Advances in Olericulture

Series Editor
Silvana Nicola, Department of Agricultural, Forest and Food Sciences
University of Turin
Grugliasco, Italy

The book series Advances in Olericulture provides a state-of-the-art account of research in olericulture, the applied life science of production and utilization of vegetable crops. The series focuses on various aspects of vegetable science and technology covering primarily but not exclusively species where the vegetative organ is the economically important component. The series of books spans current topics from sustainable fertilization to organic production; from open field cultivation to advanced soilless growing techniques; from vegetable seed and seedling physiology to vegetable quality and safety; from environmental stresses to phyllosphere communities interaction with vegetables; from postharvest biology and technology to minimally processing of vegetables. The series is designed to present the most advanced scientific information available linking basic and applied research for serving olericulturists, research workers, teachers and advanced students.

Mohammad Mahdi Jowkar
Patricia Duarte de Oliveira Paiva • Silvana Nicola
Editors

Advances in Research on Edible Flowers

From Production to Uses

Springer

Editors
Mohammad Mahdi Jowkar
Department of Biosystematics, TeMS.C.
Islamic Azad University
Tehran, Iran

Patricia Duarte de Oliveira Paiva
Department of Agriculture
Federal University of Lavras
Lavras, MG, Brazil

Silvana Nicola
Department of Agricultural, Forest
and Food Sciences
University of Turin
Grugliasco, Italy

ISSN 2367-4083 ISSN 2367-4091 (electronic)
Advances in Olericulture
ISBN 978-3-032-04560-7 ISBN 978-3-032-04561-4 (eBook)
https://doi.org/10.1007/978-3-032-04561-4

This Springer imprint is published by the registered company Springer Nature Switzerland AG
The registered company address is: Gewerbestrasse 11, 6330 Cham, Switzerland

To our Beloved Family

Human beings are members of a whole
In creation of one essence and soul

If one member is afflicted with pain
Other members uneasy will remain

If you've no sympathy for human pain
The name of human you cannot retain
Saadi (1209–1291)

Preface

Flowers have a long history of living alongside humans and being with us by improving our life quality, such as cut flowers, pot plants, and bedding plants. Considering their beautifulness (aesthetics), their major role has been improving our mental and psychological well-being besides emotional expression. This is while some flowers can be used as edible source and can improve the nutritional value of our food and improve our physiological (body) health due to the various micronutrients and bioactive compounds they contain (when consumed as food or food additives). Using flowers as edible source is not only limited to ornamental plants, they include a wide range of plants, from vegetables (such as pumpkins) to fruit trees (such as citrus).

As pollen, nectar, and petals of all plants have a significant amount of beneficial micronutrient elements and secondary metabolites such as free radical scavenging bioactive compounds, natural pigments (anthocyanins and carotenoids), vitamins, and polyphenols; they can be used as a source of food and additives, specifically as nutrients, spices, and flavor in all kinds of foods (curry, soup, pasta, and canned), drinks, desserts (cakes, jelly, and chocolate), and jam. On the other hand, by having attractive shapes and colors, edible flowers can also be used as a garnish to decorate various foods such as cakes, desserts, and even ice cubes. In addition to improving the nutritional value and taste of food prepared with edible flowers, their pleasant aroma can have a therapeutic effect (aromatherapy) in some cases.

Although some flowers such as Damascus Rose and Persian Saffron have a long history of being used as food additives and spices, changing the life trend toward natural and organic style, in addition to human welfare and moving to a more luxurious life, floriphage and consuming flowers as food is getting more popular and receiving more attention and accordingly have flourished the market of edible flowers in recent years.

From an economic point of view, the share of edible flowers in the agricultural market is rising significantly and they have become more common and well-known. In fact, edible flowers have changed the market trend in the food industry due to their unique characteristics, especially their effect on the beauty of the product and the health of the consumer, along with economic innovation.

Another issue that has caused the rapid growth of the market of this group of plants was the crisis that hit the ornamental flower market as a result of the epidemic of COVID-19 disease, which reduced the global demand for ornamental flowers and changed the producers' approach, to grow and produce specific ornamental flowers for consumption as edible flowers.

Unlike the market, research on these kinds of flowers has not grown in the same way as the rate for their demand. Conducted research on edible flowers has mostly focused on them as a new rich nutritional food source and other related aspects have not been explored. This has unfortunately resulted in the lack of comprehensive scientific books on related topics covering various aspects related to flowers as an edible source (edible flowers). The present scientific book is the first and its goal is to provide comprehensive scientific literature on various agronomical aspects of edible flowers, from cultural practices through postharvest. Every effort has been made to make this book a comprehensive resource on this subject. We believe that the present book will be of great interest to the related scientific research community; students in a wide range of biological, biomedical, agricultural, and food fields; educators; growers; and even households around the world.

The book is divided into two sections. Part I deals with edible flowers' production chain, where topics such as biodiversity, domestication, breeding, soil/soilless cultivation, precision production technologies, bioactive compounds, nutraceutical aspects, industrial food exploitation, adulterant detection, biotechnology, postharvest handling physiology, and postharvest technologies are discussed. In addition, future trends and goals for an innovative and sustainable food chain of edible flowers are presented in this section of the book. Part II deals with applications of edible flowers, where topics such as infusion, non-alcoholic beverages, liquors, food preparation, food coloring, pastry, and candies are discussed. A second volume has been prepared with the aim of dealing with specific families of edible flowers where extensively cultivated edible flowers are explored by their family members.

We hope that besides providing useful information to various communities, this book will focus the attention of researchers on other scientific aspects of edible flowers' production and consumption chain and the implementation of new and advanced technologies in their supply chain, topics such as biofortification, integration of AI, IOT, and machine learning for predictive analytics, resulting in the development of more efficient and affordable sensing technologies intended for automation, precision farming, and exploring alternative sustainable substrates to gain eco-friendly and energy-efficient production chain.

We are highly thankful to the authors for their valuable contribution and dedication in the preparation process of each chapter and to Springer publishing team for the opportunity to publish the work.

Tehran, Iran — Mohammad Mahdi Jowkar
Lavras, MG, Brazil — Patricia Duarte de Oliveira Paiva
Grugliasco, Italy — Silvana Nicola

Contents

Contributors

Awais Ali Department of Agricultural and Environmental Sciences, Università Degli Studi Di Milano, Milan, Italy

Elka Fabiana Aparecida Almeida Universidade Federal de Minas Gerais, Montes Claros, MG, Brazil

Lillian Barros CIMO, LA SusTEC, Instituto Politécnico de Bragança, Campus de Santa Apolónia, Bragança, Portugal

Mantana Buanong Division of Postharvest Technology, School of Bioresources and Technology, King Mongkut's University of Technology Thonburi, Bangkok, Thailand

Cristina Caleja CIMO, LA SusTEC, Instituto Politécnico de Bragança, Campus de Santa Apolónia, Bragança, Portugal

Matteo Caser Department of Agricultural, Forest and Food Sciences (DISAFA), University of Torino, Grugliasco, Italy

Rey Castañeda-Rodríguez Instituto Universitario de Ingeniería de Alimentos-Food UPV, Universitat Politècnica de València, Valencia, Spain

Ariane Castricini Empresa de Pesquisa Agropecuária de Minas Gerais, Caldas, MG, Brazil

Viviana Cavallaro Department of Agricultural and Environmental Sciences-Università Degli Studi Di Milano, Milan, Italy

Giacomo Cocetta Department of Agricultural and Environmental Sciences, Università Degli Studi Di Milano, Milan, Italy

Andrea Copetta CREA - Research Centre for Vegetable and Ornamental Crops, Sanremo, IM, Italy

Juliana Pinto de Lima Universidade Federal de Minas Gerais, Montes Claros, MG, Brazil

Patricia Duarte de Oliveira Paiva Department of Agriculture, Federal University of Lavras, Lavras, MG, Brazil

Maria Inês Dias CIMO, LA SusTEC, Instituto Politécnico de Bragança, Campus de Santa Apolónia, Bragança, Portugal

Nicole Mélanie Falla Department of Agricultural, Forest and Food Sciences (DISAFA), University of Torino, Grugliasco, Italy

Rhuanito Soranz Ferrarezi Department of Horticulture, University of Georgia, Athens, GA, USA

Davide Guffanti Department of Agricultural and Environmental Sciences, Università Degli Studi Di Milano, Milan, Italy

Norhayati Hussain Faculty of Food Science and Technology, Department of Food Technology, Universiti Putra Malaysia, Selangor, Malaysia

Safoura Jabbari Department of Biochemistry and Biophysics, TeMS.C., Islamic Azad University, Tehran, Iran

Mohammad Mahdi Jowkar Department of Biosystematics, TeMS.C., Islamic Azad University, Tehran, Iran

Gabriel Sthefano Lourenço Universidade de Campinas, Campinas, SP, Brazil

Carlo Mascarello CREA - Research Centre for Vegetable and Ornamental Crops, Sanremo, IM, Italy

Giovanni Minuto Centro di Sperimentazione e Assistenza Agricola – Regione Rollo, Albenga, SV, Italy

Nursabrina Munawar Faculty of Food Science and Technology, Department of Food Technology, Universiti Putra Malaysia, Selangor, Malaysia

Department of Food Technology, Faculty of Applied Sciences, Negeri Sembilan Branch, Universiti Teknologi MARA, Kuala Pilah, Malaysia

Suyun Nam Department of Horticulture, University of Georgia, Athens, GA, USA

Silvana Nicola Department of Agricultural, Forest, and Food Sciences, University of Turin, Grugliasco, Italy

Cláudia Novais CIMO, LA SusTEC, Instituto Politécnico de Bragança, Campus de Santa Apolónia, Bragança, Portugal

César Ozuna Departmento de Alimentos, División de Ciencias de la Vida, Campus Irapuato-Salamanca, Universidad de Guanajuato, Irapuato, Mexico

Beatriz Helena Paschoalinotto CIMO, LA SusTEC, Instituto Politécnico de Bragança, Campus de Santa Apolónia, Bragança, Portugal

Carla Pereira CIMO, LA SusTEC, Instituto Politécnico de Bragança, Campus de Santa Apolónia, Bragança, Portugal

Eliana Pereira CIMO, LA SusTEC, Instituto Politécnico de Bragança, Campus de Santa Apolónia, Bragança, Portugal

Alice Petrini Department of Agricultural and Environmental Sciences, Università Degli Studi Di Milano, Milan, Italy

Sara Petrucci AIAB Liguria, Varese Ligure, SP, Italy

Eleomar de O. Pires CIMO, LA SusTEC, Instituto Politécnico de Bragança, Campus de Santa Apolónia, Bragança, Portugal

Nutrition and Bromatology Group, Department of Food and Analytical Chemistry, Faculty of Food Science and Technology, University of Vigo, Ourense, Spain

Laura Pistelli Department of Agriculture, Food, and Environment, University of Pisa, Pisa, Italy

Kuan Qin Department of Horticulture, University of Georgia, Athens, GA, USA

Liqing Qiu School of Perfume and Aroma Technology, Shanghai Institute of Technology, Shanghai, China

Husnain Rauf Department of Horticulture, University of Georgia, Athens, GA, USA

Barbara Ruffoni CREA - Research Centre for Vegetable and Ornamental Crops, Sanremo, IM, Italy

Valentina Scariot Department of Agricultural, Forest and Food Sciences (DISAFA), University of Torino, Grugliasco, Italy

Hossein Sheikhi Horticultural Sciences Department, College of Agriculture and Natural Resources, University of Tehran, Karaj, Iran

Federico Tinivella Centro di Sperimentazione e Assistenza Agricola – Regione Rollo, Albenga, SV, Italy

Alessandro Triantafyllidis AIAB Liguria, Varese Ligure, SP, Italy

Chalermchai Wongs-Aree Division of Postharvest Technology, School of Bioresources and Technology, King Mongkut's University of Technology Thonburi, Bangkok, Thailand

Min Zhang State Key Laboratory of Food Science and Technology, Jiangnan University, Wuxi, Jiangsu, China

Part I
Production Chain

Chapter 1
Market Outlook and Role of Research in an Innovative and Sustainable Food Chain of Edible Flowers

Patricia Duarte de Oliveira Paiva, Mohammad Mahdi Jowkar, and Silvana Nicola

Abstract Edible flowers are those meeting essential requirements such as being non-toxic, and nutritional beneficial. Its popularity is growing as they enhance the flavour and appearance in many culinary applications. The Edible Flowers market has grown in recent years, driving new investments in this sector as its compound annual growth rate is proceeding more than five percent. To support producers and the market, scientific research helps minimize risks, optimize usage, and enhance consumer acceptance, ensuring safe and effective applications. Agronomic research plays a key role in optimizing cultivation techniques, for high-quality yields with minimal environmental impact. Sustainable practices, including hydroponics, efficient postharvest management, rational product application, organic methods, and disease or pest control, improve availability and affordability. Moreover, biotechnological tools such as RAPD and SSR markers, aid in assessing genetic diversity in ornamental crops. Standardized Agrobacterium-mediated transformation protocols have been developed to enhance quality attributes for industrial and commercial use. In addition, progress in improving edible flower crops includes mutagenesis and tissue culture. Furthermore, genome editing, chemical and radiation-induced mutations, and omics technologies are emerging in edible ornamental breeding but remain in the early stages. At the same time, sustainable production practices are essential for enhancing plant quality while minimizing environmental impact.

P. D. de Oliveira Paiva
Department of Agriculture, Federal University of Lavras, Lavras, MG, Brazil
e-mail: patriciapaiva@ufla.br

M. M. Jowkar (✉)
Department of Biosystematics, TeMS.C., Islamic Azad University, Tehran, Iran
e-mail: mm.jowkar@iau.ac.ir

S. Nicola
Department of Agricultural, Forest, and Food Sciences, University of Turin, Grugliasco, Italy
e-mail: silvana.nicola@unito.it

M. M. Jowkar et al. (eds.), *Advances in Research on Edible Flowers*, Advances in Olericulture, https://doi.org/10.1007/978-3-032-04561-4_1

Keywords Biotechnology · Industry developments · Innovation · Market outlook · Mutagenesis · Omics · Sustainability

Abbreviations

CAGR	compound annual growth rate
CRISPR/Cas systems	Precision correction of Disease-Causing Mutations
EMS	ethylmethanesulphonate
GC-MS	gas chromatography-mass
GWAS	genome-wide association studies
ISO	International Organization for Standardization
LCA	Life Cycle Assessment
LC-MS	liquid chromatography-mass
NMR	nuclear magnetic resonance
PANCs	*Plantas Alimentícias Não Convencionais*
QTLs	quantitative trait loci
RAPD	Random Amplified Polymorphic DNA
RNA-seq	RNA sequencing
SNPs	single nucleotide polymorphisms
SSR markers	Simple Sequence Repeat
TALENs	Transcription Activator-Like Effector Nucleases
TLC	Thin layer chromatography
ZFNs	Zinc Finger Nucleases

1.1 Introduction

Flowers have the power to engage all our senses, and edible flowers can elevate the sensory experience by providing attractive visual elements, aromas, and flavours while delivering valuable nutrients and medicinal properties (Sarode et al. 2024). However, not all flowering plant species are considered edible. Edible flowers are defined as those meeting essential requirements such as being non-toxic, innocuous, and possessing important nutritional properties beneficial to humans (Fernandes et al. 2017). Despite their potential, only a subset of these flowers has been thoroughly studied for their natural composition and suitability as food ingredients or medicines. This highlights the need for scientific data to minimize potential risks, determine appropriate usage, and increase consumer acceptability (Lu et al. 2016). Global research has identified approximately 180 flower species suitable for human consumption (Purohit et al. 2021; Lu et al. 2016; Fernandes et al. 2017).

Edible flowers are rich in various compounds including antioxidants, flavonoids, and anthocyanins, which contribute to their nutritional and functional properties (Falla et al. 2020). While certain flowers on their parts are commonly used as vegetables, such as artichoke (*Cynara scolymus*), cauliflower (*Brassica oleracea* var.

botrytis), and broccoli (*Brassica oleracea* var. *italica*) (Santos and Reis 2021) this discussion focuses on a different category of flowers used for diverse purposes, including beverages, colorants, salads, desserts, and patisserie.

Edible flowers are gaining popularity among food enthusiasts, professional chefs, and home bakers as a unique and visually striking addition to enhance the flavours and appearance of various dishes and baked goods. Recipes featuring edible flowers have seen growing attention as bakers share their visually appealing creations on social media platforms. Incorporating edible flowers enhances aesthetics and introduces exciting flavours to baked treats. This trend, driven by the emphasis on food presentation and the influence of social media, is expected to significantly boost the baking segment in the coming years (Technavio 2023).

The Edible Flowers market is projected to grow approximately 5% annually from 2023 to 2028. Key factors contributing to this growth include the appeal of visually striking food items, especially on social media, recommendations from health and nutritional experts emphasizing the benefits of edible flowers, and advancements in postharvest and packaging technologies (Technavio 2023). Marigold (*Calendula officinalis*), a versatile edible flower used in various culinary applications, also possesses nutraceutical properties. It is a source of bioactive flavonoids with potential efficacy against neglected tropical diseases such as *Leishmania major* promastigotes and amastigotes (Al-Huqail et al. 2023).

These findings underscore the necessity of further research to uncover additional bioactive compounds and expand the potential applications of edible flowers, ensuring their safe and effective use in diverse contexts.

Also important is research into agronomic aspects, which optimizes cultivation techniques, including soil management, water usage, and organic production, ensuring high-quality yields with minimal environmental impact. Agronomic research can contribute to cost-effective and scalable cultivation practices, ultimately enhancing the availability and affordability of edible flowers in global markets.

1.2 Brief History and Evolution

Flowers have been consumed since the Palaeolithic Age (Rodrigues and Spence 2023), reflecting a legacy of culinary traditions across cultures worldwide (Santos and Reis 2021). This rich history of flower consumption is documented in ancient civilizations such as Greece, Rome, medieval France, and Europe, as well as Victorian England, the Middle Eastern region, and Asian countries like China and Japan (He et al. 2015; Lu et al. 2016), (Rop et al. 2012).

In ancient Greece and Rome, flowers, along with herbs, roots, and vegetables were used to prepare pickles stored in cylindrical vases and held in high esteem (Rodrigues and Spence 2023). Romans incorporated roses (*Rosa* spp.), chamomile (*Matricaria chamomilla* L.), and violets (*Viola odorata* L.) into dishes and beverages, while lavender flowers were used in sauces (Cunningham 2015; Chen and Wei 201). A popular legend recounts that Emperor Nero's banquets featured fountains of rose water, wine flavoured with roses, and decorations with a single rose.

In Iran, roses and saffron flowers have enriched culinary traditions for over 3000 years (Rodrigues and Spence 2023). In China, the chrysanthemum (*Chrysanthemum morifolium* Ramat.) was a staple in traditional recipes. Meanwhile, ancient Brazilian cookbooks recorded the use of orange blossom water in soups, sauces, and desserts, as well as *Myristica fragans* in dishes like "Veal with Sardines" (Cozinheiro Nacional 1882; cf. Aki and Perosa 2002). Portuguese influences are also documented in flower waters in *Manual do Distillador e Licorista* [Distiller and Liqueurist's Manual] (Rodrigues and Spence 2023), emphasizing their use in distillation and liqueur production.

In medieval Europe, marigold flowers (*Calendula officinalis* L.) were commonly used in salads, particularly in France, where violets (*Viola odorata* L.) were also widely consumed. Other species, such as carnation (*Dianthus caryophyllus* L.), dandelion (*Taraxacum officinale* Weber), and elder (*Sambucus nigra* L.) flowers, were integrated into culinary practices. Many of these traditions originated in rural populations and have persisted to the present day. Recently, they have experienced a revival, driven by researchers and chefs seeking innovative culinary insights (Chen and Wei 2017).

The consumption of flowers is deeply rooted in cultural practices and rituals (Takahashi et al. 2020). In India, edible flowers from trees were associated with beliefs in human immortality, for example, consuming lotus flowers (*Nelumbo nucifer*a) was thought to promote balance and longevity (Pinakin et al. 2019). Similarly, the lotus flower featured prominently in ancient Greek mythology, where it was consumed for both its food value and its narcotic properties, inducing a peaceful, apathetic state (Murtola 2010).

In Africa, ancient hunter-gatherer communities traditionally consumed *Aponogeton distachyos*, a wild aquatic plant during winter, as a source of carbohydrates, minerals, and vitamins. This species was later introduced to the Netherlands in the 1700s, becoming a key ingredient in soups (Rodrigues and Spence 2023). In North America, indigenous peoples used pumpkin flowers (*Curcubita* spp.) in soups while hibiscus flowers (Hibiscus spp.) were similarly incorporated into traditional South American cuisine (Cunningham 2015; Pires Jr et al. 2021).

Flowers have been consumed across all regions of the world, including the tropics, subtropics, and even the Arctic Circle. For example, Eskimos in Alaska traditionally ate flowering axes of *Senecio congestus* and *Petasites frigiqys*. Similarly, in southern Greenland, the flowering stems of *Angelica archangelica* were consumed (Rodrigues and Spence 2023).

Although the consumption of flowers dates back centuries, often as visually appealing additions to meals, their primary use today seems to be decorative (Rodrigues and Spence 2023). Edible flowers can be consumed in various ways but are most commonly incorporated into diets in their fresh, natural form (*in natura*) (Matyjaszczyk and Smiechowska 2019). Researchers have also developed effective methods to preserve the structure, nutritional composition, and sensory characteristics of edible flowers through innovative conservation techniques (Pires et al. 2019).

In the twentieth and twentyfirst centuries, a growing movement has emerged to promote the value of wild and native plants, including lesser-known edible flowers. This trend has been strongly supported by media such as TV programs, magazines, websites, and cookbooks, further increasing awareness and appreciation of these plants (Santos and Reis 2021).

Recent studies have explored various aspects of edible flowers, addressing both their benefits and risks (Matyjaszczyk and Smiechowska 2019). Nutraceutical properties have been a particularly significant area of focus, with numerous publications emphasizing their potential health benefits (Fernandes et al. 2017; Lu et al. 2016; Rivas-García et al. 2021; Pires et al. 2019); Other research topics include post-harvest handling and preservation techniques (Fernandes et al. 2019); traditional and contemporary culinary applications (Takahashi et al. 2020; Motti et al. 2022; Santos and Reis 2021) as well as marketing perspectives (Fernandes et al. 2020a, b, c).

In Brazil, edible flowers are classified under the broader category of "PANCs" (*Plantas Alimentícias Não Convencionais*), translated as "Unconventional Food Plants" (Jacob and Albuquerque 2021). In Mexico, the consumption of edible flowers is often associated with poverty, representing a marginal aspect of agriculture and traditional cuisine (Mulík and Ozuna 2020).

1.3 Edible Flower Market outlook

In recent years, as the living level has improved, applications of edible flowers have increased and the demands for edible flower has increased. Despite the growth, market information on edible flowers (production, sales, Import and export) is rare and only available by some producers and wholesalers (Fernandes et al. 2019). This is while, producers and sale point are increasing yearly. In general, edible flower market consists of four major segments: unpacked flowers (which is mostly provided in batch to local markets, and no exact data is available), packed flowers, saffron and by products such as rose water.

1.4 Market Value

With demands growth, value of edible flowers market has increased in recent years. Statistical data show that by 2020, edible flowers market value increased by more than 2 folds compared to its previous years' value. The main reason for such growth is increasing popularity of international cousins, healthier lifestyle (due to nutritional and medicinal value), and changing consumer eating behaviour and taste. Among different market analyzers, Market research future has reported the value of edible flower market as 421 million USD in 2023 and 422 million USD in 2024. They anticipated a compound annual growth rate (CAGR) of 5.20% during 2024–2032 (Marketresearchfuture 2025). This is while, a higher growth rate, but with a different value has been forecasted by other rivals. Research and markets have reported that the market value of packaged edible flowers has increased from 124.92 million USD in 2023 to 132.23 million USD in 2024. They have predicted a growth of 6.63%, leading the value to 195.83 million USD by 2030 (Researchandmarkets 2024).

Saffron by itself had a market value of 1.5 billion USD in 2024 and its anticipated CAGR during 2024–2032 is 6.8%, reaching the value of 2.6 billion USD by 2033 (Custommarketinsights 2025). On the other hand, rose water had a market value of 477 million USD in 2024 and its anticipated CAGR during 2024–2032 is 7.6%, reaching the value of 992.30 million USD by 2034 (expertmarketresearch 2025). This is while the global flower market of ornamental flowers has been reported 31.95 billion USD (Mohamed 2024).

As the market value is influenced by increasing request for edible flowers and growing demand for delightful appealing food; high cost, product hygiene, freshness and potential allergy are a limiting factor for the market of some edible flower which should be overwhelmed by application of advanced biotechnological methods in production and innovative cultivation techniques (Jowkar and Moshashaei 2024).

1.5 Market Segmentation

Edible flowers market can be segmented according to various aspects such as: flower species, application and region as below.

1.5.1 Flower Species

In general, in different cultures and cousin, 180 species of 100 genera belonging to 90 families are consumed as edible flowers throughout the world (Lu et al. 2016). In the bulk and packed edible flower supply, rose, hibiscus, jasmine, marigold, lavender, violet, marigold, orchids hold the highest market segments of edible flowers (Marketresearchfuture 2025; Researchandmarkets 2024).

The hibiscus segment accounts for 35% of market revenue. This share is due to hibiscus medicinal and culinary application besides the plentiful applications that hibiscus has (such as tea, jam, salads, and delights). In between hibiscus species, roselle (*Hibiscus sabdariffa*) is the most common edible specie (Marketresearchfuture 2025).

1.5.2 Flower Applications

Various segments of edible flower application are: food, beverage (infusion, alcoholic and non-alcoholic), garnish and, medicine (Jowkar 2024). The foods category generated the most income (45%). After which beverage and garnish are placed. In the food segment, edible flowers are used as decorating, spice (flavouring and colouring agent) pickle and, bakery. In this segment, flowers are delivered as dry, fresh, processed, and preserved depending on their conventional or organic nature of production (Marketresearchfuture 2025).

1.5.3 Market Region

Based on the geography, edible flower market can be divided into eight segments: North America (US and Canada); South Ameriça (Brazil, Mexico, Argentina, Chile, and Peru); Europe (Germany, Italy, France, UK, Spain, Belgium, Bulgaria, and The Netherlands); South Asia (India, Thailand, Indonesia, and Malaysia); East Asia (China, Japan, and S. Korea); Middle East (Turkey, Iran, and UAE); Africa (mainly South Africa); and Oceania (Australia and New Zealand).

North America, Europe and Asia hold the most market shares, respectively. Various regions differ in growth rate depending of the application and cultivated flower species. As living level is increasing in Asia, it is estimated that Asian markets will have the highest growth rate due to new applications.

1.6 Edible Flowers Industry Developments

Although the application of edible flowers in food and medicine dates back to early civilization, and they are used in different part of the world in their local cuisines (Rodrigues and Spence 2023), but edible flowers production to sale chain has not been recognised as an industry until recent years which their market value has increased due to the rapid rise of their demand and consumption.

At the present time, edible flowers have changed the market trend in the food industry due to their unique characteristics, especially their effect on the beauty of the product and the health of the consumer, as well as economic innovation. From the nutritional aspect, most of the edible flowers have a significant amount of useful secondary metabolites such as free radical scavenging compounds (anthocyanins and carotenoids), polyphenols and micronutrient elements and are rich in fiber. Today, due to the many natural properties that edible flowers have on human health and nutrition, as well as the results of published research, the share of this group of flowers in the market of agricultural products has grown significantly and they have become more common and well-known.

Another factor that has caused the rapid growth of the market of this group of plants was the crisis that hit the market of ornamental flowers as a result of the epidemic of Covid-19 disease; which reduced the global demand for ornamental flowers and changed the approach of ornamental plant's producers to grow and produce ornamental flowers to be used as edible flowers (Jowkar and Nicola 2022).

Nowadays, development of edible flower's industry can flourish by focusing on various strategies such as:

1.6.1 Production Strategies

Improving conventional products and producing new products; commercial by products (such as health and cosmetic products); packaging improvement by various new technologies (to increase self-life and protect active ingredient and colour); development and innovation in the confectionery and bakery sector;

introducing new products (such as instant infusions, chocolate dips, smoothies, fresh boxes and delights);edible's assortment of gifting and their online and in-store sales.

1.6.2 Market Strategies

Market also needs attention in terms of customers' expectations. Technology trends and market regulatory factors are essential to expand the market sector. Special attention should be given to trends in the consumer demands, supply changes and its dynamic. Lack of legislation and related information is a challenging issue in setting strategies for market, specially at local markets. Future edible flower market strategies will depend on actionable intelligence to grasp emerging opportunities and drive investments.

1.6.3 Sustainability

In order to achieve a sustainable market for edible flowers and their byproducts, marketing should focus on the impact that the products have on the environment and consumers, besides targeting a long term value for consumers, the environment and the company. This can be achieved by various means at different aspects such as the product itself and its supply chains and market. For products, special attention should be given to value-added processing, green products, waste reduction and eco-friendly packaging (usage of recyclable and biodegradable materials). For supply chain and market, sustainable sourcing should be demonstrated by various means even by integrating it into the company's name, brand and logo. Special attention should be given to consumer-focused marketing and eco-friendly practice throughout the supply chain. Implementing green marketing strategies such as usage of organic products, nontoxic materials, smart water and energy consumption should also be in the spotlight.

1.7 Innovative Cultivation Techniques

The sensory properties of food, such as appearance, aroma, texture, and flavour, play a vital role in consumer perception, directly influencing product quality and acceptance. For edible flowers, quality is closely tied to the intrinsic characteristics of the species and the agronomic practices employed during cultivation (Demasi et al. 2021).

Among modern practices, hydroponic cultivation stands out as a promising alternative and innovative for the production of edible flowers. This method enables precise control of environmental factors, such as temperature, humidity, and light, ensuring uniform production and maintaining visual attributes like vibrant colours. Additionally, hydroponics supports continuous year-round production, reducing dependence on seasonal variations. The efficient use of water and nutrients makes this system highly sustainable while contributing to the economic viability of large-scale production (Pires Jr et al. 2021).

Other important agronomic practices include postharvest management, which is critical for preserving the quality and extending the shelf life of edible flowers. Techniques such as applying preservative solutions, controlled refrigerators, and radiation treatments (electron beams and gamma rays) have shown positive results in reducing spoilage and maintaining the sensory and nutritional properties of flowers (Pires Jr et al. 2021).

It is important to highlight that edible flower production requires intensive and specialized agronomic management. From selecting the most suitable cultivars for specific environmental conditions to adopting integrated farming practices, such as the rational application of fertilizers and effective pest control, every step must be carefully planned to maximize quality and ensure food safety. Furthermore, sustainable practices, including the use of organic fertilizers and integrated pest management are increasingly incorporated into production systems to align with the growing demand for eco-friendly agriculture.

The success of edible flower production depends on implementing innovative and sustainable agronomic practices. These practices optimize resources, ensure high-quality standards, and meet the expectations of an increasingly demanding consumer market.

1.8 Breeding and Biotechnology

For ornamental plants, the development of new varieties emphasizes the importance of plant architecture, focusing on traits such as compact growth habits, short internodes, and sturdy stems (Chandler and Sanchez 2012), which are critical for aesthetic appeal and marketability.

Biotechnology has significantly transformed the floriculture industry, particularly through advancements regarding *in vitro* cloning techniques for large-scale propagation of ornamental crops (Rout et al. 2006; Mehbub et al. 2022). These methods have enabled the generation of genetic variability, improvement of plant health, and incorporation of desirable traits through transgenics. Additionally, biotechnological approaches such as protoplast fusion, anther culture, microspore culture, ovule and embryo culture, root culture, organ culture, suspension culture, single-cell culture, and ovule layer cell culture have facilitated the production of haploid and somaclonal variants (Sarode et al. 2024).

In vitro techniques have also proven invaluable for the rapid multiplication of chimeric mutant plants developed through chemical mutagenesis, allowing for the preservation and proliferation of unique genetic traits.

While biotechnological tools were initially focused on food, feed, and herbicide/pesticide tolerance, recently shifted attention toward enhancing the quality attributes of ornamental crops for industrial and commercial purposes (Noman et al. 2017; Sarode et al. 2024). Biotechnology has enabled the introduction of genes from other species into ornamental plants, facilitating trait-specific improvements and addressing limitations posed by conventional breeding methods (Chandler and Brugliera 2011; Kamthan et al. 2016).

These innovations underscore the transformative potential of biotechnology in ornamental horticulture, providing solutions to challenges in propagation, trait enhancement, and the development of novel varieties designed to meet consumer and industry demands. In the realm of edible flowers, recent advancements have introduced biotechnological tools such as mutation breeding, molecular marker technology, plant tissue culture, transgenic approaches, omics technologies, and genome editing (Sarode et al. 2024).

While significant progress has been achieved in improving edible flower crops through mutagenesis, tissue culture, and biotechnological techniques, the application of cutting-edge technologies such as genome editing and omics for breeding in edible ornamentals remains in the initial stages (Sarode et al. 2024). These advanced methodologies hold immense potential to revolutionize the field by enabling precise trait modifications and unlocking new possibilities for crop improvement.

Edible flowers are valued not only for their aesthetic appeal but also for their role as functional food ingredients, offering essential nutrients, unique flavours, and medicinal properties. With the rising global demand for natural, functional, and healthy food products, research on edible flowers represents a promising opportunity for applying advanced biotechnological tools (Sarode et al. 2024). These innovations could further enhance the quality, yield, and sustainability of edible flower production, positioning them as a key component of the future of functional and ornamental horticulture.

Resistance to Biotic and Abiotic Stress
Biotechnological approaches play a critical role in enhancing plant resistance to biotic and abiotic stresses, a necessity given the increasing challenges posed by climate change. Developing effective strategies for crop improvement requires a comprehensive understanding of plant responses to environmental stresses to minimize crop losses and ensure sustainability.

In the context of edible flower crops, significant progress has been made. For instance, transgenic carnation plants expressing a bacterial chitinase gene from *Serratia marcescens* demonstrated improved tolerance to Fusarium wilt, a severe disease that affects crop yield and quality (Brugliera et al. 2000). Similarly, *Rosa chinensis* Jacq. plants genetically modified with the Medicago gene MtDREB1C exhibited enhanced performance under freezing temperatures (Chen et al. 2010), showcasing the potential of genetic modifications to confer tolerance to abiotic stress.

Such advancements underscore the potential of genetic engineering to fortify edible flower crops against diverse stresses, including pathogens, pests, extreme temperatures, drought, and salinity. These technologies not only ensure crop resilience but also contribute to reducing reliance on chemical pesticides and fertilizers, aligning with sustainable agricultural practices.

Shelf Life and Post-harvest Quality
Edible flowers play a significant role in the horticulture industry, and maintaining their quality and freshness is essential for their commercial value. Extending the shelf life of these flowers is particularly important for breeders. However, extending their shelf life while preserving their desirable traits such as aroma, taste, and colour poses a challenge.

Genetic engineering approaches provide an alternative solution by reducing ethylene-regulated senescence (Shantamma et al. 2021). Genes involved in the ethylene signal-transduction pathway can be manipulated to engineer longer-lasting flowers. By enhancing ethylene resistance or downregulating the expression of ethylene biosynthesis genes, the vase life and overall quality of edible flowers can be significantly improved (Sarode et al. 2024).

1.8.1 Mutation Breeding

The application of chemical and radiation-based mutations has proven particularly effective in generating ornamental crop mutants with novel flower colours, especially in (Datta 2023). For instance, the chemical mutagen EMS (ethylmethanesulphonate) demonstrated its potential in hybrid roses, enabling the development of new cultivars through induced in vitro mutagenesis (Senapati and Rout 2008). Similarly, gamma irradiation has been successfully employed for in vitro mutagenesis, resulting in the production of reduced-stature, scented rose plants (Baig et al. 2012).

1.8.2 Molecular Markers

Advancements in molecular markers, such as RAPD and SSR markers, along with quantitative trait loci (QTLs), have become invaluable tools for studying genetic diversity in ornamental crops. These applications include assessing plant genetic diversity, exploring interrelationships between genotypes, establishing DNA barcodes, developing linkage maps, facilitating marker-assisted selection, and ensuring the genetic fidelity of tissue culture-raised ornamental plants (Sarode et al. 2024).

Additionally, genome-wide association studies (GWAS) have also been successfully utilized in chrysanthemums to identify single nucleotide polymorphisms (SNPs) in genes associated with traits such as ray floret type, cultivated type, and flower shape (Chong et al. 2016).

1.8.3 Genetic Engineering and Genome Editing

For several important edible flower crops, standardized Agrobacterium-mediated transformation protocols have been established, often employing reference marker genes to evaluate transformation efficiency. However, only a few genetically engineered edible flower crops have been field-tested and released for commercialization. Notable examples include modified rose and carnation varieties, primarily developed for altered flower colours rather than their edibility (Tanaka and Brugliera 2013).

The advent of genome editing technologies, such as CRISPR/Cas systems, has opened new possibilities for improving agronomic traits in edible flower crops. These tools enable precise genetic modifications to enhance traits such as flavour nutritional content, pest, and disease resistance, tolerance to abiotic stresses, and shelf life alongside aesthetic attributes like colour and fragrance. Genome editing can also be applied to optimize growth habits, flowering time, and postharvest longevity, ensuring higher yields and extended marketability.

For instance, target edits in genes regulating plant architecture could result in more compact growth forms that are easier to cultivate and harvest, particularly in controlled environments such as greenhouses or vertical farms.

The application of genetic engineering and genome editing to edible flowers presents challenges. Research focusing on agronomic improvements in edible flowers is still in its early stages compared to ornamental varieties, highlighting the need for expanded investment and collaboration in this field to unlock its full potential.

1.8.4 Omics Approaches

The use of "omics" technologies such as genomics, transcriptomics, proteomics, and metabolomics has become an essential technique in research (Sarode et al. 2024).

1.8.4.1 Genomics

Advances in plant genomics have enabled the generation of whole-genome sequence data, providing efficient insights into genomic variations linked to high-value traits (Zheng et al. 2021). This approach supports advanced breeding strategies improving the potential to enhance complex traits and develop new varieties (Zheng et al. 2021).

1.8.4.2 Transcriptomics

Transcriptomics studies the complete set of RNA transcripts in an organism, offering a detailed profile of gene expression and identifying differentially expressed genes across tissues and various conditions. Techniques such as microarray analysis, RNA sequencing (RNA-seq), and single-cell transcriptomics are commonly used (Su et al. 2019).

1.8.4.3 Proteomics

Proteomics focuses on the structure, function, and interactions of proteins, providing critical insights into biological systems. It enables the com comparative analysis of proteins in cells, tissues, or organisms, helping identify disease biomarkers, study cellular pathways, and discover potential drug targets (Vanderschuren et al. 2013).

1.8.4.4 Metabolomics

Metabolomics analyses the full spectrum of metabolites produced in a plant system, offering valuable insights into plant growth, stress responses, and metabolic markers for crop improvement. Analytical methods like gas chromatography-mass (GC-MS), liquid chromatography-mass (LC-MS), and nuclear magnetic resonance (NMR) spectroscopy; and thin layer chromatography (TLC) are used to isolate and identify specific metabolites (Sarode et al. 2024).

1.8.5 Genome Editing Approaches

Genome editing technologies, such as Zinc Finger Nucleases (ZFNs), Transcription Activator-Like Effector Nucleases (TALENs), and CRISPR/Cas9 have emerged as powerful tools for making precise and targeted modifications of the genome (Dheer et al. 2020). These techniques enable researchers to improve various traits in plants, including resistance to stresses, yield, and quality, by directly altering specific genes.

Studying the impact of agronomy and processing on flower metabolites is essential for establishing quality standards in the horticultural industry (Zhao et al. 2019). Understanding how cultivation practices affect the biochemical composition of flowers can help optimize their quality and nutritional value.

NMR metabolite analysis plays a crucial role in assessing preservation methods that extend the shelf life of flower-derived products. By evaluating the effects of different preservation techniques, NMR analysis helps improve the quality standards, and ensures the authenticity of flower ingredients, promoting the consumption of high-quality, pure flowers (Shantamma et al. 2021).

1.9 Sustainability

Sustainable production results from management practices that enhance plant quality while reducing negative environmental impacts (Ingram et al. 2018). The Life Cycle Assessment (LCA) is a standardized methodology defined by a normative reference (Beccaro et al. 2014; Lazzerini et al. 2014), as outlined in ISO (International Organization for Standardization) standards 14040:2006 and 14044:2018. This methodology is used to investigate the environmental impact of a product, production process, or system (Lazzerini et al. 2016; Wandl and Haberl 2017). Through LCA, it is possible to objectively and transparently identify and quantify energy and material usage inputs and outputs. By associating these factors with environmental impacts, targeted activities can be developed to mitigate them (Russo and Zeller 2008).

LCA studies in the floricultural sector remain limited, with applications to edible flowers being nearly non-existent (Lazzerini et al. 2016; Wandl and Haberl 2017). One LCA analysis examined the environmental impacts associated with using large versus small containers for wax begonias and violas. The study revealed lower emissions values for both species when grown in large plastic containers compared to smaller ones. The adoption of biodegradable plastic packaging was suggested as a strategy to further reduce environmental impact, particularly at the end of the product lifecycle (Jariyasakoolroj et al. 2019).

Conducting LCA for small nurseries or farms poses challenges, as these operations often do not collect data unless required by immediate necessity or legal obligation. A potential solution lies in the development of smartphone applications to facilitate the collecting data on inputs and outputs. This approach would enable nursery and farm owners to manage their operations more effectively while simultaneously generating valuable data for research purposes (Falla et al. 2020).

Expanding data collection on the environmental pressures associated with the production and commercialization of edible flower products could broaden environmental analysis. Such efforts would help identify opportunities to reduce emissions and empower consumers to make more informed and sustainable purchasing decisions (Falla et al. 2020).

1.10 Conclusions

Research in edible flowers still requires further development, despite some promising results already being available. Optimized agronomic techniques, such as hydroponics and efficient postharvest management, ensure high-quality yields while minimizing environmental impact. Scientific research is essential for market growth and safety, enhancing consumer acceptance, minimizing risks, and supporting the expansion of the edible flowers market globally. Moreover, biotechnological advancements hold significant potential for improving quality through new technologies and the development of enhanced varieties.

References

Aki A, Perosa JMY (2002) Aspectos da produção e consumo de flores e plantas ornamentais no Brasil [Production and consumption aspects of Brazilian flowers and ornamental plants]. Ornam Hortic 8:13–23

Al-Huqail AA, Bekhit AA, Ullah H, Ayaz M, Mostafa NM (2023) Antimalarial and antileishmanial flavonoids from *Calendula officinalis* flowers. Agronomy 13:2765. https://doi.org/10.3390/agronomy13112765

Baig MMQ, Hafiz IA, Abbasi NA, Yaseen M, Akram Z, Donnelly DJ (2012) Reduced-stature Rosa species through in vitro mutagenesis. Can J Pl Sci 92:1049–1055

Beccaro GL, Cerutti A, Vandecasteele I, Bonvegna L, Donno D, Bounous G (2014) Assessing environmental impacts of nursery production: Methodological issues and results from a case study in Italy. J Clean Prod 80:159–169

Brugliera F, Kalc-Wright G, Hyland C, Webb L, Herbert S, Sheehan B, Mason J (2000) Improvement of Fusarium wilt tolerance in carnations expressing chitinase. Int Plant Mol Biol Rep 18:522–529

Chandler SF, Brugliera F (2011) Genetic modification in floriculture. Biotechnol Lett 33:207–214

Chandler SF, Sanchez C (2012) Genetic modification; the development of transgenic ornamental plant varieties. Pl Biotech J 10:891–903

Chen NH, Wei S (2017) Factors influencing consumers' attitudes towards the consumption of edible flowers. Food Qual Pref 56:93–100

Chen JR, Lü JJ, Liu R, Xiong XY, Wang TX, Chen SY, Guo LB, Wang HF (2010) DREB1C from *Medicago truncatula* enhances freezing tolerance in transgenic *M. truncatula* and China Rose. *Rosa chinensis* Jacq. Pl Growth Regul 60:199–211

Chong X, Zhang F, Wu Y, Yang X, Zhao N, Wang H, Guan Z, Fang W, Chen F (2016) A SNP-enabled assessment of genetic diversity, evolutionary relationships and the identification of candidate genes in chrysanthemum. Gen Biol Evolut 8:3661–3671

Cunningham E (2015) What nutritional contribution do edible flowers make? J Acad Nutr Diet 115:856. https://doi.org/10.1016/j.jand.2015.03.002

Custommarketinsights (2025) Global saffron market 2024–2033. https://www.custommarketinsights.com/report/saffron-market. Accessed 23 Feb 2025

Datta SK (2023) Induced mutagenesis in Chrysanthemum. In: Penna S, Jain SM (eds) Mutation breeding for sustainable food production and climate resilience. Springer Nature, pp 565–598

Demasi S, Mellano MG, Falla NM, Caser M, Scariot V (2021) Sensory profile, shelf life, and dynamics of bioactive compounds during cold storage of 17 edible flowers. Horticulturae 7:166. https://doi.org/10.3390/horticulturae7070166

Dheer P, Rautela I, Sharma V, Dhiman M, Sharma A, Sharma N, Sharma MD (2020) Evolution in crop improvement approaches and future prospects of molecular markers to CRISPR/Cas9 system. Gene 753:144795

Expertmarketresearch (2025) Global rose water market report and forecast 2025–2034. https://www.expertmarketresearch.com/reports/rose-water-market. Accessed 24 Feb 2025

Falla NM, Contu S, Demasi S, Caser M, Scariot V (2020) Environmental impact of edible flower production: a case study. Agronomy 10:579. https://doi.org/10.3390/agronomy10040579

Fernandes L, Casal S, Pereira JA, Saraiva JA, Ramalhosa E (2017) Edible flowers: a review of the nutritional, antioxidant, antimicrobial properties and effects on human health. J Food Compos Anal 60:38–50. https://doi.org/10.1016/j.jfca.2017.03.017

Fernandes L, Casal S, Pereira JA, Saraiva JA, Ramalhosa E (2019) Post-harvest technologies applied to edible flowers: a review. Food Rev Int 35:132–154. https://doi.org/10.1080/87559129.2018.1473422

Fernandes L, Casal S, Pereira JA, Pereira EL, Saraiva JA, Ramalhosa E (2020a) Freezing of edible flowers: effect on microbial and antioxidant quality during storage. J Food Sci 85:1151–1159. https://doi.org/10.1111/1750-3841.15097

Fernandes L, Casal S, Pereira JA, Saraiva JA, Ramalhosa E (2020b) Borage, camellia, centaurea and pansies: nutritional, fatty acids, free sugars, vitamin E, carotenoids and organic acids characterization. Food Res Int 132:109070. https://doi.org/10.1016/j.foodres.2020.109070

Fernandes L, Casal S, Pereira JA, Saraiva JA, Ramalhosa E (2020c) An overview on the market of edible flowers. Food Rev Int 36:258–275. https://doi.org/10.1080/87559129.2019.1639727

He J, Yin T, Chen Y, Cai L, Tai Z, Li Z, Liu C, Wang Y, Ding Z (2015) Phenolic compounds and antioxidant activities of edible flowers of Pyrus pashia. J Func Foods 17:371–379

Ingram DL, Hall CR, Knight J (2018) Global warming potential, variable costs, and water use of a model greenhouse production system for 11.4-cm annual plants using Life Cycle Assessment. HortSci 53:441–444

Jacob MCM, Albuquerque UP (eds) (2021) Local food plants of Brazil. Springer Nature

Jariyasakoolroj P, Leelaphiwat P, Harnkarnsujarit N (2019) Advances in research and development of bioplastic for food packaging. J Sci Food Agric 100:5032–5045. https://doi.org/10.1002/jsfa.9497

Jowkar MM (2024) A new look at flowers as a rich source of bioactive compounds effective on human health. Paper presented at the 6th National and International Congress on Flower and Ornamental Plants, University of Mohagheh Ardabili, Ardabil, Iran, 20–21 August 2024

Jowkar MM, Moshashaei A (2024) Biotechnological applications for improving quality and marketability of edible flowers. Paper presented at the 6th National and International Congress on Flower and Ornamental Plants, University of Mohagheh Ardabili, Ardabil, Iran, 20–21 August 2024

Jowkar MM, Nicola S (2022) Edible flowers: the neglected sector of Iran's flower and ornamental plants industry. Paper presented at the 5th National and International Congress on Flower and Ornamental Plants, Shiraz University, Shiraz, Iran, 13–15 September 2022

Kamthan A, Chaudhuri A, Kamthan M, Datta A (2016) Genetically modified (GM) crops: milestones and new advances in crop improvement. Theoret Applied Genet 129:1639–1655

Lazzerini G, Lucchetti S, Nicese FP (2014) Analysis of greenhouse gas emissions from ornamental plant production: a nursery level approach. Urban For Urban Green 13:517–525

Lazzerini G, Lucchetti S, Nicese FP (2016) Green House Gases (GHG) emissions from the ornamental plant nursery industry: a Life Cycle Assessment (LCA) approach in a nursery district in central Italy. J Clean Prod 112:4022–4030

Lu B, Li M, Yin R (2016) Phytochemical content, health benefits, and toxicology of common edible flowers: a review (2000–2015). Crit Rev Food Sci Nutr 56(Suppl.1):S130–S148. https://doi.org/10.1080/10408398.2015.1078276

Marketresearchfuture (2025) Global edible flowers market overview. https://www.marketresearchfuture.com/reports/edible-flowers-market-6634. Accessed 23 Feb 2025

Matyjaszczyk E, Smiechowska M (2019) Edible flowers. Benefits and risks pertaining to their consumption. Trends Food Sci Technol 91:670–674. https://doi.org/10.1016/j.tifs.2019.07.017

Mohamed A (2024) Economic analysis of the global flower industry. https://blogs.oregonstate.edu/nexus/2024/11/06/economic-analysis-of-the-global-flower-industry/. Accessed 24 Feb 2025

Motti R, Paura B, Cozzolino A, Falco B (2022) Edible flowers used in some countries of the Mediterranean basin: an ethnobotanical overview. Plants 11:3272. https://doi.org/10.3390/plants11233272

Mulík S, Ozuna C (2020) Mexican edible flowers: cultural background, traditional culinary uses, and potential health benefits. Int J Gastron Food Sci 21:100235. https://doi.org/10.1016/j.ijgfs.2020.100235

Murtola AM (2010) Commodification of utopia: the lotus eaters revisited. Cult Organ 16:37–54. https://doi.org/10.1080/14759550903558078

Noman A, Aqeel M, Deng J, Khalid N, Sanaullah T, Shuilin H (2017) Biotechnological advancements for improving floral attributes in ornamental plants. Front Pl Sci 8:530

Pinakin DJ, Kumar V, Suri S, Sharma R, Kaushal M (2019) Nutraceutical potential of tree flowers: a comprehensive review on biochemical profile, health benefits, and utilization. Food Res Int 127:108724. https://doi.org/10.1016/j.foodres.2019.108724

Pires EO Jr, Caleja C, Garcia CC, Ferreira ICFR, Barros L (2021) Current status of genus *Impatiens*: Bioactive compounds and natural pigments with health benefits. Trends Food Sci Technol 117:106–124. https://doi.org/10.1016/j.tifs.2021.01.074

Pires TC, Barros L, Santos-Buelga C, Ferreira IC (2019) Edible flowers: emerging components in the diet. Trends Food Sci Technol 93:244–258. https://doi.org/10.1016/j.tifs.2019.09.020

Purohit SR, Rana SS, Idrishi R, Sharma V, Ghosh P (2021) A review on nutritional, bioactive, toxicological properties and preservation of edible flowers. Fut Foods 4:100078

Researchandmarkets (2024) Packaged edible flower market by product, type, nature of production-global forecast 2025–2030. https://www.researchandmarkets.com/report/edible-flower. Accessed 23 Feb 2025

Rivas-García L, Navarro-Hortal MD, Romero-Marquez JM, Forbes-Hernandez TY, Varela-Lopez A, Llopis J (2021) Edible flowers as a health promoter: an evidence-based review. Trends Food Sci Technol 117:46–59

Rodrigues H, Spence C (2023) Looking to the future, by studying the history of edible flowers. Int J Gastron Food Sci 34:100805. https://doi.org/10.1016/j.ijgfs.2023.100805

Rop O, Mlcek J, Jurikova T, Neugebauerova J, Vabkova J (2012) Edible flowers – a new promising source of mineral elements in human nutrition. Molecules 17:6672–6683

Russo G, Zeller BDL (2008) Environmental evaluation by means of LCA regarding the ornamental nursery production in rose and sowbread greenhouse cultivation. Acta Hortic 801:1597–1604

Santos ICD, Reis SN (2021) Edible flowers: traditional and current use. Ornam Hortic 27:438–445. https://doi.org/10.1590/2447-536X.v27i4.2392

Sarode DK, Pagariya MC, Jadhav PR, Patil SA, Devarumath RM, Shingote PR, Prasad KV, Jain SM, Penna S, KawaR PG (2024) Edible flowers: biotechnological interventions for improving bioactives of food and health significance. J Food Compos Anal 134:106506. https://doi.org/10.1016/j.jfca.2024.106506

Senapati SK, Rout GR (2008) In vitro mutagenesis of rose with ethylmethanesulphonate (EMS) and early selection using RAPD Markers. Adv Hortic Sci 22:218–222

Shantamma S, Vasikaran EM, Waghmare R, Nimbkar S, Moses JA, Anandharamakrishnan (2021) Emerging techniques for the processing and preservation of edible flowers. Fut Foods 4:100094. https://doi.org/10.1016/j.fbio.2022.102188

Su J, Jiang J, Zhang F, Liu Y, Ding L, Chen S, Chen F (2019) Current achievements and future prospects in the genetic breeding of chrysanthemum: a review. Hortic Res 6. https://doi.org/10.1038/s41438-019-0193-8

Takahashi JA, Rezende FAGG, Moura MAF, Dominguete LCB, Sande D (2020) Edible flowers: bioactive profile and its potential to be used in food development. Food Res Int 129:108868. https://doi.org/10.1016/j.foodres.2019.108868

Technavio (2023) Packaged edible flower market analysis Europe, North America, APAC, South America, Middle East and Africa – US, Japan, Germany, UK, France – size and forecast 2024–2028. Available at: https://www.technavio.com/report/packaged-edible-flower-market-industry-analysis. Accessed 26 Dec 2024

Vanderschuren H, Lentz E, Zainuddin I, Gruissem W (2013) Proteomics of model and crop plant species: status, current limitations and strategic advances for crop improvement. J Proteome 93:5–19

Wandl M-T, Haberl H (2017) Greenhouse gas emissions of small scale ornamental plant production in Austria-A case study. J Clean Prod 141:1123–1133

Zhao L, Fan H, Zhang M, Chitrakar B, Bhandari B, Wang B (2019) Edible flowers: review of flower processing and extraction of bioactive compounds by novel technologies. Food Res Int 126:108660

Zheng T, Li P, Li L, Zhang Q (2021) Research advances in and prospects of ornamental plant genomics. Hort Res 8:65

Chapter 2
Biodiversity, Wild Harvesting, Domestication, Selection, and Breeding of Edible Flowers

Andrea Copetta and Barbara Ruffoni

Abstract Some flowers and inflorescences are commonly eaten in several region of the world. The biodiversity is large: bulbous, climbers, shrubs and trees can produce edible flowers. We have compiled a list of edible flower plants belonging to 98 different families, 320 genera and over 500 species and varieties, however most likely the number of edible flowers is higher. The edible flowers mainly belong to the Asteraceae (42 genera), Fabaceae (36 genera) and Lamiaceae (24 genera) families, while pansy (Violaceae), roses (Rosaceae), nasturtiums (Tropaeolaceae), begonia (Begoniaceae) and carnation (Caryophyllaceae) are the most cited flowers as ingredient or dish garnish. Many of the edible flowers consumed especially in Central America, in the Mediterranean and Asia area are wild species and they are collected and consumed locally raw or cooked during blooming season. Wild harvesting is widely carried out in many regions; however, the collectors must have skill and experience to recognize the edible flowers. Apart from the case of artichokes, broccoli, cauliflower, turnip greens, and zucchini, domestication or breeding programs with the aim of obtaining edible flower varieties are scarce. However, many varieties selected for ornamental purposes can also have food use. Finally, breeding programs to select new edible flower varieties to design new products are desirable.

Keywords Asteraceae · *Brassica* · *Hibiscus* · *Moringa oleifera* · *Paeonia lactiflora* · *Rosa* · *Sesbania grandiflora*

Abbreviations

Af	Africa
All	All world
Am	America

A. Copetta (✉) · B. Ruffoni
CREA – Research Centre for Vegetable and Ornamental Crops, Sanremo, IM, Italy
e-mail: andrea.copetta@crea.gov.it

M. M. Jowkar et al. (eds.), *Advances in Research on Edible Flowers*, Advances in Olericulture, https://doi.org/10.1007/978-3-032-04561-4_2

As	Asia
EFSA	European Food Safety Authority
Eu	Europe
FAO	Food and Agriculture Organization of the United Nations
FDA	Food and Drug Administration
Oc	Oceania
WHO	World Health Organization

2.1 Introduction

Edible flower represents a niche market in the food service and business industry. Some flowers and inflorescences are commonly eaten by consumers who consider them (wrongly) vegetables. This is the case of artichokes, broccoli, cauliflower, capers, turnip greens and courgette. Globally, edible flower has been increasing in popularity, though many consumers are wary of edible flower that they classify exclusively as cut flowers or garden flowers. For some people, edible flowers offer a unique combination of aroma and taste. For other persons the nutritional value of unprocessed or minimally processed edible flowers represents an important reason for their consumption.

2.1.1 Traditional Uses

The gastronomic use of ornamental flowers as edible flower is frequently a rediscovery of ethnobotanical traditions or a curiosity about the taste of other cultures. Indeed, since the time of Greeks and Roman civilizations in Europe (see Apicius also cited in Grocock and Grainger 2006; and in Motti et al. 2022) and in other societies throughout the World, e.g. Chinese, Indian, African, and South American cultures (Kirker and Newman 2016), edible flowers have been used to add nutritional or medicinal value, flavour, fragrance, and aesthetic appearance (Fernandes et al. 2017).

Every culture employs in traditional preparations edible flower species which grow spontaneously on its territory or characteristic habitat. The acceptance of unfamiliar ingredients also vary among countries. Many authors present edible flower lists of regions or countries giving indications about the traditional/commercial use or the nutritional properties and the content of secondary compounds (Reddy et al. 2007; Li et al. 2014; Zheng et al. 2018; Benvenuti and Mazzoncini 2021; Marchioni et al. 2019; Copetta et al. 2021a, b; Janarny et al. 2021; Kandylis 2022; Motti et al. 2022).

2.1.2 The Supply Chain

The differences among the distinct cultures are the flower species consumed. The edible flower market, though global, is niche because the production is difficult, laborious, and linked to seasonality. The shelf life of the product is short limited and packaging is not easy (Fernandes et al. 2019). The edible flower mainly marketed are those deriving from annual or perennial herbaceous plants that can be easily grown in pots such as calendula, nasturtium, pansies, and violets; however, there are many bulbous species (such as garlic and tulips) lianas, shrubs, and trees that produce edible flowers (Copetta et al. 2021b). The assortment of edible flower includes wild species, aromatic plants, and fruit tree, whose flowering is often linked to seasonality and consequently also their consumption if no transformation processes (sauces, jams, syrups) or drying processes are applied. The production of edible flowers from shrubs or trees requires larger spaces than potted herbaceous plants because these species must be grown in the ground to be fully productive, and cultivation in a protected environment (greenhouse) is hardly practicable. All these considerations influence the selection that producers of edible flowers implement when they decide to cultivate a new species of edible flower.

The "foraging" (harvesting in nature) of edible flowers from herbaceous, shrubby, or tree species can be carried out, but this activity requires systematic botany skills and knowledge of the place and environment in which this activity is carried out. In order to avoid the depletion of natural populations, it is possible to implement the collection of part of the seeds for the propagation of interesting species.

This chapter intends to illustrate the variety of edible flower available and traditionally used in various cultures, the possibility of domestication of wild edible flower species, and the news about the breeding activities of edible flower.

2.2 Biodiversity

Many species produce edible flowers, but it is difficult to have a complete list because many species (some of which are endemic) are little studied, and some have many varieties or hybrids. Actually, no organization (FAO—Food and Agriculture Organization of the United Nations; WHO—World Health Organization; EFSA—European Food Safety Authority; FDA—Food and Drug Administration) has drawn up an official list of edible and non-edible flowers (Fernandes et al. 2017). Lu et al. (2016) reported that 180 species, 100 genera, and 97 families produce edible flowers; based on the bibliography found about ethnobotanical overview, nutritional research, and commercial analysis. We have compiled a list of edible flower plants belonging to 98 different families, 320 genera, and over 500 species and varieties (Table 2.1). However, this list could be not entirely complete because:

(a) an Australian website (https://aussiegreenthumb.com/edible-flowers-australia/) lists genera and species such as *Crocosmia*, *Freesia* (Iridaceae), *Gypsophila*

Table 2.1 List of edible flowers eaten on different continents

Family	Genera/species	Eaten in	Reference
Acanthaceae	*Acanthus hirsutus* Boiss.	Eu	Simsek et al. (2004)
	Barleria prionitis L.	As	Fakhri et al. (2022)
	Ecbolium ligustrinum (Vahl) Vollesen	As	Sandhyarani Devi et al. (2009)
	Phlogacanthus tubiflorus Nees	As	Angami et al. (2006)
	Rhinacanthus nasutus (L.) Kurz	As	Wongwattanasathien et al. (2010)
	Ruellia simplex C.Wright	Am	Kritsi et al. (2022)
	Thunbergia alata Bojer ex Sims	Af	Kritsi et al. (2022)
Aizoaceae	*Carpobrotus edulis* (L.) N.E.Br.	Af, As	Kandylis (2022)
Amaranthaceae	*Amaranthus tricolor* L.	As, Oc	Sandhyarani Devi et al. (2009)
	Gomphrena globosa L.	As	Zheng et al. (2018)
Amaryllidaceae	*Allium* spp. L.	All	Kucekova et al. (2013), Demasi et al. (2021)
	Tulbaghia spp. L.	Af	Drava et al. (2020), Marchioni et al. (2023)
Anacardiaceae	*Rhus chinensis* Mill.	As	Sandhyarani Devi et al. (2009)
Annonaceae	*Cananga odorata* (Lam.) Hook.f. & Thomson	As	Kresnapati et al. (2022)
	Cymbopetalum penduliflorum (Dunal) Baill.	Am	Mulik and Ozuna (2020)
Apiaceae	*Anethum graveolens* L.	Eu, Oc	Motti et al. (2022)
	Anthriscus nemorosa (M.Bieb.) Spreng.	As	Nadiroglu and Behçet (2018)
	Centella asiatica (L.) Urb.	Eu, As	Sandhyarani Devi et al. (2009)
	Echinophora tenuifolia L.	Eu	Motti et al. (2022)
	Ferula communis L.	As	Motti et al. (2022)
	Foeniculum vulgare Mill.	As, Eu, Oc	Motti et al. (2022)
	Oenanthe javanica (Blume) DC.	As	Sandhyarani Devi et al. (2009)
	Tetrataenium wallichii (DC.) Manden.	As	Ghimeray et al. (2010)
Apocynaceae	*Allamanda cathartica* L.	Af	Li et al. (2014)
	Calotropis gigantea (L.) W.T.Aiton	As	Janarny et al. (2021)
	Catharanthus roseus (L.) G.Don	Af	Janarny et al. (2021)
	Echites panduratus A.DC.	Am	Morton et al. (1990)
	Plumeria obtusa L.	Am	Kaisoon et al. (2011)
	Telosma cordata (Burm.f.) Merr.	As	Kaisoon et al. (2011)
Aponogetonaceae	*Aponogeton crispus* Thunb.	As	Janarny et al. (2021)
Araliaceae	*Panax pseudoginseng* Wall.	As	Zheng et al. (2018)

(continued)

Table 2.1 (continued)

Family	Genera/species	Eaten in	Reference
Arecaceae	*Chamaedorea tepejilote* Liebm.	Am	Cáceres and Cruz (2019)
	Cocos nucifera L.	As	Janarny et al. (2021)
	Epipremnum aureum (Linden & André) G.S.Bunting	As	Li et al. (2014)
	Lasia spinosa (L.) Thwaites	As	Janarny et al. (2021)
Asparagaceae	*Agave* spp. L.	Am	Figueredo-Urbina et al. (2022)
	Aphyllanthes monspeliensis L.	Eu	Motti et al. (2022)
	Asparagus horridus L.	Eu	Motti et al. (2022)
	Dasylirion acrotrichum (Schiede) Zucc.	Am	Figueredo-Urbina et al. (2022)
	Muscari neglectum Guss. ex Ten.	As, Eu	Eroglu-Ozkan et al. (2018)
	Nolina parviflora (Kunth) Hemsl.	Am	Figueredo-Urbina et al. (2022)
	Thysanotus patersonii R.Br.	Oc	Hansen and Horsfall (2019)
	Yucca spp. L.	Am	Juarez-Truillo et al. (2018)
Asphodelaceae	*Aloe vera* (L.) Burm.f.	Am	Figueredo-Urbina et al. (2022)
	Asphodelus spp. L.	Eu	Motti et al. (2022)
	Hemerocallis spp. L.	As, Oc	Mlcek and Rop (2011)
Asteraceae	*Achillea* spp. L.	Eu, Oc	Vitalini et al. (2015), Lucchetti et al. (2019)
	Acmella oleracea (L.) R.K.Jansen	Am, Af	Uthpala and Navaratne (2021)
	Ageratum houstonianum Mill.	Am	Drava et al. (2020)
	Antennaria dioica (L.) Gaertn.	Eu	Motti et al. (2022)
	Arctium lappa L.	Eu	Vitalini et al. (2015)
	Artemisia spp. L.	Eu	Kaval ˙ et al. (2015), Łuczaj and Dolina (2015)
	Bellis perennis L.	All	Kucekova et al. (2013)
	Bidens spp. L.	Eu, As, Oc	Motti et al. (2022)
	Calendula officinalis L.	All	Demasi et al. (2021)
	Carduus spp. L	Eu	Dop et al. (2020), Baldi et al. (2022)
	Carlina spp. L	Eu	Dop et al. (2020), Baldi et al. (2022)
	Centaurea cyanus L.	All	Rop et al. (2012)
	Chiliadenus glutinosus (L.) Fourr.	Eu	Tardío et al. (2006)
	Chrysanthemum spp. L.	As, Oc	Ukiya et al. (2001)
	Cichorium intybus L.	Eu, Oc	Kucekova et al. (2013)
	Cirsium spp. Mill.	Eu	Motti et al. (2022)

(continued)

Table 2.1 (continued)

Family	Genera/species	Eaten in	Reference
	Coreopsis tinctoria Nutt.	As	Zheng et al. (2018)
	Cosmos sulphureus Cav.	As, Oc	Kaisoon et al. (2011)
	Cota spp. J.Gay	Eu, As	Kaval ˙ et al. (2015), Geraci et al. (2018)
	Cynara spp. L.	Eu	Idolo et al. (2010), Günes (2017)
	Dahlia spp. Cav.	Am, Oc	Pires et al. (2018)
	Echinacea spp. Moench	Am	Chen and Wei (2017)
	Gerbera jamesonii Adlam	Af	Li et al. (2014)
	Gundelia tournefortii L.	As	Motti et al. (2022)
	Helianthus annuus L.	Am	Li et al. (2014)
	Helichrysum spp. Mill.	As	Tardío et al. (2006), Zheng et al. (2018)
	Jasonia tuberosa (L.) DC.	Eu	Tardío et al. (2006)
	Leontopodium nivale (Ten.) A.Huet ex Hand.-Mazz.	Eu	Pieroni and Giusti (2009)
	Leucanthemum vulgare Lam.	Eu	Demasi et al. (2021)
	Leuzea conifera DC.	Eu	Tardío et al. (2006)
	Matricaria spp. L.	Eu	Singh et al. (2011), Ranfa and Bodesmo (2017)
	Pentanema salicinum (L.) D.Gut. Larr., Santos-Vicente, Anderb., E. Rico & M.M.Mart.Ort.	Eu	Tardío et al. (2006)
	Santolina spp. L.	Eu	Tardío et al. (2006)
	Scolymus hispanicus L.	Eu	Kaval ˙ et al. (2015)
	Pseudopodospermum undulatum (Vahl) Zaika, Sukhor. & N.Kilian	Af	Motti et al. (2022)
	Silybum marianum (L.) Gaertn.	Eu,Af	Mautone et al. (2019)
	Tagetes spp. L.	Am, As, Oc	Benvenuti et al. (2016), Marchioni et al. (2020b)
	Tanacetum spp. L.	Eu	Tardío et al. (2006), Lucchetti et al. (2019),
	Taraxacum sect. *Taraxacum* F.H.Wigg.	Eu, Af	Kucekova et al. (2013)
	Tragopogon spp. L.	All	Kucekova et al. (2013), Geraci et al. (2018)
	Tripleurospermum parviflorum (Willd.) Pobed.	Af	Simsek et al. (2004)
	Tussilago farfara L.	Eu	Jakubczyk et al. (2022)
Balsaminaceae	*Impatiens walleriana* Hook.f.	Af, Oc	Rop et al. (2012)
Begoniaceae	*Begonia cucullata* Willd.	Af, Oc	Drava et al. (2020)
	Tecomaria capensis (Thunb.) Spach	Af	Kritsi et al. (2022)

(continued)

Table 2.1 (continued)

Family	Genera/species	Eaten in	Reference
Berberidaceae	*Berberis vulgaris* L.	Eu	Vitalini et al. (2015)
Betulaceae	*Betula pendula* Roth	Eu	Polat et al. (2017)
	Corylus spp. L.	Eu	Polat et al. (2017)
Bignoniaceae	*Campsis grandiflora* (Thunb.) K. Schum.	As	Zheng et al. (2018)
	Oroxylum indicum (L.) Kurz	As	Zheng et al. (2018)
	Millingtonia hortensis L.f.	As	Wongwattanasathien et al. (2010)
Boraginaceae	*Anchusa* spp. L.	Eu, As	Pieroni et al. (2005), Kaval ˙ et al. (2015)
	Borago officinalis L.	Eu, As, Oc	Demasi et al. (2021)
	Cerinthe major L.	Eu	Motti et al. (2022)
	Echium spp. Tourn. ex L.	Eu	Pieroni et al. (2005), Tardío et al. (2006)
	Lithodora fruticosa (L.) Griseb.	Eu	Tardío et al. (2006)
	Mertensia maritima (L.) Gray	Eu, As	Copetta et al. (2021a)
	Moltkia coerulea (Willd.) Lehm.	As	Simsek et al. (2004)
	Myosotis sylvatica Ehrh. ex Hoffm.	As	Zheng et al. (2018)
	Onosma spp. L.	Eu	Pieroni et al. (2005)
Brassicaceae	*Brassica* spp. L.	Eu	Geraci et al. (2018)
	Capsella bursa-pastoris (L.) Medik.	Eu, As	Tardío et al. (2006)
	Cardamine hirsuta L.	As	Sandhyarani Devi et al. (2009)
	Coriandrum sativum L.	As, As, Oc	Janarny et al. (2021)
	Diplotaxis catholica (L.) DC.	Eu	Tardío et al. (2006)
	Eruca vesicaria (L.) Cav.	As, Eu, Oc	Rahim et al. (2022)
	Iberis amara L.	As, Oc	Fakhri et al. (2022)
	Isatis tinctoria L.	Eu	Dop et al. (2020)
	Matthiola incana (L.) W.T.Aiton	Eu, As, Oc	Li et al. (2014)
	Raphanus raphanistrum L.	Eu	Tardío et al. (2006)
	Sinapis spp. L.	Eu	Geraci et al. (2018)
	Sisymbrium officinale (L.) Scop.	Eu	Geraci et al. (2018)
Cactaceae	*Selenicereus undatus* (Haw.) D.R.Hunt	As	Zheng et al. (2018)
	Myrtillocactus geometrizans (Mart. ex Pfeiff.) console	Am	Figueredo-Urbina et al. (2022)
	Opuntia ficus-indica (L.) Mill.	Eu, Am	Tardío et al. (2006)

(continued)

Table 2.1 (continued)

Family	Genera/species	Eaten in	Reference
Calycanthaceae	*Chimonanthus praecox* (L.) Link	As	Zheng et al. (2018)
Campanulaceae	*Campanula* spp. L.	Eu, Oc	Ranfa and Bodesmo (2017)
	Platycodon grandiflorus (Jacq.) A. DC.	As	Li et al. (2014)
Cannabaceae	*Humulus lupulus* L.	Eu, As	Nadiroglu and Behçet (2018)
Cannaceae	*Canna indica* L.	As	Zheng et al. (2018)
Capparaceae	*Capparis spinosa* L.	Eu, As, Oc	Polat et al. (2017)
Caprifoliaceae	*Lonicera* spp. L.	Eu, As, Oc	Tardío et al. (2006)
Caricaceae	*Carica papaya* L.	Am	Janarny et al. (2021)
Caryophyllaceae	*Dianthus* spp. L.	All	Kritsi et al. (2022)
Celastraceae	*Celastrus paniculatus* Willd.	As	Ekka and Ekka (2016)
Convolvulaceae	*Convolvulus arvensis* L.	Eu	Tardío et al. (2006)
	Ipomoea cairica (L.) Sweet	As	Li et al. (2014)
Crassulaceae	*Sedum goldmanii* (Rose) Moran	Am	Figueredo-Urbina et al. (2022)
Cucurbitaceae	*Cucurbita* spp. L.	All	Aquino-Bolaños et al. (2013)
	Siraitia grosvenorii (Swingle) C. Jeffrey ex A.M.Lu & Zhi Y.Zhang	As	Zheng et al. (2018)
Cytinaceae	*Cytinus hypocistis* (L.) L.	Eu	Tardío et al. (2006)
Dilleniaceae	*Dillenia indica* L.	As	Sandhyarani Devi et al. (2009)
Elaeagnaceae	*Elaeagnus angustifolia* L.	As	Pieroni et al. (2005)
Ericaceae	*Arbutus xalapensis* Kunth	Am	Sotelo et al. (2007)
	Arctostaphylos pungens Kunth	Am	Figueredo-Urbina et al. (2022)
	Calluna vulgaris (L.) Hull	Eu	Polat et al. (2017)
	Rhododendron spp. L.	As	Sandhyarani Devi et al. (2009), Li et al. (2014)
Euphorbiaceae	*Cnidoscolus multilobus* (Pax) I.M.Johnst.	Am	Figueredo-Urbina et al. (2022)
	Jatropha integerrima Jacq.	As	Li et al. (2014)
Fabaceae	*Acacia* spp. Mill.	Eu, Oc	Casas et al. (2019)
	Albizia julibrissin Durazz.	As, Oc	Zeng et al. (2014)
	Anagyris foetida L.	Eu	Motti et al. (2022)
	Bauhinia variegata L.	As	Villavicencio et al. (2018)
	Cajanus cajan (L.) Huth	As	Sandhyarani Devi et al. (2009)
	Calliandra haematocephala Hassk.	As	Li et al. (2014)

(continued)

Table 2.1 (continued)

Family	Genera/species	Eaten in	Reference
	Cassia spp. L.	As	Kaisoon et al. (2011), Janarny et al. (2021)
	Cercis spp. L.	All	Amer et al. (2019), Lucchetti et al. (2019)
	Clitoria ternatea L.	As, Oc	Kaisoon et al. (2011)
	Crotalaria juncea L.	As	Janarny et al. (2021)
	Erythrina spp. L.	Am	Angami et al. (2006), Sotelo et al. (2007)
	Genista tridentata L.	Eu	Tardío et al. (2006)
	Gliricidia sepium (Jacq.) Kunth	Am	Cáceres and Cruz (2019)
	Kennedia prostrata R.Br.	Oc	Hansen and Horsfall (2019)
	Lathyrus sylvestris L.	Eu	Tardío et al. (2006)
	Leptosema aphyllum (Hook.) Crisp	Oc	Hansen and Horsfall (2019)
	Leptospron adenanthum (G.Mey.) A.Delgado	Am	Figueredo-Urbina et al. (2022)
	Leucaena leucocephala (Lam.) de Wit	Am, As	Kaisoon et al. (2011)
	Onobrychis humilis (Loefl.) G.López	Eu	Tardío et al. (2006)
	Neltuma laevigata (Humb. & Bonpl. ex Willd.) Britton & Rose	Am	Figueredo-Urbina et al. (2022)
	Ononis viscosa L.	Eu	Tardío et al. (2006)
	Parkia timoriana (DC.) Merr.	As	Sandhyarani Devi et al. (2009)
	Phaseolus coccineus L.	Am, Oc	Figueredo-Urbina et al. (2022)
	Robinia pseudoacacia L.	Eu, Am	Motti et al. (2022)
	Selloa spp. Kunth	Oc	Hansen and Horsfall (2019)
	Sesbania grandiflora (L.) Poir.	As	Loganayaki et al. (2012)
	Styphnolobium japonicum (L.) Schott	As	Zheng et al. (2018)
	Sulla coronaria (L.) B.H.Choi & H.Ohashi	Eu	Loizzo et al. (2016)
	Tamarindus indica L.	As, Af	Janarny et al. (2021)
	Templetonia retusa (Vent.) R.Br.	Oc	Hansen and Horsfall (2019)
	Trifolium spp. Tourn. ex L.	Eu	Vitalini et al. (2015), Danna et al. (2022)
	Vachellia farnesiana (L.) Wight & Arn.	Af	Ranfa and Bodesmo (2017)
	Wisteria floribunda (Willd.) DC.	As, Oc, Eu	Oh et al. (2008)
Fagaceae	*Castanea sativa* Mill.	Eu	Motti et al. (2022)
Gentianaceae	*Gentiana* spp. Tourn. ex L.	Eu	Motti et al. (2022)

(continued)

Table 2.1 (continued)

Family	Genera/species	Eaten in	Reference
Geraniaceae	*Pelargonium* spp. L'Hér. ex Aiton	As, Oc	Benvenuti et al. (2016), Marchioni et al. (2020b)
	Erodium cygnorum Nees	Oc	Hansen and Horsfall (2019)
Hypericaceae	*Hypericum perforatum* L.	Eu, As	Nadiroglu and Behçet (2018)
Iridaceae	*Crocus* spp. L.	Eu	Marchioni et al. (2019), Motti et al. (2022)
	Gladiolus (hybrids) Tourn. ex L.	Af, As, Eu	Li et al. (2014)
	Iris spp. Tourn. ex L.	As	Pieroni et al. (2005)
Lamiaceae	*Clinopodium alpinum* (L.) Kuntze	Eu	Tardío et al. (2006)
	Agastache spp. Clayton ex Gronov.	Am, As, Oc	Najar et al. (2019)
	Elsholtzia spp. Willd.	As	Sandhyarani Devi et al. (2009)
	Gmelina asiatica L.	As	Janarny et al. (2021)
	Lamium spp. L.	Eu	Baldi et al. (2022)
	Lavandula spp. L.	Eu, As	Tardío et al. (2006), Zeng et al. (2014)
	Melissa officinalis L.	Eu, As	Mattalia et al. (2020)
	Mentha spp. L.	As	Günes (2017), Mattalia et al. (2020)
	Micromeria juliana (L.) Benth. ex Rchb.	Eu, As	Motti et al. (2022)
	Monarda spp. L.	Am	Marchioni et al. (2020a)
	Nepeta spp. L.	Eu, Oc	Marchioni et al. (2020a)
	Ocimum spp. L.	Eu, As, Oc	Marchioni et al. (2020a)
	Origanum spp. L.	Eu, As	Tardío et al. (2006)
	Periploca laevigata Aiton	Eu	Tardío et al. (2006)
	Phlomis spp. L.	Eu	Tardío et al. (2006)
	Prunella spp. L.	Eu	Tardío et al. (2006), Zeng et al. (2014)
	Rotheca serrata (L.) Steane & Mabb.	Af, As	Sandhyarani Devi et al. (2009)
	Salvia spp. L.	All	Tardío et al. (2006), Marchioni et al. (2020b)
	Satureja montana L.	Eu	Łuczaj and Dolina (2015)
	Scutellaria discolor Wall. ex Benth.	As	Sandhyarani Devi et al. (2009)
	Sideritis spp. L.	Eu, Af	Mautone et al. (2019)
	Stachys lavandulifolia Vahl	Af	Motti et al. (2022)

(continued)

Table 2.1 (continued)

Family	Genera/species	Eaten in	Reference
	Teucrium spp. L.	Eu, Af	Łuczaj and Dolina (2015), Geraci et al. (2018)
	Thymbra capitata (L.) Cav.	Eu, Af	Dop et al. (2020)
	Thymus spp. L.	Af	Günes (2017)
Lauraceae	*Litsea cubeba (L. citrata)* (Lour.) Pers.	As	Sandhyarani Devi et al. (2009)
Liliaceae	*Fritillaria* spp. Tourn. ex L.	Eu	Motti et al. (2022), Tardío et al. (2006)
	Lilium spp. Tourn. ex L.	Eu, As	Zeng et al. (2014), Kritsi et al. (2022)
	Tulipa spp. L.	As	Mlcek and Rop (2011)
Loranthaceae	*Nuytsia floribunda* (Labill.) R.Br. ex G.Don	Oc	Hansen and Horsfall (2019)
Lythraceae	*Woodfordia fruticosa* (L.) Kurz	As	Arya et al. (2015)
	Punica granatum L.	Eu, As	Wongwattanasathien et al. (2010)
Magnoliaceae	*Magnolia* spp. Plum. ex L.	As, Am, Oc	Li et al. (2014), Mulik and Ozuna (2020)
Malvaceae	*Abutilon* spp. Mill.	Am	Facciola (1990), Reddy et al. (2007)
	Alcea rosea L.	Eu, As, Oc	Kalemba-Drozdz and Cierniak (2019)
	Alyogyne huegelii (Endl.) Fryxell	Oc	Hansen and Horsfall (2019)
	Bombax ceiba L.	As	Rameshwar et al. (2014)
	Callianthe spp. Donnell	Am	Facciola (1990), Reddy et al. (2007)
	Durio zibethinus L.	As	Janarny et al. (2021)
	Hibiscus spp. L.	As, Oc	Kalemba-Drozdz and Cierniak (2019), Park et al. (2022)
	Malva spp. Tourn. ex L.	Eu, Am, Oc	Loizzo et al. (2016)
	Malvaviscus arboreus Dill. ex Cav.	Eu	Kritsi et al. (2022)
	Quararibea funebris (La Llave) Vischer	Am	Mulik and Ozuna (2020)
	Tilia spp. L.	Eu	Nowicka and Wojdyło (2019)
Marcgraviaceae	*Souroubea exauriculata* Delpino	Am	Figueredo-Urbina et al. (2022)
Meliaceae	*Aglaia odorata* Lour.	As	Kandylis (2022)
	Azadirachta indica A.Juss.	As	Janarny et al. (2021)

(continued)

Table 2.1 (continued)

Family	Genera/species	Eaten in	Reference
Moraceae	*Artocarpus* spp. J.R.Forst. & G. Forst.	As	Gupta et al. (2020)
Moringaceae	*Moringa oleifera* Lam.	As	Arulmozhi and Wilson (2015)
Musaceae	*Ensete superbum* (Roxb.) Cheesman	As	Angami et al. (2006)
	Musa spp. L.	As	Kennedy (2009)
Myrtaceae	*Feijoa sellowiana* (O.Berg) O.Berg	Am, Oc	Magri et al. (2020), Vatrano et al. (2022)
	Kunzea muelleri (K. ericifolia) Benth.	Oc	Hansen and Horsfall (2019)
	Melaleuca spp. L.	Oc	Hansen and Horsfall (2019)
	Syzygium spp. Gaertn.	As, Af, Oc	Wongwattanasathien et al. (2010)
Nelumbonaceae	*Nelumbo nucifera* Gaertn.	As	Sandhyarani Devi et al. (2009)
Nyctaginaceae	*Bougainvillea* spp. Comm. ex Juss.	Am	Kaisoon et al. (2011)
Nymphaeaceae	*Nymphaea* spp. L.	As	Zheng et al. (2018)
Oleaceae	*Forsythia* × *intermedia* Zabel	Eu, Oc	Jakubczyk et al. (2022)
	Jasminum spp. L.	Eu, As, Oc	Zeng et al. (2014), Kritsi et al. (2022)
	Nyctanthes arbor-tristis L.	As	Sandhyarani Devi et al. (2009)
	Osmanthus fragrans Lour.	As	Zeng et al. (2014)
	Syringa vulgaris L.	As	Mlcek and Rop (2011)
Onagraceae	*Fuchsia* spp. Plum. ex L.	Am, Oc	Marchioni et al. (2020b)
	Oenothera biennis L.	Eu, As, Oc	Grzeszczuk et al. (2016)
Orchidaceae	*Cymbidium sinense* (Andrews) Willd.	As	Zheng et al. (2018)
	Dendrobium spp. Sw.	As	Zhao et al. (2021)
Orobanchaceae	*Bellardia trixago* (L.) All.	Eu	Tardío et al. (2006)
	Castilleja moranensis Kunth	Am	Figueredo-Urbina et al. (2022)
	Conopholis alpina Liebm.	Am	Figueredo-Urbina et al. (2022)
Oxalidaceae	*Oxalis debilis* Kunth	As	Li et al. (2014)
Paeoniaceae	*Paeonia* spp. L.	As, Oc	Voon et al. (2013)
Papaveraceae	*Fumaria capreolata* L.	Eu	Vitalini et al. (2013)
	Papaver rhoeas L.	Eu	Tardío et al. (2006)
Passifloraceae	*Passiflora* spp. L.	Am, As, Oc	Fernandes et al. (2017)

(continued)

Table 2.1 (continued)

Family	Genera/species	Eaten in	Reference
Penthoraceae	*Penthorum chinense* Pursh	As	Fakhri et al. (2022)
Phrymaceae	*Erythranthe* × *hybrida* (Voss) Silverside	Am, Af, As	Grzeszczuk et al. (2018)
Pinaceae	*Pinus pinaster* Aiton	Eu	Tardío et al. (2006)
Plantaginaceae	*Antirrhinum majus* L.	All	Drava et al. (2020)
	Linaria hirta (L.) Moench	Eu	Tardío et al. (2006)
	Veronica spp. L.	Eu	Pieroni and Giusti (2009)
Plumbaginaceae	*Limonium sinuatum* (L.) Mill.	As	Li et al. (2014)
Polemoniaceae	*Phlox paniculata* L.	Am, Oc, As	Gunes and Kahraman (2022)
Polygonaceae	*Antigonon leptopus* Hook. & Arn.	Am	Kaisoon et al. (2011)
	Calligonum comosum L'Hér.	As	Ranfa and Bodesmo (2017)
	Rumex spp. L.	Af, As Eu	Motti et al. (2022)
Pontederiaceae	*Pontederia* spp. L.	As	Deka and Nath (2014)
Primulaceae	*Cyclamen hederifolium* Aiton	As	Kandylis (2022)
	Primula spp. L.	Eu, Oc	Mattalia et al. (2013), Motti et al. (2022)
Proteaceae	*Banksia marginata (B. integrifolia)* Cav.	Oc	Lim (2014)
	Grevillea spp. R.Br. ex Knight	Oc	Hansen and Horsfall (2019)
Ranunculaceae	*Aquilegia vulgaris* L.	Eu	Motti et al. (2022)
	Caltha palustris L.	Eu	Polat et al. (2017)
Rosaceae	*Crataegus* spp. L.	Eu	Vitalini et al. (2015), Motti et al. (2022)
	Dryas octopetala L.	Eu	Pieroni and Giusti (2009)
	Eriobotrya japonica (R. bibas) (Thunb.) Lindl.	As	Zheng et al. (2018)
	Filipendula ulmaria (L.) Maxim.	Af, As, Oc	Barros et al. (2010)
	Malus spp. Mill.	As, Eu, Oc	Zheng et al. (2018)
	Prunus spp. L.	As, Eu	Zheng et al. (2018)
	Pseudocydonia sinensis (Dum. Cours.) C.K.Schneid.	As	Li et al. (2014)
	Pyrus spp. L.	As	He et al. (2015)
	Rosa spp. L.	Eu, As	Marchioni et al. (2021), Cui et al. (2022)
Rubiaceae	*Bouvardia longiflora* (Cav.) Kunth	Am	Figueredo-Urbina et al. (2022)
	Gardenia jasminoides J.Ellis	As, Oc	Zhang et al. (2017)
	Ixora spp. L.	As	Kaisoon et al. (2011)
	Paederia foetida L.	As	Sandhyarani Devi et al. (2009)

(continued)

Table 2.1 (continued)

Family	Genera/species	Eaten in	Reference
Rutaceae	*Aegle marmelos* (L.) Corrêa	As	Janarny et al. (2021)
	Citrus spp. L.	As, Eu, Oc	Kalemba-Drozdz and Cierniak (2019)
	Ruta graveolens L.	Eu	Motti et al. (2022)
Sapotaceae	*Madhuca longifolia* (L.) J.F.Macbr.	As	Janarny et al. (2021)
Sauraceae	*Houttuynia cordata* Thunb.	As	Sandhyarani Devi et al. (2009)
Solanaceae	*Brunfelsia brasiliensis* (Spreng.) L.B.Sm. & Downs	As	Li et al. (2014)
	Petunia spp. (hybrids) Juss.	Am	Benvenuti et al. (2016)
	Physalis coztomatl Moc. & Sessé ex Dunal	Am	Figueredo-Urbina et al. (2022)
Strelitziaceae	*Strelitzia reginae* Banks	As	Kandylis (2022)
Tamaricaceae	*Tamarix gallica* L.	Eu	Ksouri et al. (2009)
Theaceae	*Camellia japonica* L.	As, Oc	Li et al. (2014)
Thymelaeaceae	*Edgeworthia gardneri* (Wall.) Meisn.	As	Fakhri et al. (2022)
	Pimelea microcephala R.Br.	Oc	Hansen and Horsfall (2019)
Typhaceae	*Typha* spp. L.	Am, As	Pandey and Verma (2018)
Tropaeolaceae	*Tropaeolum majus* L.	Eu	Benvenuti et al. (2016)
Urticaceae	*Dendrocnide sinuata* (Blume) Chew	As	Deka and Nath (2014)
	Girardinia diversifolia (Link) Friis	As	Deka and Nath (2014)
Verbenaceae	*Lantana camara* L.	Am, As	Li et al. (2014)
	Verbena bonariensis L.	Am	Drava et al. (2020)
Viburnaceae	*Sambucus nigra* L.	Eu, Oc	Vitalini et al. (2013)
Violaceae	*Viola* spp. L.	All	Kucekova et al. (2013), Koike et al. (2015)
Zingiberaceae	*Alpinia galanga* (L.) Willd.	As	Rachkeeree et al. (2018)
	Amomum maximum Roxb.	As	Rachkeeree et al. (2018)
	Curcuma spp. L.	As	Angami et al. (2006), Rachkeeree et al. (2018)
	Etlingera elatior (Jack) R.M.Sm.	As	Rachkeeree et al. (2018)
	Hedychium forrestii Diels	As	Rachkeeree et al. (2018)
	Zingiber spp. Mill.	As	Rachkeeree et al. (2018)
Zygophyllaceae	*Zygophyllum* spp. L.	Af	Günes (2017)

Af Africa, *Am* America, *As* Asia, *Eu* Europe, *Oc* Oceania, *All* All world

(Caryophyllaceae), *Backhousia citriodora, Leptospermum liversidgei* (Myrtaceae), *Osteospermum* (Asteraceae) *Prostanthera* (Lamiaceae), *Wahlenbergia* (Campanulaceae) among the edible flowers typical of the Australian flora, but we have not found evidence of their use in the literature;

(b) probably several communities use local species for their subsistence and therefore not yet studied and not present in the literature;
(c) Ray et al. (2020) report the diversity of Indian wild edible flower mentioning only families and number of species. The authors do not report the name of the species, so we could not report their name in Table 2.1. The only information we can report is that in India, among others, EDIBLE FLOWERs belonging to the families Menyanthaceae, Basellaceae, Ophioglossaceae, Rhamnaceae, Piperaceae, Dioscoreaceae, Cycadaceae, Vitaceae and Poaceae are collected and used raw or in the preparation of traditional dishes.

Moreover, Hansen and Horsfall (2019) report that some nectar-rich flowers (*Beaufortia* spp., *Callistemon phoeniceus*, *Calothamnus* spp., *Grevillea* spp., *Kunzea ericifolia*, *Leptosema aphyllum*, *Nuytsia floribunda*) were collected by Australian aborigines for sweet drink.

The edible flowers mainly belong to the Asteraceae (42 genera), Fabaceae (36 genera) and Lamiaceae (24 genera) families, while pansy (Violaceae), roses (Rosaceae), nasturtiums (Tropaeolaceae), begonia (Begoniaceae), and carnation (Plantaginaceae) are the most cited flowers as ingredient or garnish of dishes. Many of the edible flowers listed are used for the preparation of teas, syrups and liqueurs, and for flavouring drinks (like *Artemisia* spp., *Gentiana* spp., *Matricaria* spp.).

Some flowers, while containing toxic substances, are treated to eliminate toxins before being used in food preparations, this group includes flowers belonging for example to the genus *Erythrina*, *Arbutus*, *Gliciridia*. (Sotelo et al. 2007). For many flowers only a part is edible: depending on the species, stems, sepal, petals or stamens must be removed. In the case of gladiolus or courgette the stamens are scarted; in rose only the petals are edible; while only ligulate flowers of inflorescence can be eaten in some Asteraceae and the central disc is not chewable. The pollen doesn't always have to be discarded. In early spring the green flowering spikes of *Typha* can be harvested, boiled and eaten like corn on the cob; in mid-summer, the pollen produced by the male flowers can be collected and used as a thickener or as a food supplement by mixing it with flour (Pandey and Verma 2018). Many flowers are harvested and consumed before flowering as is the case for capers and *Bombax ceiba* or the male flowers of *Musa* spp. (Angami et al. 2006; Rameshwar et al. 2014). Many perfume species such as roses, *Viola odorata* citrus fruits and tuberose are ideal for fresh consumption and the preparation of syrups and jams (Copetta et al. 2020; Marchioni et al. 2021). Flower extracts are used in confectionery or aromatherapy, and other flowers are used as additives in diets for their medicinal value especially in Asia. In this regard, Asia is the continent where the greatest variety of flowers is consumed with 150 different genera; followed by Europe (131), Oceania (64), America (61), and Africa (37). The consumption of edible flowers is not uniformly spread across all continents; for example, the areas of Central America (Mexico), the Mediterranean area, China, Southeast Asia and Indonesia are rich in traditional dishes and preparations based on flowers.

2.3 Wild Harvesting

Edible wild plants are an important part of traditional foods worldwide and in many areas allow to enrich and diversify (supplement food) the diet of communities (Reddy et al. 2007). Many of the edible flowers consumed especially in Central America, in the Mediterranean and Asia area are wild species and they are collected and consumed locally during the often-short blooming season. In Mexico edible flowers are widely used in regional gastronomy and are harvested from their natural habitats and generate income when they are sold. In particular, some flowers of arid regions such as *Agave salmiana*, *Aloe vera*, *Arbutus xalapensis Myrtillocactus geometrizans Erythrina americana*, *Erythrina caribaea*, *Euphorbia radiants* and *Yucca filifera* are in high demand and can be very expensive (Pinedo-Espinoza et al. 2020; Figueredo-Urbina et al. 2022). In Europe and Mediterranean area, with globalization and demographic growth, knowledge of the traditional and cultural use of wild edible flowers has been lost; the edible flowers mainly consumed are the cultivated ones, such as cauliflower, broccoli, and artichokes, while the consumption of wildflowers is limited to rural and mountainous areas for self-consumption and the main form of preparation is in drinks. However, in recent years there has been an attempt to rediscover the traditional use of wild plants including edible flowers (Mattalia et al. 2013) and the edible flowers of different cultures and countries (Fernandez et al. 2020). In Asia, the culture of consumption of edible wildflowers has always been widespread, especially in rural and semi-urban areas of China, India, and Southeast Asia. In these areas, wildflowers are a valuable nutritional resource as well as medicinal properties in fact they are part of traditional Chinese medicine (Li et al. 2014). Ray et al. (2020) evaluated the consumption of wildflowers in India and noted the use of 153 species (59 families) of which 27 are Fabaceae, 9 Lamiaceae, and 8 Moraceae.

The wild harvesting requires the ability and experience to recognize the edible flowers present in nature by the toxic or poisonous ones. Another possible risk related to the harvest of wild edible flowers could be represented by the possible concentration of pollutants (hydrocarbons, metals, and heavy metals) in their tissues (Drava et al. 2020); therefore, a knowledge of the collection environment is required by avoiding the areas close to the road arteries, those neighbour areas to industrial areas and cities or in any case with strong pollution. Lucarini et al. (2020) examined the information available on databases and bibliographies out of 62 edible flowers and found a very low risk of allergic reactions due to the ingestion of edible flowers because the protein content of this food is low, often the stamens with pollen are eliminated and the portions of raw consumption are low. In most cases, edible flowers should be harvested in the morning when the compounds that determine flavour, micronutrients, and bioactive substances are most concentrated. The harvest of withered flowers, with signs of infection and with insects must be avoided because they may have a bitter or unpleasant taste. Wildflowers should be used as soon as possible because they are sensitive to petal abscission, dehydration, colour change, wilting and tissue browning. Generally, they can be kept in the refrigerator from 2 days at a week by placing them in a container lined with damp paper towels to maintain their turgidity. In many cases the collection in nature is carried out for the preparation of herbal teas and therefore the edible flowers can be dried in a dehydrator or in a stove at low temperatures.

2.4 Domestication, Selection, Propagation, and Breeding

Many edible species grow wild and have not been domesticated, among them there are many species of edible flowers. This diversified range of wildflowers is consumed regularly in many areas of the world such as Central America and Southeast Asia and complete nutritional needs and expand the variety of gastronomic choice for humans. Domestication of plants is the process by which wild plants evolve into cultivated plants through artificial selection by humans. This process usually involves breeding activity followed by culture, evaluation and selection of progeny with the desired characteristics. Selection often involves exposing wild plants to new selective environments, and man controls the reproduction and dispersion of the species subject to domestication (Purugganan 2019). The systematic breeding and selection programs reserved for edible flowers are very limited.

2.4.1 In Brassica *Genus*

Among the most commonly consumed edible flowers because they are considered vegetables are broccoli (*B. oleracea* var. *italica*) and cauliflower (*B. oleracea* var. *botrytis*). Their inflorescences are harvested and cooked when the flowers are immature. The consumption of cauliflowers has been known since Roman times (Maggioni et al. 2018), over time they were selected suitable for cultivation in temperate mountain regions and hot and humid tropical climates, so their cultivation has developed in Europe, Asia, and Australia. In Italy, recent breeding programs have produced a wide range of colours and shapes (see Natalini et al. 2021). Broccoli, as we know it today, derives from breeding and selection activities that took place in southern Italy on cabbage. The hybridization process has led to the selection of heads with green, purple or white inflorescences, but also with high concentrations of chemicals considered beneficial to human health (Dixon 2017).

2.4.2 In Rosa *Genus*

The domestication process for ornamental plants subsequently used for food purposes has been more studied as in the case of roses. During the Han dynasty between 20 and 220 AD probably the Chinese began the domestication of roses in China and many of these oriental cultivars were exported to Europe; actually, Yunnan province is known for producing edible rose petals where about 20 edible cultivars like *Rosa* 'Kushui', *R.* 'Tuwei', YN01, *R.* 'Crimson Glory', *R. rugosa* 'Plena' and *R.* 'Zizhi' have been used (Cui et al. 2022). In Europe, in addition to wild species such as *R. canina* and *R. pendulina*, numerous rose hybrids and cultivars are edible, and the most commonly used are: *R. damascena*, *R. micrantha*, *R. rugosa*, *R. gallica,* and *R.* × *centifolia* (Demasi et al. 2021; Marchioni et al. 2021).

2.4.3 In Asteraceae Family

In the Asteraceae few species are so strongly domesticated and only the artichoke has been selected for the edible use of the flower; other species with edible flowers such as safflower (*Carthamus tinctorius*), endive (*Cichorium endivia*), chicory (*Cichorium intybus*), sunflower (*Helianthus annuus*) and lettuce (*Lactuca sativa*) have been domesticated for consumption of the leaves or the oils from the seeds (Dempewolf et al. 2008). Many species of Asteraceae have been heavily domesticated and hybridized mainly by selecting the aesthetic characters like *Chrysanthemum*, *Dahlia*, *Echinacea*, *Gerbera* and *Tagetes*. The hybridization and selection can actually be done on the basis of the nutritional content of the flowers in order to obtain edible flower varieties rich in nutrients and antioxidants.

2.4.4 The **Paeonia lactiflora** *Case Study*

Li et al. (2017) investigate to establish which ornamental herbaceous peony (*Paeonia lactiflora*) varieties can be selected based on nutritional qualities. Peony petals contain many chemical compounds such as vitamins, proteins, sugars, and organic acids, as well as rich mineral elements and amino acids. In addition, peony petals also contain a large number of total phenols, flavonoids, and other bioactive substances, which have a strong ability to scavenge free radicals. Among the 46 varieties analysed, the authors identify the ten most interesting cultivars with the best nutritional qualities for edible consumption, including 'Dielian Qihua', 'Zhushapan', 'Xueyuan Honghua', 'Wulong Jisheng', 'Honglou', 'Bingshan', 'Hongyan Yushuang', 'Zituo Ronghua', 'Zifengyu' and 'Fenlou Dianchun'.

2.4.5 The **Moringa oleifera** *Case Study*

Studies about the breeding of plants to produce edible flowers are scarce. Due to the high nutritional content of fruit, leaves, and flowers, *Moringa oleifera* deserves special attention for improvement, commercial cultivation, and use in India. In the northern regions of India, the edible parts of this tree, including the flower buds, are harvested from the wild; however, the cultivation in southern regions is more frequent and used wider as a vegetable crop, whereby *M. oleifera* has been domesticated and cultivated. In *M. oleifera* the selection of accessions from different regions takes place mainly by considering the abundance of leaves and fruits produced, rarely due to the abundance of flowering. Intending to increase the production and export of the commodity, cultivation practices and systematic production are necessary to increase the production on a commercial scale as an alternative crop in the northern region, especially in the dry/arid regions of India (Pandey et al. 2018).

2.4.6 The **Sesbania grandiflora** *Case Study*

Although a systematic breeding program leading to variety formation has not been organized in *Sesbania grandiflora*, these trees can be classified into four distinct groups based on flower colour: Sita with white flowers, Peeta with yellow flower, Neela with blue colour flowers and Lohita with red flower. The white and red flowering plants are used as vegetables while the other two groups are more popular for their medicinal uses. Furthermore, red-flowered *S. grandiflora* is more nutritious than white-flowered due to its high content of phenolic compounds (Karmakar et al. 2016).

2.4.7 In **Hibiscus** *Genus*

Hibiscus sabdariffa is one of the edible flowers treated in Africa, Asia, and Europe: its calyx is used as an ingredient for the preparation of drinks, syrups, and jams (Park et al. 2022). However, many species of the genus *Hibiscus* (more than 220 species distributed in tropical and subtropical regions) as well as having an ornamental value can be used as a nutritional resource. In this genus, many crosses between species have been carried out to obtain hybrids with desirable ornamental characteristics while the evaluation of their potential use as a source of nutrients is done retrospectively. Chemson et al. (2020) report the analysis of three ornamental cultivars ('Adonis', 'Artemis' and 'Rhea') of the hibiscus 'Titanbicus' to evaluate their use of flowers as a food resource. 'Titanbicus' is a hybrid between *Hibiscus moscheutos* and *Hibiscus coccineus. H. moscautos* is a shrubby, cold-hardy perennial, while the scarlet-flowered *H. coccineus* is a perennial found in swampy regions. The authors note that the flowers of the three cultivars are a source of functional compounds (anthocyanins and other phenolic compounds) and they can be used as a potential food source with antioxidant activity. Zhang et al. (2022) evaluate four different *H. syriacus* varieties by petal colour (white, pink, red, and purple) to determine which resource is better as an edible flower. The authors find that the most consumed flowers are those with pink and red petals, and by analyzing the stability of the colours and the content of antioxidant compounds they found that although the pink petals have a lower concentration of anthocyanins, the colour of their extract is beautiful after food preparation and processing. The red petals are very rich in anthocyanins and have better antioxidant properties, but the colour tends to become darker than that of fresh flowers. The purple petals are also rich in anthocyanins, but the colour is difficult to control during processing and for this reason it is less consumed by people.

2.4.8 Domestication of Wildflower Species

Many wildflower species grow in degraded soils that cannot be grown with conventional crops. The enhancement and domestication of these often underutilized and neglected species could be a means of using arable land in marginal areas that cannot be used by conventional agricultural production, allowing the diversification of crops, products and providing new sources of food with better nutritional properties (Corrêa et al. 2020). Furthermore, the use of such species into farming systems reduces the risk of genetic erosion in agroecosystems due to intensified cropping systems. In a recent article, the opportunity to cultivate with different fertilization regimes *Scolymus hispanicus* a highly appreciated wild herb for the consumption of both flowers and leaves was evaluated. The authors note that low to moderate fertilizer inputs can increase the fresh yield of *S. hispanicus* without compromising the nutrient profile of the leaves (Paschoalinotto et al. 2023). Unfortunately, the authors did not analyse the flowers. Safflower (*Carthamus tinctorius* L.) is a species belonging to the Asteraceae family known as one of the typical plants of the Middle East, which is its centre of origin and biodiversity. Safflower was initially domesticated and cultivated for medicinal uses for its flowers used as a food colouring and grown as an oilseed crop. This species has interesting agronomic characteristics considering climate change scenarios: it is resistant to drought and has adapted to arid and semi-arid conditions (Ali et al. 2020).

2.5 New and Future Perspectives

The cultivation and sale of edible flowers is still a niche market even though it is constantly growing. For this reason, the cultivation of edible flowers is neglected in the floriculture industry, although it can be considered an alternative source of income for the floriculture industry. The methods for the propagation of edible flower plants are the classic ones adopted in floriculture, i.e. using seeds, cuttings, offshoots, and layering. However, the cultivation method cannot be the same as used to produce cut flowers (where a large use of pesticides is foreseen) because the alimentary use foresees the absence of pesticide residues. Among the various products on the market, there are trays in plastic or biodegradable material with edible flowers and bags containing a mix of salads and edible flowers. For these two types of products, a flower shelf-life of at least 7–10 days from the moment of collection is required; then the edible flower shelf life means that only some species can be used for these products (Marchioni et al. 2020b). Producers of mixed salads with edible flowers would like the flower to have certain characteristics:

- all present and cultivable throughout the year (no seasonality);
- soilless cultivable;
- with rapid crop cycle;
- mechanized harvesting;
- does not get damaged during washing/disinfection (then washable);
- coloured petals but which do not release the colour by staining the salad;
- without strong flavours or aromas so that the taste of the salad is predominant;
- fresh consumable;
- intact consumable (not only petals, not remove sepals, stamens, or pistils);
- long shelf life;
- high nutritional compounds (not important or essential).

To date, a flower capable of meeting all these requirements has not yet been found. Currently the most used flowers are violets and nasturtiums (non-washable), or daisies, marigolds, carnations, and snapdragons (non-washable and only the petals are edible). One of the activities on which breeders will be able to concentrate will be organizing breeding programs to select new varieties of edible flowers that are able to satisfy the wishes of salad producers.

A new area where selection needs to be made is the use of edible flowers for the aroma enhancement (reconstitution) of dealcoholized wines. Sam et al. (2023) evaluated the use of edible flowers (rose, peach, and lily) for the reconstitution of a reverse osmosis dealcoholized red wine. Reverse osmosis and the addition of water during the dealcoholisation process can also cause the loss or alteration of the original wine aroma, compromising the sensory quality and acceptability of the product. The addition of edible flowers to dealcoholized wine has led to the development of reconstituted alcohol-free wines without alterations of the chemical parameters and with a notable improvement of the volatile compounds. For this type of product, it is necessary to select very fragrant edible flowers capable of releasing volatile compounds such as esters, alcohols, acids, and terpenes and interesting aromas for the palate. In the study carried out, rose flowers were the ones that gave the most satisfactory results. Currently, the use of edible flowers for the aromatic improvement of dealcoholised wines is not yet approved by the EU also because it is a relatively new sector, however, it is very interesting for wine producers.

References

Ali F, Yilmaz A, Chaudhary HJ, Nadeem MA, Rabbani MA, Arslan Y, Nawaz MA, Habyarimana E, Baloch FS (2020) Investigation of morphoagronomic performance and selection indices in the international safflower panel for breeding perspectives. Turk J Agric For 44(2):103–120. https://doi.org/10.3906/tar-1902-49

Amer J, Jaradat N, Hattab S, Al-hihi S, Juma'a R (2019) Traditional Palestinian medicinal plant *Cercis siliquastrum* (Judas tree) inhibits the DNA cell cycle of breast cancer – antimicrobial and antioxidant characteristics. Eur J Integr Med 27:90–96. https://doi.org/10.1016/j.eujim.2019.03.005

Angami A, Gajurel PR, Rethy P, Singh B, Kalita SK (2006) Status and potential of wild edible plants of Arunachal Pradesh. Indian J Tradit Knowl 5(4):541–550

Apicius MG. De re coquinaria. [On the subject of cooking]. Online sources: (1) original: De re coquinaria/Liber IV—Pandecter. https://la.wikisource.org/wiki/De_re_coquinaria/Liber_IV

Aquino-Bolaños EN, Urrutia-Hernández TA, López Del Castillo-Lozano M, Chavéz-Servia JL, Verdalet-Guzmán I (2013) Physicochemical parameters and antioxidant compounds in edible Squash (*Cucurbita pepo*) flower stored under controlled atmospheres. J Food Qual 36:302–308. https://doi.org/10.1111/jfq.12053

Arulmozhi K, Wilson L (2015) In vitro antioxidant properties of *Moringa oleifera* and *Tagetes erecta* flower extracts. Int J Microbiol App Sci 4(10):733–740

Arya A, Al-Obaidi MMJ, Karim RB, Taha H, Khan AK, Shahid N, Sayem AS, Looi CY, Mustafa MR, Mohd MA, Ali HM (2015) Extract of *Woodfordia fruticosa* flowers ameliorates hyperglycemia, oxidative stress and improves β-cell function in streptozotocin–nicotinamide induced diabetic rats. J Ethnopharmacol 175:229–240. https://doi.org/10.1016/j.jep.2015.08.057

Baldi A, Bruschi P, Campeggi S, Egea T, Rivera D, Obón C, Lenzi A (2022) The renaissance of wild food plants: insights from Tuscany (Italy). Foods. https://doi.org/10.3390/foods11030300

Barros L, Oliveira S, Carvalho AM, Ferreira ICFR (2010) In vitro antioxidant properties and characterization in nutrients and phytochemicals of six medicinal plants from the Portuguese folk medicine. Ind Crop Prod 32:572–579. https://doi.org/10.1016/j.indcrop.2010.07.012

Benvenuti S, Mazzoncini M (2021) The biodiversity of edible flowers: discovering new tastes and new health benedible flowerits. Front Plant Sci. https://doi.org/10.3389/fpls.2020.569499

Benvenuti S, Bortolotti E, Maggini R (2016) Antioxidant power, anthocyanin content and organoleptic performance of edible flowers. Sci Hortic 199:170–177. https://doi.org/10.1016/j.scienta.2015.12.052

Cáceres A, Cruz SM (2019) Edible seeds, leaves and flowers as Maya super foods: function and composition. Int J Phytocos Nat Ingred 6(2):1–5. https://doi.org/10.15171/ijpni.2019.02

Casas MP, Ribeiro D, Fernandes E, Dominguez H, Torres MD (2019) Bioactive properties of *Acacia dealbata* flowers extracts. Waste Biomass Valori 11:2549–2557. https://doi.org/10.1007/s12649-019-00639-4

Chemson S, Shimada Y, Nakayama H, Yoshida K, Kondo T, Katsuzaki H, Mishima T (2020) Determination of anthocyanins and antioxidants in 'Titanbicus' edible flowers in vitro and in vivo. Plant Foods Hum Nutr 75:265–271. https://doi.org/10.1007/s11130-020-00813-3

Chen NH, Wei S (2017) Factors influencing consumers' attitudes towards the consumption of edible flowers. Food Qual Prefer 56:93–100. https://doi.org/10.1016/j.foodqual.2016.10.001

Copetta A, Marchioni I, Mascarello C, Pistelli L, Cambournac L, Dimita R, Ruffoni B (2020) *Polianthes tuberosa* as edible flower: in vitro propagation and nutritional properties. Int J Food Eng 6(2):57–62. https://doi.org/10.18178/ijfe.6.2.57-62

Copetta A, Bazzicalupo M, Cassetti A, Marchioni I, Mascarello C, Cornara L, Pistelli La, Ruffoni B (2021a) Plant production and leaf anatomy of *Mertensia maritima* (L.) Gray: comparison of in vitro culture methods to improve acclimatization. Horticulturae. https://doi.org/10.3390/horticulturae7050111

Copetta A, Marchioni I, Ruffoni B (2021b) The edible flowers from woody ornamental plants. Acta Hortic 1331:195–204. https://doi.org/10.17660/ActaHortic.2021.1331.27

Corrêa RCG, Di Gioia F, Ferreira ICFR, Petropoulos SA (2020) Wild greens used in the Mediterranean diet. In: Preedy V, Watson R (eds) The Mediterranean diet: an evidence-based approach. Academic, London, pp 209–228

Cui WH, Du XY, Zhong MC, Fang W, Suo ZQ, Wang D, Dong X, Jiang XD, Hu JY (2022) Complex and reticulate origin of edible roses (*Rosa*, Rosaceae) in China. Hortic Res. https://doi.org/10.1093/hr/uhab051

Danna C, Poggio L, Smeriglio A, Mariotti M, Cornara L (2022) Ethnomedicinal and ethnobotanical survey in the Aosta Valley side of the Gran Paradiso National Park (Western Alps, Italy). Plants. https://doi.org/10.3390/plants11020170

Deka K, Nath N (2014) Documentation of edible flowers of Western Assam. Am J Phytomed Clin Ther 2(10):1124–1140

Demasi S, Mellano MG, Falla NM, Caser M, Scariot V (2021) Sensory profile, shelf life, and dynamics of bioactive compounds during cold storage of 17 edible flowers. Horticulturae. https://doi.org/10.3390/horticulturae7070166

Dempewolf H, Rieseberg LH, Cronk QC (2008) Crop domestication in the compositae: a family-wide trait assessment. Genet Resour Crop Evol 55:1141–1157. https://doi.org/10.1007/s10722-008-9315-0

Dixon GR (2017) The origins of edible brassicas. Plantsman 16(3):180–185

Dop MC, Kefi F, Karous O, Verger EO, Bahrini A, Ghrabi Z, El Ati J, Kennedy G, Termote C (2020) Identification and frequency of consumption of wild edible plants over a year in Central Tunisia: a mixed-methods approach. Public Health Nutr 23:782–794. https://doi.org/10.1017/S1368980019003409

Drava G, Iobbi V, Govaerts R, Minganti V, Copetta A, Ruffoni B, Bisio A (2020) Trace elements in edible flowers from Italy: further insights into health benedible flowerits and risks to consumers. Molecules. https://doi.org/10.3390/molecules25122891

Ekka NS, Ekka A (2016) Wild edible plants used by tribals of North-East Chhattisgarh (part-I), India. Res J Recent Sci 5:127–131

Eroglu-Ozkan E, Demirci-Kayiran S, Gulsoy-Toplan G, Mataraci-Kara E, Kurkcuoglu M (2018) Identification of volatile compounds and antimicrobial activities of *Muscari neglectum* growing in Turkey. Fresenius Environ Bull 27(7):4674–4678

Facciola S (1990) Cornucopia: a source book of edible plants. Kampong Publishers, Vista

Fakhri S, Tomas M, Capanoglu E, Hussain Y, Abbaszadeh F, Lu B, Wu XJ, Zou L, Smeriglio A, Simal-Gandara J, Cao H, Xiao J, Khan H (2022) Antioxidant and anticancer potentials of edible flowers: where do we stand? Crit Rev Food Sci Nutr 62(31):8589–8645. https://doi.org/10.1080/10408398.2021.1931022

Fernandes L, Casal S, Pereira JA, Saraiva JA, Ramalhosa E (2017) Edible flowers: a review of the nutritional, antioxidant, antimicrobial properties and effects on human health. J Food Compos Anal 60:38–50. https://doi.org/10.1016/j.jfca.2017.03.017

Fernandes L, Saraiva JA, Pereira JA, Casal S, Ramalhosa E (2019) Post-harvest technologies applied to edibleflowers: A review: Edible flowers preservation. Food Rev Int 35:132–154. https://doi.org/10.1080/87559129.2018.1473422

Fernandes L, Casal S, Pereira JA, Malheiro R, Rodrigues N, Saraiva JA, Ramalhosa E (2020) An overview on themarket of edible flowers. Food Rev Int 36: 258275. https://doi.org/10.1080/87559129.2019.1639727

Figueredo-Urbina CJ, Alvarez-Rios GD, Cortez Zarraga L (2022) Edible flowers commercialized in local markets of Pachuca de Soto, Hidalgo. Mexico Bot Sci 100(1):120–138. https://doi.org/10.17129/botsci.2831

Geraci A, Amato F, Di Noto G, Bazan G, Schicchi R (2018) The wild taxa utilized as vegetables in Sicily (Italy): a traditional component of the Mediterranean diet. J Ethnobiol Ethnomed 14:14. https://doi.org/10.1186/s13002-018-0215-x

Ghimeray AK, Sharma P, Bi G, Lamsal K, Ba G, Cho DH (2010) Wild edible flowering plants of the Illam Hills (eastern Nepal) mode of use by local community. Korean J Pl Taxon 40(1):74–77. https://doi.org/10.11110/kjpt.2010.40.1.074

Grocock CW, Grainger S (2006) Apicius: a critical edition with an introduction and an english translation of the Latin recipe text Apicius. Prospect, London

Grzeszczuk M, Stedible floweraniak A, Pachlowska A (2016) Biological value of various edible flower species. Acta Sci Pol Hortoru 15(2):109–119

Grzeszczuk M, Stedible floweraniak A, Meller E, Wysocka G (2018) Mineral composition of some edible flowers. J Elem 23:151–162. https://doi.org/10.5601/jelem.2017.22.2.1352

Günes F (2017) Food plants used in meriç town from Turkey. Indian J Pharm Educ Res 51(3):S271–S275. https://doi.org/10.5530/ijper.51.3s.28

Gunes Z, Kahraman O (2022) Edible ornamental plants used in landscaping areas: the case of Canakkale city center. AgroLife Sci J 11(2):66–72. https://doi.org/10.17930/AGL202228

Gupta AK, Rather MA, Kumar Jha A, Shashank A, Singhal S, Sharma M, Pathak U, Sharma D, Mastinu A (2020) *Artocarpus lakoocha* Roxb. and *Artocarpus heterophyllus* Lam. Flowers: new sources of bioactive compounds. Plants. https://doi.org/10.3390/plants9101329

Hansen V, Horsfall J (2019) Noongar bush tucker: bush food plants and fungi of the south-west of Western Australia. UWA Publishing, Perth, Australia

He J, Yin T, Chen Y, Cai L, Tai Z, Li Z, Liu C, Wang Y, Ding Z (2015) Phenolic compounds and antioxidant activities of edible flowers of *Pyrus pashia*. J Funct Foods 17:371–379. https://doi.org/10.1016/j.jff.2015.05.045

Idolo M, Motti R, Mazzoleni S (2010) Ethnobotanical and phytomedicinal knowledge in a long-history protected area, the Abruzzo, Lazio and Molise National Park (Italian Apennines). J Ethnopharmacol 127:379–395. https://doi.org/10.1016/j.jep.2009.10.027

Jakubczyk K, Koprowska K, Gottschling A, Janda-Milczarek K (2022) Edible flowers as a source of dietary fibre (total, insoluble and soluble) as a potential athlete's dietary supplement. Nutrients. https://doi.org/10.3390/nu14122470

Janarny G, Ranaweera KKDS, Gunathilake KDPP (2021) Antioxidant activities of hydro-methanolic extracts of Sri Lankan edible flowers. Biocatal Agric Biotechnol. https://doi.org/10.1016/j.bcab.2021.102081

Juarez-Truillo N, Monribot-Villanueva JL, Jimenez-Fernandez VM, Suarez-Montaño R, Aguilar-Colorado AS, Guerrero-Analco J, Jimenenez M (2018) Phytochemical characterization of izote (*Yucca elephantipes*) flowers. J Appl Bot Food Qual 91:202–210. https://doi.org/10.5073/JABFQ.2018.091.027

Kaisoon O, Siriamornpun S, Weerapreeyakul N, Meeso N (2011) Phenolic compounds and antioxidant activities of edible flowers from Thailand. J Funct Foods 3:88–99. https://doi.org/10.1016/j.jff.2011.03.002

Kalemba-Drozdz M, Cierniak A (2019) Antioxidant and genoprotective properties of extracts from edible flowers. J Food Nutr Res 58:42–50. https://doi.org/10.13140/RG.2.2.17195.49442/1

Kandylis P (2022) Phytochemicals and antioxidant properties of edible flowers. Appl Sci 12:9937. https://doi.org/10.3390/app12199937

Karmakar P, Singh V, Yadava RB, Singh B, Singh R, Kushwaha M (2016) Agathi [*Sesbania grandiflora* L. (Agast)]: current status of production, protection and genetic improvement. In: national symposium on vegetable legumes for soil and human health, pp 153-161

Kaval ˙ I, Behçet L, Çakilcioglu U (2015) Survey of wild food plants for human consumption in Geçitli (Hakkari, Turkey). J Tradit Knowl 14(2):183–190. https://doi.org/10.1016/j.jep.2014.05.014

Kennedy J (2009) Bananas and people in the homeland of genus *Musa*: not just pretty fruit. Ethnobot Res Appl 7:179–197. https://doi.org/10.17348/era.7.0.179-197

Kirker CL, Newman M (2016) Edible flowers: a global history. Reakition Books, London

Koike A, Barreira JC, Barros L, Santos-Buelga C, Villavicencio AL, Ferreira IC (2015) Edible flowers of *Viola tricolor* L. as a new functional food: antioxidant activity, individual phenolics and effects of gamma and electronbeam irradiation. Food Chem 179:6–14. https://doi.org/10.1016/j.foodchem.2015.01.123

Kresnapati BA, Ramandha MEP, Indriani N (2022) Familiar edible flowers in Indonesia. Pharm Clin J Nusantara 1(1):63–70

Kritsi E, Tsiaka T, Ioannou AG, Mantanika V, Strati IF, Panderi I, Zoumpoulakis P, Sinanoglou VJ (2022) In vitro and in silico studies to assess edible flowers' antioxidant activities. Appl Sci 12:7331. https://doi.org/10.3390/app12147331

Ksouri R, Falleh H, Megdiche W, Trabelsi N, Mhamdi B, Chaieb K, Bakrouf A, Magné C, Abdelly C (2009) Antioxidant and antimicrobial activities of the edible medicinal halophyte *Tamarix gallica* L. and related polyphenolic constituents. Food Chem Toxicol 47:2083–2091. https://doi.org/10.1016/j.fct.2009.05.040

Kucekova Z, Mlcek J, Humpolicek P, Rop O (2013) Edible flowers – antioxidant activity and impact on cell viability. Cent Eur J Biol 8:1023–1031. https://doi.org/10.2478/s11535-013-0212-y

Li AN, Li S, Li HB, Xu D, Xu XR, Chen F (2014) Total phenolic contents and antioxidant capacities of 51 edible and wild flowers. J Funct Foods 6:319–330. https://doi.org/10.1016/j.jff.2013.10.022

Li W, Yang S, Cui H, Ha Y, Tao J, Zhou C (2017) Nutritional evaluation of herbaceous peony (*Paeonia lactiflora* Pall.) petals. Emir J Food Agric 29(7):518–531. https://doi.org/10.9755/ejfa.2017-05-1061

Lim TK (2014) Banksia integrifolia. In: Edible medicinal and non medicinal plants. Springer, Dordrecht, pp 658–660. https://doi.org/10.1007/978-94-017-8748-2_52

Loganayaki N, Suganya N, Manian S (2012) Evaluation of edible flowers of agathi (*Sesbania grandiflora* L. Fabaceae) for in vivo anti-inflammatory and analgesic, and in vitro antioxidant potential. Food Sci Biotechnol 21(2):509–517. https://doi.org/10.1007/s10068-012-0065-6

Loizzo MR, Pugliese A, Bonesi M, Tenuta MC, Menichini F, Xiao J, Tundis R (2016) Edible flowers: a rich source of phytochemicals with antioxidant and hypoglycemic properties. J Agric Food Chem 64:2467–2474. https://doi.org/10.1021/acs.jafc.5b03092

Lu B, Li M, Yin R (2016) Phytochemical content, health benefits, and toxicology of common edible flowers: a review (2000-2015). Crit Rev Food Sci Nutr 56:S130–S148. https://doi.org/10.1080/10408398.2015.1078276

Lucarini M, Copetta A, Durazzo A, Gabrielli P, Lombardi-Boccia G, Lupotto E, Santini A, Ruffoni B (2020) A snapshot on food allergies: a case study on edible flowers. Sustainability. https://doi.org/10.3390/su12208709

Lucchetti L, Zitti S, Taffetani F (2019) Ethnobotanical uses in the Ancona district (Marche region, Central Italy). J Ethnobiol Ethnomed 15(9):1–33. https://doi.org/10.1186/s13002-0190288-1

Łuczaj Ł, Dolina K (2015) A hundred years of change in wild vegetable use in Southern Herzegovina. J Ethnopharmacol 166:297–304. https://doi.org/10.1016/j.jep.2015.02.033

Maggioni L, von Bothmer R, Poulsen G, Lipman E (2018) Domestication, diversity and use of *Brassica oleracea* L., based on ancient Greek and Latin texts. Genet Resour Crop Evol 65:137–159. https://doi.org/10.1007/s10722-017-0516-2

Magri A, Adiletta G, Petriccione M (2020) Evaluation of antioxidant systems and ascorbate-glutathione cycle in feijoa edible flowers at different flowering stages. Foods. https://doi.org/10.3390/foods9010095

Marchioni I, Pistelli L, Ferri B, Pl C, Copetta A, Pistelli L, Ruffoni B (2019) Preliminary studies on edible saffron bio-residues during different post-harvest storages. Bulg Chem Commun 51A:131–136

Marchioni I, Najar B, Ruffoni B, Copetta A, Pistelli Lu, Pistelli La (2020a) Bioactive compounds and aroma profile of some Lamiaceae edible flowers. Plants. https://doi.org/10.3390/plants9060691

Marchioni I, Pistelli L, Ferri B, Copetta A, Ruffoni B, Pistelli L, Najar B (2020b) Phytonutritional content and aroma profile changes during postharvest storage of edible flowers. Front Plant Sci. https://doi.org/10.3389/fpls.2020.590968

Marchioni I, Pistelli L, Copetta A, Dimita R, Descamps S, Cambournac L, Ruffoni B (2021) Edible rose as novel food with healthy value. Acta Hortic 1331:239–244. https://doi.org/10.17660/ActaHortic.2021.1331.32

Marchioni I, Najar B, Copetta A, Ferri B, Ruffoni B, Pistelli Lu, Pistelli La (2023) Phytonutritional and aromatic profiles of *Tulbaghia simmleri* Beauv. edible flowers during cold storage. Adv Hort Sci 37(1):25–32. https://doi.org/10.36253/ahsc-13857

Mattalia G, Quave CL, Pieroni A (2013) Traditional uses of wild food and medicinal plants among Brigasc, Kyé, and Provençal communities on the Western Italian Alps. Genet Resour Crop Evol 60:587–603. https://doi.org/10.1007/s10722-012-9859-x

Mattalia G, Sõukand R, Corvo P, Pieroni A (2020) Blended divergences: local food and medicinal plant uses among arbëreshë, occitans, and autochthonous calabrians living in Calabria, Southern Italy. Plant Biosyst 154(5):615–626. https://doi.org/10.1080/11263504.2019.1651786

Mautone M, de Martino L, de Feo V (2019) Ethnobotanical research in Cava de' Tirreni area, Southern Italy. J Ethnobiol Ethnomed 15:50. https://doi.org/10.1186/s13002-019-0330-3

Mlcek J, Rop O (2011) Fresh edible flowers of ornamental plants–A new source of nutraceutical foods. Trends Food Sci Tech 22:561–569. https://doi.org/10.1016/j.tifs.2011.04.006

Morton JF, Alvarez E, Quiñonez C (1990) Loroco, *Fernaldia pandurata* (Apocynaceae): a popular edible flower of Central America. Econ Bot 44:301–310. https://doi.org/10.1007/BF03183911

Motti R, Paura B, Cozzolino A, Falco BD (2022) Edible flowers used in some countries of the Mediterranean basin: an ethnobotanical overview. Plants. https://doi.org/10.3390/plants11233272

Mulik S, Ozuna C (2020) Mexican edible flowers: cultural background, traditional culinary uses, and potential health benefits. Int J Gastron Food Sci 21:100235. https://doi.org/10.1016/j.ijgfs.2020.100235

Nadiroglu M, Behçet L (2018) Traditional food uses of wild plants among the Karlıova (Bingöl-Turkey). Int J Nat Sci 2(2):57–71

Najar B, Marchioni I, Ruffoni B, Copetta A, Pistelli L, Pistelli L (2019) Volatilomic analysis of four edible flowers from *Agastache* genus. Molecules. https://doi.org/10.3390/molecules24244480

Natalini A, Acciarri N, Cardi T (2021) Breeding for nutritional and organoleptic quality in vegetable crops: the case of tomato and cauliflower. Agriculture. https://doi.org/10.3390/agriculture11070606

Nowicka P, Wojdyło A (2019) Anti-hyperglycemic and anticholinergic effects of natural antioxidant contents in edible flowers. Antioxidants 8(8):308–319. https://doi.org/10.3390/antiox8080308

Oh WG, Jang IC, Jeon GI, Park EJ, Park HR, Lee SC (2008) Antioxidative activity of extracts from *Wisteria floribunda* flowers. J Korean Soc Food Sci Nutr 37(6):677–683. https://doi.org/10.3746/jkfn.2008.37.6.677

Pandey A, Verma RK (2018) Taxonomical and pharmacological status of *Typha*: a review. Ann Plant Sci 7(3):2101–2106. https://doi.org/10.21746/aps.2018.7.3.2

Pandey A, Pradheep K, Gupta R, Roshini Nayar E, Bhandari DC (2018) 'Drumstick tree' (*Moringa oleifera* Lam.): a multipurpose potential species in India. Genet Resour Crop Evol 58:453–460. https://doi.org/10.1007/s10722-010-9629-6

Park Y, Kwon SH, Jang YL, Lee DH, Yang SO, Eo HJ, Park GH, Kwon HY (2022) Nutritional composition and phytochemical screening in different parts of *Hibiscus syriacus* L. Food Sci Nutr 10:3034–3042. https://doi.org/10.1002/fsn3.2899

Paschoalinotto BH, Polyzos N, Compocholi M, Rouphael Y, Alexopoulos A, Dias MI, Barros L, Petropoulos SA (2023) Domestication of wild edible species: the response of *Scolymus hispanicus* plants to different fertigation regimes. Horticulturae. https://doi.org/10.3390/horticulturae9010103

Pieroni A, Giusti ME (2009) Alpine ethnobotany in Italy: traditional knowledge of gastronomic and medicinal plants among the Occitans of the upper Varaita valley, Piedmont. J Ethnobiol Ethnomed 5:1–13. https://doi.org/10.1186/1746-4269-5-32

Pieroni A, Dibra B, Grishaj G, Grishaj I, Maçai SG (2005) Traditional phytotherapy of the Albanians of Lepushe, northern Albanian alps. Fitoterapia 76:379–399. https://doi.org/10.1016/j.fitote.2005.03.015

Pinedo-Espinoza JM, Gutiérrez-Tlahque J, Santiago-Saenz YO, Aguirre-Mancilla CL, Reyes-Fuentes M, López-Palestina CU (2020) Nutritional composition, bioactive compounds and antioxidant activity of wild edible flowers consumed in semiarid regions of Mexico. Plant Foods Hum Nutr 75:413–419. https://doi.org/10.1007/s11130-020-00822-2

Pires TC, Dias MI, Barros L, Calhelha RC, Alves MJ, Oliveira MB, Santos-Buelga C, Ferreira ICFR (2018) Edible flowers as sources of phenolic compounds with bioactive potential. Food Res Int 105:580–588. https://doi.org/10.1016/j.foodres.2017.11.014

Polat R, Güner B, Babacan EY, Çakılcıoglu U (2017) Survey of wild food plants for human consumption in Bingöl (Turkey). J Tradit Knowl 16(3):378–384. http://nopr.niscair.res.in/handle/123456789/42023

Purugganan MD (2019) Evolutionary insights into the nature of plant domestication. Curr Biol 29:R705–R714. https://doi.org/10.1016/j.cub.2019.05.053

Rachkeeree A, Kantadoung K, Suksathan R, Puangpradab R, Page PA, Sommano SR (2018) Nutritional compositions and phytochemical properties of the edible flowers from selected Zingiberaceae found in Thailand. Front Nutr 5:3. https://doi.org/10.3389/fnut.2018.00003

Rahim MA, Tanni ST, Islam MN, Robbani MG (2022) Genetic variability studies of a novel health promoting leafy vegetable Rucola (*Eruca sativa*). J Bangladesh Agric 47(2):53–63. https://doi.org/10.3329/bjagri.v47i2.63326

Rameshwar V, Kishor D, Tushar G, Siddharth G, Sudarshan G (2014) A pharmacognostic and pharmacological overview on *Bombax ceiba*. Sch Acad J Pharm 3(2):100–107

Ranfa A, Bodesmo M (2017) An ethnobotanical investigation of traditional knowledge and uses of edible wild plants in the Umbria Region. Central Italy. J Appl Bot Food Qual 90:246–258. https://doi.org/10.5073/JABFQ.2017.090.031

Ray A, Ray R, Sreevidya EA (2020) How many wild edible plants do we eat—their diversity, use, and implications for sustainable food system: an exploratory analysis in India. Front Sustain Food Syst 4:56. https://doi.org/10.3389/fsufs.2020.00056

Reddy KN, Pattanaik C, Reddy CS, Raju VS (2007) Traditional knowledge on wild food plants in Andhra Pradesh. Indian J Tradit Knowl 6(1):223–229

Rop O, Mlcek J, Jurikova T, Neugebauerova J, Vabkova J (2012) Edible flowers – a new promising source of mineral elements in human nutrition. Molecules. https://doi.org/10.3390/molecules17066672

Sam FE, Ma T, Wang J, Liang Y, Sheng W, Li J, Jiang Y, Zhang B (2023) Aroma improvement of dealcoholized Merlot red wine using edible flowers. Food Chem. https://doi.org/10.1016/j.foodchem.2022.134711

Sandhyarani Devi KH, Sanatombi Devi Y, Singh PK (2009) A census of edible flowers found in the valley districts of Manipur. J Econ Taxon Bot 33(1):232–239. https://doi.org/10.1016/j.foodchem.2022.134711

Simsek I, Aytekin F, Yesilada E, Yildirimli S (2004) An ethnobotanical survey of the Beypazari, Ayas, and Güdül District Towns of Ankara Province (Turkey). Econ Bot 58:705–720. https://doi.org/10.1663/0013-0001(2004)058[0705:AESOTB]2.0.CO;2

Singh O, Khanam Z, Misra N, Srivastava MK (2011) Chamomile (*Matricaria chamomilla* L.): an overview. Pharmacogn Rev 5:82–95. https://doi.org/10.4103/0973-7847.79103

Sotelo A, López-García S, Basurto-Peña F (2007) Content of nutrient and antinutrient in edible flowers of wild plants in Mexico. Plant Foods Hum Nutr 62:133–138. https://doi.org/10.1007/s11130-007-0053-9

Tardío J, Pardo-de-Santayana M, Morales R (2006) Ethnobotanical review of wild edible plants in Spain. Bot J Linn Soc 152:27–71. https://doi.org/10.1111/j.1095-8339.2006.00549.x

Ukiya M, Akihisa T, Yasukawa K, Kasahara Y, Kimura Y, Koike K, Nikaido T, Takido M (2001) Constituents of Compositae plants. 2. Triterpene diols, triols, and their 3-O-fatty acid esters from edible chrysanthemum flower extract and their anti-inflammatory effects. J Agric Food Chem 49:3187–3197. https://doi.org/10.1021/jf010164e

Uthpala TGG, Navaratne SB (2021) *Acmella oleracea* plant; identification, applications and use as an emerging food source–review. Food Rev Int 37(4):399–414. https://doi.org/10.1080/87559129.2019.1709201

Vatrano T, Amenta M, Copetta A, Guardo M, Nunziata A, Strano MC, Petriccione M (2022) Multifunctional role of *Acca sellowiana* from farm management to postharvest life: a review. Agronomy. https://doi.org/10.3390/agronomy12081802

Villavicencio AL, Heleno SA, Calhelha RC, Santos-Buelga C, Barros L, Ferreira IC (2018) The influence of electron beam radiation in the nutritional value, chemical composition and bioactivities of edible flowers of *Bauhinia variegata* L. var. candida alba Buch.-Ham from Brazil. Food Chem 241:163–170. https://doi.org/10.1016/j.foodchem.2017.08.093

Vitalini S, Iriti M, Puricelli C, Ciuchi D, Segale A, Fico G (2013) Traditional knowledge on medicinal and food plants used in Val San Giacomo (Sondrio, Italy) – an Alpine ethnobotanical study. J Ethnopharmacol 145:517–529. https://doi.org/10.1016/j.jep.2012.11.024

Vitalini S, Puricelli C, Mikerezi I, Iriti M (2015) Plants, people and traditions: ethnobotanical survey in the lombard Stelvio National Park and neighbouring areas (Central Alps, Italy). J Ethnopharmacol 173:435–458. https://doi.org/10.1016/j.jep.2015.05.036

Voon HC, Bhat R, Karim AA, Rosma A (2013) Composition of tree peony (*Paeonia suffruticosa*) and Chinese apple flower (*Malus* spp.) buds. Int Food Res J 20(3):1173–1179

Wongwattanasathien O, Kangsadalampai K, Tongyonk L (2010) Antimutagenicity of some flowers grown in Thailand. Food Chem Toxicol 48:1045–1051. https://doi.org/10.1016/j.fct.2010.01.018

Zeng Y, Deng M, Lv Z, Peng Y (2014) Evaluation of antioxidant activities of extracts from 19 Chinese edible flowers. Springerplus 3:1–5. https://doi.org/10.1186/2193-1801-3-315

Zhang H, Feng N, Xu YT, Li TX, Gao XM, Zhu Y, Song YS, Wang YN, Wu HH (2017) Chemical constituents from the flowers of wild *Gardenia jasminoides* J. Ellis. Chem Biodivers 14(5):e1600437. https://doi.org/10.1002/cbdv.201600437

Zhang P, Li Y, Chong S, Yan S, Yu R, Chen R, Si J, Zhang X (2022) Identification and quantitative analysis of anthocyanins composition and their stability from different strains of *Hibiscus syriacus* L. flowers. Ind Crop Pro 177:114457. https://doi.org/10.1016/j.indcrop.2021.114457

Zhao M, Fan J, Liu Q, Luo H, Tang Q, Li C, Zao J, Zhang X (2021) Phytochemical profiles of edible flowers of medicinal plants of *Dendrobium officinale* and *Dendrobium devonianum*. Food Sci Nutr 9(12):6575–6586. https://doi.org/10.1002/fsn3.2602

Zheng J, Yu X, Maninder M, Xu B (2018) Total phenolics and antioxidants profiles of commonly consumed edible flowers in China. Int J Food Prop 21(1):1524–1540. https://doi.org/10.1080/10942912.2018.1494195

Chapter 3
Soil Cultivation of Edible Flowers

Andrea Copetta, Carlo Mascarello, Federico Tinivella, Giovanni Minuto, Alessandro Triantafyllidis, Sara Petrucci, and Barbara Ruffoni

Abstract Edible flowers are now quite popular and their use is common in all the continents with some differentiations. In some countries the use is traditional but in some others is emerging. There are peculiar systems to provide fresh material depending on the market claim and on the type of species and the season. Many species of edible flowers are produced as vegetables since years or are MAPs (Medicinal and aromatic plants) or, again, they are species with industrial uses for transformed products as liqueurs, teas, jams, and others. They have standardized production protocols that can change according to the country of production or the final product. In this chapter the theme of the new flowers considered as "edible" are taken into consideration; they follow strictly the rules for food cultivation and the new orientation of the social and environmental approach of the sustainability of the land use. Owing to their emerging value, up to now the land used by farmers to grow edible flowers is often a remaining strip of other agricultural crops, usually vegetables or fruits or aromatic plants. Through edible flower cultivation, the growers optimize land use, both in a greenhouse and out in the field. The focal point is the reduced or absent use of non-organic pesticides. Instead, edible flowers should be grown organically to maintain high sanitary and nutritional value combined with the aesthetic requirements. In this chapter the needs and systems to grow edible flowers are described and discussed together with some specific examples.

Keywords Biocontrol · Biological control · Cultivation practices · Integrated pest management · Organic culture · Symbiosis

A. Copetta · C. Mascarello · B. Ruffoni (✉)
CREA – Research Centre for Vegetable and Ornamental Crops, Sanremo, IM, Italy
e-mail: barbara.ruffoni@crea.gov.it

F. Tinivella · G. Minuto
Centro di Sperimentazione e Assistenza Agricola – Regione Rollo, Albenga, SV, Italy

A. Triantafyllidis · S. Petrucci
AIAB Liguria, Varese Ligure, SP, Italy

M. M. Jowkar et al. (eds.), *Advances in Research on Edible Flowers*, Advances in Olericulture, https://doi.org/10.1007/978-3-032-04561-4_3

Abbreviations

AMF	Arbuscular mycorrhizal fungi
B. D747	Bacillus amyloliquefaciens subsp. plantarum strain D747
Ca	Calcium
EU	European Union
IFOAM	International Federation of Organic Agriculture Movements
IPM	Integrate pest management
MAP	Medicinal and aromatic plant
Mg	Magnesium
NOP	National Organic Program
NPK	Nitrogen, phosphorus, potassium
PGPB	Plant growth-promoting bacteria

3.1 Introduction

Over the past few years edible flowers have become a popular decoration and garnishes on food dishes and cocktails. In recent years the edible flowers acquired high importance in diet owing the numerous reports about the useful metabolite composition; furthermore, the organoleptic analysis carried out also from professional taste panel, permitted to recognize them as innovative ingredients of the food preparations (Fernandes et al. 2020). Edible flowers are used all over the world and in each continent there are peculiar systems to provide fresh material depending on the market claim and on the type of species and the season. Many of them belong to woody or semiwoody species, in Europe, it is possible to list plants belonging to Fabaceae (acacia, mimosa, Judas tree, wisteria) and Malvaceae (hibiscus and linden) along with fruit species as orange, apple, and pear tree and elder, jasmine, lilac that, having interesting taste, are still source of edible flowers for traditional and innovative food preparations. In other parts of the world, several exotic flowers of climbing plants and trees are consumed as supplements in the diet (Copetta et al. 2022). Among these species *Clitoria ternatea*, *Erythrina americana*, *Erythrina berteroana*, *Sesbania grandiflora* (Fabaceae), *Bombax ceiba* (Malvaceae), *Arbutus xalopensis* (Ericaceae), *Acca sellowiana* (Mirtaceae), *Madhuca indica* (Sapotaceae), *Moringa oleifera* (Moringaceae), *Musa acuminata* (Musaceae), *Yucca* sp. (Asparagaceae) are an important source of proteins, carbohydrates, vitamins, mineral substances and antioxidants. The flowers of these species can be collected in the wild or in gardens not really specialized or can be cultivated in soil and/or in pot depending on the delivery form.

Many species of edible flowers are produced as vegetables since years as artichoke, cauliflower, or are MAPs (medicinal and aromatic plants) as chamomile, mallow, crocus, echinacea, lavender, calendula or, again, they are species with industrial uses for transformed products as liqueurs, teas, jams, and others as, among others, *Ibiscus sabdariffa*, rose (Pires et al. 2023). They have still agronomic production protocols standardized over years that can change following the country of production or the final product.

Different is the situation of the "new "edible flowers, used itself fresh or dehydrated for new use and for new consumers; they follow strictly the rules for food cultivation and the new orientation of the social and environmental approach of the sustainability of the land use. Some popular edible flowers include chrysanthemum, daylily, lilac, mint, nasturtium, pansy, rose, tulip, and violet. Blossoms from various vegetable and fruit crops are also enjoyed for culinary purposes.

For farmers, it all starts on a flower field, where the soil is tested for optimal pH levels and fertility. The land used by farmers to grow edible flowers is often a remaining strip of other agricultural crops, usually vegetables or fruits or aromatic plants. Through edible flower cultivation, the growers optimize land use, both in a greenhouse and out in the field. The golden rule in edible flower farming is the reduced or absent use of non-organic pesticides. Instead, edible flowers should be grown organically with just four ingredients: the sun, soil, organic fertilizers and water.

In this chapter the cultivation in the open field of the annual or plurennial edible flower species is reported with some examples and the different systems for safe growth of edible flowers are presented and discussed.

3.2 The Distribution of the Use in the World

As reported in Chap. 1 the use of edible flowers is mainly concentrated in three geographic areas: Asia, Central and south America and Europe, even if some edible examples could be founded also in Oceania. Each area has different traditional uses and this reflects the cultivation needs and procedures. When the material is used to be mainly cooked it is possible to perform collection in the wild as in Asia and America without considering the cultivation system. When the most important use is with fresh materials it is important to verify that the cultivation protocol could ensure the requested sanitary conditions. In Table 3.1 the most important species all over the world are reported together with information about the utilization.

3.3 Propagation

For the cultivated species the starting point is the propagation that can be done either by seed or in vegetative way that can maintain better the characters of the selected accession. It must be noticed that for the edible flowers not yet grown for industrial uses, genetic improvement or breeding is lacking and it is possible to speak about accessions of selected seed.

Table 3.1 Examples of important edible flower species with specific information about origin, main use, cultivation, packaging

Edible species	Origin	Use	Annual/ pluriannual	Cultivation	Packaging	Type
Anthirrinum majus	Europe	Salad	Annual	Pots or full ground	Bag	Herbaceous
Begonia cucullata	South America	Various	Annual	Pots	Bag	Herbaceous
Borago officinalis	Europe	Decoration	Annual	Full ground	Bag	Herbaceous
Calendula officinalis	Europe	Salad	Annual	Pots or full ground	Bag	Herbaceous
Centaurea cyanus	Europe	Salad	Annual	full ground	Bag	Herbaceous
Dianthus spp.	Asia and Europe	Salad	Pluriannual	Pots or full ground	Bag	Herbaceous
Hibiscus spp.	Asia and Europe	Dried for tea	Annual/ pluriannual	Full ground	Bag	Herbaceous or woody
Lavandula spp.	Europe	Dried	Pluriannual	Full ground	Bag	Woody
Rosa spp.	Asia and Europe	Fresh and dried	Pluriannual	Full ground	Bag	Woody
Tagetes spp.	America	Salad	Annual	Pots or full ground	Bag	Herbaceous
Tropaeolum majus	Europe	Salad	Annual	Pots or full ground	Bag	Herbaceous
Viola spp.	Europe	Salad	Annual	Pots	Bag	Herbaceous

3.3.1 Propagation by Seed

Seed propagation permitted to produce material with low cost but with a certain degree of genetic variability, depending on the species and or accessions, or variety. Some of them are rather stable but others can be very variable leading to a certain degree of discrepancy. Commonly cultivated annuals and perennials can be raised for their edible flowers. In general, large cultivation of annual species can be obtained in field directly from sowing (Falla et al. 2020). In countries with a continental climate, sometimes seeds are pre-sowed and cultivated indoor in the plastic or glass greenhouse. In this case, once the last frost is over, the seeds are transplanted outside. During the sowing period, the same fertilizers used for vegetable production are also used on edible flowers. Once the sowing is over, the seeds are regularly watered.

Most edible flowers including chamomile, chives, cornflowers, courgettes and marigolds can be grown from seed. They can be sown directly in the ground or in a pot. They are hardy so they can be sown directly outside. The annual plants, like cornflower and marigold will flower and set seed within a year, while the perennials, as chamomile and chives may not flower until the following year.

3.3.2 Vegetative Propagation

Vegetative propagation is often used when the crop plants either do not produce seeds or when the seeds produced are not viable or are of long dormancy. It is a type of asexual reproduction in plants, where a new plant grows from a fragment or cutting of the parent plant or specialized reproductive structures, which are sometimes called vegetative propagules. This process can occur naturally in many plants, but it can also be induced artificially. Horticulturists have developed asexual propagation techniques that use vegetative propagules to replicate plants both in vivo or in vitro (micropropagation). Vegetatively reproduced plants are clones, with the genetic background identical to that of the mother plant. Vegetative propagation can happen through the use of any vegetative part of the plant, such as leaves, stems, and roots to produce new plants or through growth from specialized vegetative plant parts in geophytes such as stolons, rhizomes, tubers, corms, or bulbs (Yadad and Singh 2018). In the hands of the grower, vegetative propagation has been a powerful device to prevent genetic segregation and to "fix" desired genotypes. By discarding sexual reproduction and inventing clonal propagation, the farmer was able to select, in a single act, exceptional individuals with desirable traits from among a large number of variables, inferior plants. If the environment in which they are growing does not change, vegetatively propagated plants will always grow true to type with regard to agronomic characteristics like size and shape of harvestable components, quality (nutrient content, taste and smell, and other chemical compositions), and storage qualities. The use of vegetative propagation regarding edible flowers depends from the relationship between cost and benefit. The seed has low cost, and the cuttings have a cost depending on the mother plant and the labour cost. In case of large cultivation seed sowing is preferable but in case of particular varieties to be grown in greenhouses, in pot, the cloned plants are indicated for the homogeneity of the production. For instance, *Monarda*, lavender and pinks are easy to grow from plug plants or young plants produced by vegetative propagation as cutting from mother selected plants.

3.4 Cultivation Practices of Edible Flowers

Annual species complete their life cycle from seed to vegetative plant, to bloom, to setting seed, to death of the plant in one growing season. They perform quickly, especially if transplants are used, and provide relatively long periods of bloom.

Biennials species are typically planted in June–July and complete their growing season the following spring. Perennial plants live more than 2 years and, once established, bloom each year. Some die back to the ground in their off- season; others retain foliage year-round.

Perennials species require more maintenance than annuals, like shrubs and trees, grow for many years, flowering each year.

The cultural requirements for edible flowers are very similar to those of ornamental flowers. In general, edible flowers prefer fertile, well-drained soil and full sun throughout the day. Growing plants in raised beds will improve drainage, reduce diseases and increase ease of harvest.

Harvest and storage: Fully opened edible flowers are harvested in the cool of the day during the peak of bloom. Only flowers free of insect and disease problems should be selected. To maintain freshness, flowers should be cooled immediately after harvest. The stems, sepals, pistils, and stamens of most flowers are removed prior to use. Pollen may detract from the flower's flavour and may cause allergies in some people. The sepals should be removed from all flowers except violas, pansies, and *Viola tricolour*. In many flowers (including rose, lavender, tulip, calendula, and chrysanthemum) only the petals are edible. If the petals have a white base, this area should be removed as it may have a bitter taste. For example, chrysanthemum, dianthus, marigold, and rose have bitter petal bases.

Including edible flowers in the diet can add a dash of flavour while at the same time enhancing the nutritional value of the food. However, caution is advised in picking the right kind of flowers under appropriate conditions.

3.5 Cultivation Generalities

Considering that talking about edible flowers means considering the most disparate types of species, it is necessary to know the characteristics of each crop and above all the type of use, which can range from a well-known industrial use to little specific consumption for special high level food preparations. In the case of industrial use it is still possible to plan the set-up of large mechanizable extensions. If we consider edible flowers for direct sale or artisanal use, we must envisage a more collected management, with an association of different species and with a gradual collection which involves more complex cultivation (Falla et al. 2020) (Fig. 3.1).

As an example, chamomile that is used mainly dry for herbal teas and has simple management, the best exposure is full sun. The plant prefers dry soil, even poor ones, and tolerates a certain level of limestone. At the time of tillage it is good to incorporate 3 or 4 kg per square meter of mature compost, because even if chamomile adapts to poor soils, maintaining a good level of organic matter in the soil is always important regardless of the crop this hosts. But later, no other particular fertilizations are needed for this species.

We take then into consideration the most important factors that can condition the extensive cultivation of plants for edible flowers production.

Fig. 3.1 Field of edible flowers for direct sale: different species at different stages of development are grown simultaneously, the plants are arranged in parallel rows alternating with a meadow. (Photo by Katrin Mamberto—La Rutabaga farm, Finale Ligure, Italy)

3.5.1 Climate

The climate is determined by various factors that exist in a specific geographical region, it depends not only on the latitude and longitude but also on the altitude and on the typology and characteristics of the reliefs. In recent years the climate has been changing due to well-known environmental problems and this creates greater uncertainty in dealing with crops. The climatic elements that can condition a crop are represented by the wind which on the one hand allows pollination but if of strong and constant intensity it can condition the humidity of the environment and generate problems. Important elements are solar radiation and the atmospheric temperature correlated to it, which, depending on the environment and the aforementioned variables, can vary greatly from day to night, influencing times and types of cultivation. The photoperiod is in fact what must be controlled to determine not only the growth of the plant but, what is really important in the case of edible flowers, the flowering in terms of efficiency and duration. Important factors are also the composition and consistency of the soil.

The plants that produce edible flowers belong to numerous genera and species and come from areas with different climates; species originating from the

Table 3.2 Optimal thermal levels of development of some species with edible flowers

Species	Optimal growing temperature (°C)	Exposure
Primula acaulis	12–15	Half shade
Petunia x hybrida	20–22	Full sun
Tagetes patula	18–20	Full sun
Viola tricolour	16–18	Full sun
Anthirrinum majus	18–20	Full sun
Ageratum spp.	14–16	Full sun
Dahlia variabilis	15–18	Full sun
Fuchsia spp.	20–25	Half shade
Polianthes tuberosa	18–22	Full sun
Mertensia maritima	15–18	Half shade
Dianthus chinensis	18–22	Full sun
Tropaeolum majus	18–22	Full sun

Mediterranean area such as *Anthirrinum, Borago, Calendula, Crocus, Dianthus, Nepeta, Viola*, or others which originate from areas characterized by a Mediterranean-type climate (*Pycnanthemum, Pelargonium, Tulbaghia*) can be easily cultivated in land with this climate, without particular adaptive needs. Species such as *Hibiscus sabdariffa, Fuchsia, Petunia*, on the other hand, have high and more constant thermal requirements during the year, while others such as primroses and *Tagetes lemmonii*, have low thermal requirements. Table 3.2 shows the optimal thermal levels of development and the exposure of some edible flowers.

3.5.2 Light

Light is very important for edible flowers cultivation because, in addition to the plant growth, it conditions the flower colour intensity. Except for the fuchsias which need constant shade, the other species generally display their bright colours and adequate development when placed in full sun. In summer and in the hottest hours, however, a slight shade may be useful to avoid leaf burns or breaks in the flower colour. It is necessary to evaluate carefully the crop density as it conditions the uniform and harmonious development of the plant. Then there are some species whose flowering is conditioned by the day length: *Fuchsia* and *Petunia*, for example, are species with an obliged long day, therefore the plants have the induction to flower only if the day is longer than 14 hours; the long day also positively influences the flowering of *Ageratum, Dianthus chinensis, Tropaeolum* and *Anthirrinum.*

3.5.3 Water Quality and Irrigation Management

To set up a correct irrigation of cultivated fields for edible flowers, first of all it is necessary to verify the water requirement of the species which depends on the physiological characteristics as well as on the place of origin. In a scenario of climate change, it is increasingly common to verify the presence of periods of drought followed by strong and concentrated rainfall. Water management takes place considering the cultivation site and its characteristics, the water capacity of the soil, the quality and constant availability of water, the water requirement of the plant also defined according to the different stages of development.

Irrigation techniques include various possibilities that largely depend on the availability of water, in case of water surplus a furrow watering system can be used or, in case of need for water saving, a drip system which can be more controlled (Fig. 3.2). A system used for aromatic and medicinal plants (MAPs) is controlled water stress which allows irrigation in different quantities according to the phenological phase of the plant. In MAPs this system has a physical elicitation effect by affecting the presence of metabolites produced in response to stress. Also in edible species with a high content of secondary metabolites and antioxidants, the

Fig. 3.2 Organic cultivation of mixed edible flowers equipped with drip irrigation system. (Photo by Katrin Mamberto)

application of this system, in addition to obtaining significant water savings, can contribute to increasing the active substances contained in the flower (Liu et al. 2023). Given the irregularity of the climate, it may be useful to have evapotranspiration data management systems that automatically define water interventions.

3.5.4 Nutrients and Fertilization

Watering must be combined with fertilizing. The nutrients supply must be calibrated related to the species and respecting the crop planning. It is always good practice to carry out a preliminary analysis of the soil, both physical and chemical, which allows the precise determination of the structural and nutritional conditions useful to define corrective interventions that can be carried out (Kumar and Chaudhary 2018). The soil evaluation must be done together with the water analysis; the two data are interconnected. Irrigation water with a high saline content has to be evaluated to avoid reduced plant development for high osmotic pressure around the roots (Barbieri and De Pascale 1992). It is therefore important to know the total salt concentration of the water; up to 500 mg/L of total salts and a conductivity of 0.8 mS/cm good quality water is available (Giardini et al. 1993; Blasi and Giorgioni 1994). It is also important to know the composition of the individual salts, both to know the concentration of any compound harmful to the plant (such as sodium) and to appropriately calibrate the fertilization interventions without supplying elements already guaranteed by the water. In relation to the cultivated species, the irrigation method, the irrigation volume and the intervention interval must then be managed correctly (Bauder et al. 2011).

Excessively clay soils can be improved by adding sand and organic matter; the latter, in addition to improving the structure, restores microflora and microfauna and provides nutrients even if to a lesser extent than chemical fertilizers. The organic substance can be supplied through manure (bovine, equine or sheep), compost, cornunghia and green manure. Organic compounds to be used as a cover are bone meal, dried blood, fish meal which, like the previous ones, are allowed in organic farming. It should always be kept in mind that all organic compounds never have a certain composition, so the action cannot be calibrated in a targeted way as with chemical fertilizers. If the latter are used, a balanced fertilization is recommended in the early stages of growth (for example with a 15-10-15) to then unbalance the NPK ratio towards phosphorus and potassium so as to guarantee conspicuous flowering of the plants (for example using a 8-16-24). It is better to carry out frequent (weekly) fertigation at low doses (0.5–1‰) especially in the early stages and then possibly raise it to 1.5‰. It is very important to check the pH of the soil because its level affects the availability of nutrients, the life of the plant itself and the development of soil microorganisms. Very often chlorosis occurs on the plant which is the symptom of the lack of nutritional elements which in reality are not absent but unavailable due to immobilization of others or due to inadequate pH. Quicklime can be used to raise the pH, while sulphur or iron sulphate or gypsum can be used to lower it. Table 3.3

Table 3.3 Optimal pH values of some species with edible flowers

Species	pH
Anthirrinum majus	6.0–7.0
Dianthus chinensis	6.5–7.5
Rosa spp.	6.0–7.0
Begonia cucullata	5.5–6.5
Calendula officinalis	6.5–7.5
Primula acaulis	5.5–6.0
Petunia hybrida	5.5–6.0
Pelargonium odoratissimum	6.0–7.0
Tagetes patula	6.0–6.5
Viola tricolour	5.5–6.0
Hemerocallis fulva	6.5–7.0

shows the optimal pH values of some edible flowers (Tesi 1985). In order to improve the nutrition of edible flowers in organic farming, bio-stimulants can also be used, products that provide substances that favour or regulate the absorption of nutritional elements or correct certain physiological anomalies. Among these we can mention algae extracts, humic substances, protein hydrolysates and mycorrhizae.

3.6 Beneficial Microorganisms of Soil

The application of soil microorganism benefits for plants (fungi and bacteria) can lead to economic, environmental and human health benefits in relation to greater sustainability and ecosystem services provided by the same microbes. Among these, the most interesting are the arbuscular mycorrhizal fungi (AMF) and plant growth-promoting bacteria (PGPB).

The arbuscular mycorrhizae are symbiotic associations between plants and soil fungi belonging to the phylum Glomeromycota (Schüβler et al. 2001). AM fungi are ubiquitous soil inhabitants and establish symbiotic associations with almost 80% of plant species and among these also many agricultural and ornamental crops. The AMF are able to colonize the cortical parenchyma of the roots (not the central cylinder) where they form intercellular hyphae, coils and arbuscules. The arbuscules are the main exchange sites between host and symbiont: the plant supplies photosynthate in exchange for minerals and water. This symbiosis positively affects not only in the target organ (root), but also the entire plant physiology with detectable effects on leaves, flowers, fruits and seeds (Sohn et al. 2003; Copetta et al. 2006, 2021; Berta et al. 2014; Bona et al. 2018).

Several papers underline the role of mycorrhizal symbioses about improvement of plant mineral nutrition especially P, but also N, K, Mg, Ca and some trace elements (Sohn et al. 2003), biocontrol and resistance against pathogens (Perez-de-Luque et al. 2017), tolerance to water stress (Kaya et al. 2003; Chitarra et al. 2016) or to adverse environmental conditions (Cicatelli et al. 2014), and the fitness of host plants by influencing seed size and the competitiveness of offspring during the first stages of growth (Poulton et al. 2002). However, only few experiments reported the application of AMF about the effects on the edible flowers. Some AMF improved the number of flowers and flowering in different ornamental plant with edible flower like *Petunia hybrida*, *Impatiens balsamina* and *Chrysanthemum morifolium* (Gaur et al. 2000; Sohn et al. 2003). AMF can increase the amount and quality of essential oils in basil leaves (Copetta et al. 2006) and the production of sugars and healthy compounds in fruits including tomato, strawberry and melon (Copetta et al. 2011, 2021; Bona et al. 2015); few paper evaluate the content of nutritional and antioxidant compounds in edible flowers from mycorrhized plants compared to uninoculated ones.

Both in greenhouse and in field, artichoke plants (*Cynara cardunculus*) inoculated with a mix of AMF showed flower heads content of antioxidants (mainly phenolic compounds) increases compared with uninoculated plants (Ceccarelli et al. 2010). *Funnelliformis mosseae* (an AMF) have a significant and positive effect on the zinnia crop: in addition to supporting more efficient use of irrigation, the application of fungal inoculum increases flower dry weight, their antioxidant capacity and their total phenol content (Heidari and Nazari Deljou 2014). Caser et al. (2018) observed that the use of AMF positively affected saffron cultivation, increasing flower production and saffron yield, finally *Rhizophagus intraradices* applied alone increase the content of picrocrocin, quercitin, crocin and antioxidant activity of the spice. Kheyri et al. (2020) test the effect of eleven AMF species on *Calendula officinalis* plant development, flower production, and also the content of flavonoids and carotenoids in flowers. The authors have highlighted that the AMF species have different effects on the growth of the marigold and on the production of secondary metabolites: *F. mosseae* is the AMF that mostly increases the development of the plant and like *G. etunicatum* and *G. geosporum* it increases the content of flavonoids and carotenoids in the calendula petals.

PGPB are free-living microorganisms often present in the rhizosphere that by producing chelating compounds and phytohormones improving mineral nutrition, and synthesizing antibiotics help suppress plant diseases (Gamalero and Glick 2011). These bacteria belonging to genera such as *Pseudomonas*, *Bacillus*, *Rhizobium*, *Klebsiella*, *Pantoea*, *Serratia*, *Streptomyces*, adhere to the root epidermis (and sometimes become endophytes) and they feed on root exudates and release compounds useful for the plant. Dal'Rio et al. (2022) explore the bacterial community of the rhizosphere of *Tropaeolum majus* to identify and select PGPBs that can be used as biofertilizers to promote the growth of nasturtium plants and protect this species widely used as an edible flower from phytopathogens. Through metagenomic sequencing of 16S rRNA amplicon, they identified among the 236 isolates 76 strains belonging to 11 different genera (mainly *Serratia*, *Raoultella* and *Klebsiella*) showing positive results for nasturtium growth promotion. Often these bacteria are used alone or with mycorrhizal inoculum. Flores et al. (2007) test the application of *Bacillus subtilis* and *Glomus fasciculatum* to evaluate their effects on growth and quality of *Tagetes erecta* flowers. They observe that the two microorganisms together or alone increase the number and weight of inflorescences, while only the plants treated with *B. subtilis* had brighter and clearer yellow flowers and only the plants inoculated with *G. fasciculatum* produced flowers with a higher content of xanthophylls.

3.7 Plant Protection

Consumers demand pest and disease-free edible flowers. This could present a challenge since edible flowers must be grown without the use of any chemical pesticide. Manual control of insects by hand picking and also the use of beneficial insects help decrease insect populations. Following good cultural practices and diversifying planting will keep away insect and disease problems. Weeding should be done regularly to keep plants healthy and obtain quality blooms.

3.7.1 Integrated Pest Management for Growing Edible Flowers

Edible flowers represent a new type of agri-food product that arises from the convergence between two apparently distant sectors: floriculture and horticulture. Therefore, edible flowers can be considered as edible vegetables like traditional vegetables e.g. potatoes or tomatoes and as such they must be considered when it comes to assure and preserve consumers in terms of food safety. Moreover, they need to be qualitatively "beautiful to look at" and must have a shelf life such as to guarantee its adequate use in the kitchen.

Depending on the minimal surfaces on which they are grown and the minimum contribution to the daily diet, edible flowers - as well as many other crops that grow on surfaces that are not comparable, for example, to cereals or vines—are classified as a minor crop.

Lack of active principles registered for the application on minor crops and especially on "edible flowers" and the present legislative framework (i.e. Directive 128/2009 regarding the sustainable use of pesticides) had pushed towards management strategies against pest and pathogens alternative to the use of pesticides. Organic farming can represent an alternative considering that no chemicals are allowed for crop production and therefore no residues are expected to be found on plants. On the other hand, the complexity in applying control strategies that can be effective and, more in general, the overall crop management in an organic farming context may represent a constraint. For this reason, the adoption of integrate pest management (IPM) as the main strategy for pathogens and pest control on such niche crops can be the winning choice. IPM is not a single pest control method but, rather, a series of pest management evaluations, decisions and controls (Grasswitz 2019). The main actions to be taken are hereby summarized:

- Set action thresholds: before taking any pest control action, IPM first sets an action threshold, a point at which pest populations or environmental conditions indicate that pest control action must be taken. Sighting a single pest does not always mean control is needed. The level at which pests will become an economic threat is critical to guide future pest control decisions.
- Monitor and identify pests: not all insects, plant pathogens, weeds, and other living organisms require control. Many organisms are innocuous, and some are even beneficial. IPM programs work to monitor for pests and identify them accurately, so that appropriate control decisions can be made in conjunction with action thresholds. This monitoring and identification removes the possibility that pesticides will be used when they are not really needed or that the wrong kind of pesticide will be used.
- Prevention: As a first line of pest control, IPM programs work to manage the environment (e.g. crop, lawn, or indoor space) in order not to build conditions that can be favorable for the survival and the multiplication of plant parasites, i.e. to prevent pests from becoming a threat. In an agricultural crop, this may mean using cultural methods, such as rotating between different crops, adopting correct fertilization, selecting pest-resistant varieties, planting pest-free rootstock, irrigating in the most efficient way and maintaining biodiversity in order to support the reproduction of natural predators or parasitoids. These control methods can be very effective and cost-efficient and present little to no risk to people or the environment (Orre et al. 2007).
- Control: Once monitoring, identification, and action thresholds indicate that pest control is required, and preventive methods are no longer effective or available, IPM programs then evaluate the proper control method both for effectiveness and risk. Effective, less risky pest controls are chosen first, including highly targeted chemicals, such as pheromones to disrupt pest mating, or mechanical control, such as trapping or weeding. If further monitoring, identifications and action thresholds indicate that less risky controls are not working, then additional pest control methods would be employed, such as targeted spraying of registered pesticides (Orre et al. 2007; Grasswitz 2019).

3.7.2 Plant Protection Through Biological Control

Biological control is a technique approved in organic culture but useful also in integrated cultivation. Some helpful insects (like ladybugs) can also be used as natural fertilizers to get rid of pests and any wilted flowers should be removed by hand to allow fresh ones to grow. This way, the plant can maximise its energy use for the growth of new flower buds, otherwise lost in seed production of wilted flowers.

Every living being, in an environment in which other individuals of the same species and of different species live, develops interactions determining relationships that can be: mutualism, commensalism, parasitism or neutralism. Every environment in a state of climax is in equilibrium so that every living being has its own limiter organism which counteract the excessive numerical increase of individuals. The anthropic activity of transforming the natural environment into an agro-ecosystem, even with the introduction of alien species, causes upheavals in the relationships between the individuals who live in it and causes the uncontrolled swarming of parasites that no longer have their natural limiter. This is why it is important rather than creating the biological vacuum with chemical treatments, trying to restore the balance by introducing organisms that constitute the natural limiters of the parasite. Chemical intervention has an immediate effect of numerical reduction of the parasite but in the long run it causes alteration of the environmental balance, induction of resistance in the parasites, suppression of useful individuals, environmental pollution and risks for human health (Orre et al. 2007). The edible flowers have among their main parasites aphids, thrips, whiteflies, scale insects, lepidopteran larvae and the red spider (Table 3.4). The aphids (generally belonging to the genera *Myzus* and *Aphis*) in addition to sucking the sap, lead to the arrest of development, the formation of honeydew and can transmit virosis (Hammond et al. 2023). Nasturtiums, *Hibiscus*, *Rosa*, *Salvia*, *Viola*, *Dahlia* are particularly subject to attacks by these parasites. The aleuroids (*Trialeurodes vaporariorum* and *Bemisia tabaci*) especially in greenhouses and in the warm periods of the year can be particularly dangerous on *Salvia*, *Pelargonium*, *Petunia*, *Calendula*, *Tropaeolum*, *Agastache*, and *Anthirrinum*. Scale insects (*Icerya purchasi* and *Planococcus* sp.) are more common in hot and dry environments and particularly affect roses and

Table 3.4 Main organisms useful for the control of edible flower parasites

Parasite	Restrictor insect	Antagonist fungus or bacterium	Useful plant derivatives
Aphids	*Adalia bipunctate Aphidius colemani, Chrysoperla carnea, Aphelinus mali, Syrphus ribesii*	*Verticillium leucanii Beauveria bassiana*	*Urtica dioica* (nettle), *Artemisia absinthium* (wormwood), *Tanacetum vulgare* (tansy), *Allium sativum* (garlic) and *Quassia amara* (quassia)
Thrips	*Amblyseius cucumeris, Orius laevigatus, Chrysoperla carnea*	*B. bassiana, Metarhizium anisopliae*	Quassia
Whiteflies	*Amblyseius swirskii, Encarsia formosa, Macrolophus pigmaeus, Eretmocerus eremicus*	*B. bassiana, M. anisopliae*	Garlic
Mealybugs	*Nephus coniunctus, Rodolia cardinalis, Anagyrus pseudococci, Leptomastix dactylopii*	*V. leucanii*	Quassia
Lepidoptera	*Trichogramma brassicae*	*Bacillus thuringiensis* kurstaki	Tansy, wormwood, *Azadirachta indica*
Mites	*Amblyseius californicus, Phytoseiulus persimilis*	*B. bassiana*	*A. indica*, tansy, *Urtica dioica* (nettle)

sage. Thrips (*Frankliniella occidentalis* and *Thryps* tabaci) with their sudden pullulations cause leaf and flower spots, deformation and decolouration of flowers, virus transmission. They are especially dangerous on *Begonia*, *Calendula*, *Tagetes lemmonii*, *Dianthus*, *Viola*, *Dahlia*, and *Petunia*. Lepidoptera are dangerous when they are in the larval stage when they gnaw portions of leaves and flowers with their chewing mouthparts. Rose, *Dahlia*, *Viola*, *Petunia* are particularly palatable. The red spider (*Tetranichus urticae*) is extremely polyphagous, despite its small size it is capable of causing serious damage such as yellowing and browning, reduction of vegetation growth and very quickly. Table 3.4 shows all the organisms that can counteract the action of the main parasites of edible flowers.

In protected cultivation in greenhouses it is very useful to use insect-proof systems which prevent the entry of parasitic insects and keep the population of useful substances inside the environment unchanged. The use of blue adhesive chromotropic traps to attract thrips, well distributed near the plants in cultivation, allow the trend of the populations to be monitored; the yellow traps are less specific and attract both harmful insects (aphids and whiteflies) but also useful ones and therefore their use should be evaluated in case of use of biological control with predatory insects and parasitoids (Mahr et al. 2001).

Among fungi, those that have shown the greatest problems on the aerial part of the edible flowers are white mallow, *Botrytis* and downy mildew, while the most common that attack the roots are *Rhizoctonia*, *Phytophtora*, *Pythium* and *Fusarium*. The most effective method of contrast is prevention by avoiding creating the optimal conditions for their development. It is therefore necessary to avoid, where possible, wetting the leaves, always maintaining good ventilation in the greenhouses, choosing varieties resistant to the main pathogens of the cultivated species, eliminating the sources of infection, adopting insect-proof greenhouses, using crop rotation, using of preparations which are capable of inducing resistance in the host such as phosphates and calcium oxide. Table 3.5 shows some microorganisms that appear to be antagonists of the main fungal pathogens that can attack species with edible flowers.

3.7.3 Organic Culture

European regulation defines an increasing deletion of pesticide that can be applied to agricultural systems, moreover there are any active principle allowed specifically for edible flowers. It is obvious to use integrated cultivation that ensure the "null residues" limit. If it is necessary to certificate good practices that led to a pesticide-free food the Organic culture is the only way to follow.

The term "organic" comprises increasingly similar standards for a global vision method declined locally Organic agriculture, in all its forms and variants, is an agri-food production system that finds a common starting point and mission all over the world, having as its basic assumptions the good of the earth, of man, of plants and animals. From such a wide-ranging global vision, the countless movements,

Table 3.5 Main microorganisms that can be used in the biological control of fungal pathogens

Patogen	Antagonistic microorganisms	Natural compounds
White powdery mildew (*Erysiphaceae*)	*Ampelomyces quisqualis*	sulphur, silicon + calcium, sweet orange essential oil
Botrytis cinerea	*Bacillus amyloliquefaciens* subsp. *plantarum* strain D747 (B. D747)	Soluble zinc + peracetic acid + acetic acid + hydrogen peroxide, horsetail macerate and dandelion
Peronospora spp.	*B. subtilis* ceppo QST 713	sweet orange essential oil
Rhizoctonia solani	*Trichoderma harzianum* T 22, *T. asperellum* ICC 012, *B. subtilis, Streptomyces griseoviridis,* B. D747	
Phytophtora spp.	*T. harzianum* T 22, *S. griseoviridis,* B. D747	
Pythium spp.	*T. harzianum* T 22, B. D747	
Fusarium oxysporum	*B. subtilis, Pseudomonas chloroaphis, S. griseoviridis,* B. D747	

associations, foundations and cooperatives existing in all corners of the Earth were born over the decades. IFOAM, the International Federation of Organic Agriculture Movements, acts as an umbrella for everyone, suggests guidelines, acts as a spokesperson and promotes important requests for the sector. It is precisely from IFOAM that the four general principles on which organic is based arise: 1. Health: Organic agriculture should support and enhance the health of soil, plants, animals, humans and the planet as one and indivisible. 2. Ecology: Organic agriculture should build on, work with, emulate, and sustain living ecological systems and cycles. 3. Equity: Organic agriculture should be based on relationships that ensure equity with regard to the common environment and life opportunities. 4. Care: Organic agriculture should be managed in a precautionary and responsible way to protect the health and well-being of current and future generations and the environment. In the European Union the legislation is constantly evolving, it was officially regulated for the first time in 1991, with Regulation 2092, subsequently repealed by EU Reg 834/07. Then finally replaced by the recent EU Reg 848/2018, as a response to changing times, news and pressure from public opinion. All organic operators operating within the European Union are united by the same basic rules, therefore there is a guarantee of compliance with common principles and free movement of products, recognizable throughout the EU by the well-known green logo in the shape of leaf composed of white stars.

Organic culture is mainly related to every type of food and for this it is very important to note that there are many organic certification systems in the world, and for the purposes of trade, the various countries evaluate the correspondence of other systems with their own. In this regard, in Reg 848/2018 Chapter VII is dedicated, related to Europe, "Exchanges with Third Countries" and, in Article 47, reference is made to equivalence, defining the countries recognized as "third countries that the Union, in within the framework of a trade agreement, recognized as having a production system which meets the same objectives and principles as those of the Union by applying standards which ensure the same level of assurance of conformity." In order to allow commercial exchanges of certified organic products with non-European countries, the European Union has stipulated equivalence agreements with various countries: Argentina, Australia, Canada, South Korea, Costa Rica, Japan, India, Israel, United States (NOP, National Organic Program) Switzerland, Tunisia, New Zealand. The agreements assure consumers and operators that, from whatever country they come from, the rules adopted for organic products are applied and verified at every stage of the production, transformation and marketing chain and are therefore equivalent to EC Regulation 834/2007. In the event that the company intends to import organic products from third countries not included in the previous list, it is necessary that the control body of the exporter, and of all the players in the supply chain, has been recognized as equivalent by the EU Commission. The list of organizations recognized as equivalent is indicated in Annex IV of EC Reg. 1235/08 and subsequent amendments or additions.

With these premises it is possible to understand that the Organic culture of edible flowers is a regime that can assure the quality of the product and the cultivation protocols suitable for the environment. The shared regulations permitted to consider this system as a real alternative to the traditional growing systems.

3.8 Cultivation Protocols for Sample Species

3.8.1 Cultivation of Anthirrinum majus

Antirrhinum majus, (snapdragon), is a herbaceous erect perennial plant belong to *Plantaginaceae* with annual or biennial cycle. This species is very rustic and often grows spontaneously and spreads even in arid, sunny places and roadsides, therefore it is not very demanding from a nutritional point of view. Generally, the plants reach a variable height of 0.6–1.8 m depending on the pedological and environmental conditions. Leaves are lanceolate to oblong-lanceolate to 7.6 cm long. The flowers are zygomorphic, born in terminal racemes very showy and they bloom in most areas from late spring into fall. There are many cultivars available, classified by size and colour groupings; the length of a single flower can vary from 3 cm to more than 5 cm. Snapdragons have a long been used in gardens, in pot and in the cut flower industry because they are easy plants to grow, whether in container or in the landscape (Al-Snafi 2015). Moreover, snapdragons have been classified into four groups based on temperature and photoperiodic response: Group I cultivars require the shortest period for flowering, followed by Group II cultivars and so on (Inaba and Ohshiro 2005). Studies on the cultivation of A. majus are mainly aimed at the production of cut flowers with racemes full of flowers, large flowers (features also suitable to produce edible flowers) and with long and robust stems. Abdullah et al.

(2020) report that the best planting density to obtain racemes with a greater number of flowers is 30 cm × 30 cm; the same planting density allows to obtain flowers with a larger diameter than those collected from plants with a higher planting density.

3.8.2 *Cultivation of* Calendula officinalis

Calendula officinalis L. (Fig. 3.3) is a herbaceous plant native to the Mediterranean basin and belonging to the Asteraceae family called marigold. The flowers are grown all over the world for ornamental purposes and for their therapeutic properties (Khalid and Da Silva 2012). This species is very rustic, and the wild species disseminates itself. Marigold cultivars have wide morphological variability especially in flower size, colour of the ligulate flowers and size of the central disc (tubular flowers). The marked anatomical variation of the plant and the different forms of cultivation (in the field, in pots, in the greenhouse) mean that the quality and quantity of bioactive products in this plant are highly variable. In spring, the seeds can be sown in open fields in rows. Row distance had significant effect about the production: increasing the row distance (from 40 to 70 cm), fresh and dry flower and petal yield increases (Crnobarac et al. 2011). The sowing date, the year and the density of plants per m^2, fertilization and irrigation practices can influence the production of the flowers, and the secondary compounds produced (Mirzaei et al. 2016; Elhindi et al. 2020).

Fig. 3.3 Marigold flower head showing the orange, edible outer long ligulate florets and a central disk of short tubular florets. A pollinating insect is present on the central disk. (Photo by Katrin Mamberto)

3.8.3 *Cultivation of* Clitoria ternatea

Clitoria ternatea is a tall slender climbing plant belong to Fabaceae family. This plant originated from tropical area of Asia and naturalized in South America, Australia and Africa. The flowers (Fig. 3.4) are axillary, mostly solitary, showy, resupinate, usually papilionaceous with purplish blue edible and widely used in Asia for the production of blue tea; however, several accessions have flower with different colour (mauve, pale blue, white) and accessions with doble flower or semi-double. Standard petal is obovate, 2.5–5 cm long, 2–4 cm wide, white or yellow at base (Havananda and Luengwilai 2019). The seeds are normally sown from the beginning to the middle of the rainy season. Its roots are deep, it adapts well to a variety of soil types (pH 5.5–8.9) including calcareous soils. It is capable of surviving both in rainy regions and in prolonged periods of drought. This species has excellent regrowth in a short time after cutting and produces many flowers during late spring and summer. *Clitoria ternatea* is also adapted to heavily fissured clay soils in northern Australia. It is also used as a cover crop, manure green and forage (Abreu et al. 2014). Butterfly pea plant commonly prefers full sunlight or partially shaded. Propagation is by seed which can be germinated by soaking them in water overnight; germination occurs within 1–2 weeks while flowering occurs within 4 weeks. The propagation of *C. ternatea* includes the stems which are fine twinning as it can grow to 3 m long. (Gomez and Kalamani 2003). Temperature is a key factor for plant development: if the seasonal temperature is too low or too high, pollination

Fig. 3.4 *Clitoria ternatea* flower: the plant was grown in a greenhouse at CREA institute in Sanremo

may not take place and fruits may not form also if self-pollination is prevalent. Morris (2009) describes the cultivation of 18 accessions. He sowed the pods into jiffy pots from April to June in a greenhouse at 21–26 °C or directly field-sown by hand in June. After about 30 days, butterfly pea plants were transplanted into the field in rows of 6 m with 3–6 m between rows.

3.8.4 *Cultivation of* Dahlia *spp.*

Belonging to the Asteraceae family, the dahlia is native to the mountainous areas of Mexico and Central America. The name "Dahlia" was given in honour of Dr. Andreas Dahl a Swedish botanist and a student of Linnaeus. The dahlias are tuberose herbaceous perennial plants; the stems are empty, generally erect and branched, glabrous or scabrous varies from 30 to 180 cm depending upon the cultivar. The leaves are pinnatifid, pinnate or bipinnate; the vegetative apex produces flower heads comprising of a few hundred individual florets of variable size from 10 to 30 cm in breadth. Tubular flowers (or disc flowers) are sometimes replaced by ligulate flowers (ray flowers) which can be fertile or sterile. The multitude of colours present, the great variation in flower size, the wide profusion of flowering and the easy cultivation have made them immensely popular as garden flowers (Thakur et al. 2022) (Fig. 3.5). The more cultivated species are *D. variabilis*, *D. pinnata* and

Fig. 3.5 Heavily blooming varieties of dahlias with different colours grown at La Rutabaga farm (Finale Ligure, Italy). (Photos by Katrin Mamberto)

D. coccinia. The National Dahlia Society of England has classified dahlias into 15 groups based on the appearance of their ray flowers and disc flowers (Mariña 2015; Thakur et al. 2022). The dahlia propagates by seed, division of the tuberous roots and cuttings, the latter method being the one mainly used for the multiplication of commercial varieties. For the propagation by cuttings, in winter the tuberous roots are put on hot beds at 15–20 °C in a humid substrate (peat: sand 1:1) to rapidly induce budding; then the shoots are cut and placed in pots for them to take root. In May, the new plants can be definitively transplanted in soil, and they flower already during the first year. In November, the tubers should be removed and kept in a cool, dry place.

To induce branching and consequently the production of more flowers, the central end of the stem must be cut in spring. The planting distance is 75 cm for large flowered, 60 cm for medium sized flowers and 30–35 cm for small-flowered varieties. Dahlia is a rustic plant, however it prefer sunny areas, temperatures between 18 and 23 °C, relative humidity of 75–78% and little wind action; the tubers prefer a good drainage non-calcareous soils with a pH around 7, and enriched with mature manure and avoiding the excess of nitrogen which causes the growth of the foliage to the detriment of the growth of the tuber. The basic fertilization generally used consists of 100 kg ha-1 P2O5 and 200 kg ha-1 K2O. Phosphorus and potassium make the stems rigid and improve the colour of the flowers. Furthermore, potassium promotes the formation of reserves and the production of new tubers. In the summer months, to improve the development of the flowers, nitrogen fertilizations are carried out (50 kg ha-1 N) combined with frequent and regular irrigation (Mariña 2015). As reported by Ciobanu et al. (2021) and Thakur et al. (2022), most of the studies on the dahlias cultivation in pots focusing on the use of organic, foliar or chemical fertilizers of various origins. in an intercropping experiment using *D. variabilis* 'Glory of India' with Damask rose, the maximum weight of the flower was produced applying an basal fertilization with poultry manure 30 t ha-1 combined with a NPK fertilizer per plant (Sabah et al. 2014). Generally, to produce flowers of good size, buds and secondary buds placed in the axil of the two pairs of leaves which are found immediately under the apical bud are removed; this operation is carried out and the sprouts are still tender (Mariña 2015).

3.8.5 Cultivation of **Hibiscus sabdariffa**

Hibiscus sabdariffa (Malvaceae) is an annual plant also known under other different names e.g., roselle, viz. roselle, razelle, sorrel, red sorrel, Jamaica sorrel, Indian sorrel, Guinea sorrel, sour-sour, and queens land jelly plant and it is used to produce karkade (red inflated edible calyces) (Ilodibia et al. 2019). Moreover, roselle is usually cultivated for its stem fibres, leaves, petals and seeds. The cultivation of this plant is carried out in the tropical and sub-tropical regions of Asia (mainly India and South-East Asia), in America, Australia, throughout Africa and in the Mediterranean areas of Europe. *H. sabdariffa* can grow easily in well-drained soil and can tolerate poor soil and drought; for good growth and flower development, this species requires 13 hours of sunlight and high minimum night temperatures (20 °C) for at least 4–8 months. For the cultivation of rosella plants, the soil is worked up to a depth of about 20 cm. In India, sowing is done at the beginning of the rainy season (approximately mid-April), planting the seeds at a depth of about 0.5 cm, with a density very variable from 15 to 70 cm both along and between rows (Shalaby and Razin 1989; Al-Sayed et al. 2020), while irrigation is carried out every 2–6 days depending on the soil and rainfall, leaving the soil to dry out between one application and the next (Hewidy et al. 2018). For small plantings, seeds are sown in protected seedbeds and the seedlings are transplanted to field. Fertilizing practices vary widely: Roselle (Fig. 3.6) responds favourably to applications of nitrogen applied both in the form of compost and as a mineral fertilizer in combination with phosphate applied in low doses (Dahmardeh 2012; Besharati et al. 2022) or with biological bio-fertilizers (Al-Sayed et al. 2020). To avoid the spread of *Heterodera radicola* (a root parasitic nematode) crop rotation is applied with a green manure legume and maize.

Fig. 3.6 *Hibiscus sabdariffa* flowers cream-coloured with red centre and petals edges, grown at CREA institute in Sanremo

3.8.6 Cultivation of Tropaeolum majus

Tropaeolum majus L. (nasturtium), had large use such as medicinal and honey plant, natural colouring or vegetable and ornamental crop. This species belongs to the Tropaeolaceae family is a rustic herb biennial with prostrate and climbing stems, green and round leaves and trumped-shaped flowers with yellow, orange or red colours (Fig. 3.7).

Its leaves and flowers are rich in health-beneficial polyphenols and glucosinolates. Nasturtium is also considered a phyto-protector plant because it can attract lepidopterans and repel flea beetles and beetles dangerous for apples and cucurbits. *T. majus* seeds contain great quantity of fatty acid like erucic acid the main ingredient of Lorenzo oil. Nasturtium prefers plain sun and moderate water. Sowing can be done in a definitive place, nether less, the multiplication by cuttings can be applied (Xu et al. 2021). The nasturtium-seed germination is large applied then, it is hazardous to produce uniform seedlings of the same traits by means of seed propagation. Nasturtium seeds need to be soaked in tap water for 24 hours before sowing (Hensel et al. 1991), and the time required for seeds to sprout is long (15–20 days). In case of sowing in trays, the plants are transplanted when the sprouts had about six leaves.

Fig. 3.7 *Tropaeolum majus* plant grown in open fields: both the flowers and the leaves are edible. (Photo by Katrin Mamberto)

Xu et al. (2021) observed that the nasturtium cuttings from the apical bud position rooted earliest and produced the most roots; cuttings from other node positions produced longer roots and heavier root fresh and dry weights than those from the apical bud position. Carbonari et al. (2006) compared triple superphosphate and organic (semi-decomposed chicken manure) fertilization; the highest yields of fresh and dried masses of nasturtium flowers can be obtain incorporating to the soil 75 kg ha-1 of phosphate and 10,800 kg ha-1 of chicken manure. Silva et al. (2021) obtain similar results: the greatest yield terms of in number (15.4 million ha-1) and fresh (10.45 t ha-1) and dry weights (0.849 t ha-1) of flowers were found by using doses close to 10 t ha-1 in the soil incorporated chicken manure and 10 t ha-1 cover chicken manure. Period of the year, hour of the day and development stage affect variation of quality and quantitative compound of medicinal plants: in the case of nasturtium, the shadow reduces the production of erucic acid (Carbonari et al. 2006).

References

Abdullah B, Ahmad I, Dol JM (2020) Optimal sowing time and planting density for selected field grown cut *antirrhinum* and *Matthiola*. Sci Hortic 26. https://doi.org/10.1016/j.scienta.2019.108909

Abreu MLC, RAM V, Rocha NS, Araujo RP, Glória LS, Fernandes AM, de Lacerda PD, Gesualdi AJ (2014) *Clitoria ternatea* L. as a potential high quality forage legume. Asian Australas J Anim Sci 27(2):169–178. https://doi.org/10.5713/ajas.2013.13343

Al-Sayed HM, Hegab SA, Youssef MA, Khalafalla MY, Almaroai YA, Ding Z, Eissa MA (2020) Evaluation of quality and growth of roselle (*Hibiscus sabdariffa* L.) as affected by bio-fertilizers. J Plant Nutr 43(7):1025–1035. https://doi.org/10.1080/01904167.2020.1711938

Al-Snafi AE (2015) The pharmacological importance of *Antirrhinum majus*-A review. Asian J Pharm Sci Tech 5(4):313–320

Barbieri G, De Pascale S (1992) Salinità dell'acqua di irrigazione e colture orto-floricole. Colture Protette 2:75–81

Bauder TA, Waskom RM, Sutherland PL, Davis JG (2011) Irrigation water quality criteria. Colourado State University Extension Publication, Crop series/irrigation. Fact sheet no. 0.506,4 pp

Berta G, Copetta A, Gamalero E, Bona E, Cesaro P, Scarafoni A, D'Agostino G (2014) Maize development and grain quality are differentially affected by mycorrhizal fungi and a growth-promoting pseudomonad in the field. Mycorrhiza 24:161–170. https://doi.org/10.1007/s00572-013-0523-x

Besharati J, Shirmardi M, Meftahizadeh H, Ardakani MD, Ghorbanpour M (2022) Changes in growth and quality performance of roselle (*Hibiscus sabdariffa* L.) in response to soil amendments with hydrogel and compost under drought stress. S Afr J Bot 145:334–347. https://doi.org/10.1016/j.sajb.2021.03.018

Blasi R, Giorgioni ME (1994) Qualità delle acque di irrigazione e loro uso in serra. Colture Protette 5:27–36

Bona E, Lingua G, Manassero P, Cantamessa S, Marsano F, Todeschini V, Copetta A, D'Agostino G, Massa N, Avidano L, Gamalero E, Berta G (2015) AM fungi and PGP pseudomonads increase flowering, fruit production, and vitamin content in strawberry grown at low nitrogen and phosphorus levels. Mycorrhiza 25:181–193. https://doi.org/10.1007/s00572-014-0599-y

Bona E, Todeschini V, Cantamessa S, Cesaro P, Copetta A, Lingua G, Gamalero E, Berta G, Massa N (2018) Combined bacterial and mycorrhizal inocula improved tomato quality at reduced fertilization. Sci Hortic 234:160–165. https://doi.org/10.1016/j.scienta.2018.02.026

Carbonari VB, Vieira MC, Heredia NAZ, Marchetti ME (2006) Phosphorus and chicken manure on development and yield of *Tropaeolum majus* L. Braz J Med Plants 8:71–77

Caser M, Victorino IMM, Demasi S, Berruti A, Donno D, Lumini E, Bianciotto V, Scariot V (2018) Saffron cultivation in marginal alpine environments: how AMF inoculation modulates yield and bioactive compounds. Agronomy 9(1). https://doi.org/10.3390/agronomy9010012

Ceccarelli N, Curadi M, Martelloni L, Sbrana C, Picciarelli P, Giovannetti M (2010) Mycorrhizal colonization impacts on phenolic content and antioxidant properties of artichoke leaves and flower heads two years after field transplant. Plant Soil 335:311–323. https://doi.org/10.1007/s11104-010-0417-z

Chitarra W, Pagliarani C, Maserti B, Lumini E, Siciliano I, Cascone P, Schubert A, Gambino G, Balestrini R, Guerrieri E (2016) Insights on the impact of arbuscular mycorrhizal symbiosis on tomato tolerance to water stress. Plant Physiol 171(2):1009–1023. https://doi.org/10.1104/pp.16.00307

Cicatelli A, Torrigiani P, Todeschini V, Biondi S, Castiglione S, Lingua G (2014) Arbuscular mycorrhizal fungi as a tool to ameliorate the phytoremediation potential of poplar: biochemical and molecular aspects. I Forest 7(5):333–341. https://doi.org/10.3832/ifor1045-007

Ciobanu EC, Bădulescu L, Badea ML, Toma F (2021) Studies and research on the species and varieties of *Dahlia* in cultivation. SciPapers Ser B Hortic 65(1):683–689

Copetta A, Lingua G, Berta G (2006) Effect of three AM fungi on growth, distribution of glandular hairs, and essential oil production in *Ocimum basilicum* L. var. Genovese. Mycorrhiza 16:485–494. https://doi.org/10.1007/s00572-006-0065-6

Copetta A, Bardi L, Bertolone E, Berta G (2011) Fruit production and quality of tomato plants (*Solanum lycopersicum* L.) are affected by green compost and arbuscular mycorrhizal fungi. Plant Biosyst 145:106–115. https://doi.org/10.1080/11263504.2010.539781

Copetta A, Todeschini V, Massa N, Bona E, Berta G, Lingua G (2021) Inoculation with arbuscular mycorrhizal fungi improves melon (*Cucumis melo*) fruit quality under field conditions and plant performance in both field and greenhouse. Plant Biosyst 155(5):1063–1074. https://doi.org/10.1080/11263504.2020.1813831

Copetta A, Marchioni I, Ruffoni B (2022) The edible flowers from woody ornamental plants. Acta Hortic 1331:ISHS 2021. https://doi.org/10.17660/ActaHortic.2021.1331.27

Crnobarac J, Marinkovič B, Jačimovič G, Latkovič D, Balijagič J (2011) The effect of cultivar and stand density on yield components and yield of pot marigold. Acta Fytotech et Zootech 14(1):6–8

Dahmardeh M (2012) Effect of mineral and organic fertilizers on the growth and calyx yield of roselle (*Hibiscus sabdariffa* L.). Afr J Biotechnol 11(48):10899–10902. https://doi.org/10.5897/AJB12.203

Dal'Rio I, Mateus JR, Seldin L (2022) Unraveling the *Tropaeolum majus* L. (nasturtium) root-associated bacterial community in search of potential biofertilizers. Microorganisms 10(3). https://doi.org/10.3390/microorganisms10030638

Elhindi KM, Al-Mana FA, Algahtani AM, Alotaibi MA (2020) Effect of irrigation with saline magnetized water and different soil amendments on growth and flower production of *Calendula officinalis* L. plants. Saudi J Biol Sci 27(11):3072–3078. https://doi.org/10.1016/j.sjbs.2020.09.015

Falla NM, Contu S, Demasi S, Caser M, Scariot V (2020) Environmental impact of edible flower production: a case study. Agronomy 10(4). https://doi.org/10.3390/agronomy10040579

Fernandes L, Casal S, Pereira JA, Saraiva JA, Ramalhosa E (2020) An overview on the market of edible flowers. Food Rev Int 36(3):258–275. https://doi.org/10.1080/87559129.2019.1639727

Flores AC, AAE L, Portugal VO (2007) Yield and quality enhancement of marigold flowers by inoculation with *Bacillus subtilis* and *Glomus fasciculatum*. J Sustain Agric 31:21–31. https://doi.org/10.1300/J064v31n01_04

Gamalero E, Glick B (2011) Mechanisms used by plant growth-promoting bacteria. In: Maheshwari DK (ed) Bacteria in 10 Mycorrhiza (2017) 27:1–11 Agrobiology: plant nutrient management. Springer, Berlin, Heidelberg, pp 17–47. https://doi.org/10.1007/978-3-642-21061-7_2

Gaur A, Gaur A, Adholeya A (2000) Growth and flowering in *Petunia hybrida*, *Callistephus chinensis* and *Impatiens balsamina* inoculated with mixed AM inocula or chemical fertilizers in a soil of low P fertility. Sci Hortic 84(1–2):151–162. https://doi.org/10.1016/S0304-4238(99)00105-3

Giardini L, Borin M, Grigolo U (1993) La qualità delle acque per l'irrigazione. L'Informatore Agrario 20:29–77

Gomez SM, Kalamani A (2003) Butterfly pea (*Clitoria ternatea*): a nutritive multipurpose forage legume for the tropics–an overview. Pak J Nutr 2(6):374–379

Grasswitz TR (2019) Integrated pest management (IPM) for small-scale farms in developed economies: challenges and opportunities. Insects 10(6). https://doi.org/10.3390/insects10060179

Hammond J, Huang Q, Jordan R, Meekes E, Fox A, Vazquez-Iglesias I, Vaira AM, Copetta A, Delmiglio C (2023) International trade and local effects of viral and bacterial diseases in ornamental plants. Annu Rev Phytopathol 61(1):73–95. https://doi.org/10.1146/annurev-phyto-021621-114618

Havananda T, Luengwilai K (2019) Variation in floral antioxidant activities and phytochemical properties among butterfly pea (*Clitoria ternatea* L.) germplasm. Genet Resour Crop Evol 66:645–658. https://doi.org/10.1007/s10722-018-00738-6

Heidari Z, Nazari Deljou MJ (2014) Improvement of morpho-physiological traits and antioxidant capacity of zinnia (*Zinnia Elegance* 'Dreamland Redí') by arbuscular mycorrhizal fungi (*Glomus mosseae*) inoculation. Int J Adv Biol Biomed Res 2(10):2627–2631

Hensel A, Brummell DA, Hanna R, Maclachlan G (1991) Auxin-dependent breakdown of xyloglucan in cotyledons of germinating nasturtium seeds. Planta 183:321–326. https://doi.org/10.1007/BF00197728

Hewidy M, Elsayed MLM, Sultan E (2018) Water schedule of roselle (*Hibiscus sabdariffa* L.) under organic fertilization. Egypt J Hortic 45(1):53–64. https://doi.org/10.21608/ejoh.2018.2913.1048

Ilodibia CV, Achebe UA, Arubalueze CU, Ibeh E (2019) A comparative study on morphology, mineral and vitamin compositions of two variants of *Hibiscus sabdariffa* (L.). Arch Agri Environ Sci 4(1):50–56. https://doi.org/10.26832/24566632.2019.040108

Inaba Z, Ohshiro M (2005) Effects of planting density and methods of raising seedlings on flowering, yield and quality of cut flowers in snapdragons (*Antirrhinum majus* L.). Environ Control Biol 43(3):201–210. https://doi.org/10.2525/ecb.43.201

Kaya C, Higgs D, Kirnak H, Tas I (2003) Mycorrhizal colonisation improves fruit yield and water use efficiency in watermelon (*Citrullus lanatus* Thunb.) grown under well-watered and water-stressed conditions. Plant Soil 253(2):287–292. https://doi.org/10.1023/A:1024843419670

Khalid KA, Da Silva JT (2012) Biology of *Calendula officinalis* Linn.: focus on pharmacology, biological activities and agronomic practices. Med Aromat plant Sci Biotechnol 6(1):12–27

Kheyri Z, Moghaddam M, Moradi M (2020) Study the effect of different mycorrhizal fungi on some growth indices, photosynthetic pigments, flavonoids and carotenoid content of pot marigold flower. Hortic Plants Nutr 3(1):37–50. https://doi.org/10.22070/hpn.2020.5007.1061

Kumar M, Chaudhary V (2018) Effect of integrated sources of nutrients on growth, flowering, yield and soil quality of floricultural crops: a review. Int J Curr Microbiol App Sci 7(3):2373–2404

Liu X, Zhang Y, Jiang Z, Yue X, Liang J, Yang Q, Li J, Li N (2023) Micro-moistening irrigation combined with bio-organic fertilizer: an adaptive irrigation and fertilization strategy to improve soil environment, edible rose yield, and nutritional quality. Ind Crop Prod 196. https://doi.org/10.1016/j.indcrop.2023.116487

Mahr SER, Cloyd RA, Mahr DL, Sadof CS (2001) Biological control of insects and other pests of greenhouse crops, vol 581. North Central Regional Publication, p 100

Mariña LJ (2015) Cultivation of the *Dahlia*. Cultivos Tropicales 36(1):103–110

Mirzaei M, Zehtab-Salmasi S, Nassab ADM, Shaker-Kouhi S (2016) Effects of sowing date and plant density on marigold (*Calendula officinalis*) morphology and flower yield. J Med Plants 4(3):229–232

Morris JB (2009) Characterization of butterfly pea (*Clitoria ternatea* L.) accessions for morphology, phenology, reproduction and potential nutraceutical, pharmaceutical trait utilization. Genet Resour Crop Evol 56:421–427. https://doi.org/10.1007/s10722-008-9376-0

Orre GS, Tompkins JM, Jonsson M, Jacometti M, Wratten SD (2007) Provision of floral resources for biological control: restoring an important ecosystem service. Funct Ecosyst Communities 1(2):86–94

Perez-de-Luque A, Tille S, Johnson I, Pascual-Pardo D, Ton J, Cameron DD (2017) The interactive effects of arbuscular mycorrhiza and plant growth-promoting rhizobacteria synergistically enhance host plant defences against pathogens. Sci Rep 7(1):16409. https://doi.org/10.1038/s41598-017-16697-4

Pires JEDO, Di Gioia F, Rouphael Y, García-Caparrós P, Tzortzakis N, Ferreira IC, Barros L, Petropoulos SA, Caleja C (2023) Edible flowers as an emerging horticultural product: a review on sensorial properties, mineral and aroma profile. Trends Food Sci Technol 137:31–54. https://doi.org/10.1016/j.tifs.2023.05.007

Poulton JL, Bryla D, Koide RT, Stephenson AG (2002) Mycorrhizal infection and high soil phosphorus improve vegetative growth and the female and male functions in tomato. New Phytol 154:255–264. https://doi.org/10.1046/j.1469-8137.2002.00366.x

Sabah SS, Prasad VM, Saravanan S (2014) Effect of different organic and inorganic manure on flower yield and tubers yield of dahlia (*Dahlia variabilis*) cv. Glory of India as intercropping with Damask rose. Eur Acad Res 2(3):4265–4273

Schüβler A, Schwarzott D, Walker C (2001) A new fungal phylum, the Glomeromycota: phylogeny and evolution. Mycol Res 105:1413–1421. https://doi.org/10.1017/S0953756201005196

Shalaby AS, Razin AM (1989) Effect of plant spacing on the productivity of Roselle (*Hibiscus sabdariffa* L.) grown in newly reclamed land. J Agron Crop Sci 162(4):256–260. https://doi.org/10.1111/j.1439-037X.1989.tb00715.x

Silva OBD, Goelzer A, Santos FHMD, Carnevali TDO, Vieira MDC, Zárate NAH (2021) Increased yield and nutrient content of *Tropaeolum majus* L. with use of chicken manure. Revista Ceres 68:379–389. https://doi.org/10.1590/0034-737X202168050002

Sohn BK, Kim KY, Chung SJ, Kim WS, Park SM, Kang JG, Rim YS, Cho JS, Kim TH, Lee JH (2003) Effect of different timing of AMF inoculation on plant growth and flower quality of chrysanthemum. Sci Hortic 98(2):173–183. https://doi.org/10.1016/S0304-4238(02)00210-8

Tesi R (1985) Colture protette – ortoflorovivaismo. Edagricole, Bologna

Thakur P, Shah AH, Adhikari Y, Kumar M, Verma S (2022) Dahlia cultivation in India and abroad: A review. Int. J. Plant Soil Sci, 34, 240–251

Xu W, Lu N, Kikuchi M, Takagaki M (2021) Effects of node position and electric conductivity of nutrient solution on adventitious rooting of nasturtium (*Tropaeolum majus* L.) cuttings. Agronomy 11(2). https://doi.org/10.3390/agronomy11020363

Yadav D, Singh SP (2018) Vegetative methods of plant propagation: I-cutting layering and budding. J Pharmacogn Phytochem 7(2):3267–3273

Chapter 4
Soilless Cultivation of Edible Flowers

Hossein Sheikhi and Silvana Nicola

Abstract Edible flowers, as the name suggests, are flowers that can be ingested by humans and do not pose any harmful effects after consumption. There are numerous health benefits associated with edible flowers due to their bioactive components. Edible flowers are in high demand around the world, which requires new cultivation methods such as soilless cultivation in contrast to traditional cultivation, for producing more and better-quality flowers. In this chapter, soilless farming systems are illustrated and compared with conventional farming methods in order to identify and understand their applications. Furthermore, a description of different types of hydroponic and substrate-dependent cultures is provided, including their advantages and disadvantages. The environmental conditions required for growth in soilless cultivation systems as well as the effect of various environmental factors on the growth of secondary metabolites production are discussed in this chapter. This chapter also addresses the protocols of some edible flowers in the conditions of soilless cultivation and their uses and nutritional value.

Keywords Aeroponic · Deep water culture · Drip systems · Ebb-and-flow technique · Hydroponic · Nutrient film technique · Substrate

Abbreviations

C	Carbon
CO_2	Carbon dioxide
EC	Electrical conductivity
HPS	High-pressure sodium

H. Sheikhi
Horticultural Sciences Department, College of Agriculture and Natural Resources, University of Tehran, Karaj, Iran

S. Nicola (✉)
Department of Agricultural, Forest and Food Sciences, University of Turin, Grugliasco, Italy
e-mail: silvana.nicola@unito.it

M. M. Jowkar et al. (eds.), *Advances in Research on Edible Flowers*, Advances in Olericulture, https://doi.org/10.1007/978-3-032-04561-4_4

K	Potassium
LED	Light-emitting diode
N	Nitrogen
NH_4^+	Ammonium
NO_3^-	Nitrate
P	Phosphorus

4.1 Introduction

Soil, which is a readily available medium for growing plants, is capable of growing nearly all kinds of plants. As a general rule, soil serves two basic functions—it acts as a storage system for retaining nutrients and water and gives the plant physical support through its root system to be able to grow and develop (Kalaivanan et al. 2023). Despite this, there are some serious limitations to the growth of plants in the soil due to its degradation and erosion, a poor drainage, sometimes an inappropriate soil pH or improper soil reactions, often the presence of disease-causing organisms and nematodes, among many others (Kimbonguila et al. 2019; Fussy and Papenbrock 2022). Additionally, conventional crop cultivation, also known as open-field cultivation, involves many different resources, such as space, irrigation, and large amounts of technical equipment, and is very susceptible to environmental conditions (Fussy and Papenbrock 2022). There was a way of eliminating these limitations when soilless system was introduced in agriculture, a system of cultivation which was specifically designed to eliminate the mentioned limitations posed by soil. As a modern cultivation technology, soilless culture has received a lot of attention over the past 30–40 years, mainly as a method of growing plants in greenhouses (Gruda 2022). The popularity of soilless agricultural methods such as hydroponic greenhouses has increased due to their advantages over traditional soil-based agricultural methods. In high-tech greenhouses with year-round production, soilless systems provide high yields, high-quality produce, and greater quality control over nutrient and plant protection product emissions (Tzortzakis et al. 2020). Despite the benefits of soilless systems, they also have some disadvantages. In soilless systems, not all plant species are equally suited, and some substrates can adversely affect the environment. It is also important to consider the economic efficiency of these systems, since not all are equally efficient or applicable to all areas and locations (Fussy and Papenbrock 2022). Furthermore, soilless systems require significant initial investments in terms of equipment and infrastructure (Asaduzzaman et al. 2015). While soilless systems may have limitations, they have the potential to be an important part of sustainable agriculture. Researchers are actively pursuing ecological sustainability in soilless systems from a circular economy perspective (Gonnella and Renna 2021). Some of the strategies being explored are using renewable energy sources, such as solar energy, and developing closed-loop systems to minimize waste and resource use (Lund 2007; Owusu and Asumadu-Sarkodie 2016; Gielen et al. 2019). In general, soilless farming approaches offer a promising alternative to traditional soil-based agriculture, but their advantages and limitations must be carefully considered in each specific circumstance. This chapter provides an overview of soilless cultivation methods, their characteristics, and how to cultivate edible flowers in them. In addition, some edible flowers cultivated in these systems have been introduced.

4.2 Soilless Cultivation Systems

4.2.1 Different Soilless Cultivation Systems

Soilless culture method is one of the most common methods of providing support and nutrients to plants in greenhouses, where it allows better control over the growing environment. In soilless culture, two types of production systems can be used: substrate culture and hydroponics. In hydroponics, the roots of the plants are cultured with an inert substrate or grown in the medium without any aggregates, while in substrate culture, the roots are cultured on substrates that support their roots (Fussy and Papenbrock 2022).

4.2.1.1 Substrate Cultivation Systems

In the last few decades, various inorganic and organic materials (or a mixture of them) have been tested in soilless cultivation to find media that are equally effective at supporting root systems and storing water and nutrients, while also being recyclable, renewable, and economically and environmentally efficient (Gonnella and Renna 2021). Substrates such as rockwool, perlite, vermiculite, coconut coir, peat, and biochar are some of the substrates used in the commercial hydroponic production of cut flowers and vegetables (Fig. 4.1). Almost all inorganic substrates are derived from natural sources and are partly pre-processed industrially (Fussy and Papenbrock 2022). Rockwool is one of the most common materials used in greenhouses where plants are grown. However, rockwool is expensive and difficult to dispose of, since it is not biodegradable or environmentally friendly (Asaduzzaman et al. 2015). Throughout Europe, perlite has become a well-established substrate for hydroponic applications, largely as a result of its low cost and the fact that it is readily available (Grillas et al. 2001). Vermiculite is a soilless substrate widely used in horticulture and often used in combination with other substrates like peat or coconut coir to enhance its properties. It has a relatively high cation exchange capacity, lightweight, sterile, and high water absorption capacity (Jagtap et al. 2022). However, due to its tendency to break down over time, it is preferred for short-term crops. Among the many substrates available, coconut coir is of particular importance to the ornamental industries, especially in the Netherlands and Canada, with the product marketed as an alternative to rockwool in greenhouse vegetable production in recent years (Asaduzzaman et al. 2015). The

Fig. 4.1 Some soilless cultivation substrates. Perlite (**a**), Vermiculite (**b**), Coconut coir (**c**), Rockwool (**d**), Expanded clay (**e**), and Biochar (**f**)

popularity of peat can be attributed to its relatively low cost, and superior chemical, physical, and biological characteristics, including low nutrient content, low pH, high air space, high water-holding capacity, and lightweight, which makes it very popular among soilless substrates (Marshall and Saunby 1954). However, in recent years there has been strong criticism of the use of peat in horticulture due to environmental concerns and the threat of climate change (Gruda 2022). The turnover of peatlands is not coping with the amount of peat that is presently used in agriculture and other anthropic industries (e.g. heating systems). It is interesting to observe the current trend toward using biochars and hydrochars, solid products rendered from hydrothermal carbonization (Fussy and Papenbrock 2022). The physical properties of biochar are generally characterized by its high porosity, high surface area, high water-holding capacity, and high ion exchange capacity, while the chemical properties of biochar depend mostly on the raw organic material used to produce it (Atzori et al. 2021). However, biochar has heterogeneous properties, while both its pH value and production costs are high (Fussy and Papenbrock 2022). It should be kept in mind that salt accumulation on the soilless substrates, e.g. Na^+ and Cl^-, is common in soilless culture, negatively affecting salt-sensitive plants (Incrocci et al. 2006). For that reason, it is necessary to develop a collection and sterilization technique for the drainage nutrient solutions, so that they can be reused (Runia et al. 1988). Commercial companies often use mixes of substrates, such as peat and perlite (Fig. 4.2). The advantages and disadvantages of some other organic and inorganic substrates for soilless cultivation are well summarized in the review of Fussy and Papenbrock (2022).

Fig. 4.2 Commercial production of edible flowers grown in peat and perlite substrates by Ortogourmet (Mola di Bari, BA, Italy)

4.2.1.2 Hydroponics

Hydroponics is another soilless cultivation method. Hydroponics is a method of growing plants without soil, where roots are fed by a nutrient solution. This system allows for complete control over the nutrient content of the solution, allowing plants to grow faster and more efficiently. Typically, hydroponic systems are classified as open (i.e., the nutrient solution is not reused after delivering to the roots) or closed (i.e., excess solution is recovered, replenished, and recycled) (Fussy and Papenbrock 2022). There are several hydroponic systems, classified into different categories, based on their setting and functionality: wick system, deep water culture, nutrient film technique, ebb-and-flow technique, aeroponics, and drip system (Fig. 4.3). Aeroponics is an innovative technology that has been effectively employed to cultivate a diverse array of crops, despite its relatively recent introduction. In aeroponic systems, pumps, timers, and spray nozzles are commonly used to dispense a mist rich in oxygen, water, and nutrients to the plant roots. The mist is applied at predetermined intervals, enabling the roots to effectively absorb the necessary nutrients and moisture (Garzón et al. 2023). Aeroponics embodies a paradigm shift in agricultural practices, providing answers to urgent challenges

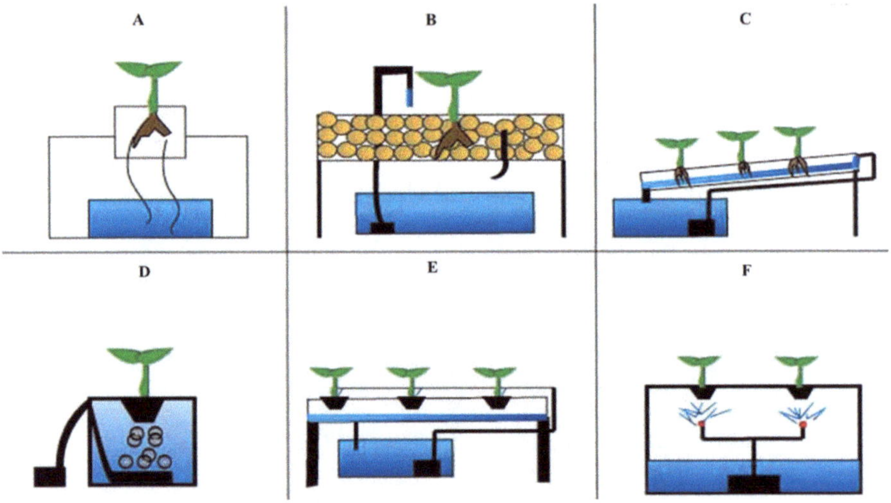

Fig. 4.3 Some soilless cultivation systems. Wick system (**a**), Ebb-and-flow technique (**b**), Nutrient film technique (**c**), Deep water culture (**d**), Drip system (**e**), and Aeroponics (**f**)

Fig. 4.4 Experimental cultivation of edible flowers (*Tropaeolum majus*) in aeroponic system at the University of Turin (Torino, Italy)

including food security, depletion of resources, and the need for environmental sustainability. As technological innovations progress, it is expected that aeroponic systems will gain greater accessibility and efficiency, leading to their broader adoption, despite at the moment they are mainly used for experimental purposes (Fig. 4.4). Various factors may affect the application of soilless systems, such as the plant type, the local climate, and the budget available for the implementation. An overview of the characteristics of each type of hydroponic system, its tolerance to pump failures and power outages, as well as its advantages and disadvantages is provided in Table 4.1.

Table 4.1 Types of hydroponic systems available and their benefits

Type of hydroponic system	Description	Tolerance to pump failure/ electricity outage	Advantages	Disadvantages
Deep water culture	Nutrient solutions constantly submerge roots	High	It is suitable for plants that require a large amount of water to grow	Delivery of oxygen is challenging
Nutrient film technique	Plants are placed in channels (gutters) fed by a film of nutrient solution	Low	There is no need for a growth medium, timers, or pumps	When nutrient flow is interrupted, plants are prone to wilting
Wick system	Pumps, aerators, and electricity are not needed	Highest	Simple hydroponic systems that are easy to maintain	It provides only a small amount of water at a time and bigger plants may suffer from water shortages
Ebb and flow	The nutrient solution is flooded into the roots, and then drained off, intermittently	High	Suitable for water-loving plants	Plants' root is susceptible to diseases
Drip system	Plants are nourished with nutrient solution dripped on their bases	Medium	As the flow rate of the nutrients can be adjusted, it can be tailored to any type of plants.	Due to pH shift, the system maintenance is difficult.
Aeroponics	Providing nutrients to roots via fog, spray, or mist	Low	Suitable for root and tuberous plants, no airing system is required	Costly, technical knowledge required, and requires high maintenance

Nutrient management and fertilization are essential components of soilless cultivation success (Pignata et al. 2017). It is possible to incorporate continuous cropping into soilless systems by using the same bags, media, and irrigation system over time, providing the media is suitable for a variety of crops. It has been estimated that soilless cultivation methods, such as hydroponics, use 90% less water than soil-based cultivation methods (Pomoni et al. 2023). The management of water is critical to the success of soilless cultivation. Although soilless cultivation is less susceptible to soilborne pests and diseases, it is imperative that potential issues are monitored and addressed (Fussy and Papenbrock 2022). In a growing environment, temperature, humidity, airflow, and light should be controlled as much as possible to enhance plant growth and productivity.

4.2.2 Different Soilless Cultivation Settings

The soilless cultivation settings provide creative solutions to the challenges faced by modern agricultural practices. Each type has its own particular strengths and is designed for different crop types, making them versatile options for sustainable food production in a variety of environments. As technology progresses, these methods are likely to become increasingly crucial in the effort to resolve global food security challenges.

4.2.2.1 Open Field Soilless Cultivation Systems

It is recognized that soilless cultivation in the open field is an efficient method that leads to a reduction in waste and an increase in productivity. Plants may be more tolerant to temperature and humidity stress when cultivating in an open field soilless cultivation system rather than growing in greenhouses or enclosed spaces (Rahim Doust et al. 2023). Advances in these fields could increase the productivity and nutritional value of horticultural crops. Several factors are required for soilless cultivation to be successful in the open field. It is important to choose the right place and avoid flood-prone and weed-filled areas (Hochmuth and Hochmuth 1969). Soilless bags should be placed on polyethylene-mulched raised beds to reduce soil contamination. There are several factors to consider when choosing a growing medium, including the type of plant, the pH of the irrigation water, the cost of the product, the life span of the medium, the type of system being used, and the grower's preference. Growstone, perlite, coconut coir, rockwool, coconut choir, peat, pine bark, mineral wool, and sand are some growing media commonly used.

4.2.2.2 Under Cover Soilless Cultivation Systems, from Basic Settings to Greenhouses

Undercover cultivation is also known as greenhouse cultivation, which provides a controlled growing environment in which factors such as temperature, humidity, and light can be readily controlled. Most greenhouse growers spend 70–80% of their energy budget on heating, and 10–15% on electricity for producing high-value intensive crops (Sanford 2009). Energy efficiency under protected cultivation is greatly influenced by greenhouse cladding and material selection. The purpose of covering a greenhouse is to allow sunlight to pass through; in this way, the energy that is trapped inside the greenhouse can be retained (Reddy 2016). Early greenhouses were covered with glass, but plastic materials e.g., acrylic, polycarbonate, fiberglass, and polyethylene have now provided several alternatives for covering greenhouses (Hanson 2020).

Temperature, light, air composition, and the type of root medium are all controlled within the greenhouse environment. Numerous climatic factors require attention in the management of a soilless cultivation systems like greenhouse. The subsequent factors are crucial for successful greenhouse cultivation.

Ventilation in greenhouses reduces greenhouse air temperatures, replaces carbon dioxide (CO_2), and modifies relative humidity (Reddy 2016). Generally, ventilation is expressed as a percentage of the floor area of a protected cropping structure. Under temperate conditions, 30% ventilation is recommended (Morgan 2021). In the autumn and spring, ventilation may be used for cooling, but in the summer, other methods must be employed (Reddy 2016).

In greenhouses, shading and fan and pad are two ways to reduce the temperature. The use of evaporation can also be used to cool greenhouses by fogging, misting, or spraying water on floors (Morgan 2021).

In many regions, the lack of natural light during the winter months is a limiting factor for crop growth. A certain percentage of light is blocked by cladding films and materials and the associated framework in protected cropping structures. Optimizing light transmission into a greenhouse is important, but artificial light can still be needed in order to maintain horticultural production throughout the year (Hemming 2011). Supplemental lighting increases light levels when natural sunlight limits growth and production; this may result in an increase in daily light integral over a 24-hour period by increasing light duration and intensity during the day (Morgan 2021). Using LED lamps can save 75% of energy per year when compared with conventional HPS lamps, as reported by Cuce et al. (2016).

Relative humidity and CO_2 enrichment are other settings that should be considered in the greenhouse environment. A relative humidity range of 50–80% is considered acceptable for most crops. Nevertheless, relative humidity of up to 90% may be suitable for plant propagation (Reddy 2016). Plants grown under protected cultivation can benefit from CO_2 enrichment, resulting in higher yields, earlier harvests, and higher-quality crops. It has been found to be economically viable to enrich CO_2 levels in various crops by 800–1200 mmol mol^{-1} (Morgan 2021).

4.2.2.3 Indoor Soilless Cultivation Systems

Controlled environment agriculture, also known as indoor farming, vertical farming, plant factory, or container farm (Mitchell 2022), involves growing crops in indoor environments using methods such as stacking plants from floor to ceiling, hydroponics, and artificial lights. All cardinal factors of plant growth (light, temperature, water, nutrients, and atmosphere) must be provided and controlled to grow crops to harvest maturity indoors. Based on the published scientific references, the settings required for indoor cultivation, particularly in indoor vertical farming, are listed in Table 4.2.

Table 4.2 A brief explanation of the settings required for indoor cultivation

Setting	Description	Reference
Light source	It is recommended to operate artificial light for 10–24 hours a day. Compared to traditional light sources such as fluorescent and HPS, LED grow lights optimize crop production and reduce energy consumption costs	Avgoustaki and Xydis (2020), Mitchell et al. (2015)
Land use and energy	Enhanced production along with improved efficiency per unit of land, water, energy, and labour.	Reddy (2016)
Fertilizer and nutrient management	Intensive sterilization and the use of supplier-application methodologies are the requirements for indoor vertical farms, which use isolated culture beds and nutrient solutions	Avgoustaki and Xydis (2020)
Environmental control	It is essential to provide and control all the cardinal factors of plant growth (light, temperature, water, nutrients, and atmosphere) in order to successfully grow crops indoors to harvest maturity	Mitchell (2022)
Pest and disease control	In order to prevent pest infestations in indoor cultivation, it is necessary to maintain a sterilized environment	Kozai et al. (2019), Reddy (2016)

4.3 Soilless Cultivation Systems for Edible Flowers

In comparison with traditional soil-based cultivation, soilless culture offers numerous advantages for growing edible flowers. Successful hydroponic flower cultivation requires choosing the right flower varieties. The use of soilless cultivation could help reduce the growing cycle used in conventional field cultivation, while simultaneously increasing biomass production, because soilless cultivation provides a balanced supply of water and nutrients, as well as allowing edible flowers to be grown year-round. Moreover, soilless cultivation provides controlled indoor environments, which makes it possible to grow edible flowers even in colder climates (Wani et al. 2023).

4.3.1 Advantages of Growing Edible Flowers in Soilless Systems

There are several advantages to this method of cultivation as compared with conventional cultivation. It is possible to control growing conditions more precisely with soilless cultivation in flowers. The control of the temperature, humidity, and nutrients of the growing solution, as well as the control of weeds, insects, and diseases can help growers to improve plant growth and flower quality (Zimmermann and Fischer 2020). Pests and diseases can be easily managed in hydroponic edible flower gardens as compared to traditional soil-based ones. Compared to conventional agriculture (using insecticides and herbicides), soilless agriculture reduces or eliminates the use of insecticides and herbicides (Savvas 2002), thereby resulting in crops that are free of pesticide residues.

There is evidence to suggest that plants grown in hydroponic systems are often of superior quality, and their bioactive compounds typically have higher levels than those grown in conventional systems (Wani et al. 2023). According to Léonhart et al. (2002), several plants (*Tanacetum parthenium*, *Taraxacum officinale*, and *Calendula officinalis*) are well suited to greenhouse hydroponic conditions and they provide abundant yields and high quality of produce within a short period of time. Stewart and Lovett-Doust (2003) reported that hydroponic greenhouse cultivation of *Calendula officinalis* under controlled conditions provided pesticide-free conditions and lacked environmental contaminants, which led to superior product quality and stability. Brechner et al. (2007) noted that growing *Hypericum perforatum* in controlled environments, such as greenhouses or growth chambers, can limit the wide differences in temperature, insect and disease pressures, as well as the water status for the plant.

An investigation was conducted to determine the effects of soil and hydroponic culture on chrysanthemum (*C. morifolium* 'wuyuanhuang') flower yields, flavonoids, and chlorogenic acid contents. Hydroponic culture systems produced higher yields of chrysanthemum flowers and contained higher levels of total flavonoids and chlorogenic acids (Ai et al. 2021). The accumulation of secondary metabolites in greenhouse hydroponic cultures has been shown to be enhanced by altering the composition of the nutrient solution in several studies (Briskin 2000; Montanari et al. 2008), as well as manipulating environmental parameters such as light (Hou et al. 2010). Both plant growth and secondary metabolism may be affected by plant mineral nutrition (Briskin 2000). In soilless cultures, growth reduction occurs as a result of nitrogen (N) starvation, which limits primary metabolism, thereby reducing biomass production. However, N deficiency may oppositely impact secondary metabolism. Bryant et al. (1983) proposed a C/N balance hypothesis to explain how C-based secondary metabolites are regulated by the C/N ratio within the plant. Since N deficiency leads to the increased formation of reactive oxygen species (Kováčik et al. 2007), phenolic compounds play a major role in plants' adaptation to N deficiency. N-deficient plants cause an increase in the production of phenolic compounds in chamomile (*Matricaria chamomilla*) (Kováčik et al. 2007). *Salvia officinalis* plants grown in soilless conditions benefited from the addition of 14.6 mM N by improving plant growth, nutrition, biochemical characteristics, and essential oil yield and composition (Khammar et al. 2021). Bioactive compounds were reported to be produced in plants grown in hydroponics using nutrient solutions containing both NH_4^+ and NO_3^- in different ratios. The accumulation of vinblastine and catharanthine in *Catharanthus roseus* was enhanced by supplying 50% total N as NH_4^+ (Guo et al. 2012).

Secondary metabolites produced by plants can be affected differently by salt stress depending on the species. Using a deep flow technique system to induce salt stress (0, 50, and 100 mM NaCl) in tagetes (*Tagetes patula* L.) plants led to increased flower polyphenols, antioxidant activity, and total carotenoids but decreased anthocyanins. This resulted in a higher antioxidant value for edible flowers at a salinity of

100 mM NaCl. It is also reported that short-term saline exposure of tagetes plants was associated with increased metabolism, as well as increased mineral accumulation on edible flowers (e.g. N, P, and Zn), which is of crucial importance to human health (Chrysargyris et al. 2018b). In another study, it was established that short-term salinity positively influences the nutritional quality of petunia (*Petunia × hybrida* L.) flower (Chrysargyris et al. 2019b). The application of 50 mmol L^{-1} NaCl was found to enhance the levels of polyphenols, antioxidants, carotenoids, and anthocyanins. On the other hand, it was reported that salinity decreased the production of essential oils in *Matricaria chamomilla* (Razmjoo et al. 2008) but did not affect the echinacoside content of *Echinacea angustifolia* (Maggini et al. 2013).

Several studies have demonstrated that supplemental lighting on plants grown hydroponically under greenhouses accumulates more bioactive molecules when compared with plants grown in the field (Pedneault et al. 2002; Brechner et al. 2007). Lighting technologies such as LEDs are proving to be very efficient for the cultivation of plants and also for the production of metabolites due to their elicitor properties. For example, in the plant factory system, Karimi et al. (2022) showed that 100% red LED light promotes the growth and flowering of *Hypericum perforatum* L. (cv. Topas), and increases secondary metabolite accumulation in the flowers, including hypericin, pseudohypericin, and hyperforin. Also, the role of combined red and blue light associated with γ-aminobutyric acid and gibberellic acid in enhancing the quality of saffron stigma (safranal, picrocrocin, and crocin) has been reported by Eftekhari et al. (2023).

4.3.2 Advantages of Applied Systems

Various hydroponic systems have been used to produce ornamental flowers in the commercial sector, including deep flow technique, nutrient film technique, troughs, and grow bags (Atherton and Li 2023). While potted and cut flowers are typically grown on solid media (Fussy and Papenbrock 2022), it is also possible to cultivate hydroponically flowers for nutritional purposes (Atherton and Li 2023). In contrast to the production of flowers for ornamental purposes, liquid culture methods such as deep flow technique (Chrysargyris et al. 2019a, b; Menegaes et al. 2020) and nutrient film technique (Melo and dos Santos 2011; Menegaes et al. 2020) are generally preferred. However, some commercial producers grow edible flowers in pots using traditional media, such as peat (Fig. 4.5). In order to grow flowers hydroponically, the floating technique is one of the most common systems used (Fig. 4.6).

This type of hydroponic system is primarily used for experimental purposes, rather than for commercial purposes (Atherton and Li 2023) (Fig. 4.7). Plants that have a low growth habit and those that require periodic harvesting of their flowers in order to promote growth of the plant are especially suited to nutrient film technique. In addition to being ideal for growing crops that grow quickly, this system can be adapted for growing flowers if they are properly maintained. Root and shoot yields were higher when deep flow technique was used in the cultivation of

Fig. 4.5 Commercial production of *Viola* edible flowers in pots in an indoor vertical farm by Potager Farm (Berlin, Brandenburg, Germany); the left picture is showing one shelf for the nursery stage, with pots and substrates at the sowing time; the right picture is showing the flowers at harvest time

Fig. 4.6 Commercial production of *Viola* edible flowers in floating system in an indoor farm by Sananbio (Xiamen, Fujian Province, China); the left picture is showing one shelf of the rack system; the right picture is showing the indoor room where the racks are installed

dandelion (*Taraxacum officinale*) plants when compared to nutrient film technique (Gill 2015). *According to* Murch et al. (2002), *hypericin, hyperforin, and pseudohypericin concentrations of Hypericum perforatum* flowers *were greater in* nutrient film technique-*grown flowers than in field-grown plants.* There is some evidence that capillary technique is effective for the cultivation of flowers (Waiba et al. 2020), but studies that use the method are extremely rare. Possibly this is due to the difference in characteristics that are desired in both cut and edible flowers when it comes to their characteristics. Growing chrysanthemums (*Dendranthema grandiflora* (Ramat.) Kitam.) in a capillary mat system induced a higher height, stronger stems, and longer vase life than growing them in a deep flow technique (Sakamoto et al. 1998). Application of ebb and flow systems has also been shown to be effective,

Fig. 4.7 Experimental cultivation of edible flowers (*Viola wittrockiana*) in floating system at the University of Turin (Torino, Italy)

which resulted in a higher phytochemical content in *H. perforatum* (Zobayed and Saxena 2004). Different soilless cultivation systems used for some edible flowers are shown in Table 4.3 (Based on Atherton and Li 2023; Pires et al. 2021; Takahashi et al. 2020).

4.3.3 Soilless Cultivation Practices for Edible Flowers

It is becoming increasingly necessary for farmers to employ skilled and trained labour who are capable of identifying edible flowers, controlling insects, caring for plants during their post-harvest and removing any parts that have an undesirable taste (for example, parts with a bitter taste) that are grown for edible purposes (Takahashi et al. 2020). Edible flowers are particularly well suited to soilless greenhouse culture (Dorais et al. 2001), as it prevents plant contamination problems linked to agricultural soil, while ensuring plant nutrition by using nutrient solutions containing all the necessary macro and micro elements. The nutritional and environmental conditions for both hydroponics and substrate culture can be tailored to meet

Table 4.3 Soilless cultivation systems used for some edible flowers

Cultivation system	Scientific name
Nutrient film	*Taraxacum officinale* L. *Crocus sativus* L.
Deep flow	*Taraxacum officinale* L. *Chrysanthemum morifolium* Ramat.
Floating	*Borago officinalis* L. *Calendula officinalis* L. *Crocus sativus* L. *Chrysanthemum morifolium* Ramat. *Lonicera japonica* Thunb. *Taraxacum officinale* L.
Ebb and flow	*Crocus sativus* L.
Capillary action	*Chrysanthemum morifolium* Ramat.
Trough	*Crocus sativus* L.
Pot	*Lavandula angustifolia* Mill. *Lonicera japonica* Thunb.

plant growth needs and optimize secondary metabolite content (Chrysargyris and Tzortzakis 2021). Recently, three cultivars of *Tagetes patula* L. were examined for morpho-physiological performance, phenolic content, and essential oil composition under different cultivation conditions (hydroponics, poly-house, and open-field) (Sharma et al. 2024). In their study, vanillic acid, caffeic acid, quercetin, as well as the number of flowers produced per plant were maximally observed under hydroponic conditions, although volatile components and phenolic content were maximally observed under open-field conditions. Open-field cultivated plants have been hypothesized to experience more stress than those in polyhouses and fields (Partap et al. 2020). Plant metabolite accumulation and biosynthesis are regulated by biotic and abiotic stresses, cultivation conditions, irrigation, substrate, and climate conditions. It can be concluded that the hydroponic system in general produces consistently healthy, disease-free plants throughout the year, regardless of seasonal variations, which is vital for meeting market demand (Gwynn-Jones et al. 2018). Furthermore, it is very important for soilless growers to understand edible flowers' nutritional needs and what elements they need. Edible flowers require the correct amount of N, P, K, and other micro and macro elements. Growers should adjust the nutrient levels in the nutrient solution accordingly to ensure healthy growth and flowering. Regular inspections and fertilization are essential for successful edible flower production. According to Stewart and Lovett-Doust (2003), *Calendula officinalis*, for example, needs low P levels to produce optimal inflorescence (the target tissue containing the bioactive constituents). It should also be noted that in light of the xerophytic characteristics of this species, the most promising results in terms of flowerhead tissue production were obtained when relatively low ratios of P relative to N and K were coupled with intermittent watering regimes. Harvesting edible flowers in the conditions of soilless cultivation is also one of the important operations in the production of these products. It is important to ensure that the flowers are harvested at the right stage of maturity. It is ideal to pick edible flowers early in the morning, after the dew has evaporated, or just before sunset in the evening, once the dew has melted. High humidity level can cause the flowers to discolour and even lose their flavour. Flowers can also be dried and reduced in taste by the heat and rays of the sun. Additionally, proper postharvest handling techniques should be employed to ensure that the flowers are free from any contaminants. The flowers should also be stored properly to ensure their freshness and flavour. Finally, the flowers should be processed in a timely manner to maximize their shelf life.

4.4 Soilless Cultivation Protocols for Selected Species

Calendula officinalis belongs to the Asteraceae family and has long been recognized as an edible flower. Given that the flowers, or more accurately the capitula within this family, represent the most valuable part of the plant, this characteristic renders the species both practical and appealing as a greenhouse crop, allowing for a continuous harvest throughout the year (Stewart and Lovett-Doust 2003). Floating technique and pot technique systems are suitable for growing this plant (Atherton and Li 2023). It has been found that peat can be substituted by up to 30% of paper waste for *C. officinalis* as a suitable substrate (Chrysargyris et al. 2019a). Moreover, it was also found that up to 30% of olive seed powder could be utilized as a substitute for peat in the cultivation of *C. officinalis* flowers (Chrysargyris et al. 2018a). Typically, *C. officinalis* likes to grow in cooler temperatures, and for this reason, rarely blooms in extreme temperatures. It is recommended to remove the old flowers (deadheading) in order to encourage reblooming. *C. officinalis* necessitates a minimal photoperiod of 6 hours and 30 minutes within a 24-hour cycle to initiate inflorescence, as noted by Sawhney et al. (1981).

Tagetes patula L. belongs to the Asteraceae family and is an essential oil-bearing annual flowering ornamental plant and cultivated throughout the world (Sharma et al. 2024). *T. Patula* has a number of natural constituents including flavonoids, carotenoids, lupeol, luteolin, thiophenes, benzofurans, phenols, lutein, and terpenoids (Marotti et al. 2010; XU et al. 2012; Riaz et al. 2020). It is grown using a variety of cultivation systems, including greenhouses, fields, and deep flow technique systems. Hydroponic conditions led to an increase in flower production for *T. patula*, surpassing the results achieved in both open-field and poly-house (Sharma et al. 2024). Hydroponic cultivation of this plant requires temperatures of 25 °C and relative humidity of 65%, as well as electrical conductivity (EC) of 2.5 dS m^{-1} and pH of 6.5 (C).

Taraxacum officinale L. (dandelion) is an edible flower belonging to the Asteraceae family. In order to consume the blossoms, they are batter-dipped and deep-fried (Takahashi et al. 2020). There is a reference to *Taraxacum officinale* L. in the Bible as one of the bitter herbs which were used in salads as well as in unleavened breads (Takahashi et al. 2020). It contains nutrients, antioxidants, and a wide variety of bioactive compounds, including phenolics, tocopherols, organic acids, and sugars (Atherton and Li 2023). Deep flow technique, nutrient film technique, and floating technique systems are suitable for growing this plant (Atherton and Li 2023). Although pH 5.5 is more favourable, this plant can adapt to nutrient solutions with acidic or neutral pH levels (Atherton and Li 2023). In order to harvest dandelion, it is recommended to clip a few leaves off the dandelions before they bloom. To determine whether the dandelion greens are too bitter, taste them once the yellow flower has bloomed.

Crocus sativus L. (saffron), a member of the Iridaceae family, is commonly grown throughout Iran and other countries such as India and Greece. *This plant contains a wide variety of chemicals in its petals, including mineral agents, anthocyanins, flavonoids, proteins, starch, glycosides, alkaloids, kaempferol, and vitamins such as riboflavin and thiamine* (Naemi and Meshkini 2023). This spice is widely used in food and beverages for its colouring, flavouring, and odour properties related to its primary metabolites *including* crocin, picrocrocin, and safranal (Hamidian et al. 2023). There are several soilless systems that can be used for saffron, including nutrient film technique, floating technique, ebb and flow technique, pot technique and trough technique (Atherton and Li 2023). In soilless cultivation, saffron with a corm size of 3.2–3.5 cm has the highest values of vegetative growth parameters and flowering, as well as cormlet production (Souret and Weathers 2000). The size of the corms plays an important role in saffron flowering in a hydroponic system. It was discovered that only corms with a diameter greater than 2.6 cm bloomed (Souret and Weathers 2000). The report also showed that small corms <2.6 cm presented poor flowering under aeroponics, hydroponics, as well as in soil. In terms of corm weight, Eftekhari et al. (2023) used corms weighing between 10 and 12 grams in an indoor cultivation system, which resulted in harvestable flowers. The harvesting of the saffron flowers starts as soon as the first flower appears and continues for 10–15 days.

Borago officinalis L. is an annual plant that belongs to the Boraginaceae family that is commonly known as the borage plant. Borage is an annual flowering herbaceous plant that is native to the Mediterranean region, including Italy, Spain, and the Middle East. In recent years, borage flowers have gained a lot of attention from consumers due to the renewed interest in edible flowers (Moliner et al. 2022). A variety of therapeutic, ornamental, and cosmetic uses are attributed to this species, as well as consumption of its leaves and flowers (Ramezani et al. 2020). It has been found that edible flowers of borage contain bioactive compounds with health-promoting properties in addition to their aesthetic value (Moliner et al. 2022). Floating technique system is suitable for growing this plant (Atherton and Li 2023). The suitable EC and pH levels in floating culture conditions for *B. officinalis* are 2–3 dS m^{-1} and 5.5–6, respectively. The flowers of borage can be harvested as soon as they open by snipping them individually or in clusters.

Chrysanthemum morifolium Ramat belongs to the Asteraceae family, *which originated in China, and is commonly referred to as a "double-use plant" with dual uses in food and drug* (Boase et al. 1996; Guo et al. 2008). Flowers of *C. morifolium* Ramat are mainly known for their carotenoids, flavonoids, and caffeoylquinic acids (Chen et al. 2021). Even though chrysanthemums have been mostly consumed as a tea, few studies have studied the health benefits that can be obtained from the ingredients that are found in chrysanthemum extract when it is dissolved in hot water (Chen et al. 2021). It has been found that deep flow technique, floating technique, and capillary action technique system are suitable for growing chrysanthemum (Atherton and Li 2023). There is a recommendation that the EC and pH of the nutrient solution be maintained between 1.3–1.8 dS m^{-1} and 5.5–6.5, respectively. Pruning, pinching, deadheading, and disbudding are necessary to encourage flowering in chrysanthemums. When the petals of the flower just start curling at the edges and the blooms are fully open, chrysanthemums can be harvested.

Lavandula angustifolia Mill. (lavender) belongs to the Lamiaceae family and is native to the mountainous regions of the Mediterranean, mostly at high altitudes. Fresh and dried lavender flowers are used for decorative purposes as well as flavouring in baking, in jams, jellies, candies, infusions, sauces, and infusions in culinary preparations (Caser et al. 2023). Essential oils and phenolic compounds are abundant in the flowers of lavender (Crişan et al. 2023; Caser et al. 2023). Pot technique system is suitable for growing this plant (Atherton and Li 2023). From the perspective of ornamental-horticulture, peat and compost contribute to plants that have improved characteristics as a result of their use (Crişan et al. 2023). For this plant, pruning is also recommended as a part of the maintenance process.

4.5 Conclusion and Perspective

Eating edible flowers has numerous health benefits including proteins, carbohydrates, key fatty acids, and dietary fiber, among many others. In the eyes of consumers, edible flowers are a healthier alternative that can be processed without any negative environmental effects by means of nonthermal methods. There is evidence to suggest that the synthesis and accumulation of nutritional compounds in edible flowers can be achieved through cultivating without soil by regulating the mineral nutrition and controlling the growth conditions under which the flowers grow. This means that correct management of these parameters is imperative in order to establish proper growing practices in the field. It was found that different soilless cultivation systems can be used for many types of edible flowers, so that some of these plants respond positively to several soilless cultivation methods. Additionally, these systems are also more environmentally friendly and sustainable than traditional soil-based methods. The literature on soilless cultivation of edible flowers is relatively sparse, so it is expected that researchers will focus on other species and optimize their growth conditions in conjunction with high flower quality. This allows for a better understanding of the potential of soilless cultivation of edible flowers and the potential to optimizing their production, as well as the potential to produce better flower qualities. In general, wick system, deep water culture, nutrient film technique, ebb-and-flow technique, drip, and aeroponics systems have been used in the cultivation of edible flowers. Aeroponics signifies a revolutionary method in agriculture, showcasing significant promise as a leading technique in soilless cultivation systems. Its benefits, including enhanced water efficiency, optimal use of space, accelerated growth rates, and diminished environmental impact, establish it as a crucial solution for tackling global food security issues. As technological advancements progress, aeroponics is expected to play a vital role in sustainable agricultural practices across the globe. The substrates most commonly used in the cultivation of this type of product include perlite, vermiculite, coconut coir, rockwool, expanded clay, and biochar. Edible flowers such as *Calendula officinalis, Tagetes patula, Taraxacum officinale, Crocus sativus, Borago officinalis, Chrysanthemum morifolium,* and *Lavandula angustifolia* are mostly planted in soilless cultivations. We believe this chapter will be very useful to the reader in terms of gaining a broader and more rounded understanding of edible flowers that can contribute to ensuring food security around the world, which is of great importance to us.

References

Ai P, Liu X, Li Z, Kang D, Khan MA, Li H, Shi M, Wang Z (2021) Comparison of chrysanthemum flowers grown under hydroponic and soil-based systems: yield and transcriptome analysis. BMC Plant Biol 21:517. https://doi.org/10.1186/s12870-021-03255-4

Asaduzzaman M, Saifullah M, Mollick AKMSR, Hossain MM, Halim GMA, Asao T (2015) Influence of soilless culture substrate on improvement of yield and produce quality of horticultural crops. In: Soilless culture – use of substrates for the production of quality horticultural Crops. https://doi.org/10.5772/59708

Atherton HR, Li P (2023) Hydroponic cultivation of medicinal plants-plant organs and hydroponic systems: techniques and trends. Horticulturae 9:349. https://doi.org/10.3390/horticulturae9030349

Atzori G, Pane C, Zaccardelli M, Cacini S, Massa D (2021) The role of peat-free organic substrates in the sustainable management of soilless cultivations. Agronomy 11:1–29. https://doi.org/10.3390/agronomy11061236

Avgoustaki DD, Xydis G (2020) How energy innovation in indoor vertical farming can improve food security, sustainability, and food safety? Adv Food Secur Sustain 5:1–51. https://doi.org/10.1016/bs.af2s.2020.08.002

Boase MR, Miller R, Deroles SC (1996) Chrysanthemum systematics, genetics, and breeding. Plant Breed Rev 14:321–361

Brechner ML, Albright LD, Weston LA (2007) Impact of a variable light intensity at a constant light integral: effects on biomass and production of secondary metabolites by *Hypericum perforatum*. In: International Symposium on Medicinal and Nutraceutical Plants. Acta Hortic. Georgia (USA) 756:221–228. https://doi.org/10.17660/ActaHortic.2007.756.23

Briskin DP (2000) Medicinal plants and Phytomedicines. Linking plant biochemistry and physiology to human health. Plant Physiol 124:507–514. https://doi.org/10.1104/pp.124.2.507

Bryant JP, Chapin FS, Klein DR (1983) Carbon/nutrient balance of boreal plants in relation to vertebrate herbivory. Oikos 40:357. https://doi.org/10.2307/3544308ertebrate

Caser M, Falla NM, Demasi S, Scariot V (2023) From fresh to dried lavender flower: changes in phytochemical profile according to drying method. Horticulturae 9:700. https://doi.org/10.3390/horticulturae9060700

Chen S, Liu J, Dong G, Zhang X, Liu Y, Sun W, Liu A (2021) Flavonoids and caffeoylquinic acids in *Chrysanthemum morifolium* Ramat flowers: a potentially rich source of bioactive compounds. Food Chem 344:128733. https://doi.org/10.1016/j.foodchem.2020.128733

Chrysargyris A, Tzortzakis N (2021) Tailor-made hydroponic recipes for the production of high quality medicinal and aromatic plants. Acta Hortic 1321:87–94. https://doi.org/10.17660/ActaHortic.2021.1321.11

Chrysargyris A, Antoniou O, Tzionis A, Prasad M, Tzortzakis N (2018a) Alternative soilless media using olive-mill and paper waste for growing ornamental plants. Environ Sci Pollut Res 25:35915–35927. https://doi.org/10.1007/s11356-017-1187-4

Chrysargyris A, Tzionis A, Xylia P, Tzortzakis N (2018b) Effects of salinity on tagetes growth, physiology, and shelf life of edible flowers stored in passive modified atmosphere packaging or treated with ethanol. Front Plant Sci 9:1765. https://doi.org/10.3389/fpls.2018.01765

Chrysargyris A, Stavrinides M, Moustakas K, Tzortzakis N (2019a) Utilization of paper waste as growing media for potted ornamental plants. Clean Techn Environ Policy 21:1937–1948. https://doi.org/10.1007/s10098-018-1647-7

Chrysargyris A, Tzionis A, Xylia P, Nicola S, Tzortzakis N (2019b) Physiochemical properties of petunia edible flowers grown under saline conditions and their postharvest performance under modified atmosphere packaging and ethanol application. J Sci Food Agric 99:3644–3652. https://doi.org/10.1002/jsfa.9586

Crişan I, Ona A, Vârban D, Muntean L, Vârban R, Stoie A, Mihăiescu T, Morea A (2023) Current trends for lavender (*Lavandula angustifolia* Mill.) crops and products with emphasis on essential oil quality. Plants 12:357. https://doi.org/10.3390/plants12020357

Cuce E, Harjunowibowo D, Cuce PM (2016) Renewable and sustainable energy saving strategies for greenhouse systems: a comprehensive review. Renew Sust Energ Rev 64:34–59. https://doi.org/10.1016/j.rser.2016.05.077

Dorais M, Papadopoulos AP, Luo X, Leonhart S, Gosselin A, Pedneault K, Angers P, Gaudreau L (2001) Soilless greenhouse production of medicinal plants in north eastern Canada. Acta Hortic 554:297–304. https://doi.org/10.17660/ActaHortic.2001.554.32

Eftekhari M, Javid MG, Aliniaeifard S, Nicola S (2023) Alteration of flower yield and phytochemical compounds of saffron (*Crocus sativus* L.) by application of different light qualities and growth regulators. Horticulturae 9. https://doi.org/10.3390/horticulturae9020169

Fussy A, Papenbrock J (2022) An overview of soil and soilless cultivation techniques—chances, challenges and the neglected question of sustainability. Plants 11:1153. https://doi.org/10.3390/plants11091153

Garzón J, Montes L, Garzón J, Lampropoulos G (2023) Systematic review of technology in aeroponics: introducing the technology adoption and integration in sustainable agriculture model. Agronomy 13(10):2517. https://doi.org/10.3390/agronomy13102517

Gielen D, Boshell F, Saygin D, Bazilian MD, Wagner N, Gorini R (2019) The role of renewable energy in the global energy transformation. Energy Strateg Rev 24:38–50. https://doi.org/10.1016/j.esr.2019.01.006

Gill R (2015) Nutrient management for growing Dandelion (*Taraxacum officinale* L.) in nutrient film and deep flow hydroponics. Dissertation, University of Arkansas, Fayetteville, Arkansas, USA. https://scholarworks.uark.edu/etd/1505

Gonnella M, Renna M (2021) The evolution of soilless systems towards ecological sustainability in the perspective of a circular economy. Is it really the opposite of organic agriculture? Agronomy 11:950. https://doi.org/10.3390/agronomy11050950

Grillas S, Lucas M, Bardopoulou E, Sarafopoulos S, Voulgari M (2001) Perlite based soilless culture systems: current commercial applications and prospects. Acta Hortic 548:105–114. https://doi.org/10.17660/ActaHortic.2001.548.10

Gruda NS (2022) Advances in soilless culture and growing media in today's horticulture – an editorial. Agronomy 12:2773. https://doi.org/10.3390/agronomy12112773

Guo Q-S, Wang T, Cheng L-T, Wang T-Y, Liang Y-N (2008) Study on dynamic accumulation of flavone in various *Chrysanthemum morifolium* for medicine. Zhongguo Zhong Yao Za Zhi 33:1237–1239

Guo X-R, Zu Y-G, Tang Z-H (2012) Physiological responses of *Catharanthus roseus* to different nitrogen forms. Acta Physiol Plant 34:589–598. https://doi.org/10.1007/s11738-011-0859-9

Gwynn-Jones D, Dunne H, Donnison I, Robson P, Sanfratello GM, Schlarb-Ridley B, Hughes K, Convey P (2018) Can the optimisation of pop-up agriculture in remote communities help feed the world? Glob Food Sec 18:35–43. https://doi.org/10.1016/j.gfs.2018.07.003

Hamidian M, Movahhedi-Dehnavi M, Sayyed RZ, Almalki WH, Gafur A, Fazeli-Nasab B (2023) Author correction: co-application of Mycorrhiza and methyl jasmonate regulates morpho-physiological and antioxidant responses of *Crocus sativus* (Saffron) under salinity stress conditions. Sci Rep 13:8032. https://doi.org/10.1038/s41598-023-35118-3

Hanson AA (2020) Horticultural crops. In: Practical handbook of agricultural science. CRC, pp 273–283

Hemming S (2011) Use of natural and artificial light in horticulture – interaction of plant and technology. Acta Hortic 907:25–35. https://doi.org/10.17660/ActaHortic.2011.907.1

Hochmuth G, Hochmuth R (1969) Open-field soilless culture of vegetables. Edis 2003. https://doi.org/10.32473/edis-hs176-2003

Hou J, Li W, Zheng Q, Wang W, Xiao B, Xing D (2010) Effect of low light intensity on growth and accumulation of secondary metabolites in roots of *Glycyrrhiza uralensis* Fisch. Biochem Syst Ecol 38:160–168. https://doi.org/10.1016/j.bse.2009.12.026

Incrocci L, Malorgio F, Della Bartola A, Pardossi A (2006) The influence of drip irrigation or subirrigation on tomato grown in closed-loop substrate culture with saline water. Sci Hortic (Amsterdam) 107:365–372. https://doi.org/10.1016/j.scienta.2005.12.001

Jagtap P, Profile S, Lakhawat SS, Singh PK, Jagtap PP, Bhakar SR, Kothari M (2022) Present status and future prospective of hydroponics technique: hope and hype for future welfare. J Postharvest Technol 10:65–77

Kalaivanan D, Selvakumar G, Carolin Rathinakumari A (2023) Soilless cultivation to secure the vegetable demand of urban and peri-urban population. In: Recent research and advances in soilless culture. IntechOpen

Karimi M, Ahmadi N, Ebrahimi M (2022) Red LED light promotes biomass, flowering and secondary metabolites accumulation in hydroponically grown *Hypericum perforatum* L. (cv. Topas). Ind Crops Pro 175:114239. https://doi.org/10.1016/j.indcrop.2021.114239

Khammar AA, Moghaddam M, Asgharzade A, Sourestani MM (2021) Nutritive composition, growth, biochemical traits, essential oil content and compositions of *Salvia officinalis* L. grown in different nitrogen levels in soilless culture. J Soil Sci Plant Nutr 21:3320–3332. https://doi.org/10.1007/s42729-021-00608-8

Kimbonguila A, Matos L, Petit J, Scher J, Nzikou J-M (2019) Effect of physical treatment on the physicochemical, rheological and functional properties of yam meal of the cultivar "Ngumvu" from Dioscorea Alata L. of Congo. Int J Recent Sci Res 10:30693–30695. https://doi.org/10.24327/IJRSR

Kováčik J, Klejdus B, Bačkor M, Repčák M (2007) Phenylalanine ammonia-lyase activity and phenolic compounds accumulation in nitrogen-deficient *Matricaria chamomilla* leaf rosettes. Plant Sci 172:393–399. https://doi.org/10.1016/j.plantsci.2006.10.001

Kozai T, Niu G, Takagaki M (2019) Plant factory: an indoor vertical farming system for efficient quality food production. 2nd edn

Léonhart S, Pedneault K, Gosselin A, Angers P, Papadopoulos AP, Dorais M (2002) Diversification of greenhouse crop production under supplemental lighting by the use of new cultures with high economic potential. Acta Hortic 580:249–254. https://doi.org/10.17660/ActaHortic.2002.580.33

Lund H (2007) Renewable energy strategies for sustainable development. Energy 32:912–919. https://doi.org/10.1016/j.energy.2006.10.017

Maggini R, Galluzzo F, Pardossi A (2013) Effect of nitrogen nutrition on growth and accumulation of caffeic acid derivatives in hydroponically-grown *Echinacea angustifolia* DC. var angustifolia. Agrochimica 57:22–30

Marotti I, Marotti M, Piccaglia R, Nastri A, Grandi S, Dinelli G (2010) Thiophene occurrence in different tagetes species: agricultural biomasses as sources of biocidal substances. J Sci Food Agric 90:1210–1217. https://doi.org/10.1002/jsfa.3950

Marshall HS, Saunby T (1954) Soilless culture. Kew Bull 9:56. https://doi.org/10.2307/4108931
Melo EFR, dos Santos OS (2011) Growth and production of nasturtium flowers in three hydroponic solutions. Hortic Bras 29:584–589. https://doi.org/10.1590/s0102-05362011000400023
Menegaes JF, Fiorin TT, Backes FAAL, Lidório HF, Franzen F d L, Zini PB, Sousa NA (2020) Germinação e produção de flores comestíveis de nastúrcio em cultivo hidropônico [Germination and production of edible flowers of nasturtium in hydroponic cultivation]. Braz J Dev 6:13381–13394
Mitchell CA (2022) History of controlled environment horticulture: indoor farming and its key technologies. HortScience 57:247–256. https://doi.org/10.21273/HORTSCI16159-21
Mitchell CA, Dzakovich MP, Gomez C, Lopez R, Burr JF, Hernández R, Kubota C, Currey CJ, Meng Q, Runkle ES, Bourget CM, Morrow RC, Both AJ (2015) Light-emitting diodes in horticulture. In: Horticultural reviews, vol 43. Wiley. Hoboken, NJ, USA. pp 1–88. https://doi.org/10.1002/9781119107781.ch01
Moliner C, Cásedas G, Barros L, Finimundy TC, Gómez-Rincón C, López V (2022) Neuroprotective profile of edible flowers of borage (*Borago officinalis* L.) in two different models: Caenorhabditis elegans and neuro-2a cells. Antioxidants 11:1244. https://doi.org/10.3390/antiox11071244
Montanari M, Degl'Innocenti E, Maggini R, Pacifici S, Pardossi A, Guidi L (2008) Effect of nitrate fertilization and saline stress on the contents of active constituents of Echinacea angustifolia DC. Food Chem 107:1461–1466. https://doi.org/10.1016/j.foodchem.2007.10.001
Morgan L (2021) Hydroponics and protected cultivation: a practical guide. CABI. Wallingford, Oxfordshire, UK. https://doi.org/10.1079/9781789244830.0000
Murch SJ, Vasantha Rupasinghe HP, Saxena PK (2002) An in vitro and hydroponic growing system for hypericin, pseudohypericin, and hyperforin production of St. John's wort (*Hypericum perforatum* CV new stem). Planta Med 68:1108–1112. https://doi.org/10.1055/s-2002-36352
Naemi S, Meshkini A (2023) Phytosynthesis of graphene oxide encapsulated selenium nanoparticles using *crocus Sativus* petals' extract and evaluation of their bioactivity. J Drug Deliv Sci Technol 81:104286. https://doi.org/10.1016/j.jddst.2023.104286
Owusu PA, Asumadu-Sarkodie S (2016) A review of renewable energy sources, sustainability issues and climate change mitigation. Cogent Eng 3:1167990
Partap M, Kumar P, Kumar A, Joshi R, Kumar D, Warghat AR (2020) Effect of elicitors on morpho-physiological performance and metabolites enrichment in *Valeriana jatamansi* cultivated under aeroponic conditions. Front Plant Sci 11:1263. https://doi.org/10.3389/fpls.2020.01263
Pedneault K, Léonhart S, Gosselin A, Papadopoulos AP, Dorais M, Angers P (2002) Variations in concentration of active compounds in four hydroponically-and field-grown medicinal plant species. Acta Hortic 580:255–262. https://doi.org/10.17660/ActaHortic.2002.580.34
Pignata G, Casale M, Nicola S (2017) Water and nutrient supply in horticultural crops grown in soilless culture: resource efficiency in dynamic and intensive systems. In: Advances in research on fertilization management of vegetable crops. Springer. Cham, Zug, CH. pp 183–219. https://doi.org/10.1007/978-3-319-53626-2_7
Pires E d O, Di Gioia F, Rouphael Y, Ferreira ICFR, Caleja C, Barros L, Petropoulos SA (2021) The compositional aspects of edible flowers as an emerging horticultural product. Molecules 26:6940. https://doi.org/10.3390/molecules26226940
Pomoni DI, Koukou MK, Vrachopoulos MG, Vasiliadis L (2023) A review of hydroponics and conventional agriculture based on energy and water consumption, environmental impact, and land use. Energies 16:1690. https://doi.org/10.3390/en16041690
Rahim Doust J, Nazarideljou MJ, Arshad M, Ferrante A (2023) Comparison of the growth, physio-biochemical characteristics, and quality indices in soilless-grown strawberries under greenhouse and open-field conditions. Horticulturae 9. https://doi.org/10.3390/horticulturae9070774
Ramezani M, Amiri MS, Zibaee E, Boghrati Z, Ayati Z, Sahebkar A, Emami SA (2020) A review on the phytochemistry, ethnobotanical uses and pharmacology of Borago species. Curr Pharm Des 26:110–128. https://doi.org/10.2174/1381612825666191216152733

Razmjoo K, Heydarizadeh P, Sabzalian MR (2008) Effect of salinity and drought stresses on growth parameters and essential oil content of *Matricaria chamomila*. Int J Agric Biol 10:451–454

Reddy PP (2016) Chrysanthemum. In: Sustainable crop protection under protected cultivation. Springer Singapore, pp 363–374

Riaz M, Ahmad R, Rahman NU, Khan Z, Dou D, Sechel G, Manea R (2020) Traditional uses, phyto-chemistry and pharmacological activities of *Tagetes Patula* L. J Ethnopharmacol 255:112718. https://doi.org/10.1016/j.jep.2020.112718

Runia WT, van Os EA, Bollen GJ (1988) Disinfection of drainwater from soilless cultures by heat treatment. Neth J Agric Sci 36:231–238. https://doi.org/10.18174/njas.v36i3.16674

Sakamoto Y, Nakashima T, Okano K (1998) Growth and quality of Chrysanthemum (*Dendranthema grandiflora*) grown under various types of hydroponic systems. Environ Control Biol 36:77–84. https://doi.org/10.2525/ecb1963.36.77

Sanford S (2009) Reducing greenhouse energy consumption – an overview (A3907-01). Energy 3907:1–16

Savvas D (2002) Nutrient solution recycling. In: Hydroponic Production of Vegetables and Ornamentals. Embryo Publications. Athens, 299–343

Sawhney S, Kamlesh K, Kohli RK (1981) *Calendula officinalis* L., a long day plant with an exceptionally low photoperiodic requirement for flowering. Indian J Plant Physiol 24:299–303

Sharma D, Partap M, Warghat AR, Bhargava B (2024) Hydroponic cultivation enhances the morpho-physiological traits and quality flower production in three cultivars (marigold scarlet red, marigold orange, and marigold yellow) of French marigold (*Tagetes patula* L.). Sci Hortic (Amsterdam) 327:112803

Souret FF, Weathers PJ (2000) The growth of saffron (*Crocus sativus* L.) in Aeroponics and hydroponics. Int J Geogr Inf Syst 7:25–35. https://doi.org/10.1300/J044v07n03_04

Stewart CL, Lovett-Doust L (2003) Effect of phosphorus treatment on growth and yield in the medicinal herb *Calendula officinalis* L. (Standard Pacific) under hydroponic cultivation. Can J Plant Sci 83:611–617. https://doi.org/10.4141/P02-132

Takahashi JA, Rezende FAGG, Moura MAF, Dominguete LCB, Sande D (2020) Edible flowers: bioactive profile and its potential to be used in food development. Food Res Int 129:108868. https://doi.org/10.1016/j.foodres.2019.108868

Tzortzakis N, Nicola S, Savvas D, Voogt W (2020) Editorial: soilless cultivation through an intensive crop production scheme. Management strategies, challenges and future directions. Front Plant Sci 11:10–12. https://doi.org/10.3389/fpls.2020.00363

Waiba KM, Sharma P, Sharma A, Chadha S, Kaur M (2020) Soil-less vegetable cultivation: a review. J Pharmacogn Phytochem 9:631–636

Wani MA, Din A, Nazki IT, Rehman TU, Al-Khayri JM, Jain SM, Lone RA, Bhat ZA, Mushtaq M (2023) Navigating the future: exploring technological advancements and emerging trends in the sustainable ornamental industry. Front Environ Sci 11:1188643. https://doi.org/10.3389/fenvs.2023.1188643

XU L w, CHEN J, QI H y, SHI Y p (2012) Phytochemicals and their biological activities of plants in Tagetes L. Chinese Herb Med 4:103–117. https://doi.org/10.3969/j.issn.1674-6384.2012.02.004

Zimmermann M, Fischer M (2020) Impact assessment of water and nutrient reuse in hydroponic systems using Bayesian belief networks. J Water Reuse Desalin 10:431–442. https://doi.org/10.2166/wrd.2020.026

Zobayed SSPK, Saxena PK (2004) Production of St. John's wort plants under controlled environment for maximizing biomass and secondary metabolites. In Vitro Cell Dev Biol Plant 40:108–114

Chapter 5
Precision Production Technologies for Edible Flowers

Kuan Qin, Suyun Nam, Husnain Rauf, and Rhuanito Soranz Ferrarezi

Abstract Edible flowers have been widely used due to their nutritional, medicinal, and aesthetic values. In recent years, advanced production techniques, particularly used in controlled environment agriculture (e.g., greenhouses and vertical farms) have been introduced and initiated in edible flowers production to reduce production cost, increase yield and optimize resource use efficiency. This review illustrates recent advances in control systems for microclimatic conditions (e.g., artificial light, temperature and humidity control), IoT for real-time data collection (e.g., sensor network, and image-based machine learning data analysis), greenhouse automation and control system (e.g., precision fertilization and irrigation, artificial intelligence guided harvesting and storage technologies). Sustainability and environmental impact of edible flowers is highlighted, and emerging trends and innovations in precision edible flower farming are discussed.

Keywords Artificial light · Automation and control systems · Controlled environment agriculture · Image analysis · IoT sensors · Machine learning · Resource use efficiency

Abbreviations

AP	Acidification potential
AI	Artificial intelligence
CO_2	Carbon dioxide
CEA	Controlled environment agriculture
DLI	Daily light integral
EC	Electrical conductivity
EP	Eutrophication potential

K. Qin · S. Nam · H. Rauf · R. S. Ferrarezi (✉)
Department of Horticulture, University of Georgia, Athens, GA, USA
e-mail: ferrarezi@uga.edu

M. M. Jowkar et al. (eds.), *Advances in Research on Edible Flowers*, Advances in Olericulture, https://doi.org/10.1007/978-3-032-04561-4_5

GIS	Geographic information systems
GPS	Global positioning system
GWP	Global warming potential
HVAC	Heating and cooling air conditioning
HHP	High hydrostatic pressure
HPS	High-pressure sodium
ISO	International Organization for Standardization
IoT	Internet of things
LEDs	Light-emitting diodes
LCA	Life cycle assessment
ML	Machine learning
MAP	Modified atmosphere packaging
NDAI	Normalized difference anthocyanin index
NDVI	Normalized difference vegetation index
NUE	Nutrient use efficiency
POCP	Photochemical Ozone creation potential
PPFD	Photosynthetic photon flux density
PSATs	Precision smart agriculture technologies
RGB	Red, green, and blue
RH	Relative humidity
UV	Ultraviolet
VWC	Volumetric water content
WUE	Water use efficiency

5.1 Introduction: Uses and Economics of Edible Flowers

Edible flowers are known for enhancing the visual appeal of dishes and increasing the nutritional and medicinal value of diets and have gained popularity in recent years. They fulfil the function of being consumable and augment the aesthetic quality of dishes, contributing to the visual dimension of culinary experiences. Edible flowers have a long history of worldwide consumption with traditional and medicinal uses due to their active compounds such as carotenoids, anthocyanins, flavonoids, and phenolic acids. For example, rose (*Rosa* spp.), Chinese rose (*Hibiscus rosasinensis*), Arabian jasmine (*Jaminum sambac*), Indian chrysanthemum (*Chrysanthemum* spp.), and gerbera (*Gerbera jamesonii*) have been traditionally used as tea, and medicinally used as anti-inflammations to treat cancer, inflammation, ageing, heart diseases, colitis, pneumonia, hypertension, coughs, bronchitis, disorders of the liver and kidneys, swelling, ulceration, uterine bleeding, skin diseases and wound healing; while day lily (*Hemerocallis fulva*), carnation (*Dianthus caryophyllus*), lotus (*Nelumbo nucifera*), marigold (*Tagetes erecta*, *Tagetes patula*), gladiolus (*Gladiolus* spp.), lavender (*Lavandula multifida*, *Lavandula stoechas*, *Lavandula angustifolia*), saffron (*Crocus sativus*), damask rose (*Rosa* × *damascena*) are traditionally used as garnishing agents, and flavouring ingredients in different food dishes such as vegetables, salads, cakes, flavour extracts, food colourant, and also served as a delightful base for sweet or savoury dishes or extracted as fragrance oil, and medicinally used to treat human diseases, infections, and allergies (Sarode et al. 2024).

Regarding the global market for edible flowers, it was valued at $318 million in 2022 and is projected to have a $503 million value by 2032 (Factmr 2022). Another survey also showed that an overall increase of $90.5 million is projected between 2020 and 2026. Annual growth rates show a steady rise, with increases of 4.24% in 2021 and 4.40% in 2022, gradually climbing to 4.85% by 2026. Additionally, the cumulative growth over the period is anticipated to reach 31.17% by 2026 compared to the baseline in 2020, where East Asia (China, India) has higher production and consumption of edible flowers (39%) than Western (31%) world due to cultural aspects, favourable climate conditions, and relatively low labour costs, while Europe (England, France, Spain, Italy, and Portugal) has shown the potential of an increasing edible flowers market (Pires et al. 2023). Although the global regular flower trade value is over $100 billion, increasing competition, labour cost, climate change, and environmental degradation post challenges for edible flower production with reduced profit margins and economic returns for growers (Wani et al. 2023).

5.2 Introduction to Precision Production Technologies for Edible Flowers

To reduce production cost, increase yield and resource use efficiency, precision agriculture technologies with economic advantages have been introduced and initiated in edible flowers production, especially considering the huge benefits in productivity from precision agriculture for cost effectiveness in high-value edible crops. Precision production techniques, particularly used in indoor farming systems like greenhouses and vertical farms, exhibit less reliance on climatic and meteorological fluctuations, need little or no pest control, may be extensively mechanized, and provide continuous production throughout the year while enhancing food quality (Mempel et al. 2021). Precision production systems also enhance spatial efficiency by promoting vertical plant growth or closely spaced rows of plants (Stanghellini 2013). By using artificial light sources and intensive HVAC systems, and sensors to control the temperature, RH, and other environmental factors, precision production for edible flowers can have a fast turnover rate, high light and fertilizer use efficiency, and reduced fungicide and pesticide use by maintaining optimal environmental parameters for rapid plant growth [e.g., optimal light intensity, spectrum, photoperiod, temperature, CO_2, water, and RH] (Al-Chalabi 2015; Al-Kodmany 2018; Beacham et al. 2019; van Delden et al. 2021), yielding long-term economic and environmental advantages. Components in precision edible flower production include automated robotics for cultivation, maintenance (planting, pruning, weeding), and harvesting; imaging technology (visible, multispectral and hyperspectral remote sensing, thermal infrared imaging, fluorescence imaging, 3D imaging, tomographic imaging) using cameras and sensors to detect surrounding environments, plant growth, development, disease, pest, and health; and the

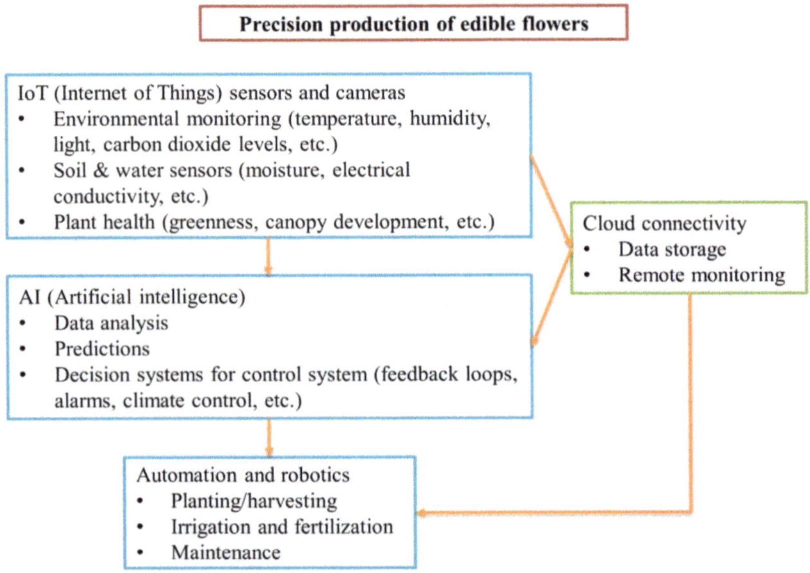

Fig. 5.1 The use of precision technologies (e.g., IoT sensors, robotics, AI, etc.) in edible flowers production under a controlled environment setup

combined use of data analytics (machine learning, artificial intelligence) and IoTs to build the computerized monitoring systems to automatically optimize the environmental conditions for plant growth as well as precisely controlling system (Wani et al. 2023) (Fig. 5.1).

5.2.1 Control Systems for Microclimatic Conditions

During the last decade, many management systems have been used in greenhouses for sustaining ideal growth conditions, enhancing energy efficiency, and reducing energy requirements (Dieleman and Hemming 2011). These systems have also been adapted to edible flower production with a key component that uses sensors for controlling and monitoring the environment. The effective real-time monitoring and regulation of the microclimatic conditions (humidity, temperature, CO_2, lighting, nutrients) inside a greenhouse is crucial for the preservation of crops against adverse surroundings, especially when the human oversight and regulation of such large systems are very challenging and impracticable (Bhujel et al. 2020). Sensors are extensively used in precision agriculture for monitoring crop growth and health, delivering essential data for enhancing the growing environment and mitigating biotic and abiotic challenges. Sensors can be used for various plant parts (including leaves, stems, and roots) and use a large variety of physical, chemical, and biological sensors to get real-time physiological data, using both invasive and non-invasive methods for on-site assessment of plant health (Yan et al. 2024).

As detailed in Table 5.1, greenhouse control systems offer distinct advantages and disadvantages. Temperature control, which utilizes heaters, fans, and vents, is beneficial for enhancing plant growth and protecting crops from excessive heat. However, a primary challenge with this system is balancing the temperature with humidity levels. This challenge directly relates to the need for humidity control,

Table 5.1 Examples of control systems used in controlled environment agriculture (greenhouses and vertical farms)

Control systems	Explanation	Advantages	Disadvantages	References
Humidity	Utilizing ventilation systems, humidifiers, and dehumidifiers, the humidity level is monitored and managed.	Protect against damage caused by high humidity levels. Prevents low humidity from causing water stress in plants	Result in an increase in energy expenses. Consistent monitoring is necessary to ensure that humidity levels are at an appropriate level.	Ghani et al. (2019)
Temperature	Utilizes radiators, vents and ventilation to monitor the temperature. Guarantees that plants are in optimal condition.	Mitigate damage resulting from high temperature	It is challenging to maintain a balance between temperature and humidity control.	Iddio et al. (2020)
CO_2	Utilizes CO_2 generators and sensors to guarantee optimal CO_2 levels within the enclosure.	Enhances growth and the yield	The electricity consumption results in an increase in energy costs. Has the potential to generate heat, thereby increasing greenhouse temperatures.	Vatistas et al. (2022)
Nutrients	Utilizes fertilizers and nutrient monitoring devices to regulate the quantity and composition of nutrients that crops obtain.	Enhances growth and yield Minimize waste by supplying solely the essential nutrients.	Needs regular monitoring to avoid nutrient deficiencies. Requires supplemental devices to monitor and regulate the amounts of nutrients.	Yuan et al. (2016)
Lighting	Utilizes artificial light, shade covers, and curtains to regulate the duration and total amount of light which plants receive.	Promote consistent light conditions. By extending the flowering season, the crop yield is increased	The electricity consumption results in an increase in energy costs. Has the potential to generate heat, thereby increasing greenhouse temperatures.	Afzali et al. (2021), Weaver and van Iersel (2020)

which uses humidifiers, dehumidifiers, and ventilation systems. While this second system effectively protects plants from the stresses of both excessive humidity and insufficient moisture, it requires continuous supervision to maintain proper levels and often leads to high energy costs. (Ghani et al. 2019; Maher et al. 2016). The lighting management system controls artificial lighting, shading, and curtain illumination to optimize plant growth. By extending the growing season and maintaining optimal light levels, it enhances overall agricultural productivity. However, operating this system can lead to higher energy costs, and the heat produced by the lighting may contribute to increased greenhouse temperatures (Afzali et al. 2021; Weaver and van Iersel 2020). Generators and sensors work together to optimize CO_2 levels in greenhouses, conserving water while improving plant growth and overall agricultural productivity. However, operating these advanced systems requires energy, which may raise electricity costs. The resulting heat can also increase greenhouse temperatures, and a monitoring system failure could pose risks or cause damage (Vatistas et al. 2022). Fertilizers and monitoring systems regulate the supply of nutrients to plants, ensuring they receive the proper amounts to enhance yield and agricultural productivity while minimizing waste. Maintaining optimal nutrient balance requires regular monitoring and the use of additional equipment (Bergstrand 2022). Moreover, crop sap flow serves as a crucial physiological indication of the usage of water, nutrients, and the overall strength status of the crop (Miner et al. 2017). The transpiration rate of branches or tillers may be assessed by monitoring the rate of plant sap flow. Li et al. (2022) developed a novel moisture sensor by utilizing flexible graphene oxide. This sensor demonstrated a high level of sensitivity, rapid response, and excellent biological compatibility.

5.2.2 IoT for Real-Time Data Collection

Edible flower production requires careful management of challenges such as diseases, insect infestations, and the use of chemical fertilizers. The implementation of advanced technologies such as IoT and AI in edible flower production, combined with the quantification of the cultivation process for thorough analysis, is crucial (Morii et al. 2023). Technological advancements have significantly influenced flower crop production, allowing growers to enhance yields, minimize losses, and improve flower quality. The application of modern technologies, including high-throughput phenotyping, facilitates the identification and selection of enhanced cultivars. Precision agricultural techniques, such as sensors and remote monitoring devices, facilitate thorough monitoring of the development of plants and optimization of inputs like fertilizers and water, resulting in enhanced resource efficiency and increased yields (Wani et al. 2023).

The integration of AI and IoT has enabled the floral business to realize the benefits of precision agriculture. Employs various drones, GPS maps, and sensors to enhance agricultural inputs, reduce waste, and optimize yields via technological advancements. The use of AI, robots, and automation substantially reduces labour costs and improves efficiency in flower production. The use of automated farming and harvesting systems may reduce the need for human labour since these systems may perform tasks more swiftly and accurately than manual labour (Jha et al. 2019). Evaluating plant health and growth enables farmers to enhance their fertilizer, irrigation, and pest management techniques, therefore mitigating the negative environmental effects on edible flower production (Belal et al. 2021; Ferroukhi et al. 2023).

Saini et al. (2025) illustrate several IoT-based agricultural solutions. These technologies rapidly digitize the environment and collect the resulting data. Mendon et al. (2022) stated that the cultivation of premium spinach is fully automated, with efforts underway to increase output capacity. Moreover, technologies have been developed to autonomously provide circumstances favourable for certain plant species (Kondaka et al. 2023). From a greenhouse perspective, traditional systems have heavily relied on the expertise of cultivators. Nevertheless, IoT technologies are used to reduce manpower, while advancements in monitoring various climatic elements and developing new applications are ongoing. This trend meets food demand while becoming more popular as an agricultural strategy that advantages farmers (Farooq et al. 2022). Moreover, IoT factors, ML, and electrical usage have been included in the production for hydroponically grown crops including edible flowers (Bhargava et al. 2022). By modifying factors including the amount of water and intensity of light, energy usage has been decreased, and the acquired data may be enhanced via analysis.

5.2.3 Greenhouse Automation and Control System

Precision smart agriculture technologies (PSATs) are transforming the agricultural industry by improving production, efficiency, and sustainability (Majumder et al. 2021). These technologies use sophisticated instruments such as GPS technology, drones, IoT, and info regarding data analysis to oversee and optimize agricultural production with unparalleled precision. IoT is an application of technology enabling remote connectivity among equipment that allows for smart farming practices (Patil and Kale 2016). IoT refers to a network of devices that are equipped with software, sensors, and other technologies that intelligently link and share data across devices and networks over the Internet. In smart farming, automated systems use sensors to monitor horticultural crops by measuring environmental factors like solar radiation, RH, supplemental light intensities, soil moisture and temperature, EC, wind speed, and direction, and subsequently uploaded to cloud information support systems (Bhat and Huang 2021; Javaid et al. 2022). To achieve a successful IoT system, AI, ML, and robotics need to be synergistically used.

AI demonstrated the machine's capability to execute different tasks traditionally necessitating the intelligence of humans, including acquiring knowledge, solving problems, and decision-making. A number of tools related to AI and ML can be applied in floriculture by analysing extensive datasets obtained from sensors and cameras that involve environmental and plant physiological information to identify patterns and predict optimal conditions for cultivating various flower types (Benos et al. 2021). ML is utilized in the flower industry to enhance production and quality, alongside optimizing supply chain management. Moreover, ML may improve the management of supply chains through the analysis of the state and position of flowers throughout transportation (Zhang et al. 2021). ML can also be applied to enhance early disease symptom identification and avoidance in flowers, this enables growers to implement corrective actions prior to the disease's proliferation, thereby safeguarding the entire crop. ML algorithms assist growers in optimizing the allocation of water, fertilizer, and light to flowers, thereby enhancing both quality and yield (Virnodkar et al. 2020).

Robotics has become a developing technology that is advancing within the floriculture sector while reducing labour costs. The application of robotics in flower crops encompasses automated and autonomous systems designed to execute tasks associated with the cultivation, maintenance, and harvesting processes of different flowering stages while using AI-assisted cultivators in making choices on optimal harvest timing, therefore minimizing waste and enhancing efficiency (Gürel and Erden 2013). Automated maintenance equipment, including robotic mowers and trimmers, is now available for maintaining lawns and gardens (Oliveira et al. 2021). With the use of cameras, sensors, and other technology, robots can navigate fields, identify flowers, and precisely pick them while causing the least amount of harm to the plants. For example, Quick Flora is evaluating a new robot, 'Flora Bot,' designed to automate the flower arrangement process (Wani et al. 2023).

5.3 Artificial Lighting Technologies

5.3.1 Advantages of Lighting in Edible Flower Production

Artificial lighting is essential in CEA, particularly for promoting plant growth and enhancing crop quality in conditions where natural sunlight is insufficient, inconsistent, or in fully indoor production systems (Albright et al. 2000). Among artificial lighting options, LEDs have gained popularity in CEA due to their longevity, low heat emission, and higher energy efficiency (Katzin et al. 2021). A primary advantage of LED lights over traditional HPS lamps is their flexibility in control; lighting conditions, such as light intensity, spectrum, and duration, can be precisely adjusted according to environmental conditions, different crops, and growth stages (van Iersel 2017). Numerous studies have shown that specific LED configuration can improve not only vegetative growth but also influence flowering time and increase secondary metabolites concentrations, flower numbers, and petal pigmentation (Al Murad et al. 2021; Sams et al. 2016; Weeplian et al. 2018). These attributes make LED lighting especially valuable for edible flower production, where it can increase crop yield and potentially enhance aesthetic and nutritional values.

5.3.2 *Impact of LED Light Spectrum on Edible Flower Development and Quality*

Plants detect light spectrum through photoreceptors, which mediate plant growth, development, morphology, and secondary metabolite production. Five main classes of photoreceptors enable plants to respond to light across the UV to far-red range (280–750 nm) (Appolloni et al. 2022). Red and far-red light, absorbed by phytochromes, promotes vegetative growth, stem elongation, and leaf expansion. Blue and UV light, primarily detected by cryptochromes and phototropins, inhibits stem elongation while encouraging compact growth and branching, along with the accumulation of nutritional compounds like carotenoids and anthocyanins (Ouzounis et al. 2015).

Specific LED spectra combinations can regulate flowering and enhance flower quality. For example, a low red-to-far-red ratio accelerates flowering in some short-day plants (Demotes-Mainard et al. 2016), while blue light encourages flowering in long-day plants but suppresses it in short-day plants (Trivellini et al. 2023). In marigolds, 75% red and 25% blue LED lighting led to greater bloom production compared to 50% red and 50% blue LED or HPS treatments (Sams et al. 2016). Responses to LED spectrum vary by species and cultivars. For example, in *Hypericum perforatum* L. (cv. Topas), 100% red LED enhanced growth, flower number, and medicinal compound contents, contrary to studies suggesting blue light's importance for secondary metabolites (Karimi et al. 2022).

LED technology holds promise for edible flower cultivation, allowing precise spectral adjustments to enhance yield, visual quality, and nutritional value. In lemon marigold (*Tagetes tenuifolia*), 80% red and 20% blue LED increased petal carotenoids compared to other blue ratios (Kopsell et al. 2016). Supplementing far-red light at late growth stages promoted reproductive growth and antioxidant levels in edible common ice plants (*Mesembryanthemum crystallinum* L.) (Weeplian et al. 2018). Monochromatic red LED light improved stigma yield and crocin production in saffron (*Crocus sativus* L.) (Zhu et al. 2022). However, specific LED protocols for various edible flowers remain limited. Thus, future research should focus on tailoring LED spectrum recipes by species and cultivars to optimize yield, flowering, and desired secondary compounds (Appolloni et al. 2022).

5.3.3 *LED Light Intensity, DLI, and Photoperiod for Edible Flower Crops*

Higher PPFD generally increases photosynthetic rate to a certain extent, resulting in enhanced biomass accumulation (Mayorga-Gomez et al. 2024). However, DLI - the cumulative amount of photosynthetic light received by a plant over a day—plays a more significant role in determining crop yield. DLI integrates both instantaneous light intensity and photoperiod, making it particularly relevant in fluctuating light environments such as greenhouses. Many studies have shown that increased DLI enhances various crop growth and quality aspects such as compactness, flower number, flower visual quality, and antioxidant compounds, although optimal DLI requirements are highly species-specific (Dou et al. 2018; Fausey et al. 2005; Spall and Lopez 2024). Representative edible flowers, such as pansy, marigold, violet, and hibiscus, respond similarly to increased DLI, showing improvements in total biomass, flower size, bud number, as well as reductions in time to flowering (Niu et al. 2000; Pramuk and Runkle 2005; Warner and Erwin 2003). For example, saffron demonstrated increased flower number, stigma dry weight, and crocetin esters content when the total accumulated DLI during the entire corm cultivation period reached up to 150 $mol \cdot m^{-2}$ (Gao et al. 2023). Achieving optimal PPFD or DLI through sole-source or supplemental LED lighting in CEA production systems (e.g., greenhouses and vertical farms) can significantly improve the productivity and quality of edible flower crops.

In addition to DLI, photoperiod is a critical factor in regulating flowering responses. Photoperiodic crops can be manipulated with low intensity of end-of-day or night interruption lighting (1–3 $\mu mol \cdot m^{-2} \cdot s^{-1}$), which extends the photoperiod to accelerate flowering in long-day plants or inhibit it in short-day plants (Zhang and Runkle 2019). In pansy, the rate of flowering increased with longer photoperiods (Adams et al. 1997). Interestingly, sufficient DLI can override photoperiodic flowering responses, allowing flowering induction in long-day plants under short-day conditions (Whitman et al. 2022). However, research on the combined effects of DLI and photoperiod on flowering responses in edible flower crops remains limited, highlighting the need for further studies in this area.

5.3.4 Enhancing Light Use Efficiency with Advanced LED Strategies

High electrical costs remain a major limiting factor for the commercial application of LED lighting, despite its numerous advantages for improving crop growth and quality (Dou et al. 2018). In vertical farming systems, artificial lighting can account for up to 30% of production costs (Kaiser et al. 2024). Therefore, optimizing LED lighting through dynamic control strategies is crucial to reduce production expenses and maximize profitability.

Red and blue LED products generally offer higher photon efficacy than white or green LEDs, meaning they emit more photosynthetically active photons per unit of electrical energy. For instance, horticultural LED fixtures with a spectrum of 90% red and 10% blue photons can achieve an efficacy of ≈ 4.1 $\mu mol \cdot J^{-1}$, making them highly efficient (Kusuma et al. 2020). Additionally, since the quantum yield of photosystem II decreases as PPFD increases, longer photoperiods with lower light intensities can promote growth compared to shorter photoperiods with higher intensities (Palmer and van Iersel 2020). Continuous lighting, using a 24-hour photoperiod, has also been explored to maximize crop yield and light use efficiency, but some crops may experience light stress without a nighttime recovery period (Xu et al. 2021). Enriched CO_2 levels are frequently employed in greenhouses and vertical farms to accelerate growth and enhance light use efficiency. For example, supplemental lighting treatment combined with CO_2 enrichment of 1000 $\mu mol \cdot mol^{-1}$ boosted the number of axillary shoots, flower diameter, and number of petals per flower while decreasing flower disorders of roses (Naing et al. 2016). In greenhouses, supplemental lighting strategies must account for the variability of natural sunlight levels, which fluctuates with latitude, season, and weather conditions. Maintaining a consistent DLI throughout the cultivation period is essential to ensure uniform growth (Seginer et al. 2006). Adaptive lighting strategies dynamically adjust the LED intensity in real-time based on ambient sunlight levels to maintain a target PPFD (van Iersel and Gianino 2017). The timing of supplemental lighting applications also affect light use efficiency. Providing supplemental light later in the plug stage, rather than earlier in seedling development, has been shown to improve transplant biomass and subsequent flowering aspects after transplant (Oh et al. 2010). Moreover, innovative lighting control methods now integrate real-time electricity pricing and crop-specific photosynthetic performance, further optimizing energy use and crop yield (Mahdavian and Wattanapongsakorn 2017; van Iersel et al. 2016).

5.4 Irrigation and Fertigation Technologies

Water management is essential in edible flower production for improving WUE, characterized as the total biomass per unit of water used by a plant (Sonneveld and Voogt 2009). The growing pressure on water resources requires an efficient irrigation system for water management in agriculture. The amount of water present in the root zone influences agricultural productivity quantitatively. About 97% of water consumed through the roots is lost through transpiration on the surfaces of leaves, driven by the continuous requirement for CO_2 absorption (Gonçalves et al. 2023). Due to its influence on crop health and water efficiency, commercial producers are finding that managing irrigation for edible flower crops effectively and precisely is becoming more crucial. Substrate moisture monitoring systems have been evaluated in recent years to provide more information and help producers with watering choices. Those systems begin watering when the substrate reaches the VWC specified by the user (Henderson et al. 2018). Automated irrigation systems are crucial in the production of flowers by delivering a precise quantity of water to plants, improving crop output, and maximizing water resource efficiency. Irrigation technology in floriculture involves the exact delivery of water in appropriate volumes and frequency using advanced equipment and techniques (Wani et al., 2023). The primary technologies that are used most are surface systems (drip-irrigation) and subirrigation systems (ebb-and-flow, capillaries mats, through benches, and flood floors) (Reed 1996). Drip irrigation is a contemporary technology designed to provide water directly to the roots in tiny, concentrated amounts, so preserving the necessary soil moisture for plant growth. In comparison to capillary mats, ebb-and-flow systems, and manual watering, microtube irrigation in greenhouse production maintained a higher level of moisture in the medium, resulting in plants with the highest dry weights (Dole et al. 1994). Smart irrigation technology is another good example of precision edible flower production. Soil moisture sensors such as capacitance and electrical conductivity-based tensiometer, time domain reflectometers, and dielectric probes have been jointly used with automatic irrigation control systems through IoT systems to save >83% water for hydrangea (*Hydrangea arborescens*) production (Jongyun et al. 2014)—which can probably be adapted to edible flowers. A study used precision irrigation scheduling based on evapotranspiration (from 40% to 120%) with the use of mycorrhizal colonization (*Claroideoglomus etunicatum*) to grow greenhouse garden pansies (*Viola* × *wittrockiana*), results showed that there were no significant increases in the growth and production of flowers (floral dimensions, flower fresh and dry weight, and flower production) in response to inoculation with mycorrhizal fungi, while increased irrigation level increased flower quality and plant biomass with the peak happened at 100% evapotranspiration irrigation scheduling (Gonçalves et al. 2023), with a potential similar effect in edible flowers.

Fertigation enables meticulous regulation of the spatial and temporal dimensions of mineral nutrition, which is essential in the precision cultivation of edible flowers. The development of drip irrigation systems integrated with fertigation provides exceptional precision in the distribution of nutrients creates a high nutrient-use efficiency, and attains plant nutrient recovery rates of up to 90%, in contrast to the 40%–45% efficiency noted in traditional fertilizer application systems (Agostini et al. 2010). Nutrient recycling is another fundamental aspect of precision edible flower cultivation, including methods such as detecting and maintaining EC to evaluate nutrient content (Christie 2014). Hydroponic systems often recycle nutrient solutions to improve WUE and NUE while reducing environmental pollutions (Nederhoff and Stanghellini 2010). Hydroponic systems such as nutrient film technique, ebb-and-flow system, and deep-water culture are also frequently employed in the production of edible flowers.

5.5 Precision Harvesting and Storage Technologies

5.5.1 Imaging and Sensor Technologies for Flower Quality Estimation and Harvest Prediction

Imaging and remote sensing techniques combined with machine learning analyses can effectively detect crop growth and quality based on canopy size and pigmentation. For example, RGB images of rose flowers analyzed using ML models, such as back-propagation neural networks and random forest algorithms, accurately predicted anthocyanin content in petals, which was highly correlated to traditional destructive methods (Liu et al. 2024). These imaging techniques enable non-invasive and time-efficient detection of flower pigmentation. However, manual image segmentation and evaluation remain labour-intensive despite their high predictability. Low-cost imaging systems have been developed to address this limitation, utilizing multi-spectral image collection and automated analysis through Python-based programs with microcomputers embedded in the imaging system. Chlorophyll fluorescence imaging has also been employed to separate plant canopies from the background, quantify canopy size and overall growth, and calculate indices such as the NDVI and NDAI (Kim and van Iersel 2023; Wacker et al. 2024).

These imaging techniques, integrated with environmental sensors, have been tested in greenhouse systems to quantify flower quality and maturity, enabling farmers to determine optimal harvest times. For example, the number of strawberry flowers during the growing period was monitored using a fixed-position imaging system in conjunction with machine vision techniques. RGB images were captured above a circulating movable-bench system, and flower regions and color features were analysed to detect flower numbers (Naito et al. 2018). Similarly, self-supervised learning algorithms continuously analysing orchid flower images and environmental factors (e.g., soil moisture, temperature, humidity, and CO_2 concentration) have been used to monitor growth status and further, control greenhouse conditions, including fans and sprinklers, to optimize plant growth (Chen et al. 2022). These technologies can be adapted for edible flowers to monitor flower numbers, size, and growth stages. Self-supervised algorithms have the potential to optimize greenhouse conditions based on edible flower growth and developmental data, by automating adjustments based on flowering stages to improve harvest timing and crop quality.

In another application, IoT clip-type cameras installed in greenhouses detected anthesis using deep learning models. Combining image data with daily average temperature forecasts, this approach predicted harvest dates with an accuracy of ± 2 days compared to manual screening (Lee et al. 2022). Furthermore, hyperspectral sensing techniques paired with machine learning have successfully classified developmental stages and determined maturity in greenhouse crops, providing information to farmers on harvest readiness (Hassanzadeh et al. 2020). These imaging and sensor-based technologies are especially valuable for edible flower production, where visual quality, nutritional content, and shelf life are critical. However, research specific to edible flower crops in CEA environments regarding image- or sensor-based monitoring remains limited. Further studies are needed to develop automated, objective tools for monitoring flower quality, predicting harvest timing, and optimizing management practices for edible flower producers.

5.5.2 Automated and Robotic Harvesting for Edible Flowers

Automated harvesting technologies are becoming essential in floriculture, including edible flower production, as they address labour-intensive practices while improving efficiency, precision, and consistency. Robotic systems equipped with advanced cameras and sensors can autonomously navigate fields, identify flowers, and harvest them with minimal plant damage. Additionally, these robotic systems can monitor plant growth, and detect diseases and pests during the cultivation period, which can optimize overall crop management (Wani et al. 2023). However, challenges persist in achieving high success rates for delicate crops. Early systems, such as chrysanthemum pickers, caused plant damage due to unsuitable designs, and large-scale machines for cotton harvest proved inefficient for small production areas. Even robotic arms with specialized grippers for roses struggled with low efficiency especially when stem density is high in the plant canopy. Therefore, developing crop-specific systems tailored to flower size, canopy structure, and stem attachment is critical for overcoming these limitations (Wani et al. 2023; Wu et al. 2022). Edible flowers are highly delicate and valued for their visual quality, so precision robotic harvesting technologies are essential to maintain marketable quality and reduce labour dependency.

Edible flowers such as marigolds (*Tagetes patula*), snapdragons (*Antirrhinum majus*), and pansies (*Viola* × *wittrockiana*) have been tested with robotic systems using computer vision for flower detection, pose estimation, and plucking. In one study, AI-based robotic harvesting was evaluated for marigolds, snapdragons, pansies, and horned pansies (*Viola cornuta*) using the YOLOv5 ("You Only Look Once" object detection model version 5; Ultralytics, Frederick, MD, USA) framework for 2D flower detection and 3D flower localization, achieving 0.96 detection precision and identifying flower positions within one second (Taddei Dalla Torre et al. 2024). Neural networks have improved accuracy by identifying only mature flowers based on flower size and color, while depth cameras and 3D imaging have enabled precise harvesting for jasmine (Amer et al. 2023; Shree et al. 2019). Saffron harvesting robots integrated sensors, grippers, and post-harvest handling systems to maintain quality during and after harvest (Asimopoulos et al. 2013). For these highly sensitive crops, specialized post-harvest containers were used during harvesting to control light, temperature, and humidity, ensuring that the harvested flowers are not adversely affected by atmospheric conditions. Robotic technologies offer promising solutions for edible flower production by reducing labor and improving consistency. Future research should focus on adapting these systems to various edible flower species/cultivars and developing cost-effective solutions for commercial-scale adoption.

5.5.3 Innovative Approaches to Extend Edible Flower Shelf Life

Post-harvest technologies, including low-temperature storage and various drying methods, are examples of precision management for edible flowers to improve their nutritional and aesthetic qualities, ensure food safety by minimizing bacteria and mold contamination, and extend shelf life as they are highly perishable with short shelf life (2-5 days after harvest), limiting their commercial potential despite their growing popularity (Zhao et al. 2019).

Low-temperature storage remains a fundamental practice in post-harvest management. It reduces respiration rates and minimizes water loss, wilting, and microbial growth, effectively extending the shelf life of edible flowers. Additionally, it lowers ethylene production, slowing down tissue breakdown and quality deterioration. Some traditional methods include 7–10 days low temperature (−2–5 °C) storage with polyethylene bags, as cold storage reduces microbial growth and helps retain antioxidant activity in species like nasturtium (*Tropaeolum majus*) (Fernandes et al. 2019). MAP is another widely used technology that reduces oxygen levels while increasing CO_2 or N_2 concentrations within polymer-based packaging. This also slows down respiration, decreases water loss, and inhibits mold growth. For example, active MAP with 100% N_2 preserved petal colors, reduced decay, and maintained phenolic content in begonia and viola flowers, especially when coupled with cold storage (Falla et al. 2024). These technologies can be readily adapted for edible flower harvest by optimizing gaseous compositions in packaging and combining them with low-temperature storage.

Post-harvest LED treatment has shown potential in preserving edible flower metabolites. For instance, cold-stored horned violet flowers treated with 2 hours of LED light retained polyphenols and antioxidant activity while delaying depigmentation (Marchioni et al. 2020). UV light treatments also showed promise, with UV-B delaying chlorophyll degradation in vegetables and UV-C enhancing phenolic content, reducing bacterial and mold populations, and improving storage quality (Aiamla-or et al. 2010).

HHP, applying pressures of 50–1000 MPa, offers a novel approach to maintaining flower quality by inactivating enzymes and reducing microbial contamination. However, its effects highly vary by species. For example, borage (*Borago officinalis*) and rose camellia (*Camelia japonica*) flowers showed undesirable changes in appearance, while pansies (*Viola* × *wittrockiana*) and blue centaurea (*Centaurea cyanus*) exhibited enhanced or retained bioactive compound production (Chen et al. 2010; Fernandes et al. 2019).

Edible films and coatings can provide an additional layer of protection, forming a barrier against gas exchange and microbial growth. Polysaccharide-based coatings have been successfully applied to extend shelf life and microbial stability of artichokes (*Cynara cardunculus* var. scolymus), suggesting their potential for delicate edible flowers as well (Fernandes et al. 2019). Post-harvest technologies tailored for edible flowers play a vital role in ensuring their quality, food safety, and marketability. Future advancements should focus on integrating these methods to develop cost-effective, species-specific solutions for prolonging shelf life while maintaining visual and nutritional value. Drying methods using freeze-vacuum or microwave-vacuum drying and precision micro-climate control methods have been developed to maintain flower quality over time by delaying ripening or senescence through treatments such as HHP (50 and 1000 MPa) and ionizing radiation using gamma rays from the radionuclides ^{60}Co (Fernandes et al. 2019).

5.6 Sustainability and Environmental Impact of Edible Flowers

The environmental impacts of edible flowers involve several aspects. Cultivation-related chemical, water, electricity, labour, infrastructure, and growth media inputs; harvesting-related labour, mechanical, transportation, post-harvest storage, packaging, and selling costs will all impose negative effects on sustainability. At the same time, the ecological role of flowers (phytochemicals that have attractions for pollination), nutritional value (essential oil, amino acids, minerals, vitamins, carbohydrates, protein, sugars, phenolic acids), and biodiversity will also contribute to the sustainability (Benvenuti and Mazzoncini 2021). Additionally, different cropping systems will lead to different environmental impacts, as organic cropping systems are "chemical" (e.g., pesticide, fungicide, herbicide) free, and using organic fertilizers can form a recycled resource use to significantly reduce negative environmental impacts.

LCA, defined by the International Organization for Standardization (ISO), is a standardized procedure for investigating the environmental impact of a product, taking into consideration of objectively and transparently inputs and outputs of energy and material usage (Hiloidhari et al. 2017). A case study of environmental impact for LCA showed that using edible flowers, e.g., *Begonia x semperflorens – cultorum hort* and *Viola cornuta* L. as potted flowering plants with plastic vases base and consumed package flowers have GWP values ranged from 24.94 to 31.25 g CO_2 eq/g flowers, Acidification Potential (AP) ranged from 8.169×10^{-2} to 1.249×10^{-1} g SO_2 eq/g flowers, EP ranged from 3.961×10^{-2} to 5.284×10^{-2} g PO_4^{3-} eq/g flowers, and POCP ranged from 8.998×10^{-3} to 1.134×10^{-2} g C_2H_4 eq/g flowers, where the top impactful phases were the propagation and cultivation, due to the requirements of night supplemental light for facilitating rooting and high amounts of peat-based substrate (Falla et al. 2020). To reduce the environmental impact, biodegradable plastic packaging, pots, and foam could be used for edible flower production (Jariyasakoolroj et al. 2020; Razza et al. 2015; Schettini et al. 2013). Using precision agriculture under indoor facilities in urban areas can significantly reduce environmental impact by reducing greenhouse gas emissions, lowering travel costs and environmental pollution between local markets and production sites, increasing space use efficiency, and providing green impact by using green roofs or walls (Despommier 2014). At the same time, balancing energy consumption in precision systems (LED, automation), using renewable energy integration (solar, wind) and recycled resources in greenhouse operations, and minimizing resource inputs (water, nutrients) through precision farming by significantly increasing plant resource use efficiencies, and using a high-output growing design such as hydroponics will contribute considerably to the sustainability by providing a low-impact production system.

5.7 Future Trends and Innovations in Precision Edible Flower Farming

As the trend of pursuing innovative food products and formulations that combine enhanced visual appeal, superior nutritional quality, and health-promoting properties is increased, the production of edible flowers will face a promising market with the consideration of safety production using cropping system with fewer chemicals (pesticides, fungicide, etc.) (Pires et al. 2023), the precision farming using advanced AI combined with GPS and GIS for informative and location data collection and analysis, machine learning and automated robotics for flower production will reduce the chemical requirements, especially when produced in urban area. Emerging IoT devices and systems for improved precision and integration of blockchain for supply chain transparency will improve the resource use efficiency in edible flower production by optimizing crop inputs, reducing waste, increasing yield, and providing sustainability. The combined use of these technologies (IoT gateway, IoT robotics system for agriculture precision applications includes agricultural inputs, harvesting, storage, plant level sensing system) creates promising quantum-based computing for smart agriculture in the future (Maraveas et al. 2024).

Biotechnology has recently been applied to enhance the yield, stress tolerance, biochemical composition, and quality of edible flower production using micropropagation, molecular marker and omics approaches, genome editing (CRISPR-based gene editing). Advanced in vitro tissue culture techniques have been employed, including protoplast culture, anther culture, microspore culture, ovule culture, embryo culture, root culture, organ culture, suspension culture, single-cell culture, and ovular layer cell culture. These methods enable the mass multiplication of haploid and somaclonal variants, significantly improving production efficiency compared to traditional cultivation methods (Sarode et al. 2024). Precision agriculture could provide data from the environment (e.g., soil moisture, temperature) and plant (morphological and physical responses) for driving the biotechnology-based breeding and crop management programs, for example, using phenotyping and genomic data to identify target genes for editing and accelerating trait development (Steinwand and Ronald 2020). Conversely, plant materials obtained from biotechnology could be further utilized in precision production systems, especially under controlled environment to optimize the efficiency and efficacy of biotechnology outputs.

Future trends will also include the combination use of flower vision detection and a robotic arm for automated flower harvesting, converting in-lab tests to more suitable for practical deployment in real-world scenarios (Taddei Dalla Torre et al. 2024). Sustainable materials, online sale, transportation and delivery platforms, virtual reality, establishment of an efficient and fast management system, and development of low environmental and high health and educational impacts connections between consumers and growers will also become the trends for the future precision production of edible flowers (Wani et al. 2023).

Another promising technology that has been introduced to edible flowers is nanotechnology, which focuses on manipulating matter at the nanoscale (1–100 nm) and is being explored for its potential to enhance flower quality, improve disease resistance, and extend shelf life. For example, binding fertilizer or pesticide molecules to nanoparticles to deliver nutrients and other essential compounds directly to plant cells, which can increase their use efficiency, reduce the quantities required, improve cost-effectiveness and minimize environmental impact; another application is the use of nanosensors for monitoring plant growth and health by detecting fluctuations in temperature, humidity, nutrient levels, and other critical factors affecting plant development, this application can provide early detection of pests or diseases, guiding precision management practices for growers; nanomaterials have also been used in edible flowers by providing coating to improve the durability, longevity, and shelf-life of flowers (Wani et al. 2023).

5.8 Conclusion

Application of precision technologies in edible flowers production provides significant benefits, which includes optimizing resource use and improving energy use efficiency, enhancing crop production and quality, reducing waste during production, harvesting, packaging, and transportation, and improving profitability by reducing labor, energy, and chemical costs. Innovated technologies such as high accuracy monitoring and prediction systems using advanced cameras, sensors, and AI-driven analytics; automated irrigation and fertilization systems; optimized artificial light adjusting systems; precise robot-based harvesting system, and other advanced technologies (e.g., nanotechnology, biotechnology) contribute to the sustainability of edible flowers production by minimizing environmental impact and promoting the efficient use of water, energy, and nutrients. However, high implementation costs for specific devices and equipment, the need for specialized knowledge to constantly monitor and maintain the system, and the difficulties in integrating these technologies into existing practices (e.g., data compatibility and interoperability, scalability and cost, energy efficiency and reliability, cybersecurity) will still create challenges for using precision technologies. Addressing these challenges requires collaborations among researchers, technology developers, agricultural stakeholders, and policymakers. Despite the challenges, precision technology-driven edible flower production will become promised for the future of sustainable agriculture as these technologies become more accessible and user-friendly. By enhancing innovation and supporting research, the industry can unlock new opportunities for cultivating edible flowers in eco-friendly and economical ways that meet the consumers' needs and food security goals while reducing negative environmental impacts for a sustainable agricultural production.

References

Adams SR, Pearson S, Hadley P (1997) The effects of temperature, photoperiod and light integral on the time to flowering of pansy cv. Universal Violet (*Viola* × *wittrockiana* Gams.). Ann Bot 80(1):107–112. https://doi.org/10.1006/anbo.1997.0411

Afzali S, Mosharafian S, van Iersel MW et al (2021) Development and implementation of an IoT-enabled optimal and predictive lighting control strategy in greenhouses. Plants 10(12). https://doi.org/10.3390/plants10122652

Agostini F, Tei F, Silgram M et al (2010) Decreasing nitrate leaching in vegetable crops with better N management. In: Lichtfouse E (ed) Genetic engineering, biofertilisation, soil quality and organic farming, pp 147–200. https://doi.org/10.1007/978-90-481-8741-6_6

Aiamla-or S, Kaewsuksaeng S, Shigyo M et al (2010) Impact of UV-B irradiation on chlorophyll degradation and chlorophyll-degrading enzyme activities in stored broccoli (*Brassica oleracea* L. Italica group) florets. Food Chem 120(3):645–651. https://doi.org/10.1016/j.foodchem.2009.10.056

Al Murad M, Razi K, Jeong BR et al (2021) Light emitting diodes (LEDs) as agricultural lighting: impact and its potential on improving physiology, flowering, and secondary metabolites of crops. Sustainability 13(4). https://doi.org/10.3390/su13041985

Albright LD, Both AJ, Chiu AJ (2000) Controlling greenhouse light to a consistent daily integral. Trans ASAE 43(2):421–431. https://doi.org/10.13031/2013.2721

Al-Chalabi M (2015) Vertical farming: skyscraper sustainability? Sustain Cities Soc 18:74–77. https://doi.org/10.1016/j.scs.2015.06.003

Al-Kodmany K (2018) The vertical farm: a review of developments and implications for the vertical city. Buildings 8(2):24. https://doi.org/10.3390/buildings8020024

Amer MA, Rajan P, Mehanny S et al (2023) Innovative design of an experimental jasmine flower automated picker system using vertical gripper and YOLOv5. Agriculture 13(8). https://doi.org/10.3390/agriculture13081595

Appolloni E, Pennisi G, Zauli I et al (2022) Beyond vegetables: effects of indoor LED light on specialized metabolite biosynthesis in medicinal and aromatic plants, edible flowers, and microgreens. J Sci Food Agric 102(2):472–487. https://doi.org/10.1002/jsfa.11513

Asimopoulos N, Parisses C, Smyrnaios A et al (2013) Autonomous vehicle for saffron harvesting. Proc Technol 8:175–182. https://doi.org/10.1016/j.protcy.2013.11.025

Beacham AM, Vickers LH, Monaghan JM (2019) Vertical farming: a summary of approaches to growing skywards. J Hortic Sci Biotechnol 94(3):277–283. https://doi.org/10.1080/14620316.2019.1574214

Belal AA, EL-Ramady H, Jalhoum M et al (2021) Precision farming technologies to increase soil and crop productivity. In: Abu-hashim M, Khebour Allouche F, Negm A (eds) Agro-environmental sustainability in MENA regions. Springer, Cham, pp 117–154. https://doi.org/10.1007/978-3-030-78574-1_6

Benos L, Tagarakis AC, Dolias G et al (2021) Machine learning in agriculture: a comprehensive updated review. Sensors 21(11). https://doi.org/10.3390/s21113758

Benvenuti S, Mazzoncini M (2021) The biodiversity of edible flowers: discovering new tastes and new health benefits. Front Plant Sci 11. https://doi.org/10.3389/fpls.2020.569499

Bergstrand KJ (2022) Organic fertilizers in greenhouse production systems – a review. Sci Hortic 295. https://doi.org/10.1016/j.scienta.2021.110855

Bhargava YV, Chittoor PK, Bharatiraja C et al (2022) Sensor fusion based intelligent hydroponic farming and nursing system. IEEE Sensors J 22(14):14584–14591. https://doi.org/10.1109/jsen.2022.3177777

Bhat SA, Huang NF (2021) Big data and AI revolution in precision agriculture: survey and challenges. IEEE Access 9:110209–110222. https://doi.org/10.1109/access.2021.3102227

Bhujel A, Basak JK, Khan F et al (2020) Sensor systems for greenhouse microclimate monitoring and control: a review. J Biosyst Eng 45(4):341–361. https://doi.org/10.1007/s42853-020-00075-6

Chen XM, Hu C, Raghubeer E et al (2010) Effect of high pressure pasteurization on bacterial load and bioactivity of *Echinacea Purpurea*. J Food Sci 75(7):C613–C618. https://doi.org/10.1111/j.1750-3841.2010.01753.x

Chen LB, Huang GZ, Huang XR et al (2022) A self-supervised learning-based intelligent greenhouse orchid growth inspection system for precision agriculture. IEEE Sensors J 22(24):24567–24577. https://doi.org/10.1109/jsen.2022.3221960

Christie E (2014) Water and nutrient reuse within closed hydroponic systems. Thesis, Georgia Southern University

Demotes-Mainard S, Péron T, Corot A et al (2016) Plant responses to red and far-red lights, applications in horticulture. Environ Exp Bot 121:4–21. https://doi.org/10.1016/j.envexpbot.2015.05.010

Despommier D (2014) Vertical farms in horticulture. In: Thompson PB, Kaplan DM (eds) Encyclopedia of food and agricultural ethics. Springer, Dordrecht, pp 1791–1799. https://doi.org/10.1007/978-94-007-0929-4_88

Dieleman JA, Hemming S (2011) Energy saving: from engineering to crop management. In: Dorais M (ed) International symposium on high technology for greenhouse systems: GreenSys 2009, Quebec City, Canada, Apr 2011. Acta Horticulturae, vol 893. ISHS, Belgium, pp 65–73. https://doi.org/10.17660/ActaHortic.2011.893.2

Dole JM, Cole JC, Vonbroembsen SL (1994) Growth of poinsettias, nutrient leaching, and water-use efficiency respond to irrigation methods. HortScience 29(8):858–864. https://doi.org/10.21273/hortsci.29.8.858

Dou HJ, Niu GH, Gu MM et al (2018) Responses of sweet basil to different daily light integrals in photosynthesis, morphology, yield, and nutritional quality. HortScience 53(4):496–503. https://doi.org/10.21273/hortsci12785-17

Factmr (2022) Edible flowers market outlook (2022 to 2032). https://www.factmr.com/report/2979/edible-flowers-market#:~:text=The%20global%20edible%20flowers%20market,USD%20317.90%20Million%20in%202022. Accessed 18 Dec 2024

Falla NM, Contu S, Demasi S et al (2020) Environmental impact of edible flower production: a case study. Agronomy 10(4):579

Falla NM, Zadeh NS, Stelluti S et al (2024) Active modified atmosphere packaging helps preserve quality of edible flowers. Agronomy 14(10). https://doi.org/10.3390/agronomy14102409

Farooq MS, Uzair M, Raza A et al (2022) Uncovering the research gaps to alleviate the negative impacts of climate change on food security: a review. Front Plant Sci 13. https://doi.org/10.3389/fpls.2022.927535

Fausey BA, Heins RD, Cameron AC (2005) Daily light integral affects flowering and quality of greenhouse-grown *Achillea*, *Gaura*, and *Lavandula*. HortScience 40(1):114–118. https://doi.org/10.21273/hortsci.40.1.114

Fernandes L, Saraiva JA, Pereira JA et al (2019) Post-harvest technologies applied to edible flowers: a review. Food Rev Int 35(2):132–154. https://doi.org/10.1080/87559129.2018.1473422

Ferroukhi M, Saadi H, Bendib R et al (2023) Connected sensors for a smart green farm. In: Hatti M (ed) International conference on artificial intelligence in renewable energetic systems, Tamanghasset, Algeria, Nov 2022, Lecture notes in networks and systems, vol 591. Springer, Cham, pp 213–222. https://doi.org/10.1007/978-3-031-21216-1_23

Gao D, Ji XY, Yuan Q et al (2023) Effects of total daily light integral from blue and broad-band red LEDs on flowering of saffron (*Crocus sativus* L.). Sci Rep 13(1). https://doi.org/10.1038/s41598-023-34424-0

Ghani S, Bakochristou F, Elbialy EMAA et al (2019) Design challenges of agricultural greenhouses in hot and arid environments – a review. Eng Agric Environ Food 12(1):48–70. https://doi.org/10.1016/j.eaef.2018.09.004

Gonçalves J, Borges JCF, Souza FA et al (2023) Production of edible flowers: irrigation and biotechnology. Rev Ceres 70(2):1–12. https://doi.org/10.1590/0034-737X202370020001

Gürel C, Erden A (2013) Conceptual design of a rose harvesting robot for greenhouses. In: Eroğlu AK, Turhan Ç, Erden Z, Erden A (eds) 20th annual international conference on mechatronics and machine vision in practice, Ankara, Turkey. Mechatronics and Machine Vision in Practice, Australia, pp 18–20

Hassanzadeh A, Murphy SP, Pethybridge SJ et al (2020) Growth stage classification and harvest scheduling of snap bean using hyperspectral sensing: a greenhouse study. Remote Sens 12(22). https://doi.org/10.3390/rs12223809

Henderson S, Gholami D, Zheng YB (2018) Soil moisture sensor-based systems are suitable for monitoring and controlling irrigation of greenhouse crops. HortScience 53(4):552–559. https://doi.org/10.21273/hortsci12676-17

Hiloidhari M, Baruah DC, Singh A et al (2017) Emerging role of geographical information system (GIS), life cycle assessment (LCA) and spatial LCA (GIS-LCA) in sustainable bioenergy planning. Bioresour Technol 242:218–226. https://doi.org/10.1016/j.biortech.2017.03.079

Iddio E, Wang L, Thomas Y et al (2020) Energy efficient operation and modeling for greenhouses: a literature review. Renew Sust Energ Rev 117. https://doi.org/10.1016/j.rser.2019.109480

Jariyasakoolroj P, Leelaphiwat P, Harnkarnsujarit N (2020) Advances in research and development of bioplastic for food packaging. J Sci Food Agric 100(14):5032–5045. https://doi.org/10.1002/jsfa.9497

Javaid M, Haleem A, Singh RP et al (2022) Enhancing smart farming through the applications of agriculture 4.0 technologies. Int J Intell Networks 3:150–164. https://doi.org/10.1016/j.ijin.2022.09.004

Jha K, Doshi A, Patel P et al (2019) A comprehensive review on automation in agriculture using artificial intelligence. Artif Intell Agric 2:1–12. https://doi.org/10.1016/j.aiia.2019.05.004

Jongyun K, Lea-Cox JD, Chappell M et al (2014) Wireless sensors networks for optimization of irrigation, production, and profit in ornamental production. In: Son JE, Lee Ib, Oh MM (eds) International symposium on new technologies for environment control, energy-saving and crop production in greenhouse and plant factory – Greensys 2013, Jeju, Korea. Acta Horticulturae, vol 1037, ISHS, Belgium, pp 643–649. doi: https://doi.org/10.17660/ActaHortic.2014.1037.82

Kaiser E, Kusuma P, Vialet-Chabrand S et al (2024) Vertical farming goes dynamic: optimizing resource use efficiency, product quality, and energy costs. Front Sci 2. https://doi.org/10.3389/fsci.2024.1411259

Karimi M, Ahmadi N, Ebrahimi M (2022) Red LED light promotes biomass, flowering and secondary metabolites accumulation in hydroponically grown *Hypericum perforatum* L. (cv. Topas). Ind Crop Prod 175. https://doi.org/10.1016/j.indcrop.2021.114239

Katzin D, Marcelis LFM, van Mourik S (2021) Energy savings in greenhouses by transition from high-pressure sodium to LED lighting. Appl Energy 281. https://doi.org/10.1016/j.apenergy.2020.116019

Kim C, van Iersel MW (2023) Image-based phenotyping to estimate anthocyanin concentrations in lettuce. Front Plant Sci 14. https://doi.org/10.3389/fpls.2023.1155722

Kondaka L, Iyer R, Jaiswal S et al (2023) A smart hydroponic farming system using machine learning. In: Niranjan SK (ed) International conference on intelligent and innovative technologies in computing, electrical and electronics (IITCEE), Bengaluru, pp 357–362. https://doi.org/10.1109/IITCEE57236.2023.10090860

Kopsell D, Belisle C, Lowery H et al (2016) Genotype and lighting environment impact petal tissue pigmentation in *Tagetes tenuifolia*. In: Currey CJ, Lopez RG, Runkle ES (eds) 8th international symposium on light in horticulture, East Lansing, MI, USA. Acta Horticulturae, vol 1134, ISHS, Belgium, pp 103–109. doi: https://doi.org/10.17660/ActaHortic.2016.1134.14

Kusuma P, Pattison PM, Bugbee B (2020) From physics to fixtures to food: current and potential LED efficacy. Hortic Res 7(1). https://doi.org/10.1038/s41438-020-0283-7

Lee U, Islam MP, Kochi N et al (2022) An automated, clip-type, small internet of things camera-based tomato flower and fruit monitoring and harvest prediction system. Sensors 22(7). https://doi.org/10.3390/s22072456

Li DH, Li GQ, Li JZ et al (2022) Wearable crop sensor based on nano-graphene oxide for non-invasive real-time monitoring of plant water. Membranes 12(4). https://doi.org/10.3390/membranes12040358

Liu X-Y, Yu J-R, Deng H-N (2024) Non-destructive prediction of anthocyanin content of rosa chinensis petals using digital images and machine learning algorithms. Horticulturae 10(5):503. https://doi.org/10.3390/horticulturae10050503

Mahdavian M, Wattanapongsakorn N (2017) Optimizing greenhouse lighting for advanced agriculture based on real time electricity market price. Math Probl Eng 2017:6862038. https://doi.org/10.1155/2017/6862038

Maher A, Kamel E, Enrico F et al (2016) An intelligent system for the climate control and energy savings in agricultural greenhouses. Energ Effic 9(6):1241–1255. https://doi.org/10.1007/s12053-015-9421-8

Majumder D, Akhter J, Mandal A et al (2021) Precision input management for minimizing and recycling of agricultural waste. In: Bhatt R, Meena RS, Hossain A (eds) Input use efficiency for food and environmental security. Springer, pp 567–603. https://doi.org/10.1007/978-981-16-5199-1_19

Maraveas C, Konar D, Michopoulos DK et al (2024) Harnessing quantum computing for smart agriculture: empowering sustainable crop management and yield optimization. Comput Electron Agric 218:108680. https://doi.org/10.1016/j.compag.2024.108680

Marchioni I, Colla L, Pistelli L et al (2020) Different growing conditions can modulate metabolites content during post-harvest of *Viola cornuta* L. edible flowers. Adv Hortic Sci 34(1S):61–69. https://doi.org/10.13128/ahsc-8395

Mayorga-Gomez AM, van Iersel MW, Ferrarezi RS (2024) Varying light intensities affect lettuce growth and physiology in controlled indoor environments. Horticulturae 10(9). https://doi.org/10.3390/horticulturae10090931

Mempel H, Jüttner I, Wittmann S (2021) The potentials of indoor farming for plant production. Automatisierungstechnik 69(4):287–296. https://doi.org/10.1515/auto-2020-0044

Mendon A, Votavat BM, Singh S (2022) Design and construction of automated hydroponics system with remote monitoring and control. 13th international conference on computing communication and networking technologies (ICCCNT). pp 1–5. https://doi.org/10.1109/ICCCNT54827.2022.9984617

Miner GL, Ham JM, Kluitenberg GJ (2017) A heat-pulse method for measuring sap flow in corn and sunflower using 3D-printed sensor bodies and low-cost electronics. Agric For Meteorol 246:86–97. https://doi.org/10.1016/j.agrformet.2017.06.012

Morii M, Fujimoto M, Imajo A et al (2023) Design and implementation of hydroponic cultivation kit with iot for edible flowers. International conference on electrical, communication and computer engineering (ICECCE), Dubai, UAE, pp 1–6. https://doi.org/10.1109/ICECCE61019.2023.10442155

Naing AH, Jeon SM, Park JS, Kim CK (2016) Combined effects of supplementary light and CO2 on rose growth and the production of good quality cut flowers. Can J Plant Sci 96(3):503–510. https://doi.org/10.1139/cjps-2015-0304

Naito H, Yoshinaga K, Fukatsu T et al (2018) Developing techniques for counting strawberry flowers in movable-bench systems in a greenhouse. In: Li T, Yang Q (eds) International symposium on new technologies for environment control, energy-saving and crop production in greenhouse and plant factory – Greensys 2017, Beijing, China. Acta Horticulturae, vol 1227, ISHS, Belgium, pp 401–408. https://doi.org/10.17660/ActaHortic.2018.1227.50

Nederhoff E, Stanghellini C (2010) Water use efficiency of tomatoes. PHG 115:52–59

Niu GH, Heins RD, Cameron AC et al (2000) Day and night temperatures, daily light integral, and CO^2 enrichment affect growth and flower development of pansy (*Viola x wittrockiana*). J Am Soc Hortic Sci 125(4):436–441. https://doi.org/10.21273/jashs.125.4.436

Oh W, Runkle ES, Warner RM (2010) Timing and duration of supplemental lighting during the seedling stage influence quality and flowering in petunia and pansy. HortScience 45(9):1332–1337. https://doi.org/10.21273/hortsci.45.9.1332

Oliveira LFP, Moreira AP, Silva MF (2021) Advances in agriculture robotics: a state-of-the-art review and challenges ahead. Robotics 10(2). https://doi.org/10.3390/robotics10020052

Ouzounis T, Rosenqvist E, Ottosen CO (2015) Spectral effects of artificial light on plant physiology and secondary metabolism: a review. HortScience 50(8):1128–1135. https://doi.org/10.21273/hortsci.50.8.1128

Palmer S, van Iersel MW (2020) Increasing growth of lettuce and mizuna under sole-source led lighting using longer photoperiods with the same daily light integral. Agronomy 10(11):1659

Patil KA, Kale NR (2016) A model for smart agriculture using IoT. International conference on global trends in signal processing, information computing and communication (ICGTSPICC), Jalgoan, India, pp 543–545. https://doi.org/10.1109/ICGTSPICC.2016.7955360

Pires EO, Di Gioia F, Rouphael Y et al (2023) Edible flowers as an emerging horticultural product: a review on sensorial properties, mineral and aroma profile. Trends Food Sci Technol 137:31–54. https://doi.org/10.1016/j.tifs.2023.05.007

Pramuk LA, Runkle ES (2005) Photosynthetic daily light integral during the seedling stage influences subsequent growth and flowering of *Celosia*, *Impatiens*, *Salvia*, *Tagetes*, and *Viola*. HortScience 40(5):1336–1339. https://doi.org/10.21273/hortsci.40.5.1336

Razza F, Degli Innocenti F, Dobon A et al (2015) Environmental profile of a bio-based and biodegradable foamed packaging prototype in comparison with the current benchmark. J Clean Prod 102:493–500. https://doi.org/10.1016/j.jclepro.2015.04.033

Reed D (1996) Closed production systems for containerized crops. In: Reed D (ed) Water, media and nutrition for greenhouse crops. Ball Publishing, Batavia, pp 221–245

Saini AK, Yadav AK, Dhiraj (2025) A comprehensive review on technological breakthroughs in precision agriculture: IoT and emerging data analytics. Eur J Agron 163:127440. https://doi.org/10.1016/j.eja.2024.127440

Sams CE, Kopsell D, Morrow RC (2016) Light quality impacts on growth, flowering, mineral uptake and petal pigmentation of marigold. In: Currey CJ, Lopez RG, Runkle ES (eds) 8th international symposium on light in horticulture, East Lansing, MI, USA, may 2016. Acta Horticulturae, vol 1134, ISHS, Belgium, pp 139–145. https://doi.org/10.17660/ActaHortic.2016.1134.19

Sarode DK, Pagariya MC, Jadhav PR et al (2024) Edible flowers: biotechnological interventions for improving bioactives of food and health significance. J Food Compos Anal 134:106506. https://doi.org/10.1016/j.jfca.2024.106506

Schettini E, Santagata G, Malinconico M et al (2013) Recycled wastes of tomato and hemp fibres for biodegradable pots: physico-chemical characterization and field performance. Resour Conserv Recycl 70:9–19. https://doi.org/10.1016/j.resconrec.2012.11.002

Seginer I, Albright LD, Ioslovich I (2006) Improved strategy for a constant daily light integral in greenhouses. Biosyst Eng 93(1):69–80. https://doi.org/10.1016/j.biosystemseng.2005.09.007

Shree C, Kaur R, Upadhyay S et al (2019) Multi-feature based automated flower harvesting techniques in deep convolutional neural networking. 4th international conference on internet of things: smart innovation and usages (IoT-SIU), Ghaziabad, pp 1–6. https://doi.org/10.1109/IoT-SIU.2019.8777338

Sonneveld C, Voogt W (2009) Plant nutrition in future greenhouse production. In: Sonneveld C, Voogt W (eds) Plant nutrition of greenhouse crops. Springer, pp 393–403. https://doi.org/10.1007/978-90-481-2532-6_17

Spall CE, Lopez RG (2024) Daily light integral, but not the photoperiod, influences the time to flower and finished quality of dianthus specialty cut flowers. HortScience 59(7):1007–1019. https://doi.org/10.21273/hortsci17848-24

Stanghellini C (2013) Horticultural production in greenhouses: efficient use of water. In: Blok C, Os Ev, Voogt W (eds) International symposium on growing media and soilless cultivation, Leiden, Netherlands, Jun 2013. Acta Horticulturae, Vol 1034, ISHS, Belgium, pp 25–32. https://doi.org/10.17660/ActaHortic.2014.1034.1

Steinwand MA, Ronald PC (2020) Crop biotechnology and the future of food. Nat Food 1(5):273–283. https://doi.org/10.1038/s43016-020-0072-3

Taddei Dalla Torre F, Melgani F, Pertot I et al (2024) Artificial intelligence vision methods for robotic harvesting of edible flowers. Plants 13(22):3197. https://doi.org/10.3390/plants13223197

Trivellini A, Toscano S, Romano D et al (2023) LED lighting to produce high-quality ornamental plants. Plants 12(8). https://doi.org/10.3390/plants12081667

van Delden SH, SharathKumar M, Butturini M et al (2021) Current status and future challenges in implementing and upscaling vertical farming systems. Nat Food 2(12):944–956. https://doi.org/10.1038/s43016-021-00402-w

van Iersel MW (2017) Optimizing LED lighting in controlled environment agriculture. In: Dutta Gupta S (ed) Light emitting diodes for agriculture: smart lighting. Springer, pp 59–80. https://doi.org/10.1007/978-981-10-5807-3_4

van Iersel MW, Gianino D (2017) An adaptive control approach for light-emitting diode lights can reduce the energy costs of supplemental lighting in greenhouses. HortScience 52(1):72–77. https://doi.org/10.21273/hortsci11385-16

van Iersel MW, Weaver G, Martin MT et al (2016) A chlorophyll fluorescence-based biofeedback system to control photosynthetic lighting in controlled environment agriculture. J Am Soc Hortic Sci 141(2):169–176. https://doi.org/10.21273/jashs.141.2.169

Vatistas C, Avgoustaki DD, Bartzanas T (2022) A systematic literature review on controlled-environment agriculture: how vertical farms and greenhouses can influence the sustainability and footprint of urban microclimate with local food production. Atmos 13(8). https://doi.org/10.3390/atmos13081258

Virnodkar SS, Pachghare VK, Patil VC et al (2020) Remote sensing and machine learning for crop water stress determination in various crops: a critical review. Precis Agric 21(5):1121–1155. https://doi.org/10.1007/s11119-020-09711-9

Wacker K, Kim C, van Iersel MW et al (2024) Development of an automated low-cost multispectral imaging system to quantify canopy size and pigmentation. Sensors 24(17). https://doi.org/10.3390/s24175515

Wani MA, Din A, Nazki IT et al (2023) Navigating the future: exploring technological advancements and emerging trends in the sustainable ornamental industry. Front Environ Sci 11. https://doi.org/10.3389/fenvs.2023.1188643

Warner RM, Erwin JE (2003) Effect of photoperiod and daily light integral on flowering of five *Hibiscus* sp. Sci Hortic 97(3–4):341–351. https://doi.org/10.1016/s0304-4238(02)00157-7

Weaver G, van Iersel MW (2020) Longer photoperiods with adaptive lighting control can improve growth of greenhouse-grown 'Little Gem' lettuce (*Lactuca sativa*). HortScience 55(4):573–580. https://doi.org/10.21273/hortsci14721-19

Weeplian T, Yen TB, Ho YS (2018) Growth, development, and chemical constituents of edible ice plant (*Mesembryanthemum crystallinum* L.) produced under combinations of light-emitting diode lights. HortScience 53(6):865–874. https://doi.org/10.21273/hortsci12997-18

Whitman C, Padhye S, Runkle ES (2022) A high daily light integral can influence photoperiodic flowering responses in long day herbaceous ornamentals. Sci Hortic 295. https://doi.org/10.1016/j.scienta.2022.110897

Wu ZC, Li LJ, Zhao Q et al (2022) Design and research of a harvesting actuator for *Camellia oleifera* flowers during the budding period. Agriculture 12(10):1698. https://doi.org/10.3390/agriculture12101698

Xu WS, Lu N, Kikuchi M et al (2021) Continuous lighting and high daily light integral enhance yield and quality of mass-produced nasturtium (*Tropaeolum majus* L.) in plant factories. Plants 10(6):1203. https://doi.org/10.3390/plants10061203

Yan BP, Zhang F, Wang MY et al (2024) Flexible wearable sensors for crop monitoring: a review. Front Plant Sci 15. https://doi.org/10.3389/fpls.2024.1406074

Yuan H, Li L, Wang J et al (2016) Design and test of regulation and control equipment for nutrient solution of water and fertilizer integration in greenhouse. Trans Chin Soc Agric Eng 32(8):27–32. https://doi.org/10.11975/j.issn.1002-6819.2016.08.004

Zhang MZ, Runkle ES (2019) Regulating flowering and extension growth of poinsettia using red and far-red light-emitting diodes for end-of-day lighting. HortScience 54(2):323–327. https://doi.org/10.21273/hortsci13630-18

Zhang JY, Rao Y, Man C et al (2021) Identification of cucumber leaf diseases using deep learning and small sample size for agricultural internet of things. Int J Distrib Sens Netw 17(4). https://doi.org/10.1177/15501477211007407

Zhao LL, Fan HZ, Zhang M et al (2019) Edible flowers: review of flower processing and extraction of bioactive compounds by novel technologies. Food Res Int 126. https://doi.org/10.1016/j.foodres.2019.108660

Zhu J, Zhang YC, Zhou L et al (2022) Growth and flowering of saffron (*Crocus sativus* L.) with three corm weights under different LED light qualities. Sci Hortic 303. https://doi.org/10.1016/j.scienta.2022.111202

Chapter 6
Edible Flowers' Bioactive Compounds: Nutraceutical Aspect and Industrial Food Exploration

Chalermchai Wongs-Aree and Mantana Buanong

Abstract Edible flowers are increasingly in demand in the contemporary food and beverage industries due to their diverse natural flavours and colours, outstanding nutritional properties, and advantageous phytochemical compounds, beneficial to human health that go beyond their aesthetic appeal. Various floral components, including stamens, pistils, sepals, and receptacles, make up the flower's structure, but petals are generally acknowledged as the most palatable and valuable for immediate food intake. Petals produce distinctive substances, such as pigments, bioactive compounds, and volatile compounds, which greatly augment the flower's considerable value. Consequently, flowers, being the reproductive organs of plants, possess vibrant and aromatic corollas that allure both pollinators and individuals. Edible flowers serve as a delightful complement to floral salads and can be integrated into savoury preparations featuring meats, fish, and soups. Such cuisine provides a distinctive blend of flavours and colours, enhancing its overall nutritional value. Although numerous floral chemicals contain advantageous bioactive compounds and nutraceuticals with antioxidant and antimicrobial properties, it is essential to acknowledge that certain chemicals are poisonous. When harvesting ready-to-eat flowers, it is advisable to avoid wilted, dusty, aged, or discoloured specimens due to their low nutritional value and short postharvest life. Avoiding the use of pesticides is also essential when practising intensive cultural care. Furthermore, microbial contamination is another significant concern for fresh consumption. In the industrial sector, processed commodities obtained from flowers, including beverages, flavourings, and natural colourants, continue to increase in demand. In essence, flowers have been a staple of human diets for millennia and will undoubtedly remain so in the future.

Keywords Blossom · Food properties · Health benefits · Phytochemistry · Poisoning

---

C. Wongs-Aree · M. Buanong (✉)
Division of Postharvest Technology, School of Bioresources and Technology, King Mongkut's University of Technology Thonburi, Bangkok, Thailand
e-mail: mantana.bua@kmutt.ac.th

M. M. Jowkar et al. (eds.), *Advances in Research on Edible Flowers*, Advances in Olericulture, https://doi.org/10.1007/978-3-032-04561-4_6

Abbreviations

$ABTS^+$	2,2′-Azino-bis (3-ethylbenzothiazoline-6-sulfonic acid
ACE	Angiotensin-I conversion enzyme
AChE	Acetylcholinesterase
AGS	Human gastric adenocarcinoma cell line
BHT	Butylated hydroxytoluene
BL-13	The Back-Shu point of the Lung
CCD	Carotenoid cleavage dioxygenase
Cy	Cyanidin
DIPs	Dearomatized isoprenylated phloroglucinols
DM	Diabetes mellitus
DPPH	2,2-Diphenyl-1-picryhydrazyl
DW	Dry weight
F3H	Flavanone 3-hydroxylase
FRAP	Ferric-reducing antioxidant power
FW	Fresh weight
GAE	Gallic acid equivalent
HT-29	Human colorectal adenocarcinoma cell line with epithelial morphology
IC_{50}	Half maximal inhibitory concentration
KB ATCC CCL-17	Oral cancer cell
MCF-7 ATCC HTB-22	Breast cancer cell
MICs	Minimum inhibitory concentrations
MMPs	Matrix metalloproteinase
NCI-H187 ATCC CRL-5804	Small lung cancer cell
NI	No inhibition effect
ORAC	Oxygen radical absorbance capacity
RAAS	Renin-angiotensin-aldosterone system
ROS	Reactive oxygen species
TE	Trolox equivalent
UV	Ultraviolet

6.1 Introduction

6.1.1 Floral Components

Flowers are the component of a plant responsible for seed production. They are made up of reproductive organs (stamens and carpels) encircled by a colourful, fragrant corolla (petals) and a green calyx (sepals) (Streich 2007). Flowers are, without a doubt, one of the most beautiful features of plants. Four floral whorls of a flower comprise the calyx, corolla, androecium, and gynoecium (Fig. 6.1). Sepals

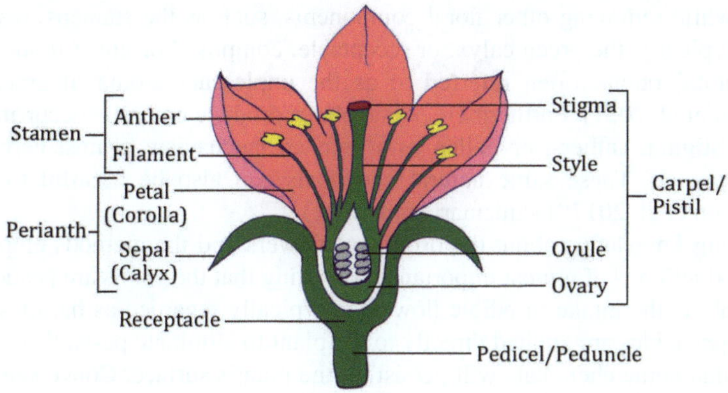

Fig. 6.1 Parts of a complete flower

are often green, leafy appendages that act as a protective shield for the flower bud. Those are the bloom's very last whorls. Corolla (petals) is typically brightly coloured and fragrant to entice pollinators like insects and animals. The corollas of different plants come in all sorts of shapes and colours. The male reproductive organ known as 'stamen' consists of filaments and anthers. Each lobe of an anther is typically divided into two separate pollen sacs. Pollen sacs are where the individual pollen grains form. Stamens are usually connected to the petals or the perianth. The female reproductive part of a flower is called 'carpel' (or pistil) and consists of stigma, style, and ovary. The ovary is linked to the stigma by the style. The stigma is the pollen-receptive surface, commonly present at the apex of the style. One or more ovules are attached to a flattened, cushion-like placenta within each ovary (Vidanapathirana and Rifna 2023). After fertilisation, the ovules transform into seeds, while the ovary and some floral accessories develop into a fruit.

6.1.2 Floral Parts for Culinary Prospects

Flowers of flowering plants primarily serve to attract pollinators through their colour, fragrance, or both, while also being widely used for decorative purposes. Flowers of plants can be consumed as food or drink in many cases. Traditionally, vegetable flowers, such as cucurbits, peas, and artichokes, are used as edible flowers. Because of their health benefits, ornamental flowers have recently gained high demand for both fresh and processed commodities. The most advisable approach is to exclusively consume the petals of edible flowers, as they contain the highest bioactive compounds (Kumari et al. 2021), pigments (Narbona et al. 2021), and aromatic compounds (Dudareva and Pichersky 2000). Consumers should not indiscriminately assume that a plant is safe for consumption just because one of its parts is edible. To ensure safe consumption, it is recommended to consume only the petals, while removing other floral components, such as the stamens, pistils, and calyx. Typically, the green calyx, or receptacle, composed of green tissue, emits a 'green note' odour, often referred to as the unpleasant aroma of green leaves (Vincenti et al. 2021). Furthermore, some plants produce naturally occurring toxins in their stigma, anthers, or pollen as a defence mechanism against herbivory or insect damage. These same natural chemicals can also be harmful to humans (Stevenson et al. 2017; Jacquemart et al. 2021).

Gaining knowledge about the origins of flowers and the methods employed in their production is of utmost importance. Ensuring that the plants are pesticide-free is crucial, as the intake of edible flowers is typically regarded as harmless. When contact pesticides are applied directly to the plant to eliminate pests, there is a possibility that some chemicals will persist on the plant's surface. Conversely, certain plant components assimilate systemic insecticides, which are subsequently distributed throughout the entirety of the plant. The chemical residues are nonremovable by washing and hence remain in the plants (Cloyd et al. 2011). Therefore, it is essential to adhere to rigorous cultivation protocols when applying pesticides to edible flowers. Flowers with good flavours, on the other hand, are adorable. Some flowers, such as lavender, have a pleasant aroma but a bitter or astringent taste. Some, like butterfly pea flowers, have a good flavour and crisp texture but have a green note (with the calyx). Furthermore, the same flower grown in different locations can have a distinct flavour and colour (Barazani et al. 2019). Sweet or floral-scented blossoms can be used to decorate meals, desserts, or garnish drinks in addition to fresh consumption as floral salads.

6.1.3 Freshly Edible Flowers

Edible flowers are widely used for culinary purposes, including exotic drinks, desserts, and raw salads, to improve the aesthetic value, food taste, and visual presentation of dishes (Matyjaszczyk and Smiechowska 2019; Pires et al. 2019). Flowers are served as garnishes for different meals, buffet food, and decorations for salads, sweet meals, fruit dishes and ice-cream, cold drinks, etc. Additionally, they contribute to a specific taste and smell of foods (Scherf 2004). The worldwide tendency to consume edible flowers has recently gained attention due to their potential as sources of bioactive compounds that promote health benefits. Studies have reported that edible flowers are composed of 70–95% water (Chensom et al. 2019; Fernandes et al. 2017; Stefaniak and Grzeszczuk 2018), and the dry matter of edible flowers is a source of active chemical constituents such as carbohydrates, proteins and lipids, vitamins, minerals, and phytochemicals (Chensom et al. 2019; Fernandes et al. 2017). Edible flowers have been consumed worldwide such as in China, Japan, Rome, Greece, Victorian England, and Europe (Jadhav et al. 2023). Figure 6.2 depicts edible flowers cultivated in the central region of Thailand and sold as mixed flowers. These flowers have recently been widely used in niche markets to furnish deliciate Thai menus. They could be included as the main ingredient in flowery

Fig. 6.2 Mixed fresh flowers with good quality are packaged and sold in Thailand.

drinks or beautifully decorated in high tea sets, which are now increasingly popular among leading restaurants.

Edible flowers comprise three main parts (pollen, nectar, and petals), each of which contributes plenty of valuable nutrients to human health. Pollen is a small part of the source of carbohydrates, proteins, lipids (saturated and unsaturated lipids) (Dobson 1988), carotenoids, and flavonoids (Lunau 1995; Wiermann and Gubatz 1992). Nectar contains sugars (glucose, fructose, and sucrose), amino acids (such as proline), proteins, and lipids. It also provides an abundant source of organic acids, phenolic compounds, alkaloids, and terpenoids (Nicolson et al. 2007). Petals, on the other hand, are rich in vitamins, minerals, and antioxidants, while exhibiting a wide range of colours and shapes (Mlcek and Rop 2011). Figure 6.3 shows some varieties of edible flowers organically grown in Thailand, which are directly supplied to restaurants for different purposes. These flowers are excellent for use in raw and mixed salads. Neugebauerova and Vabkova (2009) reported that the quality criteria of edible flowers include the capability of individual species to be grown over larger areas, and the stability of their yields, as well as their resistance to diseases and pests.

6.2 Pigments as Potential Antioxidants in Edible Flowers

Flowers, particularly their petals, display vibrant colours, which are performed by pigments. Molecular pigments absorb or reflect light at specific wavelengths (Tanaka et al. 2008). The colour perceived by humans depends on the wavelength of light that is reflected. Primary pigments reflect the important light wavelength for human sight (van der Kooi et al. 2019). Among the most prevalent pigment classes in flowers, chlorophylls, carotenoids, and flavonoids contribute to the reflectivity and visibility of pollinators (Narbona et al. 2021). In some plant species, petals can

Fig. 6.3 Edible flowers organically grown in Thailand and directly sold to restaurants. (**a**) White coral vine flower, (**b**) Coral vine flower, (**c**) Peacock flower and (**d**) Rangoon creeper flower

advertise with basic homogenous colours, while in others, they exhibit incredibly complex colour patterns. Carotenoids, flavonoids, and betalains are the three major classifications of flower pigments (Table 6.1). Certain flowers derive their colour from a combination of these pigments. Additionally, all pigments act as vital cellular antioxidants.

6.2.1 Chlorophylls

Chlorophylls play a vital function in photosynthesis since they absorb light energy and transform it into chemical energies. The molecular structure, which includes a haem group composed of tetraporphyrin rings, closely resembles the structure of haemoglobin (Nagini et al. 2015). The bioactive effects of chlorophyll molecules and their derivatives are significantly connected to their chemical structure. These compounds can boost oxygen absorption in the blood, hence increasing energy release and relieving fatigue (Kumari et al. 2017). The potential benefits of chlorophyllin as a chemopreventive agent are significant (Nagini et al. 2015). Moreover, the bioactivity of chlorophylls is attributed to their chemical ability to operate as antioxidants, antimutagens, and anticarcinogens. These properties are responsible for slowing the development of many cancerous tumours and exhibiting anti-obesogenic and anti-inflammatory effects (McQuistan et al. 2012; Hannan et al. 2020; Martins et al. 2023).

Typically, most floral buds have green sepals that shield the inside floral organs. The majority of the petals are either white or coloured. Only a few flowering plants, like Sesbania (*Sesbania javanica*), Cowslip creeper (*Telosma cordata* (Burm. f.) Merr., and Green Bird Flower (*Crotalaria cunninghamii*), perform a green corolla (a whorl of petals) containing some chlorophylls during blooming. Certain green tissues, including the calyx, receptacle, and peduncle, generally impart a strong green flavour note, commonly described as an unpleasant aroma (Vincenti et al. 2021). Therefore, removing the calyx or receptacle from flower buds before eating them fresh is recommended, as doing so improves the taste and overall enjoyment of the dish.

Table 6.1 Petal pigments and colours associated with some edible flowers

Colour	Main pigments	Edible flowers
Green/ Greenish	Chlorophylls/Carote noids	Sesbania flower, Cowslip creeper, Turmeric (inflorescence)
Yellow	Carotenoids	Rose, Marigold, Chrysanthemum
Yellow	Flavonones	Orchid
Orange	Carotenoids	Waterlily, Lotus
Red/Purple	Anthocyanins	Rose, Turmeric (bracts), Torch ginger, Red ginger, Waterlily, Lotus, Orchid, Gladiolus
Red/Purple	Betalains	Bougainvillea (bracts), *Gomphrena globose, Mirabilis jalapa*
Blue/Violet	Anthocyanins	Butterfly pea, Poppy, Waterlily, Lotus, Orchid, Carnation
White	No pigment	Bread flower, Jasmine, Orchid, *Antigonon leptopus 'alba'*

6.2.2 Carotenoids

Carotenoids are pigments characterised by a long polyene chain (35–40 carbon atoms), occasionally ended by rings, and are important for preventing autooxidation from the photosynthetic process (Maoka 2020). Carotenoids can be divided into two categories: carotene (such as lycopene, α-carotene, β-carotene, etc.) and xanthophyll (including lutein, violaxanthin, zeaxanthin, etc.). Carotenoids with a large number of connected double bonds have a unique quality: they are highly efficient at neutralising free radicals and single oxygen molecules. This ability allows them to prevent lipid oxidation in cells during the early stages of ageing. Additionally, they are powerful at removing singlet oxygen molecules when an organism is subjected to oxidative stress (Crupi et al. 2023). Carotenoids, as plentiful antioxidants, offer defence against immune system disorders, cardiovascular diseases, and cancer (Kumari et al. 2017). Consistent consumption of lycopene can help mitigate the risk of chronic ailments like type-2 diabetes (Jiang et al. 2021), and bone diseases (Sahni et al. 2021). β-carotene and other carotenoids containing at least a substituted β-ring are nutritionally important because they serve as provitamin A precursors, necessary for maintaining good eyesight. Furthermore, lutein or zeaxanthin accumulates in the inner layer of human retina as macular melanin, which protects the retinal membrane from the detrimental effects of high-energy, short-wavelength light and enhances visual acuity (Arunkumar et al. 2020).

Among the carotenes, β-carotene is the most prevalent in coloured flowers, followed by α-carotene and γ-carotene. On the other hand, in terms of xanthophylls, lutein is the most common, with minor quantities of zeaxanthin and epoxyxanthophylls, neoxanthin, violaxanthin, and luteoxanthin (Ohmiya 2011; Meléndez-Martínez et al. 2022). Pure crocetin, glycosides of the diapocarotenedioic acid, are responsible for the bright yellow hue of saffron (*Crocus sativus*) flowers (Britton 1996) that has been studied in nutraceutical research. Certain waterlilies, such as *Nymphaea* 'Mangkala Ubon', contain carotenoids in their flowers (2 mg/100 g fresh weight (FW)), which give them their characteristic yellow-orange colour (Sriyong

2010). Because of their wide variety of beautiful blooms and ease of growing, waterlilies serve as an excellent source of edible flowers. In addition, fully opening marigold flowers (*Tagetes erecta*) store up to 900 mg lutein/g FW of carotenoids (Del Villar-Martínez et al. 2005). For these reasons, fresh marigold flowers can be a valuable addition to a variety of culinary dishes and medicinal remedies. Interestingly, the oxidation of carotenoids in floral cells can produce cyclic oxidised volatiles, such as β-ionone and β-damascenone, as well as branched-chain oxidised volatiles, including 6-methyl-5-hepten-2-one and pseudoionone, depending on the carotenoid type (Winterhalter and Rouseff 2001). These aromatic substances intensify the flavour profiles of dishes.

6.2.3 Flavonoid-Related Compounds and Others

Anthocyanins, a group of flavonoids, are the most prevalent pigments found in flowers, producing red, purple, magenta, and blue hues. Anthocyanins consist of flavylium-based cations, recognised as the principal and significant category of water-dispersible pigments (Khoo et al. 2017). The major anthocyanin pigments include cyanidin, pelargonidin, delphinidin, peonidin, petunidin, and malvidin derivatives. Flavonoids are classified into UV-absorbing flavonoids, aurones-chalcones, and anthocyanins, which function as UV-absorbing substances and protect against environmental stress. The presence of ortho-hydroxylation and methoxylation on the glycosylated B-ring structure of flavylium significantly increases the antioxidant activity of anthocyanins (Wang et al. 1997). Previous studies have revealed that consuming anthocyanins can stimulate the creation of insulin, which plays a crucial role in regulating glucose levels in humans (Christison and MacKenzie 1993; He and Giusti 2010). The number of hydroxyl groups on the B-ring of anthocyanins determines their capacity to promote insulin release (Tsuda et al. 2004). Anthocyanin pigments are essential for maintaining optimal vision and preventing the decline in photoreceptor cell function, contributing to improved night vision (Miyake et al. 2012; Shim et al. 2012). Furthermore, with their potent antioxidants, anthocyanins can support cardiovascular disease, reduce the risk of cancer, and alleviate allergies in humans (Kumari et al. 2017).

Anthocyanin pigment contributions have been investigated in certain ornamental flowers with dark purple to dark red petals. Black pea (*Thermopsis barbata* (Fabaceae)) petals contain major anthocyanins of 3-*O*-glucosides of delphinidin, cyanidin, and petunidin (Murai et al. 2017). Similarly, chocolate cosmos (*Cosmos atrosanguineus* (Hook.) Voss) possesses two predominant anthocyanins: cyanidin 3-*O*-glucoside and cyanidin 3-*O*-rutinoside (Amamiya and Iwashina 2016). Hollyhock (*Alcea rosea* L. cv. 'Nigra') flowers accumulate 9 distinct anthocyanins derived from delphinidin-, cyanidin-, petunidin- and malvidin-based glycosides (Hosaka et al. 2012). Meanwhile, dahlias (*Dahlia variabilis* Hort.) exhibit a high accumulation of cyanidin (Cy)-based anthocyanins (Deguchi et al. 2016). The anthocyanins and other flavonoids in these petals may play a role in protecting plants from a variety of environmental stresses at such high altitudes. Furthermore, flowers of *Hypericum calycinum*, which appear uniformly yellow to the human eye, bear a UV pattern presumably visible to insects. Two categories of pigments [flavonoids and dearomatized isoprenylated phloroglucinols (DIPs)] are responsible for the UV demarcations observed in this flower (Gronquis et al. 2001).

The deep blue petals of butterfly pea (*Clitoria ternatea*) (Fig. 6.4a) contain many flavonoid compounds, such as *p*-coumaric acid and ferulic acid, and anthocyanins, all of which are derived exclusively from a basic delphinidin structure (Wongs-Aree et al. 2006). Anthocyanin pigment contents in orchid perianths of 3 species from the Vandeae tribe have been recorded as follows: *Aerides rosea* (22.9 mg/100 g FW, Fig. 6.4b), *Rhynchostylis retusa* (10.5 mg/100 g FW, Fig. 6.4c), and *Vanda coerulea* (47.3 mg/100 g FW, Fig. 6.4d) (Junka et al. 2007; Junka et al. 2012). The red perianth of *Aerides rosea* contains more than 80% cyanidin derivatives, followed by delphinidin derivatives, whereas *Rhynchostylis retusa* consists of only cyanidin derivatives (Junka et al. 2007). The blue perianth of *Vanda coerulea* contains 52.2% delphinidin and 47.8% cyanidin skeletons (Junka et al. 2012). Furthermore, anthocyanin characteristics were studied in 4 cultivars of waterlilies in blue- and red-toned petals (Fig. 6.4e–h). The petals of 'Red Flare' exhibited the highest amount of anthocyanins (27.1 mg/100 g FW), followed by 'Pink Capensis' (10.7 mg/100 g FW), 'King of Siam' (8.4 mg/100 g FW), and 'Blue Capensis' (5.2 mg/100 g FW). However, the highest content of anthocyanin was detected in the stamen (male gamete) of 'Red Flare' (Table 6.2). The primary anthocyanins present in these 4 varieties are cyanidin and delphinidin derivatives (Fig. 6.5a–b), referenced against extracts from Concord grape peel (Fig. 6.5c), and strawberry peel (Fig. 6.5d). Interestingly, in the blue-shaded petals of 'King of Siam' and 'Blue Capensis', unidentified

Fig. 6.4 Visual appearance of blooming flowers of *Clitoria ternatea* butterfly pea (**a**), *Aerides rosea* (**b**), *Rhynchostylis retusa* (**c**), wild *Vanda coerelea* (**d**), *Nymphaea* 'King of Siam' (**e**), *N.* 'Red Flare' (**f**), *N. capensis* var. zanzibariensis Pink (Pink Capensis) (**g**), and *N. capensis* var. zanzibariensis Blue (Blue Capensis) (**h**)

Table 6.2 Anthocyanin content in waterlily parts in 4 varieties

	Anthocyanin content (mg/100 g Fresh weight)			
Waterlily part	'King of Siam'	'Red Flare'	'Pink Capensis'	'Blue Capensis'
Sepal	0.59 ± 0.24	18.12 ± 5.78	0.56 ± 0.16	0.60 ± 0.32
Petal	8.37 ± 3.65	27.13 ± 15.52	10.66 ± 2.30	5.15 ± 2.35
Stamen	–	43.63 ± 13.32	2.04 ± 1.00	0.95 ± 0.33

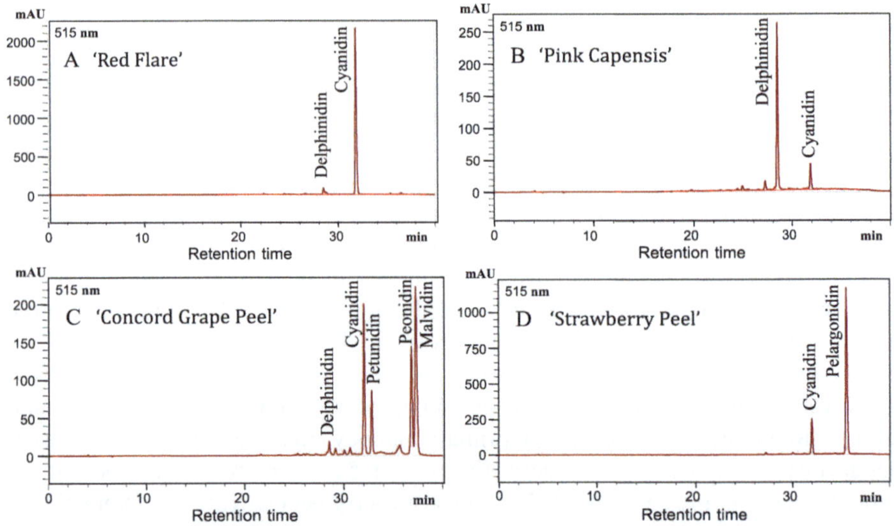

Fig. 6.5 HPLC chromatograms absorbed at 515 nm of acid-hydrolysed anthocyanins of *Nymphaea* 'Red Flare' (**a**), *Nymphaea* 'Pink Capensis' (**b**), compared to the known references from Concord grape peel (**c**) and strawberry peel (**d**)

compounds appeared in the anthocyanin chromatogram patterns, accounting for approximately 70% of the total anthocyanins (Khunmuang 2008).

Based on stability properties, anthocyanins can be spectrophotometrically classified into 2 types: typical and rare anthocyanins, depending on conditions of co-pigments, metal ions, and pH. Typical anthocyanins, commonly found in many agricultural commodities, are most stable and highly absorbed in a pH 1.0 buffer, but they degrade quickly in a buffer at pH 4.5 (Lee et al. 2005), as demonstrated by the pH-different spectral patterns in red rose petal extracts (Fig. 6.6a). In contrast, rare anthocyanins with polyacylated conjugation show high stability under various conditions, in particular across pH differences, compared to typical anthocyanins. Polyacylated anthocyanins occur in a limited number of agricultural flowers, especially in some tropical flowers, such as orchids, waterlilies, *Tradescantia pallida*, and butterfly peas. The blue butterfly pea (*Clitoria ternatea*) petals contained

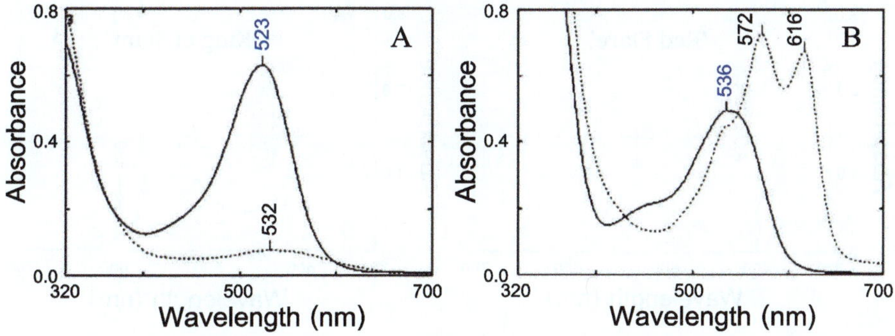

Fig. 6.6 Spectra absorbance patterns between 320 and 700 nm of anthocyanin extracts of red rose petals (**a**- left), and butterfly pea petals (**b**- right) when incubated in buffer pH 1.0 (bold lines) and pH 4.5 (dashed lines)

substantial amounts of anthocyanins, along with flavonoid compounds such as p-coumaric acid and ferulic acid. Anthocyanins in *C. ternatea* demonstrate high stability at pH 4.5, exhibiting a unique spectral behaviour. Rather than a disappearance of the maximum absorbance peak, the peak shifts to a higher absorbance, and is accompanied by an additional peak (at 572 and 616 nm) (Fig. 6.6b). This distinctive pattern disappears after saponification of the anthocyanins, highlighting polyacylation as a critical factor in the stability of *C. ternatea* anthocyanins (Wongs-Aree et al. 2006). Anthocyanins play an important role in floral colouration in *Vanda* spp. All anthocyanin extracts from 3 species within the Vandeae tribe show the light-absorption patterns of polyacylated anthocyanins, which are more stable in different pH buffers, compared to typical anthocyanins (Junka et al. 2007; Junka et al. 2012). In waterlilies, although the maximum anthocyanin absorbance of blue-shade petals, 'King of Siam' and 'Blue Capensis,' sharply decreases in buffer pH 4.5, they exhibit peak shifts and additional spectral peaks (Figs. 6.7b, d). This behaviour is different from that observed in red-shade cultivars such as 'Red Flare' (Fig. 6.7a) and 'Red Capensis' (Fig. 6.7c).

Betalains exhibit strong free radical scavenging properties (Kanner et al. 2001). These nitrogen-containing compounds, derived from tryptophan, serve as an important antioxidant and an attractive pigment in plants, imparting red to red-violet colours to flower petals. Betalains are water-soluble, nitrogen-containing compounds. Since betalains are derived from tyrosine, an amino acid common to all organisms, the presence of nitrogen in their molecule is characteristic. Betalains differ structurally in their sidechain groups, which may include amino acids or amines, and are categorised into betaxanthins (yellow) and betacyanins (red to violet). Flowers containing betalains include four O'clock plant (*Mirabilis jalapa*), *Celosia* spp., *Gomphrena* spp., *Portulaca grandiflora* (Portulacaceae), *Mirabilis jalapa* (Nyctaginaceae), *Bougainvillea glabra* (Nyctaginaceae), and *Gomphrena globosa* (Amaranthaceae) (Kumari et al. 2017; Jadhav and Bhujbal 2020).

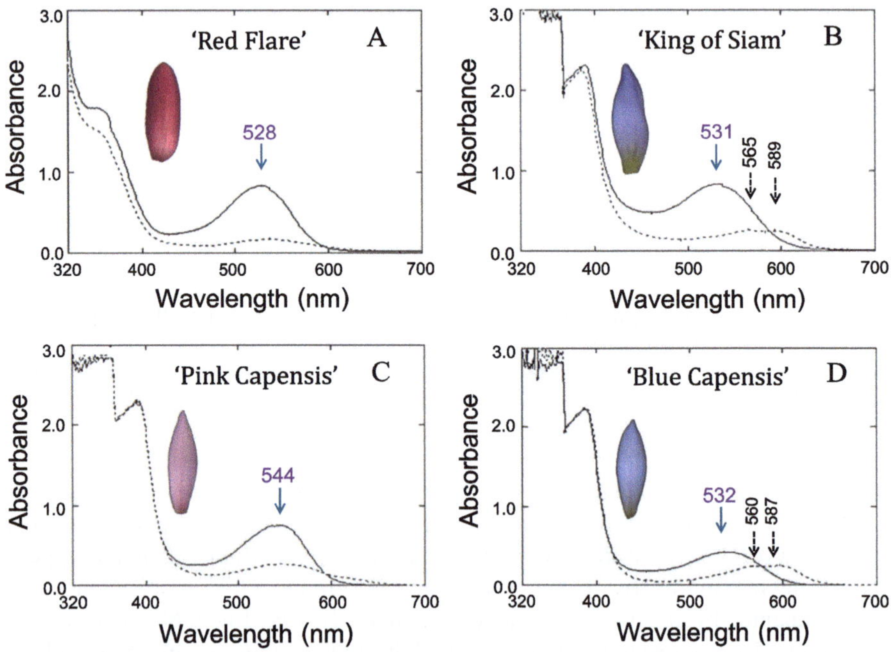

Fig. 6.7 Spectra absorbance patterns between 320 and 700 nm of petal extracts of *Nymphaea* 'Red Flare' (**a**), 'King of Siam' (**b**), 'Pink Capensis' (**c**), and 'Blue Capensis' (**d**) when incubated in buffer pH 1.0 (bold lines) and pH 4.5 (dashed lines)

Fig. 6.8 Visual appearance of open *Vanda* florets of pale-mauve (*V. teres* × *V. hookerina*) (**a**) and white (*V. teres* var. alba × *V. Cooperi* var. 'White Wing') (**b**) varieties

6.2.4 White Petals of Flowers

Flowers with white petals, unlike other flowers with coloured petals, lack pigmentation. This may be due to either the presence of flavones in the tissue (Stickland 1976) or the absence or reduction of the relevant pigment contents (Junka et al. 2011; Kazuma et al. 2003). Pale-mauve petals of *Vanda teres* × *V. hookerina* (Fig. 6.8a) contain higher levels of anthocyanins and antioxidants, compared with a white hybrid, *V. teres* var. alba × *V. Cooperi* var. 'White Wing' (Fig. 6.8b), in which anthocyanins almost disappear due mainly to significantly low expression of the flavanone 3-hydroxylase (*Va-F3H1*) gene involved in anthocyanin biosynthesis (Junka et al. 2011). Conversely, when fully bloomed Japanese chrysanthemum flowers of the wild-type 'Jimba' develop pure white petals, their immature buds initially display yellowish petals with the presence of carotenoids. The yellow carotenoid molecules are degraded by CCD (Carotenoid Cleavage Dioxygenase) enzyme at the full blooming stage. The CCD4-silencing transgenic plants are approved to restore the bright yellow petal colouration of full blooms (Ohmiya 2009).

6.3 Bioactive Compounds and Nutraceuticals in Edible Flowers

Antioxidants, free radical scavenging, and anti-ageing activities are strongly linked to the prevention and care of chronic illnesses such as cardiovascular diseases (Vivekananthan et al. 2003) and cancer (Greenlee et al. 2012). Phenolic compounds represent a particularly main group of phytochemicals, consisting of more than 10,000 compounds (Haslam 1998). Many of these compounds function as important antioxidants that effectively reduce oxidative stress and act as chemo-preventive agents (Thomasset et al. 2007). Therefore, several studies focus on the concentrations of phytochemicals, *in vitro* antioxidative properties [such as ($ABTS^{+\bullet}$ (2,2′-azino-bis(3-ethylbenzothiazoline-6-sulfonic acid) radical cation decolourisation, ORAC (Oxygen radical absorbance capacity), DPPH (2,2-Diphenyl-1-picryhydrazyl), and FRAP (Ferric-reducing antioxidant power)] as well as nutraceutical properties associated with the prevention of chronic diseases, including heart disease, cancer, diabetes, and Alzheimer's disease. However, the content of bioactive compounds varies considerably and depends on plant species, cultivar, genotype, growing conditions, harvesting time, food, and processing technologies (Naczk and Shahidi 2004).

6.3.1 Antioxidant Capacities and Bioactive Compounds

Flowers contain a great variety of natural antioxidants, such as phenolic acids (hydroxycinnamic and hydroxybenzoic acids), flavonoids (anthocyanin, chalcone, flavanones, flavones, and flavonols), and phenolic compounds, which are the majority and have a high bioactive potential in edible flowers and are extremely important for their inclusion in functional foods (Kaur et al. 2006; Youwei et al. 2008; Zheng et al. 2019; Navarro-González et al. 2015). Several factors have direct effects on the profile of bioactive compounds in flowers, including the variation in colours among cultivars of the same species, the preharvest conditions (soil and climate), the production system, the flowering stage, or even the floral organs (Fernandes et al. 2018). The bioactive compounds (total sugar, anthocyanin, phenolic contents, vitamin C, activities of DPPH and FRAP) in orange and pink flowers of *Zinnia* (*Zinnia violacea* Cav.) are rather similar, but the pink flowers expressed higher levels of DPPH than the orange flowers but lower content of anthocyanin. In *Torenia flowers* (Torenia fournieri Linden. ex E. Fourn), it was found that the pink flower showed 5-time higher content of anthocyanin than the violet *Torenia* but lower activity of FRAP assay and vitamin C content, as shown in Table 6.3 (Buanong et al. 2025).

In waterlilies, 'Red Flare' petals comprised the highest phenolic content of 1318.5 GAE mg/100 g FW, whereas 'Blue Capensis' petals contained the lowest, at 616.4 GAE mg/100 g FW (Fig. 6.9a). Antioxidant capacities as assessed by FRAP (Fig. 6.9b) and DPPH (Fig. 6.9c) were the highest in petal extracts of 'Red Flare'. When compared to 200 mg/L butylated hydroxytoluene (BHT), a standard antioxidant, the DPPH antioxidant capacity of 'Red Flare' petal extracts was equivalent to 500 mg/L, while 'Red Flare' leaf extracts demonstrated an antioxidant capacity comparable to 1000 mg/L (Wongs-Aree et al. 2017).

6.3.2 Nutraceutical Properties

The nutraceutical properties and antioxidant capacities have been extensively analysed in varied edible flowers, demonstrating positive effects on human health. It is not only the floral tissue found to have higher antioxidant powers before ingestion but also after digestion processes, underlining the prolonged bioactive effects in various phytochemicals (Chen et al. 2015). Also, flowers with high contents of phenolic compounds and high antioxidant capacities can mitigate the oxidative stress induced by various pathological conditions (Xiong et al. 2014; Lu et al. 2016). Increased phenolic consumption involves an anti-inflammatory activity and reduces a risk of cardiovascular disease and cancers (Hamaguchi et al. 2009; Kaz Owska et al. 2010; Oueslati et al. 2012). Kaisoon et al. (2012) studied the antioxidant properties of 4 varieties of edible flowers in Thailand [*Tagetes erecta*, *Cosmos sulphureus*, *Antigonon leptopus* and *Bougainvillea glabra]* and found that *T. erecta* extract contained 3- to 4-fold higher level of phenolic compounds than the other flowers and exhibited the highest activity of FRAP assay. Flowers of *T. erecta*, *A. leptopus* and *B. glabra* showed a high anti-proliferative effect on the human colon cancer lines HT-29, gastric adenocarcinoma AGS and bladder BL-13. The anti-proliferative effect was dose-dependent and mostly visible against the HT-29 cell line. In *Coreopsis* flowers, the analysis of phytochemical capacities showed approximately 591.60 ± 44.44 DPPH, 2364.20 ± 267.11 FRAP and 1974.87 ± 123.12 ABTS assays, while the ORAC assay yielded a higher antioxidant capacity of 5807.70 ± 544.51 μmol TE/g than other analysis methods. These flowers contained 1.028 ± 0.10 mg/g FW of phenolic compounds and demonstrated nutraceutical properties, showing a wide range of medicinal properties against specific pathologies such as hypertension, Alzheimer's and ageing. The inhibition of ACE, AChE and tyrosinase was recorded at 98.43 ± 0.09, 35.28 ± 1.10 and 55.12 ± 5.77%,

Table 6.3 Antioxidant properties of some edible flowers

Name	Colour	Total sugar content (mg/g FW)	Anthocyanin content (OD_{530}/g FW)	DPPH (mM Trolox/g FW)	FRAP (mg Ascorbic acid/g FW)	Phenolic compound (mg Gallic acid/g FW)	Vitamin C (mg Ascorbic acid/g FW)
Zinnia	Orange	1.46 ± 0.11	2.01 ± 0.12	1.62 ± 0.17	5.55 ± 0.35	31.37 ± 0.54	0.44 ± 0.04
	Pink	1.24 ± 0.26	1.74 ± 0.14	3.66 ± 0.48	4.68 ± 0.21	30.67 ± 0.98	0.40 ± 0.03
Torenia	Violet	2.86 ± 0.10	2.53 ± 0.40	1.79 ± 0.27	5.65 ± 0.19	31.29 ± 0.65	1.46 ± 0.17
	Pink	2.24 ± 0.22	13.25 ± 2.12	1.14 ± 0.21	4.16 ± 0.06	30.07 ± 0.92	0.37 ± 0.04

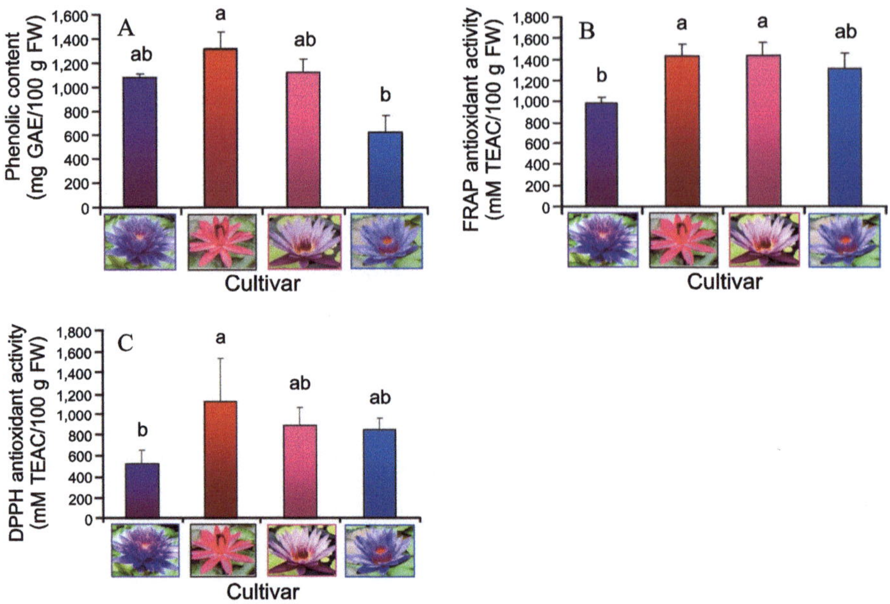

Fig. 6.9 Phenolics content (**a**), FRAP antioxidant activity (**b**), and DPPH antioxidant activity (**c**) of extracts from 4 waterlily petals: *Nymphaea* 'King of Siam', 'Red Flare', 'Pink Capensis', and 'Blue Capensis'

Table 6.4 The antioxidant activities and some nutraceutical properties in *Coreopsis tinctoria* flower after harvest

Name	Antioxidant capacities (µmol TE/g DW)			
Coreopsis tinctoria Nutt	DPPH radical scavenging assay	FRAP assay	ABTS assay	ORAC assay
	591.60±44.44	2364.20±267.11	1974.87±123.12	5807.70±544.51
	Enzyme inhibition IC_{50} (mg/ml) *			
	Hypertension: ACE inhibition (%)	Alzheimer: AChE Inhibition (%)	Anti-aging: tyrosinase inhibition (%)	Phenolic compound (mg/ g FW)
	98.43±0.09	35.28±1.10	55.12±5.77	1.028±0.10

*Inhibition of enzyme activities at sample concentration of 4.17 mg/mL

respectively (Table 6.4) (Buanong et al., 2025). Benvenuti et al. (2016) reported that most of the nutraceutical features are often due to colour of flower cultivars, strongly linked to their pigmentation. In the red varieties of *Viola × wittrockiana*, *Dianthus × barbatus*, *Pelargonium peltatum*, the high anthocyanin content was closely related to a high antioxidant activity. Similarly, red flowers of *Althaea officinalis* showed higher antioxidant activity than other cultivars with lighter-coloured cultivars. The activity slightly decreased in the pink-coloured varieties and then fell to white flowers (Sadighara et al. 2012).

Neurodegenerative illness is strongly associated with ageing. Alzheimer's disease, a general term for memory loss and cognitive damage, results from progressive and irreversible neurodegeneration and the depletion of acetylcholine levels in the brain (Coin et al. 2016). Acetylcholinesterase enzyme (AChE) facilitates the physiological cleavage of acetylcholine, causing cognitive impairment. Inhibition of AChE enzyme is a therapeutic strategy employed in some drugs (donepezil, memantine, rivastigmine, galantamine, etc.), which have recently been approved and used to control the progression of Alzheimer's disease by monotherapy or combined therapy (Birks and Harvey 2018; Takahashi et al. 2019; Tricco et al. 2018). Additionally, consuming foods with high antioxidant and anti-AChE activities can also delay Alzheimer's disease progression, inhibit AChE, and protect neurones from oxidative damage. It has been found that anti-AChE activity in flowers is related to the presence of alkaloids, terpenes, coumarins, flavonoids, and phenolic compounds (Olennikov et al. 2017). Edible flowers bearing anti-AChE properties may serve as supplemental therapies or provide a foundation for new drug development for Alzheimer's patients. Rezende et al. (2019) studied nutritional and functional values in five species of edible flowers: orange hibiscus (*Hibiscus sabdariffa*), carnation (*Dianthus caryophyllus*), sunflower (*Helianthus annuus*), purple and pink violet (*Saintpaulia ionantha*). According to the report, the purple and pink violet flowers showed efficiency in inhibiting AChE enzyme, while the orange hibiscus had high concentrations of phenolic compounds and the most active antioxidant activities in FRAP, DPPH, and ABTS assays, along with inhibiting acetylcholinesterase. Moreover, flower extracts from *Hydrocotyle umbellata* L. and *Mentha* spp. (*Mentha longifolia* L. Huds, *Mentha arvensis* L., *Mentha gentilis* L., *Mentha piperita* var. officinalis Sole, and *Mentha spicata* L.) have also been effective in inhibiting AChE (Hamdy et al. 2018).

Hypertension is a common progressive disorder resulting in several chronic diseases, such as cardiovascular disease, stroke, renal disease, and diabetes, which are often influenced by lifestyle and food consumption habits. Hypertension results from an imbalance in the renin-angiotensin system (RAAS) in which angiotensin-I conversion enzyme (ACE) is the first rate-limiting step to form angiotensin II (Nehme and Zibara 2017). Angiotensin-II conversion enzyme, the second rate-limiting step, accelerates the contraction of vascular smooth muscles and increases extracellular fluid volumes, leading to elevated blood pressure (Forrester et al. 2018). Therefore, the inhibition of ACE is a promising approach for controlling over-expression of RAAS (Balasuriya and Rupasinghe 2011). Some pharmacological studies reported that ACE inhibitors derived from plant metabolites (mostly belonging to the groups of peptides, protein hydrolysates, phenolics, flavonoids, terpenoids, and alkaloids) have the potential to regulate hypertension and other anomalies related to diabetes (Chakraborty and Roy 2021). Flavonoids extracted from the buds of *Rosa damascene* show a high cardiovascular protective activity (Kwon et al. 2018). Therefore, flavonoids have the potential to inhibit ACE *in vitro* with inhibitory property influenced by the types of sugar moiety attached at C-3 position of the flavylium structure (Balasuriya and Rupasinghe 2011).

Ageing is an irreversible and complex process related to the progressive loss of cellular functions and systems. Skin ageing can be divided into the intrinsic ageing process (natural ageing) and diverse extrinsic processes, induced by environmental factors, especially ultraviolet (UV) radiation, eventually increasing wrinkles, elastosis, laxity, aberrant pigmentation, etc. (Chen et al. 2020). The mechanism of skin ageing may regard the generation of reactive oxygen species (ROS) and oxidative stress (Gonzaga 2009; Lephart 2016). These factors stimulate the upregulation of matrix metalloproteinase (MMPs) and elastase, leading to the degradation of collagen and elastin within the dermal matrix of the skin (Tulah and Birch-Machin 2013; Li et al. 2019). Tyrosinase is the rate-limiting enzyme that controls the production of melanin. Over-production of melanin causes skin hyperpigmentation (Seo et al. 2003). Therefore, the development of antioxidants and tyrosinase inhibitors can reduce melanin biosynthesis and achieve skin whitening, effectively solving pigmentation disorders (Lall and Kishore 2014). Previous studies found that natural compounds and herbal or edible resources' extracts could act as promising candidates for anti-ageing interventions (Argyropoulou et al. 2013). For example, *R. Chinnensis* flowers cv. 'JinBian', a major raw material for rose tea in China, have demonstrated strong tyrosinase inhibitory activities and possess potential application values in the formulation of natural anti-ageing, skin-whitening, and antibacterial products (Li et al. 2021).

Inflammation is a physiological defence response in a host tissue, triggered by infections, chemical exposure, tissue damage, microbial infection, etc. (Lu et al. 2016; Kim et al. 2013; Nahar et al. 2014). Also, it can typically be induced by tissue injury or trauma, which occurs without the intervention of pathogens. For example, excessive production of inflammatory cytokines (Oh et al. 2013) and inflammatory mediators (An et al. 2011) can endanger tissues and organisms, resulting in various inflammation-related diseases, such as rheumatoid arthritis, diabetes, inflammatory bowel disease, atherosclerosis, and cancer (Shao et al. 2013). However, the process of inflammation usually leads to recovery from infection and healing (Nathan 2002). Previous studies reported that some pigments such as flavonoids, anthocyanins, and phenolic acids, are responsible for anti-inflammatory properties (Beltrán-Debón et al. 2010). For example, *Moringa oleifera* flower extract is a good source of natural antioxidants such as tannins, alkaloids, flavonoids, cardiac glycosides, and phenolic compounds. The anti-inflammatory activity is also significant and comparable with the standard drug diclofenac sodium, as it prevents protein denaturation (Alhakmani et al. 2013). In *Sesbania grandiflora* L. Fabaceae, flavonoids can suppress the formation of edema associated with inflammation (Loganayaki et al. 2012). Moreover, the anti-inflammation properties of *Hibiscus sabdariffa* extracts were evaluated using a cell modelling system. It was found that *Hibiscus* flower has a prospective anti-inflammatory effect at 500 mg/kg of body weight (El Bayani et al. 2018).

Diabetes mellitus (DM) is a non-communicable chronic disease classified into two main types: type I and type II. Type II diabetes is a non-insulin-dependent condition resulting from the body's ineffective use of insulin (Hawash et al. 2020). The long-term hyperglycaemia results in micro vascular complications (retinopathy, nephropathy, and neuropathy) and macro vascular complications (cardiovascular events and stroke) (Tahergorabi and Khazaei 2012). Pancreatic α-amylase hydrolyses large, insoluble starch molecules into absorbable forms, then intestinal α-glucosidase catalyses the last digestion step of starch and disaccharides abundant in the human diet. To reduce the levels of postprandial hyperglycaemia, inhibiting digestive enzymes such as α-amylase and α-glucosidase could be used to reduce glucose absorption by suppressing carbohydrate digestion (Loizzo et al. 2016; Hawash et al. 2020). Phenolic compounds are associated with enzyme inhibitory activity against α-glucosidase and α-amylase enzymes, which become potential as effective hypoglycaemic agents (Adisakwattana et al. 2009; Heo et al. 2009; McDougall et al. 2005; Lee et al. 2010). A study by Nowicka and Wojdyło (2019) showed a positive correlation between the content of polymeric procyanidins and the ability of hawthorn, primrose, and *Tilia cordata* flowers to inhibit α-glucosidase. Additionally, Loizzo et al. (2016) studied the hypoglycaemic effects via the inhibition of α-amylase and α-glucosidase in *Malva sylvestris* flower and found that the extract effectively inhibited both enzymes.

Petal extracts of 4 waterlily cultivars (*Nymphaea* 'Red Flare', *N.* 'King of Siam', *N. capensis* 'Pink Capensis', and *N. capensis* 'Blue Capensis') were *in vitro* studied for their inhibitory effects against some pathogenic bacterial cells and cancer cells (Yimyong 2010). From monitoring Minimum Inhibitory Concentrations (MICs), the extract from 'Red Flare' exhibited the strongest inhibitory effects on pathogenic bacterial growth (Table 6.5). Waterlily extracts show potential as disinfectants for nosocomial infections in hospitals since they effectively inhibit *Acinetobacter* spp. and *Staphylococcus* spp. (Antunes et al. 2014; Lax and Gilbert 2015). Although they conducted fewer inhibitory effects on foodborne bacteria such as *Bacillus* spp., *Escherichia coli*, and *Salmonella* spp., waterlily petals as edible flowers could provide reinforcing health benefits for consumers. Furthermore, 0.625 mg/mL and 1.25 mg/mL 'Red Flare' extracts demonstrated *in vitro* inhibition of breast cancer cells (MCF-7 ATCC HTB-22) and small lung cancer cells (NCI-H187 ATCC CRL-5804), respectively. Extracts from 'King of Siam' and 'Capensis Blue' petals at 2.5 mg/mL inhibited small lung cancer cells. However, none of the waterlily extracts exhibited an inhibitory effect (>2.5 mg/mL) on oral cancer cells (KB ATCC CCL-17) (Yimyong 2010).

6.3.3 Poisonous Flowers

Certain secondary metabolites, such as alkaloids found in flowers, can contribute to nutraceutical ability. However, in some cases or when consumed in excessive amounts, these chemicals may pose health risks to humans. A list of some poisonous flowers and their associated toxicities is shown in Table 6.6.

Table 6.5 Antimicrobials activities of some pathogenic bacteria by minimum inhibitory concentrations (MICs) in 4 waterlily petal extracts

Microbials	King of Siam	Red Flare	Pink Capensis	Blue Capensis
1. *Acinetobacter antritratus* DMST 4183	NI	250	NI	NI
2. *Acinetobacter baumanni* ATCC 19066	NI	125	NI	NI
3. *Acinetobacter calcoaceticus* ATCC 23055	NI	125	NI	NI
4. *Acinetobacter lwoffii* ATCC15309	125	31	125	500
5. *Bacillus cereus* ATCC 11778	125	125	125	500
6. *Bacillus subtilis* ATCC 6633	NI	NI	NI	NI
7. *Escherichia coli* ATCC 25922	500	NI	NI	NI
8. *Escherichia coli* O157.H7 DMST12743	NI	NI	NI	NI
9. *Pseudomonas aeruginosa* ATCC 27853	NI	250	NI	1000
10. *Psedumonas fluorescens* DMST 6034	NI	NI	NI	NI
11. *Salmonella enteritidis* ATCC 17368	NI	NI	NI	NI
12. *Salmonella typhi* DMST 5784	NI	NI	NI	NI
13. *Salmonella dysenteriae* DMST 15111	250	31	125	500
14. *Staphylococcus aureus* ATCC 25923	1000	125	500	1000
15. *Staphylococcus epidermidis* ATCC 12228	1000	31	125	500
16. *Streptococcus pyogenes* DMST 17020	250	125	125	250

Modified from Yimyong (2010)
NI no inhibition effect at the concentration of 1000 μg/mL

6.4 Industrial Food Exploration

Due to the diverse characteristics of edible flowers, varying in shape, colour, brightness, crown thickness, taste, and even nutrient content, they have been widely used to improve the sensory and nutritional qualities of food. These flowers contribute to culinary preparations by adding colour, flavour, taste and visual appeal to dishes, beverages, garnishes, baked goods, and salads (Koike et al. 2015). Chen and Wei (2017) reported that three main factors (specific curiosity, aroma and health consciousness) affect consumer attitudes towards the consumption of edible flowers. The vivid pigmentation of flowers reflects the presence of phenolic acids and flavonoids. Anthocyanins, for instance, offer significant health benefits due to their antioxidant, anti-inflammatory, anti-cancer, and anti-diabetic properties, making them greatly valuable in the food industry (Rodriguez-Amaya 2016). Benvenuti et al. (2016) reported that among 12 species of edible flowers regarding organoleptic performance, *Trapaeolum majus*, *Ageratum houstonianum* and *Viola × wittrockiana* flowers showed a high biodiversity of sensory profiles. The overlap between

Table 6.6 Some poisonous flowers of ornamental plants

Common name	Scientific name	Toxicity	Reference
Jimson Weed, Devil's trumpet, Moonflower	*Datura* spp.	The tropane alkaloids scopolamine, hyoscyamine, and atropine can lead to nausea, vomiting and hallucinations.	Kerchner and Farkas (2019)
Angel's trumpet, Brugmansia, Datura	*Brugmansia* spp.	Atropine, hyoscyamine, and scopolamine in the flowers can lead to nausea, vomiting and hallucinations.	Kerchner and Farkas (2019)
Lantana	*Lantana camara*	Methylcoumarin, salicylic acid, and lantadene A and B are poisonous to children, dogs, cats, rabbits, etc.	Sharma et al. (2007)
Mountain Laurel, Spoonwood	*Kalmia latifolia*	After consumption, diterpene grayanotoxin may cause dizziness, hypotension and atrial-ventricular block.	Jansen et al. (2012)
Lily of the Valley	*Convallaria majalis*	A saponin called convallotoxin, the primary cardiac glycoside, causes dizziness, vomiting, and diarrhoea.	Morimoto et al. (2012)
Foxglove	*Digitalis* spp.	It contains a significant toxicity of digitoxin (a cardiac glycoside), causing nausea, convulsions, delirium, and in severe cases, an arrhythmic heartbeat.	Lin et al. (2010)
Wisteria	*Wisteria frutescens*	Wisteria flowers contain a saponin known as Wisterin, causing gastroenteritis, nausea, frequent vomiting, abdominal pain and diarrhoea.	Slattery (2005)
Oleander	*Nerium oleander* L.	Eating only a small part of the flowers containing cardiac glycosides causes digitalis toxicity of drowsiness, slowed heart rate, and shaking.	Bandara et al. (2010)
Rhododendron	*Rhododendron ferrugineum* L.	Grayanotoxin in the flowers could lead to abdominal pain, difficulty breathing, dizziness, hypotension and atrial-ventricular block.	Jansen et al. (2012)
Water Hemlock	*Cicuta* spp.	White hemlock's toxin (cicutoxin) attacks the central nervous system, causing nausea, vomiting, diarrhoea, renal failure, and respiratory impairment.	Schep et al. (2009)
Aconite, Monkshood, Wolfs' bane, Devil's Helmet	*Aconitum napellus*	Aconitine, a bioactive alkaloid, could induce severe arrhythmia and neurotoxicity.	Gao et al. (2020)
Anthurium	*Anthurium* spp.	Flowers contain calcium oxalate crystals and are considered poisonous, resulting in painful burning of the lips, mouth, tongue, and throat. Sometimes acute inflammatory reactions, including blistering and swelling.	de Alcantara Guimarães et al. (2019)

(continued)

Table 6.6 (continued)

Common name	Scientific name	Toxicity	Reference
Colchicum autumnal, Autumn crocus	*Colchicum autumnale*	Toxic alkaloid colchicine can cause general gastrointestinal upset, including vomiting and diarrhoea.	Kyan et al. (2015)
Delphinium, Larkspur	*Delphinium* spp.	Larkspur containing diterpenoid alkaloids of lycotonine derivatives causes laboured breathing, rapid and irregular heartbeat, muscular weakness, and collapse. Contacting Delphinium flowers causes skin irritation or allergic reaction.	Pfister et al. (1999)
Birthwort	*Aristolochia* spp.	Aristolochic acid in the flowers causes the side effect of kidney failure.	Zhang et al. (2022)

nutraceutical value and organoleptic aspects suggests that promising edible flower species could be introduced to a potential market targeting novel foods that satisfy both taste and health. The flowers of *A. houstorianum*, *B. semperflorens*, *V. vittrockiana* and *T. majus* are particularly attractive, while meeting health requirements due to their high antioxidant levels (Friedman et al. 2005). Cultivars showing the most intensely coloured flowers, especially red and blue, become the most suited for gastronomic and nutraceutical innovation in food development. The food industrial sector has investigated the possibilities for introducing bioactive compounds from edible flowers as new foodstuffs, either by directly adding intact floral parts (petals, stems, sepals) or by incorporating their extracts or essential oils (Zheng et al. 2021). For instance, the application of flower extracts such as rose syrup and marigold powder to nutricereals and milk-based functional beverages significantly improved consumer acceptability, phytochemical profile and storage stability (Kumar et al. 2020a). A previous study from Kumar et al. (2020b) revealed that among 4 drinks (sweetened plain milk, functional drink, rose-flavoured drink and marigold-flavoured drink) cultured with 1% *Lactobacillus acidophilus* NCDC14, 5% *L. rhamnosus* NCDC347 and 5% *L. casei* NCDC297, the rose-flavoured drink cultured with *L. acidophilus* had the most desirable sensory properties. This result indicates that the products containing probiotics, millets, oats, milk, and flowers have the potential to be marketed as health food. These findings reinforce the impact of edible flowers as functional ingredients, given their beneficial phytochemical profiles and disease-preventive properties. Now, different varieties of edible flowers are usually used by gourmet chefs in their culinary dishes and are commercially available at specialised sale points worldwide.

6.4.1 Refreshing Drinks

Many edible flowers have great potential for processing into dried teas. Dried petals of flowers can be used for preparing antioxidant-rich, caffeine-free beverages, either individually or in combination with other herbal ingredients. Different drying methods: sun drying, hot-air drying, and freeze-drying, were evaluated for their effects

Fig. 6.10 Commercially dried blue *Nymphaea caerulea* petals (Left), and hot air-dried *Nymphaea* 'King of Siam' petals (Right)

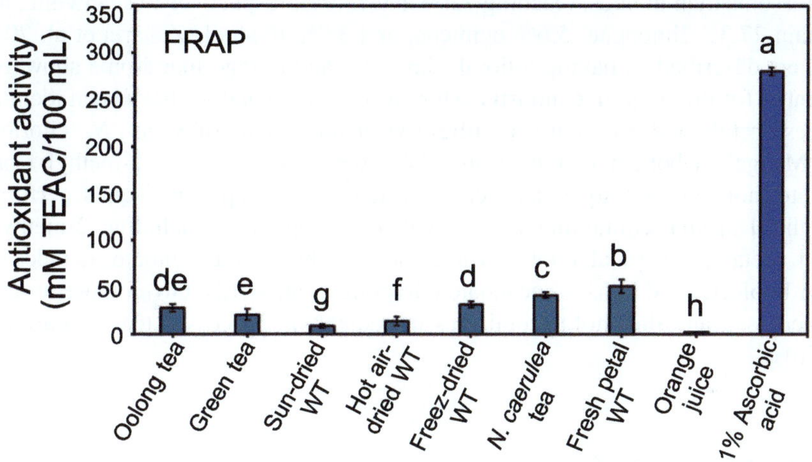

Fig. 6.11 Antioxidant activity of tea drink made from 'King of Siam' waterlily petals (Wongs-Aree et al. 2017)

on dried waterlily petals of 'King of Siam', in comparison with dried blue *Nymphaea caerulea* petals, a commercial product (Fig. 6.10). The FRAP antioxidant activity of tea drinks prepared using freeze dehydration was significantly higher than that of teas produced by other drying methods as well as the control (green tea) (Wongs-Aree et al. 2017) (Fig. 6.11). Moreover, the antioxidant ability of hot water teas made from air-dried petals of 12 selected rose cultivars investigated, using green tea as a reference antioxidant-rich drink. Rose petal teas from different cultivars exhibited ABTS scavenging capacities ranging from 713 to 1771 μM Trolox equivalents (TE)/g dry weight (DW), compared with 1228 μM TE /g DW in green tea. Total phenolic content in rose teas ranged from 50.7 to 119.5 mg gallic acid equivalents (GAE) /g DW , compared with 62.1 mg GAE/g DW in green tea. Rose teas were particularly rich in free gallic acid with the highest values of antioxidant activity, total phenols, and gallic acid contents, found in the cultivars 'San Francisco', 'Katharina Zeimet', and 'Mercedes' (Vinokur et al. 2006). Besides, Liamkaew et al. (2020) focused on sensory evaluation and consumer acceptability studies on flower beverages made from damask rose and water lily flowers. The products were found to be nutritionally acceptable and well received by consumers.

6.4.2 Flowering Flavour

Some fragrant flowers serve various purposes, including decoration, production of aromatic teas, and applications in the fragrance industry. *Citrus grandis* buds and blooms principally include mono- and sesquiterpenes, including β-myrcene, limonene, ocimene, linalool and caryophyllene. Ocimene and linalool levels are higher in blossoms than in buds, reaching 7.4% and 15.9%, respectively. Meanwhile, buds contain 27.3% limonene, 5.6% ocimene, and 3.0% linalool (Zakaria et al. 2010). Flowers described as having a floral-, jasmine- and orange-like aroma show great potential for the fragrance industry. Moreover, aroma volatiles have been identified in fresh petals of 4 different waterlilies (*Nymphaea* 'King of Siam', *N.* 'Gloriosa', *N.* 'Mangala Ubol', and 'Pink Capensis'). Across all cultivars, 2-methyl-1-butyl acetate, nonane, and squalene were detected. 'Pink Capensis' petals perform a delightful aroma, containing diverse volatile compounds, including 2-methyl-1-butyl acetate, benzyl alcohol, germacrene D, ethyl hexadecanoate, octadecanal, ethyl linoleate, and benzyl decanoate (Jirapong et al. 2012). Given these aromatic properties, waterlilies hold promise as ingredients in processed teas and aromatic additives.

6.4.3 Natural Colourants

Melastoma malabathricum (Indian, or Singapore rhododendron), a shrub belonging to the family Melastomataceae, comes with beautiful pink flowers and is widely distributed across the world. The vibrant pink flower makes it a valuable ornamental plant and a potential source for natural colourant extraction (Joffry et al. 2012). Similar to butterfly pea (*Clitoria ternatea*), extracts from the blue petals have been locally used as food and cosmetic colourant additives in Southeast Asian countries due to their high pigment content and stability (Wongs-Aree et al. 2006; Vidana Gamage et al. 2021). Edible flower extracts from rose, dahlia, and cornflower, which are potential substitutes for E163, a commercial anthocyanin-based colourant, have been added to yoghurts (Pires et al. 2018). Among these, enriched anthocyanin extracts from *Rosa damascene* demonstrated the most suitable alternative to E163, leading to the incorporation of rose petals in yoghurt production (Chanukya and Rastogi 2016). Betacyanins promote pigmentation in shades of intense violet in plants (Gonçalves et al. 2015). The addition of natural betalain colouring from *Gomphrena globosa* L. to ice creams, in comparison with beetroot extracts and commercial betalains, showed similar effects in terms of nutritional composition, colour, individual sugars and fatty acid profiles throughout storage time (Roriz et al. 2018). Further studies have evaluated the application of *Impatiens balsamina* L. flower extract as a natural colouring ingredient in pastry products (Karra et al. 2020; Pirez Jr. et al. 2021; López et al. 2019). These flowers hold significant promise as natural colourants, presenting a sustainable alternative to the use of synthetic dyes in food applications (Giusti and Wrolstad 2003).

6.5 Conclusion

The trend of consuming novel and functional foods with health benefits has significantly increased in modern gastronomy, leading to their incorporation as ingredients in various dishes, such as salads, desserts, and appetisers. Edible flowers serve as excellent sources of phytochemicals, including pigments, flavonoids, and phenolic compounds. These bioactive compounds provide numerous health benefits, including antioxidant, anti-inflammatory, anticancer, anti-diabetic, cardioprotective, and genoprotective properties. Therefore, edible flowers hold strong potential for inclusion in commercially produced foods and pharmaceuticals, contributing to the development of innovative nutraceuticals that offer potential health advantages to consumers.

References

Adisakwattana S, Chantarasinlapin P, Thammarat H et al (2009) A series of cinnamic acid derivatives and their inhibitory activity on intestinal α-glucosidase. J Enzyme Inhib Med Chem 24:1194–1200. https://doi.org/10.1080/14756360902779326

Alhakmani F, Kumar S, Khan SA (2013) Estimation of total phenolic content, *in-vitro* antioxidant and anti-inflammatory activity of flowers of *Moringa oleifera*. Asian Pac J Trop Biomed 3(8):623–627. https://doi.org/10.1016/S2221-1691(13)60126-4

Amamiya K, Iwashina T (2016) Qualitative and quantitative analysis of flower pigments in chocolate cosmos, *Cosmos atrosanguineus*, and its hydrids. Nat Prod Commun 11:77–78. https://doi.org/10.1177/1934578X1601100122

An HJ, Kim IT, Park HJ, Kim HM, Choi JH, Lee KT (2011) Tormentic acid, a triterpenoid saponin, isolated from *Rosa rugosa*, inhibited LPS-induced iNOS, COX-2, and TNF-α expression through inactivation of the nuclear factor-κb pathway in RAW 264.7 macrophages. Int Immunopharmacol 11:504–510

Antunes LC, Visca P, Towner KJ (2014) *Acinetobacter baumannii*: evolution of a global pathogen. Pathog Dis 71(3):292–301. https://doi.org/10.1111/2049-632X.12125

Argyropoulou A, Aligiannis N, Trougakos IP, Skaltsounis A-L (2013) Natural compounds with anti-ageing activity. Nat Prod Rep 30:1412–1437. https://doi.org/10.1039/C3NP70031C

Arunkumar R, Gorusupudi A, Bernstein PS (2020) The macular carotenoids: a biochemical overview. Biochim Biophys Acta Mol Cell Biol Lipids 1865:158617

Balasuriya BWN, Rupasinghe HPV (2011) Plant flavonoids as angiotensin converting enzyme inhibitors in regulation of hypertension. Func Food Health Dis 1(5):172–188. https://doi.org/10.31989/ffhd.v1i5.132

Bandara V, Weinstein SA, White J, Eddleston M (2010) A review of the natural history, toxinology, diagnosis, and clinical management of *Nerium oleander* (common oleander) and *Thevetia peruviana* (yellow oleander) poisoning. Toxicon 56:273–281. https://doi.org/10.1016/j.toxicon.2010.03.026

Barazani O, Erez T, Ogran A, Hanin N, Barzilai M, Dag A, Shafir S (2019) Natural variation in flower color and scent in populations of *Eruca sativa* (Brassicaceae) affects pollination behavior of honey bees. J Insect Sci 19(3):1–9. https://doi.org/10.1093/jisesa/iez038

Beltrán-Debón R, Alonso-Villaverde C, Aragonès G, Rodríguez-Medina I, Rull A, Micol V, Segura-Carretero A, FernándezGutiérrez A, Camps J, Joven J (2010) The aqueous extract of *Hibiscus sabdariffa* calices modulates the production of monocyte chemoattractant protein-1 in humans. Phytomedicine 17:186–191

Benvenuti S, Bortolotti E, Maggini R (2016) Antioxidant power, anthocyanin content and organoleptic performance of edible flowers. Sci Hortic 199:170–177. https://doi.org/10.1016/j.scienta.2015.12.052

Birks JS, Harvey RJ (2018) Donepezil for dementia due to Alzheimer's disease. Cochrane Database Syst Rev 6. https://doi.org/10.1002/14651858.CD001190.pub2

Britton G (1996) Carotenoids. In: Hendry GAF, Houghton JD (eds) Natural food colorants, 2nd edn. Chapman & Hall, London, pp 197–243

Buanong M, Khunmuang S, Promboon J, Penchaiya P, Promyos N, Thepouyporn A (2025) Low storage temperature maintains the quality, antioxidant activity, and nutraceutical properties of edible flowers and their anti-inflammatory effects on RAW 264.7 cells. JAFR:101938. https://doi.org/10.1016/j.jafr.2025.101938

Chakraborty R, Roy S (2021) Angiotensin-converting enzyme inhibitors from plants: a review of their diversity, modes of action, prospects, and concerns in the management of diabetes-centric complications. J Integr Med 19(6):478–492. https://doi.org/10.1016/j.joim.2021.09.006

Chanukya BS, Rastogi NK (2016) A comparison of thermal processing, freeze drying and forward osmosis for the downstream processing of anthocyanin from rose petals. J Food Process Preserv 40(6):1289–1296. https://doi.org/10.1111/jfpp.12714

Chen N-H, Wei S (2017) Factors influencing consumers' attitudes towards the consumption of edible flowers. Food Qual Prefer 56:93–100. https://doi.org/10.1016/j.foodqual.2016.10.001

Chen G-L, Chen S-G, Xie Y-Q, Chen F, Zhao Y-Y, Luo C-X, Gao Y-Q (2015) Total phenolic, flavonoid and antioxidant activity of 23 edible flowers subjected to in vitro digestion. J Func Foods 17:243–259. https://doi.org/10.1016/j.jff.2015.05.028

Chen Y, Onken B, Chen H, Zhang X, Driscoll M, Cao Y et al (2020) Healthy lifespan extension mediated by oenothein B isolated from *Eucalyptus grandis x Eucalyptus urophylla* GL9 in *Caenorhabditis elegans*. Food Funct 11:2439–2450. https://doi.org/10.1039/C9FO02472G

Chensom S, Okumura H, Mishima T (2019) Primary screening of antioxidant activity, total polyphenol content, carotenoid content, and nutritional composition of 13 edible flowers from Japan. Prev Nutr Food Sci 24(2):171. https://doi.org/10.3746/pnf.2019.24.2.171

Christison GB, MacKenzie HA (1993) Laser photoacoustic determination of physiological glucose concentrations in human whole blood. Med Biol Eng Comput 31:284–290

Cloyd R, Bethke JA, Cowles RS (2011) Systemic insecticides and their use in ornamental plant system. Floricult Ornament Biotechnol Glob Sci Book 5(1):1–9

Coin A, Pamio MV, Alexopoulos C, Granziera S, Groppa F, de Rosa G, Girardi A, Sergi G, Manzato E, Padrini R (2016) Donepezil plasma concentrations, CYP2D6 and CYP3A4 phenotypes, and cognitive outcome in Alzheimer's disease. Eur J Clin Pharmacol:72711–72717

Crupi P, Faienza MF, Naeem MY et al (2023) Overview of the potential beneficial effects of carotenoids on consumer health and well-being. Antioxidants 12:1069. https://doi.org/10.3390/antiox12051069

de Alcantara Guimarães AL, do Tanque PR, Pyrrho A, Vieira A (2019) Toxicological and anatomical study of vegetative organs of *Anthurium maricense* Nadruz and Mayo (*Araceae*). Rev Agrogeoambiental 11(2):87–106. https://doi.org/10.18406/2316-1817v11n220191289

Deguchi A, Tatsuzawa F, Hosokawa M, Doi M, Ohno S (2016) Quantitative evaluation of the contribution of four major anthocyanins to black flower coloring of dahlia petals. Hortic J 85:340–350. https://doi.org/10.2503/hortj.MI-121

Del Villar-Martínez AA, García-Saucedo PA, Cárabez-Trejo A, Cruz-Hernández A, Paredes-López O (2005) Carotenogenic gene expression and ultrastructural changes during marigold flowering. J Plant Physiol 162:1046–1056. https://doi.org/10.1016/j.jplph.2004.12.004

Dobson HEM (1988) Survey of pollen and pollenkitt lipids e chemical cues to flower visitors? Amer J Bot 75:170e182

Dudareva N, Pichersky E (2000) Biochemical and molecular genetic aspects of floral scents. Plant Physiol 122:627–633. https://doi.org/10.1104/pp.122.3.627

El Bayani GF, Marpaung NLE, Simorangkir DAS, Sianipar IR, Ibrahim N, Kartinah NT et al (2018) Anti-inflammatory effects of *Hibiscus sabdariffa* Linn. on the IL-1β/IL-1ra ratio in plasma and hippocampus of overtrained rats and correlation with spatial memory. Kobe J Med Sci 64(2):E73

Fernandes L, Casal S, Pereira JA, Saraiva JA, Ramalhosa E (2017) Edible flowers: a review of the nutritional, antioxidant, antimicrobial properties and effects on human health. J Food Compos Anal 60:38–50. https://doi.org/10.1016/j.jfca.2017.03.017

Fernandes L, Ramalhosa E, Pereira JA, Saraiva JA, Casal S (2018) The unexplored potential of edible flowers lipids. Agriculture 8:146. https://doi.org/10.3390/agriculture8100146

Forrester SJ, Booz GW, Sigmund CD et al (2018) Angiotensin II signal transduction: an update on mechanisms of physiology and pathophysiology. Physiol Rev 98(3):1627–1738

Friedman H, Vinokur Y, Rot I, Rodov V, Goldman G, Resnick N et al (2005) *Tropaeolum majus* L. as edible flowers: growth and postharvest handling. Adv Hortic Sci 19:3–8

Gao X, Hu J, Zhang X, Zuo Y, Wang Y, Zhu S (2020) Research progress of aconitine toxicity and forensic analysis of aconitine poisoning. Forensic Sci Res 5(1):25–31. https://doi.org/10.1080/20961790.2018.1452346

Giusti MM, Wrolstad RE (2003) Acylated anthocyanins from edible sources and their applications in food systems. Biochem Eng J 14:217–225

Gonçalves LCP, Marcato AC, Rodrigues ACB, Pagano APE, De Freitas BC, De Machado CO, Nakashima KK, Esteves LC, Lopes NB, Bastos EL (2015) Betalains: from the colors of beetroots to the fluorescence of flowers. Rev Virtual Quim 7:292–309. https://doi.org/10.5935/1984-6835.20150015

Gonzaga ER (2009) Role of UV light in photodamage, skin aging, and skin cancer. Am J Clin Dermatol 10:19–24. https://doi.org/10.2165/0128071-200910001-00004

Greenlee H, Kwan ML, Kushi LH, Song J, Castillo A, Weltzien E, Quesenberry CP, Caan BJ (2012) Antioxidant supplement use after breast cancer diagnosis and mortality in the life after cancer epidemiology (LACE) cohor. Cancer 118:2048–2058

Gronquis M, Bezzerides A, Attygalle A, Meinwald J, Eisner M, Eisner T (2001) Attractive and defensive functions of the ultraviolet pigments of a flower (*Hypericum calycinum*). PNAS 98(24):13745–13750. https://doi.org/10.1073/pnas.2314716

Hamaguchi T, Ono K, Murase A, Yamada M (2009) Phenolic compounds prevent Alzheimer's pathology through different effects on the amyloid-beta aggregation pathway. Am J Patho 175(6):2557–2565

Hamdy SA, El Hefnawy HM, Azzam SM, Aboutabl EA (2018) Chemical profiling, volatile oil analysis and anticholinesterase activity of *Hydrocotyle umbellata* L. aerial parts cultivated in Egypt. S Afr J Bot 115:108–112. https://doi.org/10.1016/j.sajb.2018.01.019

Hannan MA, Dash R, Sohag AAM et al (2020) Neuroprotection against oxidative stress: phytochemicals targeting TrkB signaling and the Nrf2-ARE antioxidant system. Front Mol Neurosci 13:116

Haslam E (1998) Practical polyphenolics: from structure to molecular recognition and physiological action. Cambridge University, Cambridge

Hawash M, Jaradat N, Elaraj J, Hamdan A, Saja Abu Lebdeh SA, Halawa T (2020) Evaluation of the hypoglycemic effect of seven wild folkloric edible plants from Palestine. J Comp Int Med 17(1):20190032. https://doi.org/10.1515/jcim-2019-0032

He J, Giusti MM (2010) Anthocyanins: natural colorants with health-promoting properties. Annu Rev Food Sci Technol 1:163–187

Heo SJ, Hwang JY, Choi JI et al (2009) Diphlorethohydroxycarmalol isolated from Ishige okamurae, a brown algae, a potent α-glucosidase and α-amylase inhibitor, alleviates postprandial hyperglycemia in diabetic mice. Eur J Pharmacol 615:252–256. https://doi.org/10.1016/j.ejphar.2009.05.017

Hosaka H, Mizuno T, Iwashina T (2012) Flavonoid pigments and color expression in the flowers of black hollyhock (*Alcea rosae* 'Nigra'). Bull Natl Mus Nat Sci Ser B 38:69–75

Jacquemart A, Buyens C, Herent M, Quetin-Leclercq J, Lognay G, Hance T, Quinet M (2021) Male flowers of Aconitum compensate for toxic pollen with increased floral signals and rewards for pollinators. Sci Rep 9(1):16498. https://doi.org/10.1038/s41598-019-53355-3

Jadhav RV, Bhujbal SS (2020) A review on natural food colors. Pharm Res 2(2):12–20

Jadhav HB, Badwaik LS, Annapure U, Casanova F, Alaskar K (2023) A review on the journey of edible flowers from farm to consumer's plate. App Food Res 3(2):100312. https://doi.org/10.1016/j.afres.2023.100312

Jansen SA, Kleerekooper I, Hofman ZLM, Kappen IFPM, Stary-Weinzinger A, van der Heyden MAG (2012) Grayanotoxin poisoning: 'mad honey disease' and beyond. Cardiovasc Toxicol 12(3):208–215. https://doi.org/10.1007/s12012-012-9162-2

Jiang YW, Sun ZH, Tong WW et al (2021) Dietary intake and circulating concentrations of carotenoids and risk of type 2 diabetes: a dose-response meta-analysis of prospective observational studies. Adv Nutr 12:1723–1733

Jirapong C, Inplub K, Wongs-Aree C (2012) Volatile compounds in four species of Thai waterlily (*Nymphaea* spp.). Acta Hortic 943:117–122. https://doi.org/10.17660/ActaHortic.2012.943.12

Joffry SM, Yob NJ, Rofiee MS et al (2012) *Melastoma malabathricum* (L.) Smith ethnomedicinal uses, chemical constituents, and pharmacological properties: a review. Evid Based Complement Alternat Med 2012:258434. https://doi.org/10.1155/2012/258434

Junka N, Wongs-Aree C, Kanlayanarat S, Thanomchit K (2007) Predominance of cyanidin found in flower from two species of *Aerides* orchids. Acta Hortic 755:549–556. https://doi.org/10.17660/ActaHortic.2007.755.76

Junka N, Kanlayanarat S, Buanong M, Wongchaochant S, Wongs-Aree C (2011) Analysis of anthocyanins and the expression patterns of genes involved in biosynthesis in two *Vanda* hybrids. Int J Agri Biol 13:873–880

Junka N, Kanlayanarat S, Buanong M, Wongchaochant S, Wongs-Aree C (2012) Characterisation of floral anthocyanins and their antioxidant activity in *Vanda* hybrid (*V. teres* x *V. hookeriana*). J Food Agri Env 10(2):221–226

Kaisoon O, Konczak I, Siriamornpun S (2012) Potential health enhancing properties of edible flowers from Thailand. Food Res Int 46(2):563–571. https://doi.org/10.1016/j.foodres.2011.06.016

Kanner J, Harel S, Granit R (2001) Betalains – a new class of dietary cationized antioxidants. J Agric Food Chem 49(11):5178–5185. https://doi.org/10.1021/jf010456f

Karra S, Sebii H, Bouaziz MA, Blecker C, Attia H, Besbes S (2020) Male date palm flower powder: effect of incorporation on physico-chemical, textural, and sensory quality of biscuits. J Food Process Preserv 44:1–9

Kaur G, Alamb MS, Jabbar Z, Javed K, Athar M (2006) Evaluation of antioxidant activity of *Cassia siamea* flowers. J Ethnopharmacol 108:340–348

Kaz Owska K, Hsu T, Hou C, Yang W, Tsai G (2010) Antiinflammatory properties of phenolic compounds and crude extract from *Porphyra dentata*. J Ethnopharmacol 128(1):123–130

Kazuma K, Noda N, Suzuki M (2003) Flavonoid composition related to petal color in different lines of *Clitoria ternatea*. Phytochemistry 64(6):1133–1139. https://doi.org/10.1016/S0031-9422(03)00504-1

Kerchner A, Farkas A (2019) Worldwide poisoning potential of *Brugmansia* and *Datura*. Forensic Toxicol 38(4):30–41. https://doi.org/10.1007/s11419-019-00500-2

Khoo HE, Azlan A, Tang ST, Lim SM (2017) Anthocyanidins and anthocyanins: colored pigments as food, pharmaceutical ingredients, and the potential health benefits. Food Nutri Res 61:1361779. https://doi.org/10.1080/16546628.2017.1361779

Khunmuang S (2008) Analysis of anthocyanins in some waterlily species. Dissertation, King Mongkut's University of Technology Thonburi

Kim KS, Lee DS, Bae GS, Park SJ, Kang DG, Lee HS, Oh H, Kim YC (2013) The inhibition of JNK MAPK and NF-κB signaling by tenuifoliside A isolated from polygala tenuifolia in lipopolysaccharide-induced macrophages is associated with its anti-inflammatory effect. Eur J Pharmacol 721:267–276

Koike A, Barreira JCM, Barros L, Santos-Buelga C, Villavicencio ALCH, Ferreira ICFR (2015) Edible flowers of *Viola tricolor* L. as a new functional food: antioxidant activity, individual phenolics and effects of gamma and electron-beam irradiation. Food Chem 179:6–14. https://doi.org/10.1016/j.foodchem.2015.01.123

Kumar A, Kaur A, Tomer V, Gupta K, Kaur A (2020a) Effect of rose syrup and marigold powder on the physicochemical, phytochemical, sensorial and storage properties of nutricereals and milk-based functional beverage. J Am Coll Nutr 40(2):1–8. https://doi.org/10.1080/07315724.2020.1744487

Kumar A, Kaur A, Tomer V (2020b) Process optimization for the development of a synbiotic beverage based on lactic acid fermentation of nutricereals and milk-based beverage. LWT-Food Sci Technol 131:109774. https://doi.org/10.1016/j.lwt.2020.109774

Kumari P, Panwar S, Thaneshwari T, Kaushal S, Ullas SP (2017) Pigment profiling of flower crops: a review. Eco Env Cons 23(2):1000–1008

Kumari P, Ujala, Bhargava B (2021) Phytochemicals from edible flowers: opening a new arena for healthy lifestyle. J Func Foods 78:104375. https://doi.org/10.1016/j.jff.2021.104375

Kwon E-K, Lee D-Y, Lee H, Kim D-O, Baek N-I, Kim Y-E, Kim H-Y (2018) Flavonoids from the buds of *Rosa damascena* inhibit the activity of 3-hydroxy-3-methylglutaryl-coenzyme A reductase and angiotensin I-converting enzyme. J Agric Food Chem 58:882–886. https://doi.org/10.1021/jf903515f

Kyan R, Uemura S, Tanno K, Sawamoto K, Harada K, Sakawaki E, Fujita Y, Endo S, Narimatsu E (2015) Severe accidental colchicine poisoning by the autumn crocus: a case of successful treatment. JACME 5(4):103–106. https://doi.org/10.1016/j.jacme.2015.09.002

Lall N, Kishore N (2014) Are plants used for skin care in South Africa fully explored? J Ethnopharmacol 153:61–84. https://doi.org/10.1016/j.jep.2014.02.021

Lax S, Gilbert JA (2015) Hospital-associated microbiota and implications for nosocomial infections. Trends Mol Med 21(7):427–432. https://doi.org/10.1016/j.molmed.2015.03.005

Lee J, Durst RW, Wrolstad RE (2005) Determination of total monomeric anthocyanin pigment content of fruit juices, beverages, natural colorants, and wines by the pH differential method: collaborative study. J AOAC Int 88(5):1269–1278

Lee SH, Park MH, Heo SJ et al (2010) Dieckol isolated from Ecklonia cava inhibits α-glucosidase and α-amylase *in vitro* and alleviates postprandial hyperglycemia in streptozotocin-induced diabetic mice. Food Chem Toxicol 48:2633–2637. https://doi.org/10.1016/j.fct.2010.06.032

Lephart ED (2016) Skin aging and oxidative stress: equol's anti-aging effects via biochemical and molecular mechanisms. Ageing Res Rev 31:36–54. https://doi.org/10.1016/j.arr.2016.08.001

Li MX, Bai X, Ma YP, Zhang HX, Nama N, Pei SJ, Du ZZ (2019) Cosmetic potentials of extracts and compounds from *Zingiber cassumunar* Roxb. rhizome. Ind Crop Pro 141:111764. https://doi.org/10.1016/j.indcrop.2019.111764

Li M-X, Xie J, Bai X, Du Z-Z (2021) Anti-aging potential, anti-tyrosinase and antibacterial activities of extracts and compounds isolated from *Rosa chinensis* cv. 'JinBian'. Ind Crops Pro 159:113059. https://doi.org/10.1016/j.indcrop.2020.113059

Liamkaew R, Boonpan A, Chompreeda P (2020) Sensory evaluation and consumer acceptability of ready-to-drink flower product. Prog Appl Sci Tech 10(2):1–6. https://doi.org/10.14456/past.2020.5

Lin CC, Yang CC, Phua DH, Deng JF, Lu LH (2010) An outbreak of foxglove leaf poisoning. J Chi Med As 73:97–100. https://doi.org/10.1016/S1726-4901(10)70009-5

Loganayaki N, Suganya N, Manian S (2012) Evaluation of edible flowers of agathi (*Sesbania grandiflora* L. Fabaceae) for *in vivo* anti-inflammatory and analgesic, and in vitro antioxidant potential. Food Sci Biotechnol 21:509–517. https://doi.org/10.1007/s10068-012-0065-6

Loizzo MR, Pugliese A, Bonesi M, Tenuta MC, Menichini F, Xiao J, Tundis R (2016) Edible flowers: a rich source of phytochemicals with antioxidant and hypoglycemic properties. J Agri Food Chem 64(12):2467–2474. https://doi.org/10.1021/acs.jafc.5b03092

López CJ, Caleja C, Prieto MA, Sokovic M, Calhelha RC, Barros L, Ferreira ICFR (2019) Stability of a cyanidin-3-O-glucoside extract obtained from *Arbutus unedo* L. and incorporation into wafers for colouring purposes. Food Chem 275:426–438

Lu B, Li M, Yin R (2016) Phytochemical content, health benefits, and toxicology of common edible flowers: a review (2000–2015). Crit Rev Food Sci Nutr 56:S130–S148. https://doi.org/10.1080/10408398.2015.1078276

Lunau K (1995) Notes on the color of pollen. Plant Sys Evol 198:235e252

Maoka T (2020) Carotenoids as natural functional pigments. J Nat Med 74:1–16

Martins T, Barros AN, Rosa E, Antunes L (2023) Enhancing health benefits through chlorophylls and chlorophyll-rich Agro-food: a comprehensive review. Molecules 28:5344. https://doi.org/10.3390/molecules28145344

Matyjaszczyk E, Smiechowska M (2019) Edible flowers. Benefits and risks pertaining to their consumption. Trends Food Sci Technol 91:670–674. https://doi.org/10.1016/j.tifs.2019.07.017

McDougall GJ, Shpiro F, Dobson P et al (2005) Different polyphenolic components of soft fruits inhibit α-amylase and α-glucosidase. J Agric Food Chem 53:2760–2766. https://doi.org/10.1021/jf0489926

McQuistan TJ, Simonich MT, Pratt MM et al (2012) Cancer chemoprevention by dietary chlorophylls: a 12,000-animal dose-dose matrix biomarker and tumor study. Food Chem Toxicol 50:341–352

Meléndez-Martínez AJ, Mandi'c AI, Bantis F et al (2022) A comprehensive review on carotenoids in foods and feeds: status quo, applications, patents, and research needs. Crit Rev Food Sci Nutr 62:1999–2049

Miyake S, Takahashi N, Sasaki M et al (2012) Vision preservation during retinal inflammation by anthocyaninrich bilberry extract: cellular and molecular mechanism. Lab Investig 92:102–109

Mlcek J, Rop O (2011) Fresh edible flowers of ornamental plants – a new source of nutraceutical foods. Trends Food Sci Technol 22(10):561–569. https://doi.org/10.1016/j.tifs.2011.04.006

Morimoto M, Tatsumi K, Yuui K, Terazawa I, Kudo R, Kasuda S (2012) Convallatoxin, the primary cardiac glycoside in lily of the valley (*Convallaria majalis*), induces tissue factor expression in endothelial cells. Vet Med Sci 7:2440–2444. https://doi.org/10.1002/vms3.614

Murai Y, Yangzom R, Gyeltshen C, Dorji K, Wangmo C, Iwashina T (2017) Flower pigments of black pea *Thermopsis barbata* (Fabaceae) in Bhutan. Bull Natl Mus Nat Sci Ser B 43(3):87–91

Naczk M, Shahidi F (2004) Extraction and analysis of phenolics in food. J Chrom A 1054(1–2):95–111

Nagini S, Palitti F, Natarajan AT (2015) Chemopreventive potential of chlorophyllin: a review of the mechanisms of action and molecular targets. Nutr Cancer 67:203–211

Nahar PP, Driscoll MV, Li L, Slitt AL, Seeram NP (2014) Phenolic mediated anti-inflammatory properties of a maple syrup extract in RAW 264.7 murine macrophages. J Funct Foods 6:126–136

Narbona E, del Valle JC, Arista M, Buide ML, Ortiz PL (2021) Major flower pigments originate different colour signals to pollinators. Front Ecol Evol 9:743850. https://doi.org/10.3389/fevo.2021.743850

Nathan C (2002) Points of control in inflammation. Nature 420:846–852

Navarro-González I, González-Barrio R, García-Valverde V, Bautista-Ortín AB, Periago MJ (2015) Nutritional composition and antioxidant capacity in edible flowers: characterisation of phenolic compounds by HPLC-DAD-ESI/MSn. Int J Mol Sci 16:805–822. https://doi.org/10.3390/ijms16010805

Nehme A, Zibara K (2017) Efficiency and specificity of RAAS inhibitors in cardiovascular diseases: how to achieve better end-organ protection? Hypertens Res 40:903–909

Neugebauerova J, Vabkova J (2009) Jedle kvety soucasti food stylingu. Zahradnictvi 83:22e24

Nicolson SW, Nepi M, Pacini E (2007) Nectaries and nectar. Springer, Dordrecht

Nowicka P, Wojdyło A (2019) Anti-hyperglycemic and anticholinergic effects of natural antioxidant contents in edible flowers. Antioxidants 8(8):308. https://doi.org/10.3390/antiox8080308

Oh YC, Cho WK, Jeong YH, Im GY, Lee KJ, Yang HJ, Ma JY (2013) Anti-inflammatory effect of Sosihotang via inhibition of nuclear factor-κB and mitogen-activated protein kinases signaling pathways in lipopolysaccharide-stimulated RAW 264.7 macrophage cells. Food Chem Toxicol 53:343–351

Ohmiya A (2009) Carotenoid cleavage dioxygenases and their apocarotenoid products in plants. Plant Biotechnol 26:351–358. https://doi.org/10.5511/plantbiotechnology.26.351

Ohmiya A (2011) Diversity of carotenoid composition in flower petals. JARQ 45(2):163–171. https://doi.org/10.6090/jarq.45.163

Olennikov DN, Kashchenko NI, Chirikova NK, Akobirshoeva A, Zilfikarov IN, Vennos C (2017) Isorhamnetin and quercetin derivatives as anti-acetylcholinesterase principles of marigold (*Calendula officinalis*) flowers and preparations. Int J Mol Sci 18:1685. https://doi.org/10.3390/ijms18081685

Oueslati S, Ksouri R, Falleh H, Pichette A, Abdelly C, Legault J (2012) Phenolic content, antioxidant, anti-inflammatory and anticancer activities of the edible halophyte *Suaeda fruticosa* Forssk. Food Chem 132(2):5

Pfister JA, Gardner DR, Panter KE, Manners GD, Ralphs MH, Stegelmeier B, Schoch TK (1999) Larkspur (*Delphinium* spp.) poisoning in livestock. J Nat Toxins 8:81–94. https://doi.org/10.1016/j.rala.2020.01.004

Pires EDO Jr, Pereira E, Carocho M, Pereira C, Dias MI, Calhelha RC, Ćirić A, Soković M, Garcia CC, Ferreira ICFR, Caleja C, Barros L (2021) Study on the potential application of *Impatiens balsamina* L. flowers extract as a natural colouring ingredient in a pastry product. Int J Environ Res Public Health 18:9062. https://doi.org/10.3390/ijerph18179062

Pires TCSP, Dias MI, Barros L, Barreira JCM, Santos-Buelga C, Ferreira ICFR (2018) Incorporation of natural colorants obtained from edible flowers in yogurts. LWT-Food Sci Technol 97:668–675. https://doi.org/10.1016/j.lwt.2018.08.013

Pires TC, Barros L, Santos-Buelga C, Ferreira IC (2019) Edible flowers: emerging components in the diet. Trends Food Sci Technol 93:244–258. https://doi.org/10.1016/j.tifs.2019.09.020

Rezende F, Sande D, Coelho AC, Oliveira G, Boaventura MA, Takahashi JA (2019) Edible flowers as innovative ingredients for future food development: antialzheimer, antimicrobial, and antioxidant potential. Chem Eng Trans 75:337–342. https://doi.org/10.3303/CET1975057

Rodriguez-Amaya DB (2016) Natural food pigments and colorants. Curr Opin Food Sci 7:20–26. https://doi.org/10.1016/j.cofs.2015.08.004

Roriz CL, Barreira JCM, Morales P, Barros L, Ferreira ICFR (2018) *Gomphrena globosa* L. as a novel source of food-grade betacyanins: incorporation in ice-cream and comparison with beet-root extracts and commercial betalain. LWT 92:101–107. https://doi.org/10.1016/j.lwt.2018.02.009

Sadighara P, Gharibi S, Jafari AM, Khaniki JG, Salari S (2012) The antioxidant and flavonoids contents of *Althaea officinalis* L. flowers based on their color. Avicenna J Phytomed 2(3):113–117

Sahni S, Dufour AB, Fielding RA et al (2021) Total carotenoid intake is associated with reduced loss of grip strength and gait speed over time in adults: the Framingham Offspring Study. Am J Clin Nutr 113:437–445

Schep LJ, Slaughter RJ, Becket G, Beasley DMG (2009) Poisoning due to water hemlock. JCT 47(4):270–278. https://doi.org/10.1080/15563650902904332

Scherf G (2004) Plane rostliny a jejich pouziti v kuchyni. Beta-Dobrovsky, Praha, p 8e15

Seo SY, Sharma VK, Sharma N (2003) Mushroom tyrosinase: recent prospects. J Agric Food Chem 51:2837–2853. https://doi.org/10.1021/jf020826f

Shao J, Li Y, Wang Z, Xiao M, Yin P, Lu Y, Qian X, Xu Y, Liu J (2013) A novel naphthalimide derivative exhibited anti-inflammatory effects via targeted-inhibiting TAK1 following down-regulation of ERK1/2- and p38 MAPK-mediated activation of NF-κB in LPS-stimulated RAW264.7 macrophages. Int Immunopharmacol 17:216–228

Sharma O, Sharma S, Pattabhi V, Mahato SB, Sharma PD (2007) A review of the hepatotoxic plant *Lantana camara*. Crit Rev Toxicol 37(4):313–335. https://doi.org/10.1080/10408440601177863

Shim SH, Kim JM, Choi CY et al (2012) *Ginkgo biloba* extract and bilberry anthocyanins improve visual function in patients with normal tension glaucoma. J Med Food 15:818–823

Slattery AP (2005) Wisteria. In: Wexler P (ed) Encyclopedia of toxicology, 2nd edn. Elsevier, pp 463–464. https://doi.org/10.1016/B0-12-369400-0/01018-8

Sriyong M (2010) Efficiency of preservatives on petal color stability of waterlilies (*Nympheaea* spp.) cv. 'King of Siam' and 'Mangkala Ubon' during vase period and after drying. Dissertation, King Mongkut's University of Technology Thonburi

Stefaniak A, Grzeszczuk ME (2018) Nutritional and biological value of five edible flower species. Not Bot Horti Agrobot Cluj-Napoca 47(1):128–134. https://doi.org/10.15835/nbha47111136

Stevenson PC, Nicolson SW, Wright GA (2017) Plant secondary metabolites in nectar: impacts on pollinators and ecological functions. Funct Ecol 31:65–75. https://doi.org/10.1111/1365-2435.12761

Stickland RG (1976) The nature of the white colour of petals. Ann Bot 38:1033–1037

Streich A (2007) Garden terms: reproductive plant morphology – seeds, flowers, and fruits. https://extensionpubs.unl.edu/publication/ec1257/2007/pdf/view/ec1257-2007.pdf. Accessed 25 Dec 2023

Tahergorabi Z, Khazaei M (2012) Imbalance of angiogenesis in diabetic complications: the mechanisms. Int J Prev Med 3:827. https://doi.org/10.4103/2008-7802.104853

Takahashi JA, Sande D, da Silva LG, Emoura MAF, Lima MTNS (2019) Fungal metabolites as promising new drug leads for the treatment of Alzheimer's disease. In: Atta-ur-Rahman (ed) Studies in natural products chemistry. Elsevier, pp 1–39. https://doi.org/10.1016/B978-0-444-64185-4.00001-0

Tanaka Y, Sasaki N, Ohmiya A (2008) Biosynthesis of plant pigments: anthocyanins, betalains and carotenoids. Plant J 54:733–749. https://doi.org/10.1111/j.1365-313X.2008.03447.x

Thomasset SC, Berry DP, Garcea G, Marczylo T, Steward WP, Gescher AJ (2007) Dietary polyphenolic phytochemicals promising cancer chemopreventive agents in humans? A review of their clinical properties. Int J Cancer 120:451–458

Tricco AC, Ashoor HM, Soobiah C, Rios P, Veroniki AA, Hamid JS, Blondal E (2018) Comparative effectiveness and safety of cognitive enhancers for treating Alzheimer's disease: systematic review and network metaanalysis. JAGS 66(1):170–178

Tsuda T, Ueno Y, Aoki H et al (2004) Anthocyanin enhances adipocytokine secretion and adipocyte-specific gene expression in isolated rat adipocytes. Biochem Biophys Res Commun 316:149–157

Tulah AS, Birch-machin MA (2013) Stressed out mitochondria: the role of mitochondria in ageing and cancer focusing on strategies and opportunities in human skin. Mitochondrion 13:444–453. https://doi.org/10.1016/j.mito.2012.11.007

van der Kooi CJ, Dyer AG, Kevan PG, Lunau K (2019) Functional significance of the optical properties of flowers for visual signalling. Ann Bot 123:263–276. https://doi.org/10.1093/aob/mcy119

Vidana Gamage GC, Lim YY, Choo WS (2021) Anthocyanins from *Clitoria ternatea* flower: biosynthesis, extraction, stability, antioxidant activity, and applications. Front Plant Sci 12:792303

Vidanapathirana NP, Rifna LM (2023) Plant morphology. University of Colombo Institute for Agro-technology and rural science. https://ssrn.com/abstract=4693969. Accessed 25 Dec 2023

Vincenti S, Mariani M, Alberti J, Jacopini S, de Caraffa VB, Berti L, Maury J (2021) Biocatalytic synthesis of natural green leaf volatiles using the lipoxygenase metabolic pathway. Catalysts 9(10):873. https://doi.org/10.3390/catal9100873

Vinokur Y, Rodov V, Reznick N, Goldman G, Horev B, Umiel N, Friedman H (2006) Rose petal tea as an antioxidant rich beverage: cultivar effects. J Food Sci 71(1):S42–S47. https://doi.org/10.1111/j.1365-2621.2006.tb12404.x

Vivekananthan DP, Penn MS, Sapp SK, Hsu A, Topol EJ (2003) Use of antioxidant vitamins for the prevention of cardiovascular disease: meta-analysis of randomized trials. Lancet 361:2017–2023

Wang H, Cao G, Prior RL (1997) Oxygen radical absorbing capacity of anthocyanins. J Agric Food Chem 4:304–309

Wiermann R, Gubatz S (1992) Pollen wall and sporopollenin. Int Rev Cytol 140:35e72

Winterhalter P, Rouseff RL (2001) Carotenoid-derived aroma compounds. ACS symposium series. American Chemical Society, Washington, DC. https://doi.org/10.1021/bk-2002-0802.ch001

Wongs-Aree C, Giusti MM, Schwartz SJ (2006) Anthocyanins derived only from delphinidin in the blue petals of *Clitoria ternatea*. Acta Hortic 712:437–442. https://doi.org/10.17660/ActaHortic.2006.712.51

Wongs-Aree C, Chansilpa NS, Buanong M (2017) Study of characterization and property of anthocyanins extracted from tropical waterlily and their antioxidant capacities, P_NRCT/2552 [Full research report]. National Research Council of Thailand (NRCT)

Xiong L, Yang J, Jiang Y, Lu B, Hu Y, Zhou F, Mao S, Shen C (2014) Phenolic compounds and antioxidant capacities of 10 common edible flowers from China. J Food Sci 79(4):C517–C525. https://doi.org/10.1111/1750-3841.12404

Yimyong S (2010) Stability, antioxidant capacity, antimicrobial and anticancer activities of tropical waterlily extracts. Dissertation, King Mongkut's University of Technology Thonburi

Youwei Z, Jinlian Z, Yonghong P (2008) A comparative study on the free radical scavenging activities of some fresh flowers in southern China. LWT–Food Sci Technol 41:1586–1591

Zakaria Z, Zakaria S, Ishak MM (2010) Analysis of major fragrant compounds from *Citrus grandis* flowers extracts. Sains Malaysiana 39(4):565–569

Zhang S, Wang Y, Yang J, Zhang D, Wang Y, Li S, Pan Y, Zhang H, Sun Y (2022) Comparative analysis of aristolochic acids in *Aristolochia* medicinal herbs and evaluation of their toxicities. Toxins 14:879. https://doi.org/10.3390/toxins14120879

Zheng J, Meenu M, Xu B (2019) A systematic investigation on free phenolic acids and flavonoids profiles of commonly consumed edible flowers in China. J Pharm Biomed Ana 172:268–277. https://doi.org/10.1016/j.jpba.2019.05.007

Zheng J, Lu B, Xu B (2021) An update on the health benefits promoted by edible flowers and involved mechanisms. Food Chem 340:127940

Chapter 7
Adulterant Detection Techniques in Edible Flowers and Their Products

Safoura Jabbari

Abstract Edible plants play a crucial role in promoting therapy, aiding in the recovery process, supporting a healthy lifestyle, and offering significant benefits from both nutritional and economic perspectives. However, the intentional addition of foreign substances to edible plants and their products, known as adulteration, has been a pervasive issue throughout history. This chapter provides a comprehensive review of various types of adulteration in edible plants and their products, along with the methods employed for their detection. Numerous techniques are outlined for assessing the purity of edible plants and their derivatives. The discussion delves into specific examples of edible plants and their products to illustrate the methods used in detecting adulteration. Additionally, the chapter explores the advantages and disadvantages of these detection methods, emphasizing the need to develop and implement reliable approaches for identifying adulterations. Furthermore, novel detection methods like surface plasmon resonance (SPR) biosensors are highlighted as potential tools. The chapter underscores the importance of addressing adulteration issues in edible plants and stresses the significance of utilizing advanced and dependable methods for detection. Remarkably, this content could provide a good support for studies of adulterants detection in edible flowers, and a guidance to achieving foods with higher quality, and more safe.

Keywords Biosensors · Detection methods · DNA barcoding · Molecular techniques · PCR · Plant derivatives

Abbreviations

AFLP	Restriction Fragment Length Polymorphism
C NMR	Carbon Nuclear Magnetic Resonance
C/P-IRMS	Combustion/Pyrolyze-Isotope Iratio Mass Spectrometry

S. Jabbari (✉)
Department of Biochemistry and Biophysics, TeMS.C., Islamic Azad University, Tehran, Iran
e-mail: sjabbari@iau.ac.ir

M. M. Jowkar et al. (eds.), *Advances in Research on Edible Flowers*, Advances in Olericulture, https://doi.org/10.1007/978-3-032-04561-4_7

CAM	Crassulacean Acid Metabolism
E-nose	Electronic-nose
ESI-MS	Electrospray Ionization-Mass Spectrometry
E-tongue	Electronic tongue
EVOO	Extra Virgin Olive Oil
FID	Flame-Ionization Detector
FTIR	Fourier Transform Infrared Spectroscopy
FTR	Fourier Transform Raman
GC	Gas Chromatography
GC-MS	Gas Chromatography-Mass Spectrometry
H NMR	Proton Nuclear Magnetic Resonance
HPLC	High Performance Liquid Chromatography
HPTLC	High Performance Thin Layer Chromatography
HS-E nose	Headspace-Electronic nose
IR	Infra Red
IRMS	Isotope Ratio Mass Spectrometry
ISO	International Organization for Standardization
ISSR	Inter Simple Sequence Repeats
ITS	Internal Transcribed Spacer
LC	Liquid Chromatography
LC-MS	Liquid Chromatography-Mass Spectrometry
LoD	Limit of Detection
LoQ	Limit of Quantification
LSPR	Localized Surface Plasmon Resonance
MOS	Metal Oxide Sensor
MS	Mass Spectrometry
NGS	Next Generation Sequencing
NMR	Nuclear Magnetic Resonance
P NMR	Phosphorus Nuclear Magnetic Resonance
PAGE	Poly Acrylamide Gel Electrophoresis
PANOs	Pyrrolizidine Alkaloid N-Ooxide forms
Pas	Pyrrolizidine Alkaloids
PCA	Principal Component Analysis
PCR	Polymerase Chain Reaction
PCR-RFLP	Polymerase Chain Reaction- Restriction Fragment Length Polymorphism
PLS	Partial Least Squares
QC	Quality Control
qPCR	Quantitative Polymerase Chain Reaction
RAPD	Random Amplified Polymorphic DNA
RFLP	Restriction Fragment Length Polymorphism
RTE	Ready To Eat
SEM	Scanning Electron Microscopy
SIRA	Stable Isotope Ratio Analysis

SNP	Single Nucleotide Polymorphism
SPE	Screen Printed Electrode
SPME-GCFID	Solid Phase Micro Extraction-Gas Chromatography Flame Ionization Detection
SPME-MS	Solid Phase Micro Extraction-Mass Spectrometry
SPR	Surface Plasmon Resonance
TLC	Thin Layer Chromatography
UHPLC	Ultra High Performance Liquid Chromatography
UHPLC-IT-MS	Ultra High Performance Liquid Chromatography-Ion Trap-Mass Spectrometry
UV	Ultra Violet
UV-Vis	Ultra Violet-Visible
VOO	Virgin Olive Oil
WGS	Whole Genome Sequences
WHO	World Health Organization's
μSPEed	Solid Phase microextraction

7.1 Introduction

While historically appreciated for their fragrance and visual appeal, edible flowers have gained prominence over the past decade as integral components in therapy, recovery processes, and healthy lifestyles. Recent developments highlight various biological properties associated with edible flowers, including their potential to improve ulcerative colitis (Meurer et al. 2019), exhibit anti-hyperglycemic and anti-cholinergic activity (Nowicka and Wojdyło 2019), protect against oxidative effects in erythrocytes (Yang et al. 2019), and even demonstrate anticancer activity (Nguyen et al. 2019). Simultaneously, there is a growing recognition of the value of edible flowers as food sources (Nowicka and Wojdyło 2019; Jakubczyk et al. 2021), and expanding knowledge of their chemical composition and biological properties positions them as attractive candidates for the development of functional foods enriched with naturally occurring bioactive compounds (Takahashi et al. 2020). As the commercial demand for plant-based products, particularly those with medicinal values, rises in the food and beverage industries, there is an anticipated increase in cases of adulteration that necessitates strict control. Adulterated products may compromise the biological activity of these flowers or pose health risks and adverse effects, especially if contaminated with toxic species or allergens. Researchers have dedicated considerable efforts to address the issue of adulteration by developing methods and techniques for detecting adulterants. This review aims to provide an overview of various techniques currently in use or with future potential for the detection and quantification of different adulterants.

7.2 Adulteration and Types

Adulteration is commonly defined as the process that diminishes the quality or nature of a given substance by introducing a foreign or inferior substance while removing a vital element. According to the Merriam-Webster dictionary, adulterate is described as "to corrupt, debase, or make impure by the addition of a foreign or inferior substance or element" (Simmler et al. 2018). The World Health Organization's (WHO) guidelines for the quality control of herbal materials offer a specific definition, stating that adulterated material refers to "herbal material, an herbal constituent, or other substance deliberately or non-intentionally added to an herbal material, herbal preparation, or finished herbal product (Simmler et al. 2018), whether through cross-contamination or contamination". An alternative definition characterizes adulteration as "making impure by adding foreign, extraneous, poisonous, insanitary, or inferior substances/ingredients to a (food) product" (Whitsitt et al. 2013). Furthermore, Title 21 of the U.S. Code of Federal Regulations (Simmler et al. 2018), which covers food and drugs, provides a comprehensive description of adulteration.

In a general, various classifications of adulterants exist, encompassing drug adulteration, food adulteration, herbal medicinal products adulteration, and more. However, here, we will broadly categorize adulteration into three distinct types (Fig. 7.1), which include (Manasha and Janani 2016):

7.2.1 *Natural*

Adulterants are chemicals, organic compounds or radicals that exist naturally in the food or the premium quality produce. For instance, in saffron, natural or organic impurities mainly consist of corn silk, red-dyed silk fibres, and ray florets from other flowers (Kumari et al. 2021). Moreover, saffron may be mixed with the style portion and stamens of flowers, making it difficult to distinguish. The addition of floral parts such as styles, stamens, and petals from saffron is sometimes called Auto adulteration (Kumari et al. 2021; Foschi et al. 2022).

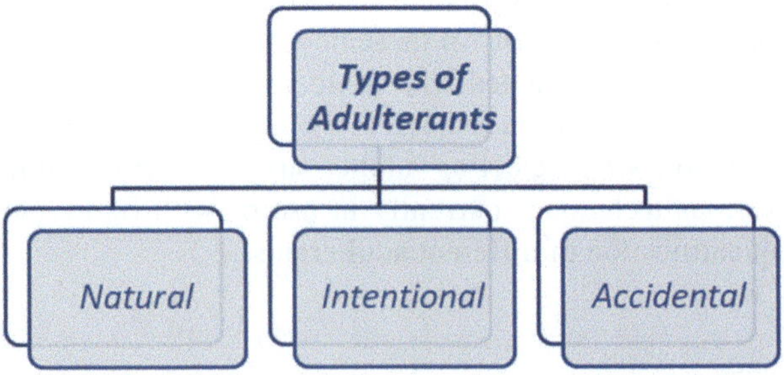

Fig. 7.1 Types of adulteration. (Manasha and Janani 2016)

7.2.2 Intentional

Intentional adulterants refer to substances intentionally added by adulterers with the aim of boosting profit margins. Examples of such deliberate additions include sand, marble chips, stones, mud, chalk powder, water, dyes, and more. These adulterants can be detrimental to the body (Manasha and Janani 2016).

7.2.3 Accidental

Adulterants in food substances may result from ignorance, negligence, or a lack of proper facilities, rather than a deliberate act by the adulterer. Examples include pesticides, droppings of rodents, and larvae found in food. In societies with limited legal controls on food quality and inadequate monitoring by authorities, the use of adulterants has been prevalent (Banti 2020). Various methods have been employed to assess the purity of products, emphasizing the need for organizations related to food, drugs, edible flowers, and products like edible oils and fats to develop and implement reliable detection methods. This is crucial to instil confidence in consumers and markets regarding the authenticity and purity of these products (Azadmard-Damirchi and Torbati 2015). Three major basic strategies are commonly used based on the presence of physical and chemical testing to distinguish original from non-original products:

I. Identification of an unusual profile of a component
II. Differentiation from the normal standard compound
III. Identification of markers or foreign materials in the product

The third strategy, focused on identifying foreign materials in food commodities, is widely employed as it is relatively straightforward and target-oriented (Negi et al. 2021).

7.3 Approaches for Adulterant Authentication

Various advanced methods for detecting adulterants, including chromatographic, spectroscopic, immunological, electrophoresis-based, molecular-based, and biosensor-based approaches, have been developed (Fig. 7.2). In this chapter, most of these methods are considered.

Physical methods for identifying adulterants rely on the shape, colour, odour, and texture of food materials, as well as microscopic examination of tissue structure and arrangement in food products (Dhanya and Sasikumar 2010). While cost-effective and straightforward, these methods do not guarantee product quality. Modern adulterations are more sophisticated, necessitating the use of advanced and

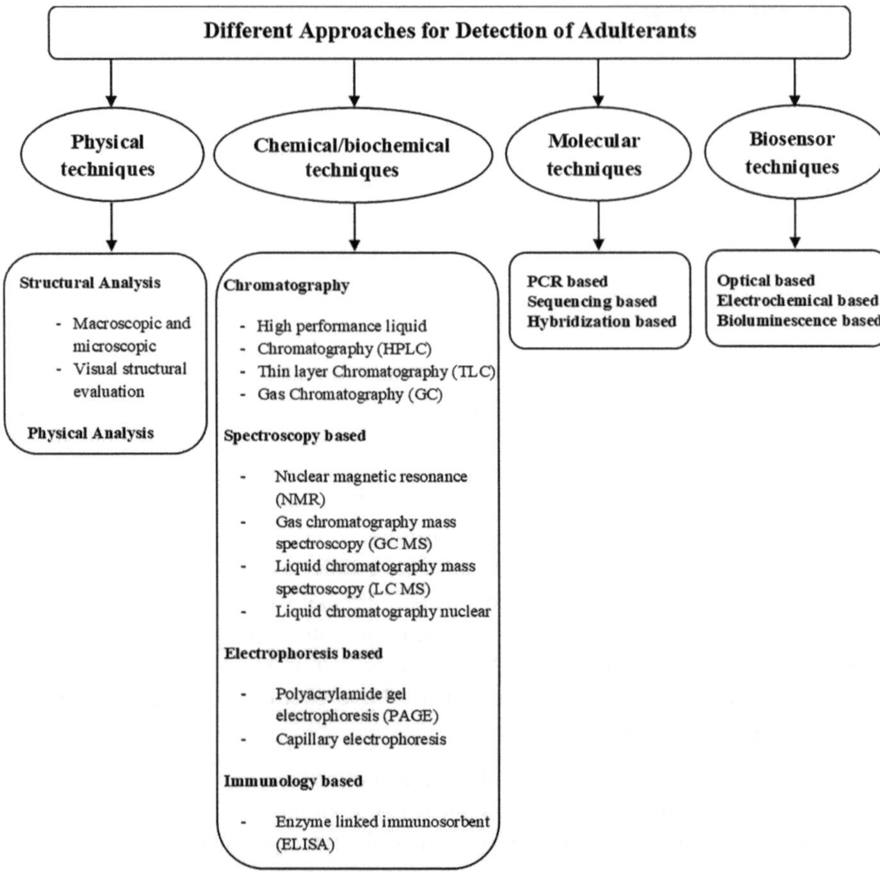

Fig. 7.2 Various analytical approaches to detect adulterants. (Nanda et al. 2022; Haji et al. 2023; Sudhakar et al. 2023)

suitable detection methods. Traditional physical properties like refractive index, viscosity, melting point, saponification, and iodine value are generally impractical for detecting adulteration (Azadmard-Damirchi and Torbati 2015).

Chemical and biochemical methods, such as chromatography, spectroscopy, electrophoresis, and immunology, offer greater precision and sensitivity than physical methods. However, their industrial applications face challenges due to high costs, technical complexities, and the need for trained personnel (Gonzálvez et al. 2008). Laboratory detection methods, including High Performance Liquid Chromatography (HPLC), High-Performance Thin-Layer Chromatography (HPTLC), Gas Chromatography-Mass Spectrometry (GC-MS), Liquid Chromatography-Mass Spectrometry (LC-MS), among others, provide accurate results based on parameters like accuracy, precision, linearity, limit of detection, limit of quantitation, specificity, and robustness. Despite their accuracy, these methods require expensive instrumentation (Churi 2021).

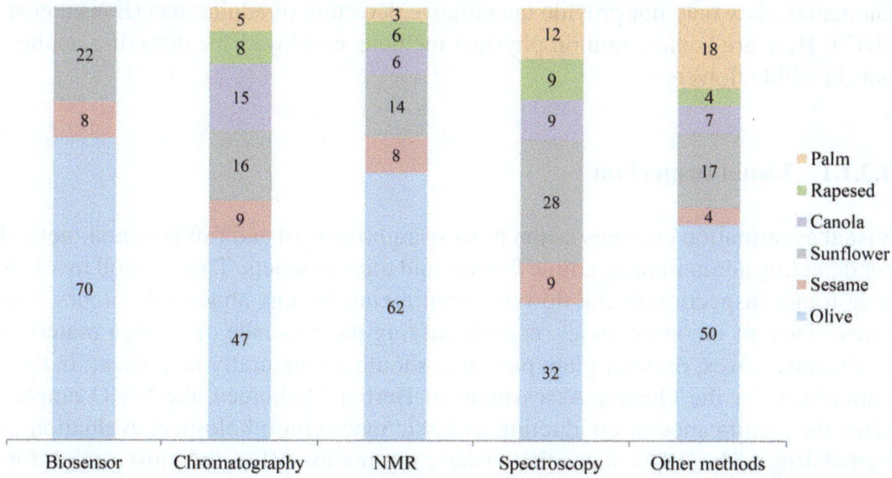

Fig. 7.3 Publications related to the study of different adulteration detection methods vis- a-vis various edible oils. (Figures on the body of the bars indicate the fraction of the whole in per cent). (Sudhakar et al. 2023)

Considering these factors, the selection of analytical techniques for quality assessment becomes a challenging decision for analysts, as each technique comes with its own advantages and disadvantages (Sudhakar et al. 2023).

As illustrated in Fig. 7.3, exemplifying the number of publications from 2010 to 2020, specific to the testing of various edible flower and vegetable oils using different methods, which all are consider in this chapter, it is evident that spectroscopy is the most extensively studied tool for detecting adulteration, constituting 46% of the research. Chromatography follows closely at 37%, while Nuclear Magnetic Resonance (NMR) accounts for 11%, and biosensors contribute 2%. Some fewer common methods, including differential scanning calorimetry, dielectric spectroscopy, etc., collectively represent the remaining 4% of the studies (Sudhakar et al. 2023).

7.3.1 Physical Methods

Physical methods are applicable for detecting adulterants in edible flowers and their products, even though they may not offer precise identification of specific adulterants. These methods serve to identify irregularities, foreign materials, or discrepancies in the physical characteristics of the flowers. Various physical methods for detecting adulteration include microscopic and macroscopic visual structural analysis, along with the analysis of physical parameters such as morphology, colour, odour, texture, solubility, bulk density, etc. (Dhanya and Sasikumar 2010). It is important to note that while these methods are useful for revealing physical anomalies, they may not provide quantitative detection of adulterants (Bansal et al. 2017). Here are some common physical methods employed for detecting adulterants in edible flowers:

7.3.1.1 Visual Inspection

Visual examination emerges as the most straightforward and fundamental method for detecting adulterants in edible flowers and their products. This method involves a thorough inspection of the flowers, scrutinizing for any abnormal colours, textures, sizes, or inconsistencies, and identifying the presence of foreign materials like stems, leaves, or other plant parts that should not naturally be present. In their Guidelines for the Quality Assessment of Herbal Medicines, the WHO emphasizes the significance of conducting a classic macro morphological evaluation of herbal drugs. The WHO states that visual examination offers the most straightforward and rapid method for confirming identity, purity, and potentially, quality. When a sample noticeably differs in terms of colour, texture, smell, or flavour from the defined criteria, it is deemed to not meet the standards (Upton et al. 2020). As an example, Figs. 7.4 and 7.5 show very similar appearance of dyed lotus stamens, corn stigmas and pure saffron (Li et al. 2020). In order practically apply this method, one can visually inspect the detection of coloured dried maize cob tendrils in saffron. This procedure entails adding a small amount of saffron to a transparent glass of water. If the saffron is impure, the artificial colour will rapidly dissolve in the water. On the other hand, genuine saffron will retain its colour as long as it lasts when dissolved in water (Shukla and Iqbal 2015). It's important to note that pure saffron doesn't break easily, unlike artificial saffron, which is produced by immersing maize cob in sugar and colouring it with coal tar. (Banti 2020).

Fig. 7.4 Flower and stigmas of saffron. (Li et al. 2020)

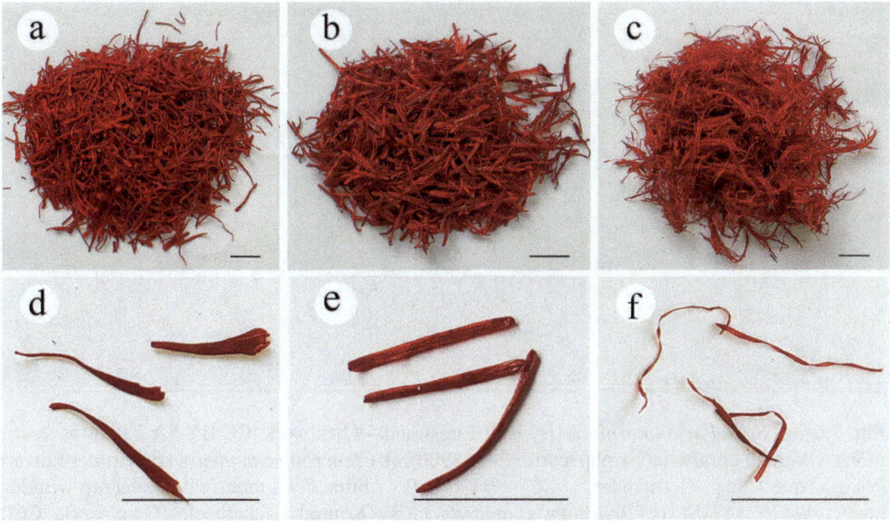

Fig. 7.5 Samples of saffron (**a**, **d**), lotus stamens (**b**, **e**), and corn stigmas (**c**, **f**) after drying. Scale bar =1 cm. (Li et al. 2020)

7.3.1.2 Microscopic Analysis

Microscopic examination proves valuable for identifying adulterants that may escape detection by the naked eye. This method employs a microscope to scrutinize the structure, morphology, and characteristics of edible flowers, enabling the observation of irregularities or foreign particles that may include non-edible flower parts or other contaminants (Kumari et al. 2021).

Moreover, microscopic techniques, such as optical microscopy, scanning electron microscopy (SEM) and fluorescence microscopy are commonly used to differentiate structural, cellular, and internal tissue features between plant species (Joshi et al. 2005; Smillie and Khan 2010; Revathy et al. 2012; Negi et al. 2021). These techniques help analyse various anatomical features, including the presence or absence of oil glands, epidermal cells, trichomes, seed and pollen structures, starch grains, and extraneous matter such as insect fragments and faeces (Negi et al. 2021).

In particular, advanced electron microscopic techniques, such as SEM, further enhance the analysis of anatomical features like the presence or absence of hairs (trichomes), starch grains, oil glands, canals, seed or pollen morphology, and vascular traces for detecting adulteration (Smillie and Khan 2010). For example, SEM has been utilized to differentiate between Illicium verum (star anise) and its adulterant Illicium anisatum by revealing differences in their epicarp morphologies (Fritz et al. 2008). Fluorescence microscopy has been employed for distinguishing star anise adulterants (Joshi et al. 2005).

Fig. 7.6 (**a**) *Scutellaria lateriflora* (By Rolf Engstrand—Own work, CC BY-SA 3.0, https://commons.wikimedia.org/w/index.php?curid=24468900), (**b**) *Teucrium canadense* (By Homer Edward Price—American Germander, CC BY 2.0, https://commons.wikimedia.org/w/index.php?curid=25214715), (**c**) *Teucrium chamaedrys* (By Konrad Lackerbeck—Own work, CC0, https://commons.wikimedia.org/w/index.php?curid=50769008)

Meanwhile, though improvements have been made in optical microscopy, scanning electron microscopy offers greater precision in detecting adulterants. However, the accuracy of microscopic methods relies on the expertise and skill in sample preparation. It is worth noting that microscopic analysis may not be cost-effective and suitable for routine examination, especially for highly processed food samples like fine powders with other plant materials (Negi et al. 2021).

Here, there is some example of microscopic analysis. For instance, Gafner et al. (2003) have utilized digital photo microscopy to distinguish between the *Scutellaria lateriflora* and its adulterants *Teucrium canadense* and *Teucrium chamaedrys* (Fig. 7.6) (Gafner et al. 2003).

The digital photo microscopy images of the raw materials from the three different plant species analysed (Figs. 7.7, 7.8, and 7.9) illustrated that this technique is a straightforward and refined method for detecting adulteration of skullcap with germanders. The simple presence of abundant strap-shaped trichomes on the stem, the bristle-like trichomes on the *T. canadense* leaf, and the waxy cuticle with numerous glandular scales on the *T. chamaedrys* leaf, which are absent in *S. lateriflora*, can differentiate adulteration between germander and skullcap. (Gafner et al. 2003).

In addition, microscopic techniques are commonly used for powdered spices. Figure 7.10, as an example, illustrates the detection of adulteration in powdered spices, turmeric (extraneous matter), chili (insect fragment), and black pepper (insect fragment), using microscopic methods.

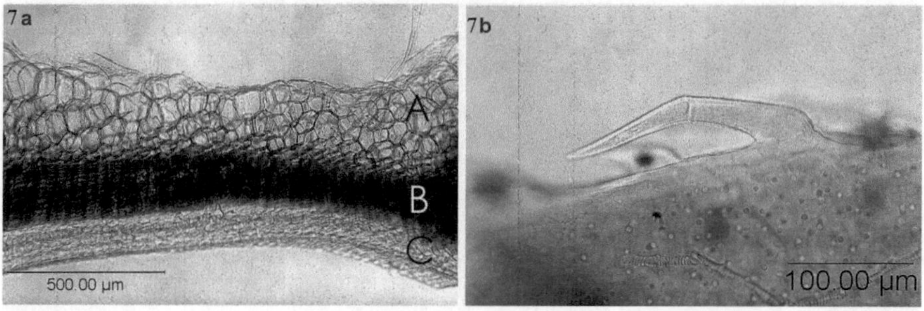

Fig. 7.7 (**a**) *Scutellaria lateriflora*, cross section of stem showing (**a**) thin-walled parenchyma of pith; (**b**) cortex; and (**c**) epidermis with flattened cells; 100× magnification. (**b**) *S. lateriflora*, cross section of leaf showing finger-like trichome with warted exine; 400× magnification. (Gafner et al. 2003)

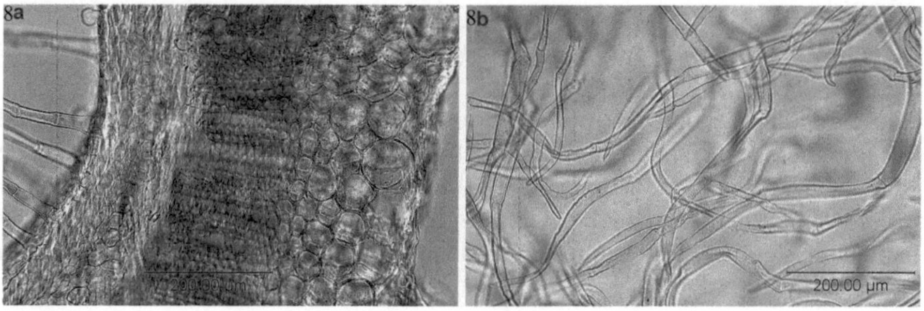

Fig. 7.8 (**a**) *Teucrium canadense*, cross section of stem showing epidermal trichomes, epidermis, cortex, and parenchyma of hollow pith; 200× magnification. (**b**) *T. canadense*, surface of stem showing strap-shaped multicellular trichomes; 200× magnification. (Gafner et al. 2003)

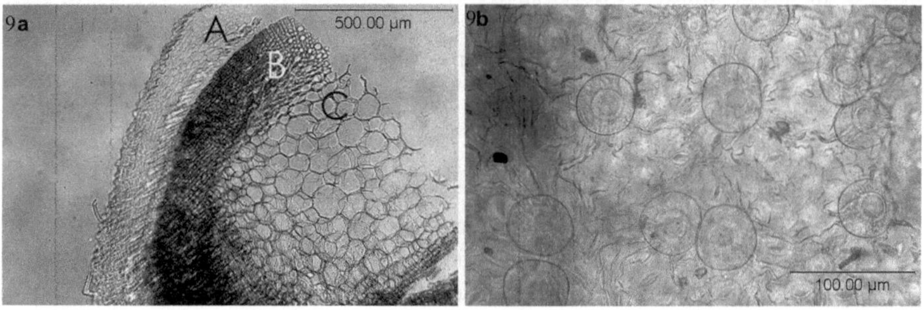

Fig. 7.9 (**a**) *Teucrium chamaedrys*, cross section of stem showing (**a**) epidermis; (**b**) cortex; (**c**) pith comprised of thin-walled parenchyma; 100× magnification. (**b**) *T. chamaedrys*, surface of leaf showing waxy cuticle with numerous stomata and glandular scales; 400× magnification. (Gafner et al. 2003)

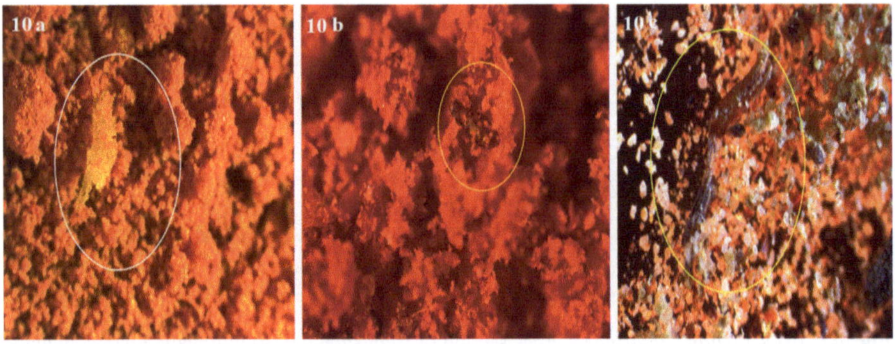

Fig. 7.10 Adulteration detection in (**a**) turmeric (extraneous matter), (**b**) chilli (Insect fragment) and (**c**) black pepper (Insect fragment) detected by physical authentication techniques. Viewed under 160 x magnification (Leica S8AP0, CMS GmbH stereo-zoom microscope LAS version 3.8.0, Singapore). (Negi et al. 2021)

7.3.1.3 Texture and Consistency Analysis

Adulterated flowers may exhibit differences in texture or consistency compared to authentic ones. Physical methods such as touch, feel, and squeeze tests can be conducted to assess the firmness, moisture content, elasticity, or brittleness of the flowers. Any significant deviations from the expected characteristics can indicate potential adulteration (Shukla and Iqbal 2015; Banti 2020; Negi et al. 2021).

7.3.1.4 Size and Shape Analysis

Edible flowers often have specific size and shape characteristics. Adulteration can involve the use of smaller or misshapen flowers to mimic authentic ones. Comparing the size, shape, and symmetry of the flowers with known standards can help identify any deviations that may indicate adulteration (Kumari et al. 2021).

7.3.1.5 Weight Analysis

Measuring the weight of the flowers can provide insights into their quality and potential adulteration. Significant variations in weight compared to standard values or expected ranges could indicate the presence of foreign materials or additives (Kumari et al. 2021; Negi et al. 2021).

7.3.1.6 Water Solubility Test

The solubility properties of certain adulterants may differ from authentic edible flowers, providing a basis for a water solubility test. This test involves immersing the flowers or their products in water and observing the extent to which they dissolve or release colour. Rapid or excessive colour bleeding or poor solubility during this test may indicate the presence of adulterants (Banti 2020; Negi et al. 2021).

For instance, a simple flotation test (bulk density) for detecting papaya seed in black pepper is widely used, followed by visual microscopic detection (Pruthi and Kulkarni 1969). Saffron, a highly expensive spice, is often adulterated with plant materials and artificial colorants in its powdered form, which helps maintain colour strength in saffron products while reducing saffron concentration (Dai et al. 2020). Table 7.1 outlines some characteristics of physical techniques used for determining adulterants in saffron.

A straightforward method to check saffron adulteration involves immersing saffron filaments in water. Authentic saffron expands into a trumpet-like shape, easily distinguishable from other biological adulterants of different floral origin (Kumari et al. 2021). To assess possible adulteration, the behaviour of adulterated samples was compared with true saffron in terms of releasing colour in water.

In Figs. 7.11 and 7.12, it is clear that the adulterated samples exhibited a more pronounced colour. In the case of sample AD.02.JM, upon water soaking, stigmas of varying colours (red and brown) were observed, indicating potential adulteration

Table 7.1 Characteristics of physical techniques used for determination of adulterants in Saffron (Kumari et al. 2021)

Major techniques/ methods	Specific methods/ techniques used for determination of adulterants	Main advantages	Limitations	Suitability for type of adulterants and compositional constituents of Saffron
Physical methods and techniques	Inspection of morphological features (size, shape, colour etc.) visually or through microscope Traditional testing of flavour and aroma by human olfaction Colorimetric technique based on colour parameters ISO test methods based on parameters (like moisture content, ash content etc.)	Simplest and easy to use Lesser or no sample preparation Good for primary examination for presence of adulteration	Lacks sensitivity, specificity, reproducibility, and accuracy for identification of adulterants Quantification of adulterants is not possible.	Plant adulterants with dissimilarity in morphological feature, Synthetic adulterants having colour variations Adulteration of Saffron floral parts (visible)

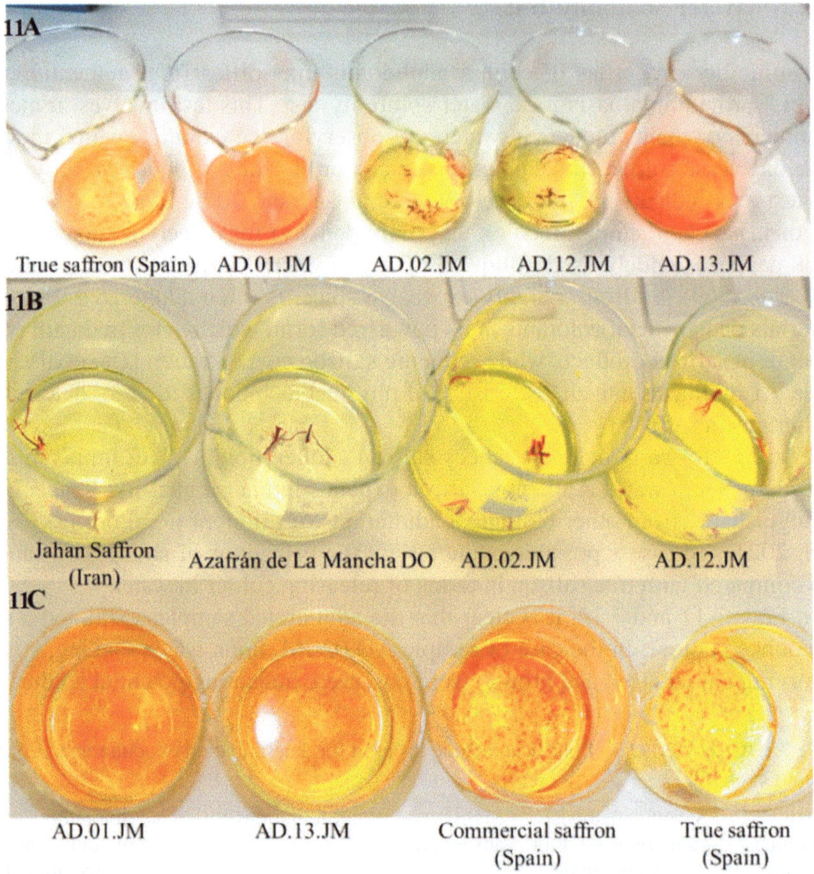

Fig. 7.11 Comparison of the different colouring properties of true and adulterated saffron samples. (**a**) Different staining capacity of the same amount of true saffron powder (true saffron from Spain) compared to adulterated powders and stigmas. The colour of true saffron is clearly less intense than the colour of the two adulterated powders (AD.01.JM and AD.13.JM) and more or less similar to the colour of the adulterated stigmas. The colouring properties of powder are stronger than the properties of stigmas. The fact that the colour intensity of true saffron and stigmas is similar is a consequence of the adulteration of stigmas; (**b**) Different staining capacity of the same amount of true saffron (Jahan saffron, from Iran, and Azafrán de La Mancha DO (Denominación de origen), from Spain) and adulterated saffron (AD.02.JM and AD.12.JM) stigmas. It is evident that true saffron has less staining capacity than adulterated saffron in cold water. The latter samples probably release some synthetic dye; (**c**) Different staining capacity of the same amount of true saffron powder (true saffron from Spain), a commercial sample (commercial saffron from Spain) and the two adulterated samples used in this work. At the same time, adulterated and commercial samples produced a stronger coloration of water than true saffron. (Reproduced with permission from Soffritti et al. 2016)

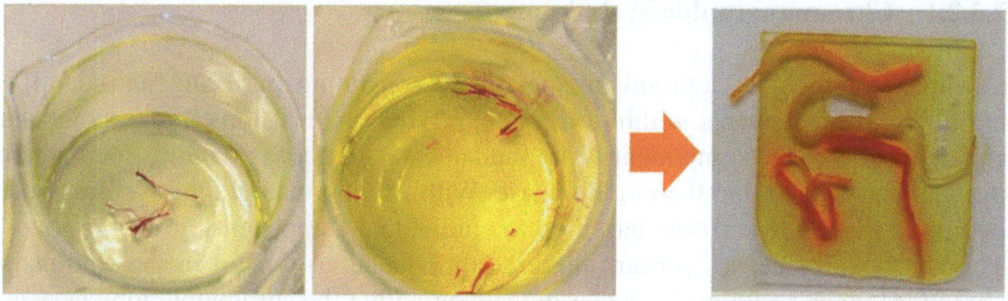

Azafrán de La Mancha DO AD.02.JM AD.02.JM (detail)

Fig. 7.12 Comparison between same amounts of true saffron stigmas (Azafrán de La Mancha DO, Spain) and adulterated stigmas (Sample AD.02.JM) after just 3 min in cold water. It is clearly visible that, at the same time, adulterated saffron (dark yellow) releases colour faster than true saffron (pale yellow). On the right, a detail of the adulterated stigmas from Sample AD.02.JM. There are clearly visible stigmas with different colours: red, pale orange, likely brown stigmas still decolouring and pale brown. The adulteration probably corresponds to the mixture and staining of two or more very old saffron stigmas with some fresh stigma lots. (Reproduced with permission from Soffritti et al. 2016)

with a blend of saffron from different years, possibly tainted with a synthetic dye. This could explain why there was no amplification with markers other than the one for crocus (Dai et al. 2020).

While physical methods can offer initial indications of adulteration, they may fall short in identifying specific adulterants or contaminants. Additionally, despite their cost-effectiveness and straightforward nature, these methods do not guarantee the overall quality of the product. For more precise identification, chemical and biochemical methods such as chromatography, spectroscopy, electrophoresis, immunology, DNA testing, and other specialized techniques are employed to detect adulterants.

Chemical or biochemical methods are considered more precise and responsive than physical methods. However, their industrial applications face challenges due to their high cost, technical complexities, and the need for trained human resources (Gonzálvez et al. 2008). In the subsequent discussion, various techniques currently in use or with futuristic potential for detecting different types of adulterants will be reviewed and explored. This comprehensive examination aims to shed light on existing methods and pave the way for future research efforts in the field of adulterant detection.

7.3.2 *Chemical Analysis*

Chemical methods of detecting adulterants can be divided into four types: those that use chromatography, spectroscopy and immunology, and those that employ electrophoresis (Haji et al. 2023). Each of these techniques has been designed with a particular chemistry or biochemistry in mind.

7.3.2.1 Chromatographic Techniques

Chromatography offers a promising method for separating, purifying, and identifying different components within a mixture, relying on variations in the molecular characteristics of the components. The fundamental components of this method are the stationary phase and the mobile phase. Within a chromatography system, these two phases begin to separate based on the individual components at varying rates. This discrepancy causes certain mixture components to remain in the stationary phase for longer, resulting in slower movement within the chromatography system, while others swiftly transition into the mobile phase, exiting the system rapidly (Sudhakar et al. 2023).

Typically, chromatographic methods are employed to assess the composition of constituents or compounds, with some techniques capable of quantifying their levels. The fundamental analytical process includes various steps such as extraction using a suitable solvent, chromatographic separation, clean up to eliminate interfering matrix components, and selective detection (Banti 2020; Nehal et al. 2021). High Performance Liquid Chromatography (HPLC), Gas Chromatography (GC), and Thin Layer Chromatography (TLC) are widely utilized chromatographic techniques for analysis. In Liquid Chromatography (LC)-based methods (TLC and HPLC), the separation of constituents relies on factors such as polarity, electrical charge, or molecular size of the compounds. At times, these analytical methods are put together with other procedures such as Mass Spectroscopy in order to obtain more accurate information about the specified substances (Kumari et al. 2021). The detection of adulterants involves following the specific procedures associated with each method. This discussion focuses on the most common types of chromatographic techniques employed for detecting adulteration.

7.3.2.2 Thin Layer Chromatography (TLC)

Thin Layer Chromatography (TLC) is a method used for the separation and analysis of various compounds within a sample. The procedure entails applying the sample extract onto a thin layer of adsorbent material and facilitating the separation of compounds by allowing them to move through the layer. Adulterants can be recognized by comparing the chromatogram of the sample to established standards, discerning distinct separation patterns (Poole 2023). Figure 7.13 provides an overview of the entire thin-layer chromatography process Flanagan et al. 2008).

In Haochen Dai et al. (2020) research, they devised a rapid screening and quantification technique that integrates TLC and Raman spectroscopy for analysing saffron powder. This method allows for the quantification of the primary colorant crocin content and the identification of potential adulterants beyond laboratory confines. The quality characteristics of saffron products are primarily influenced by the concentrations of three key constituents: crocin, picrocrocin, and safranal (Anastasaki et al. 2010; Varliklioz Er et al. 2017). These three compounds (crocin, picrocrocin, and safranal) are responsible for the colour, bitter taste, and a unique

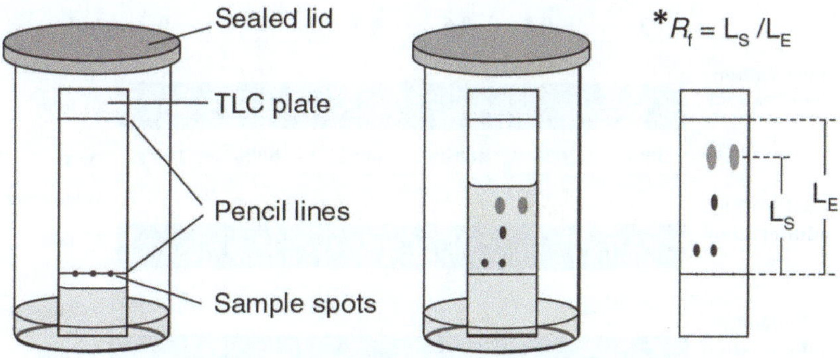

Fig. 7.13 Schematic approach for thin-layer chromatography analysis. (Flanagan et al. 2008)

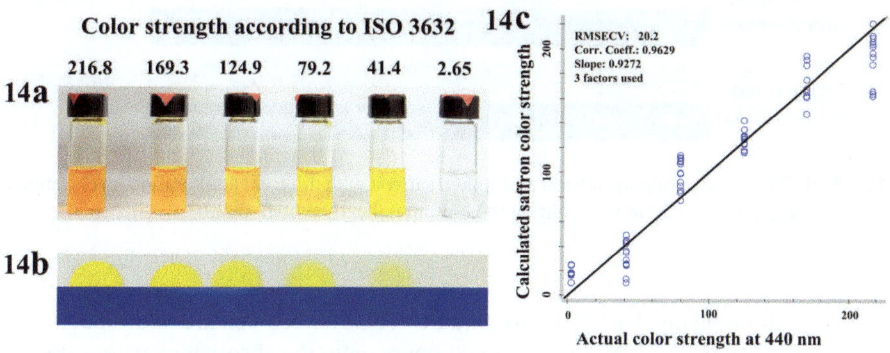

Fig. 7.14 Physical observation of pure saffron extract (**a**) and its TLC pattern (**b**) under different lighting conditions (ambient light on the top half, 365 nm UV light on the bottom half); (**c**) Predict vs. actual colour strength PLS plots for saffron specimens using L*a*b* data. (Dai et al. 2020)

aroma to saffron, respectively. The intensity of saffron's colour, often referred to as colour strength, holds significant importance in the commercial market (Anastasaki et al. 2010). In the conducted study, a dual approach was adopted, employing TLC for visual screening and subsequently utilizing Raman spectroscopy to directly measure the Raman signal of saffron water extracts on a TLC plate through a portable Raman instrument. Chemometric models were established to quantify crocin concentration and identify potential adulterants. The amalgamation of a cost-effective TLC plate with a portable Raman instrument presents a promising solution for the swift assessment of the quality and authenticity of saffron powders beyond traditional laboratory settings (Dai et al. 2020).

As depicted in Fig. 7.14, as the concentration of saffron extract increased, the colour of the extract solutions transitioned from a vibrant yellow to orange. When applied onto a TLC plate, the saffron extracts produced a consistent yellow circle without any pronounced coffee ring pattern under bright light or a fluorescence pattern under UV light (Fig. 7.14b).

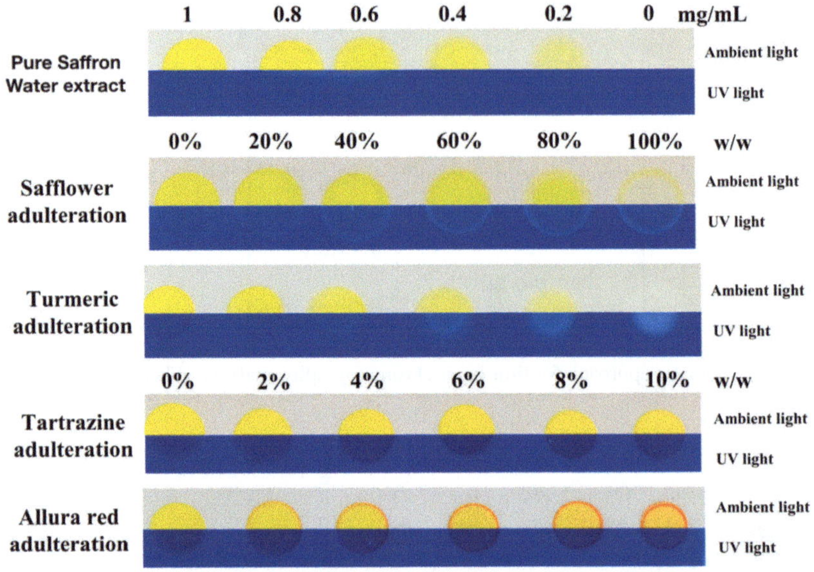

Fig. 7.15 TLC pattern of pure saffron and spiked saffron with different adulteration level (ambient light on the top half, 365 nm UV light on the bottom half). (Dai et al. 2020)

Calibrations using Partial Least Squares (PLS) were constructed, specifically PLS1, utilizing the L*a*b*2 values in comparison to the absorbance values derived from the ISO method. The Limit of Quantification (LoQ) for this method was determined to be 38.6, significantly below the threshold for category III (120) as defined by ISO 3632-1 (2011). In this investigation, premium saffron and different adulterants underwent analysis under two distinct lighting conditions, namely ambient and UV light (Fig. 7.15). Observation revealed that premium saffron produced a consistently coloured circle under both ambient and UV light. In contrast, safflower exhibited a distinctive fluorescence ring, while turmeric formed a fluorescence inner circle (Dai et al. 2020).

In addition to visual inspection, the researchers employed Principal Component Analysis (PCA) to assess the distinctions among the samples. Figure 7.16a illustrates a well-separated distribution of pure saffron and each adulterant in the PCA plot, highlighting the inherent variability among these samples captured by three principal components, totalling 77.78% of the variation. To evaluate the sensitivity of the TLC method for each individual adulterant, the Limit of Detection (LoD) was calculated from each PLS plot (Fig. 7.16b–e). The results indicated LoD values as follows: Allura red (2.59%), tartrazine (4.15%), safflower (31.01%), and turmeric (41.98%) (Dai et al. 2020).

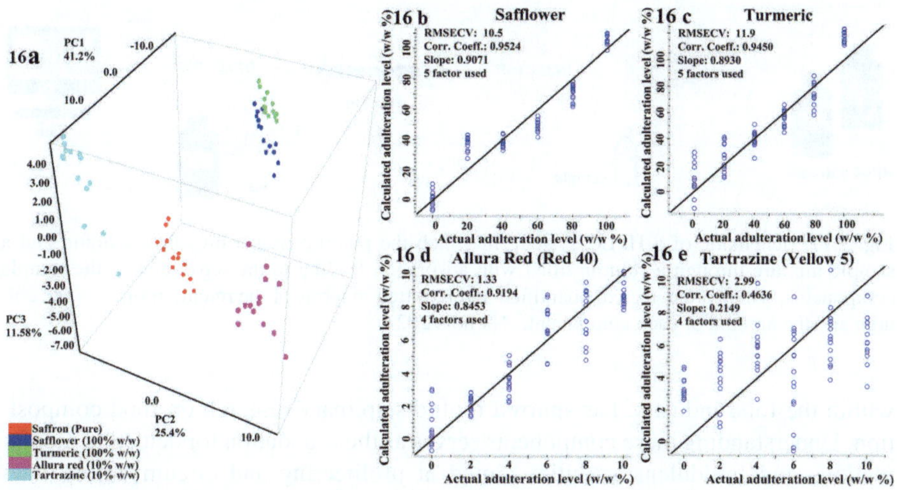

Fig. 7.16 (**a**) Principal component analysis of pure saffron and max adulterated samples based on TLC pattern L*a*b* values. Predict vs. actual colour strength plots for adulterated saffron specimens based on L*a*b* data of (**b**) safflower, (**c**) turmeric, (**d**) Allura red, tartrazine. (Dai et al. 2020)

7.3.2.3 High-Performance Liquid Chromatography (HPLC)

High-Performance Liquid Chromatography (HPLC) stands out as a robust chromatographic separation technique capable of analysing a wide range of compounds with diverse properties, spanning from low to very high molecular mass substances. The fundamental principle involves passing a dissolved mixture in a mobile phase through a material known as the stationary phase, with both phases being immiscible (Reuhs 2017; Lozano-Sánchez et al. 2018). The selection of the specific type of liquid chromatography is contingent upon the nature, chemical structure, and molecular weight of the analytes. Consequently, various forms of HPLC have been developed to provide both qualitative and quantitative insights into individual components within the sample under investigation (Lozano-Sánchez et al. 2018).

In contemporary analytical practices, continuous HPLC methods have become commonplace, wherein compounds undergo online detection following the separation process. The detectors employed in HPLC exhibit considerable diversity. HPLC is seamlessly integrated with various detection systems, including ultraviolet-visible (UV-Vis) spectroscopy, mass spectrometry (MS), nuclear magnetic resonance (NMR) spectroscopy, and Fourier transform Raman (FTR) spectroscopy (Reuhs 2017; Ali 2022). These combinations enable the analysis of complex mixtures of compounds, facilitating their separation, identification, and quantitation. The coupling of HPLC with Mass Spectrometry (MS), in particular, has proven instrumental in detecting and discovering minor and specific compounds or groups within the mixtures. Such analysers present a robust analytical solution, as evidenced by their application in recent publications characterizing food products. The chemical characterization of minor compounds, which hold significance in authenticity testing

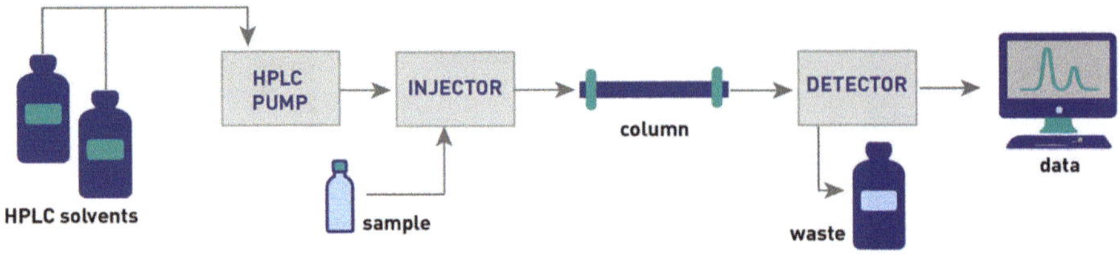

Fig. 7.17 Schematic of a HPLC system, in which the pump passes a pressurised liquid and a sample mixture through a column filled with adsorbent, leading to the separation of the sample components, which are detected, quantified and reported in terms of the retention time on the column and the amount of each component. (Nichols 2021)

within the food industry, has spurred multidisciplinary research on food composition. Understanding these components serves as the foundation for detecting manipulations and fraudulent activities aimed at profiteering and circumventing laws established to safeguard food authenticity. Authenticity in the context of food implies that the product exclusively contains the materials indicated on the packaging. Authentic food is characterized by being natural, pure, undiluted, and free from undeclared ingredients. The rules and regulations in place serve as a framework to ensure and guarantee the quality of food products (Reuhs 2017; Lozano-Sánchez et al. 2018; Ali 2022).

A contemporary HPLC system comprises essential components such as a solvent delivery system, sample injector, column, detection system, and a computer data station. Figure 7.17 provides a schematic diagram illustrating the configuration of an HPLC system (Nichols 2021).

7.3.2.4 Ultra-High-Performance Liquid Chromatography (UHPLC)

Similar to HPLC, Ultra-High-Performance Liquid Chromatography (UHPLC) is a method for separating various components of a material. Mostly used to quantify, identify, and separate mixture components by pushing solvents through a column at high pressure (Nováková et al. 2017).

In a recent study, Natalia Casado et al. (2022) used tandem mass spectrometry analysis in conjunction with UHPLC to analyse the pyrrolizidine alkaloids found in infusions of Hibiscus, Calendula, and Mallow flowers. Actually, pyrrolizidine alkaloids (PAs), which are naturally occurring toxins that may be present in food, are produced by plants in the groups Asteraceae, Boraginaceae, Fabaceae, Apocynaceae, and Orchidaceae. The goal of this work was to assess the solid phase micro extraction (μSPEed) method for the simultaneous capture of phenolic alkaloids and then analyse them using ultra-high-performance liquid chromatography coupled to ion trap tandem mass spectrometry (UHPLC-IT-MS)/MS). This was intended to create a reliable and sensitive analytical procedure to detect the presence of these alkaloids in mallow, calendula, and hibiscus infusions, for which the evaluation of phenolic

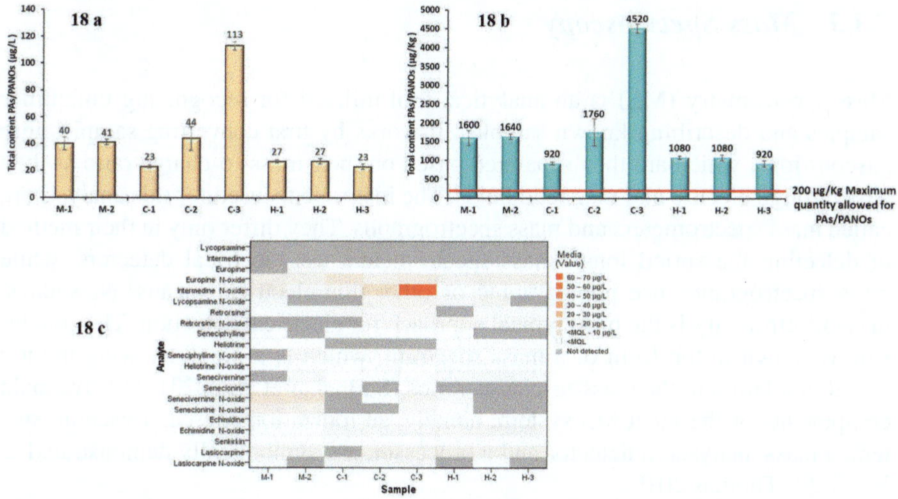

Fig. 7.18 (**a**) Total content of pyrrolizidine alkaloids in infusion samples (µg/L), (**b**) total content of pyrrolizidine alkaloids in the dry plants (µg/ kg) considering a transfer rate of 100%, and (**c**) heat map plot of the individual pyrrolizidine alkaloids found in the different edible flower infusion samples analysed with the µSPEed method proposed. In the sample identification code, the first letter indicates the type of flower (M for mallow, C for calendula and H for hibiscus). (Casado et al. 2022)

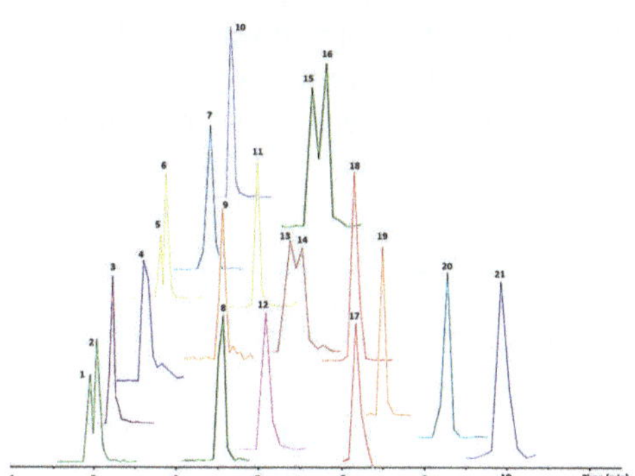

Analyte [Precursor ion (m/z)]	Retention time (min)	MS^2 Product ions (m/z)
1. Intermedine [299]	4.8	138*, 120
2. Lycopsamine [299]	5.1	138*, 120
3. Europine [329]	5.3	253*, 138
4. Europine N-oxide [345]	5.7	327*, 171.5
5. Intermedine N-oxide [315]	5.8	225, 171.5*
6. Lycopsamine N-oxide [315]	5.9	171.5*, 138
7. Retrorsine [351]	6.4	323*, 275
8. Retrorsine N-oxide [367]	6.6	339*, 245
9. Seneciphylline [333]	6.6	305*, 120
10. Heliotrine [313.5]	6.7	138*, 120
11. Seneciphylline N-oxide [350]	7.0	321*, 118
12. Heliotrine N-oxide [329]	7.1	171*, 136
13. Senecivernine [335]	7.4	307*, 120
14. Senecionine [335]	7.5	307*, 120
15. Senecivernine N-oxide [351]	7.6	323*, 219.5
16. Senecionine N-oxide [352]	7.8	220, 118*
17. Echimidine N-oxide [398]	8.2	220, 120*
18. Echimidine [413]	8.2	395*, 351
19. Senkirkin [365]	8.5	167.5*, 150
20. Lasiocarpine [411]	9.3	335*, 219.5
21. Lasiocarpine N-oxide [328]	9.9	409*, 352

** Ions used for quantification*

Fig. 7.19 Retention time, precursor ion and product ions of the analytes analysed with the UHPLC-IT-MS/MS chromatographic method used. (Casado et al. 2022)

alkaloids has previously been unexplored. Natalia Casado et al. (2022) demonstrate: "it was observed that the contamination of the infusions was mainly due to the N-oxide forms (PANOs) in the three types of flowers, which were more predominant than their corresponding PAs (Figs. 7.18c and 7.19)" (Casado et al. 2022).

7.3.3 Mass Spectroscopy

Mass spectrometry (MS) is an analytical tool utilized for recognizing unfamiliar samples and describing known samples. It works by first converting samples into gaseous ions, which are then separated based on their mass-to-charge ratio as they pass through electric and magnetic fields. The instruments used for this analysis are called mass spectrometers and mass spectrographs. They differ only in their method of detecting the sorted ions—mass spectrometers use electrical detectors, while mass spectrographs use photographic or other non-electrical means. Nowadays, mass spectrometry is the most typical approach for electrical detection. The conclusion is shown in the form of a mass spectrum, which is a graph showing the ion signal in relation to the mass-to-charge ratios (Sudhakar et al. 2023). The five main components of the most MS system, namely, an ionization source, a vacuum system, a mass analyser, a detector and a processor, are symbolically demonstrated in Fig. 7.20 (Thomas 2019).

As MS is a powerful analytical technique that detect an unknown sample in a known one, Peña et al. (2005), has developed a MS method and multivariate regression that could detect adulterants in virgin olive oil (VOO) (Peña et al. 2005). They developed a mass spectrometry method paired with multivariate regression models that was able to detect adulteration in refined olive oil at 7% and virgin olive oil at 15%. Another study conducted by Catharino et al. (2005) showed that electrospray ionization MS (ESI-MS) could clearly identify olive oil adulteration with soybean oil (Catharino et al. 2005). Matrix-assisted laser desorption ionization time-of-flight mass spectrometry was employed by Calvano et al. (2012) to investigate whether hazelnut oil was present in extra virgin olive oil (EVOO) (Calvano et al. 2012). Using a statistical tool in combination with mass spectrometry, it was found that the data indicated the presence of 1% hazelnut oil in extra-virgin olive oil, as well as other adulterants such as corn oil, soybean oil, and rapeseed oil in trace amounts

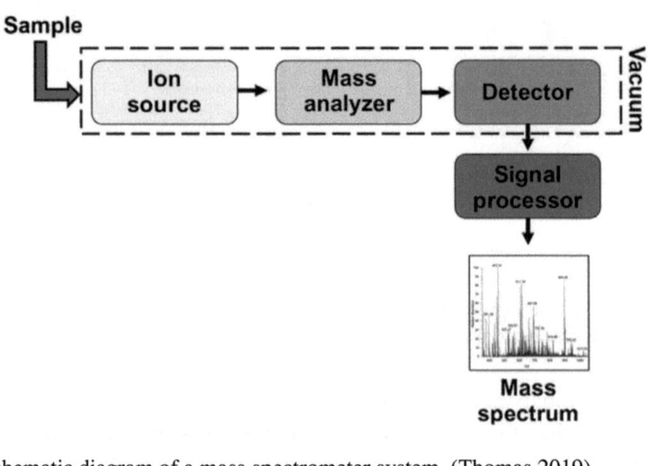

Fig. 7.20 Schematic diagram of a mass spectrometer system. (Thomas 2019)

(Zhang et al. 2016). Based on the findings from the reviewed studies utilizing mass spectrometry for edible oil authentication, it can be concluded that MS paired with either qualitative or quantitative chemometric data analysis techniques demonstrates immense capability for identifying adulterants of all types of edible oils, even at very low amount 0.5%. The chemical fingerprints generated through MS provide the potential to differentiate edible oils based on detecting patterns within the spectral data resulting in specific compound of adulterants. A major advantage of MS over techniques like vibrational spectroscopy is minimal reaction with water. However, limitations hindering broader adoption of MS for edible oil purity verification involve the longer analysis durations, higher costs of instruments and tests, as well as larger requiring area (Sudhakar et al. 2023).

7.3.3.1 Gas Chromatography–Mass Spectrometry (GC–MS)

Due to highly accurate analysis, gas chromatography and mass spectroscopy (GC-MS) is widely utilized for both qualitative and quantitative analysis within food as well as other industries, including forensics, energy, environment, and pharmaceuticals (Bubli et al. 2021).

GC/MS instrument is a highly precise tool for analysing environmental samples. It operates by separating chemical mixtures through the GC component, which relies on the principle that heating a mixture causes individual substances to separate. The separated substances are carried through a column by an inert gas, such as helium. Upon emerging from the column opening, these substances enter the MS component. In MS, compounds are identified based on the mass of the analyte molecule. A computer stores a comprehensive "library" of known mass spectra, encompassing thousands of compounds. Mass spectrometry is widely recognized as the definitive analytical detector in this context.

The GC/MS technique is utilized to detect adulteration in plant essential oils. Here we specifically focus on plant essential oils and various analytical methodologies employed for identifying adulteration, including GC/MS technique.

Essential oils are natural substances extensively employed across various global industries, and have become an essential aspect of daily life. The growing demand for these oils has led to instances of adulteration. Consequently, authentication holds significant importance for consumers and chemical companies. Essential oils, being intricate matrices, necessitate analysis through diverse techniques. To guarantee quality, consumer safety, and fair trade, a broad spectrum of instrumental techniques is employed. These encompass physical, organoleptic, chemical, chromatographic, and spectroscopic analyses (Fig. 7.21) (Panda 2003; Goodner and Rouseff 2011).

The control methods and standardization of essential oils aim to confirm adherence to quality monographs or standards. However, instances of non-compliance do not necessarily indicate adulteration. Factors such as aging, processing, or storage can lead to racemization of chiral compounds or polymerization reactions of terpenoids. These processes may alter optical activity values without involving adulteration (Mosandl 1995).

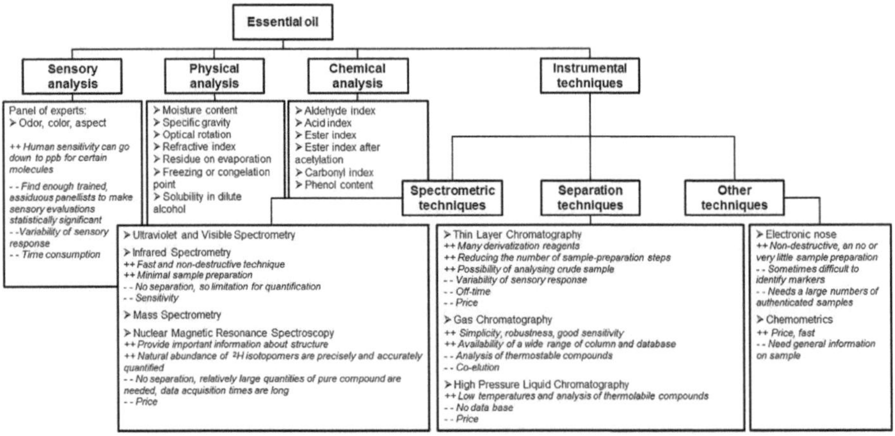

Fig. 7.21 Advantages and disadvantages of quality assessment techniques (++: Advantages; −−: Disadvantages). (Do et al. 2015)

In summary, the variety of adulteration strategies is linked to the extensive and diverse array of essential oils utilized in the production of valuable goods, rendering each case distinct from one another.

The presence of adulterants like ethanol, edible oils, or liquid paraffin can be identified through techniques such as TLC, GC, or infrared (IR) spectroscopy (Do et al. 2015). GC analysis, considered a crucial step in essential oil analysis, is recommended in certain monographs. It offers a detailed chromatogram that effectively highlights quality defects. The chromatographic profile obtained by GC equipped with a flame-ionization detector (FID) is commonly compared for quality control, given its broad linearity range and simple maintenance. This involves examining the chromatogram for the appearance or disappearance of peaks and comparing the relative percentage area to assess significant differences in the product being analysed. While this technique provides an approximation of component quantity, it is widely utilized in Quality Control (QC) for essential oils, where the product's chromatogram is compared to a standard meeting chosen specifications. In the case of various perfumery materials, GC coupled with mass spectrometry (GC-MS) is employed for both qualitative and quantitative characterization to accurately detect adulteration (Do et al. 2015). For instance, in the analysis of sandalwood oil, GC-MS reliably assesses the santalol content, serving as a valuable marker to detect adulteration with synthetic materials like Sandalore (Howes et al. 2004), Verdox, Santaliff, Vertofix Coeur, or Ebanol (Sciarrone et al. 2011).

7.3.3.2 Chiral GC Analysis

Chiral gas chromatography (GC) is an increasingly crucial method for verifying the authenticity of essential oils and detecting potential adulterants (McGorrin 2009; Thao and Satake 2010). Many plant metabolites exist as chiral molecules, and enantiomers can vary among species within the same genus. Despite having similar physicochemical properties, enantiomers can differ in biological activities, with one being harmless and the other toxic. Therefore, accurate and efficient chemical analyses are essential (Waksmundzka-Hajnos and Sherma 2010). For instance, the scent of oranges is attributed to (R)-limonene, whereas the odour of lemons is due to (S)-limonene. (S)-carvone is the predominant odorant in caraway essential oil (smelling like cumin), while (R)-carvone is found in spearmint essential oil (smelling like spearmint). In some cases, certain stereoisomers may be less active or even odourless, such as the powerful flowery note of (R)-linalool compared to the less intense (S)-linalool, or the rosy scent of (2S,4R)-cis-rose oxide versus the odourless (2R,4S)-cis-rose oxide (Do et al. 2015). Most compounds in essential oils are chiral, and their enantiomeric ratios are often specific to the oil, making them useful markers for determining the origin and detecting adulteration. Chiral analysis is instrumental in identifying adulteration with synthetic substitutes, typically in racemic form, or bulking oils from different sources. Enantiomeric purity and enantiomeric excess e-values, expressed as percentages, are crucial parameters in this analysis. For example, (−)-*trans* rose oxides are specific indicators of genuine rose oils, differentiating them from geranium oils. The chirality of the components can be used to uncover the presence of synthetic linalool and linalyl acetate that may have been added to lavender oil (Do et al. 2015). Chiral analysis is a vital part of essential oil analysis, becoming one of the most noteworthy analytical techniques in recent years. It is a cost effective, sensitive technique, but the process of developing a method can be time-consuming, particularly due to the absence of a universal chiral stationary phase (Rubiolo et al. 2010). Essential oils, undergoing processing or storage, may experience non-enzymatic reactions or racemization, potentially leading to false-positive outcomes in chiral analysis (Skaria 2007; Do et al. 2015; Sadgrove et al. 2022).

7.3.3.3 Isotope-Ratio Mass Spectrometry

To authenticate the natural origin of specific components in an essential oil, an additional type of analysis involves isotope-ratio analysis using either isotope-ratio mass spectrometry (IRMS) or stable-isotope-ratio analysis (SIRA). Discrimination among plants is achieved by examining their metabolic assimilation of atmospheric CO2, particularly through reaction intermediates involving carbon dioxide incorporation, specifically molecules with three or four carbon atoms. Most plants follow the C3 pathway through the intermediate 3-phosphoglycerate, while C4 plants utilize a malate intermediate with four carbon atoms. Some plants can select either the glycerate or malate pathway (CAM cycle) based on environmental factors. Due to distinct isotopic fractionation of these metabolites, plants can be differentiated by their isotopic ratios. Isotope-ratio analysis relies on the principle that various stable isotopes of chemical elements lead to different molecular weights. Each element has one or more isotopes present at varying levels, forming a unique distribution pattern. Stable-isotope ratios of carbon, hydrogen, oxygen, or nitrogen within

Table 7.2 Enantiomeric Ratio (%) of α-pinene, β-pinene and limonene (Do et al. 2015; Mosandl 1995)

	α-pinene		β-pinene		limonene	
	1S	1R	1S	1R	1S	4R
Oil of bergamot	72	28	94	6	14	86
Oil of bitter orange	8	92	97	3	1	99
Oil of grapefruit	–	100	34	66	tr	100
Oil of lemon	67	38	95	5	1	99
Oil of lime	76	24	97	3	2	98
Oil of orange	tr	100	46	54	tr	100
Oil of neroli	77	23	96	4	3	97
Oil of petigrain	82	18	98	2	12	88
Oil of mandrain	43	57	3	97	tr	100
Oil of citronella	23	77	tr	tr	96	4
Oil of lemongrass	96	4	tr	tr	100	tr

tr = trace <0.5%

metabolites enable the identification of accidental or intentional incorporation of synthetic products, particularly those of fossil origin. The analysis of isotopic data has become a key method for assessing the origin and natural authenticity of flavours and fragrances. IRMS IS often coupled with combustion/pyrolyze (C/P--IRMS), employed for controlling adulteration. This technique effectively distinguishes natural from synthetic mandarin essential oil by analysing C-isotope ratios for terpinen-4-ol, γ-terpinene, α-terpineol, and terpinolene. Table 7.2 shows some chiral compounds, are often specific to the essential oil (e.g., α-pinene, β-pinene and limonene), (Do et al. 2015).

Additional methods valuable for confirming the authenticity of essential oils encompass spectroscopic techniques like middle infrared spectroscopy (Bounaas et al. 2018), nuclear magnetic resonance spectroscopy (Sadgrove et al. 2022), and UV-fluorescence coupled with transmittance, demonstrated in the authentication of patchouli oil (Al Riza et al. 2019). However, it's important to note that while spectroscopic methods are useful, they may not offer immediate and conclusive evidence to challenge intricate forms of adulteration (Sadgrove et al. 2022).

7.3.3.4 Nuclear Magnetic Resonance Spectroscopy (NMR)

NMR spectroscopy provides valuable information regarding the molecular structure and composition of a sample. When applied to edible flowers, NMR can be utilized for the identification and quantification of compounds. The detection of adulterants is facilitated by comparing the NMR spectrum of the sample with known spectra. In the realm of authenticity control, NMR spectroscopy plays a crucial role by examining stable-isotope ratios, offering a method to analyse isotopic patterns within both natural and synthetic molecules for differentiation purposes (Thao and Satake 2010;

Li et al. 2012). Although NMR is a powerful technique, its application requires compound isolation, databases, an experienced operator, and substantial investment (Do et al. 2015).

NMR spectroscopy is also a straightforward, swift, and non-destructive method employed for assessing the quality of food (Fig. 7.22) (Todt et al. 2008; Santos et al. 2016).

Actually, the fundamental principle of NMR revolves around inducing energy absorption in a nucleus situated in a uniform magnetic field through simultaneous termination of radio frequency pulses at the appropriate frequency (Slichter 2013). When the absorbed energy is released, the atom of interest achieves thermal equilibrium, resulting in the emission of an NMR signal. In the magnetic field, the nucleus, possessing a positive charge, behaves like a bar magnet, causing it to spin alongside the protons. The chemical composition of the food dictates the distribution of lipids, carbohydrates, proteins, and water within the food matrix, influencing the mobility and proton distribution. These factors, in turn, impact the relaxation time of the proton to reach equilibrium. Relaxations involve the process of the nucleus returning to a stable state after being elevated to high energy levels.

Various NMR applications, such as low-field NMR, proton nuclear magnetic resonance (^{1}H NMR), diffusion NMR, ultrafast 2D NMR, and time-domain NMR, offer different approaches to the utilization of this technique. A lot of investigations have been conducted to determine the genuineness, derivation, and characterization of fats and oils utilizing their chemical components (Agiomyrgianaki et al. 2010; Guyader et al. 2018; Zhu et al. 2019; Shi et al. 2019). NMR versatile technique can be applied for both qualitative and quantitative analyses, particularly in examining

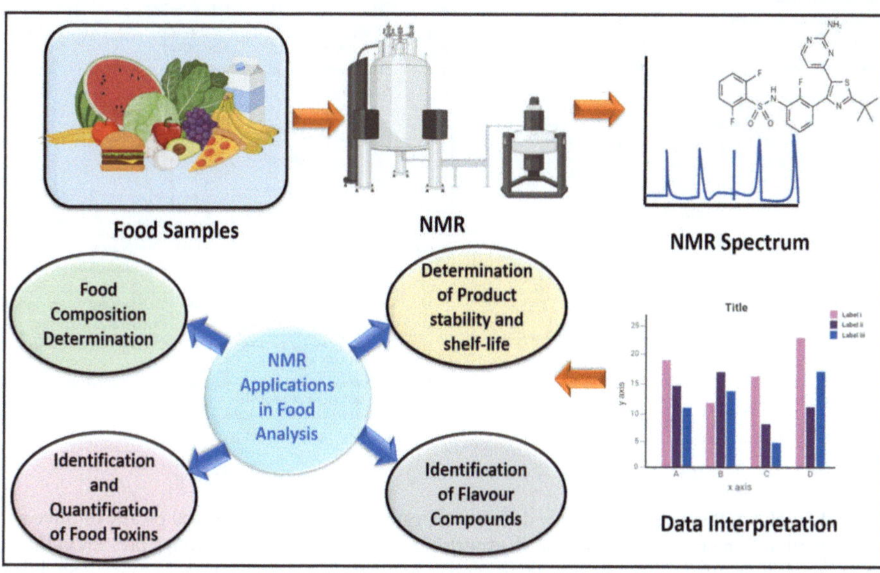

Fig. 7.22 Applications of NMR spectroscopy in food quality analysis. (Bharti et al. 2024)

edible oils composed of triglycerides with saturated (e.g., palmitic and stearic) and unsaturated (e.g., oleic, linoleic, and α-linolenic acid) fatty acids (Pariak and Güzeler 2016). The interpretation of the ^{1}H NMR spectrum of oils necessitates a comprehensive understanding of the intricate nature of free fatty acids. Key chemical signatures of free fatty acids include olefins, bis-allylics, allylic/α-CH2's, alkyls, and carboxylic acids. The concentration of fatty acids in specific edible oils accounts for variations in spectral resonance ranges (Fig. 7.23a) (Riegel 2015).

Consequently, ^{1}H NMR spectroscopy can be employed for the classification of edible oils based on fatty acid profiles. An advantage of NMR lies in its ability to analyse samples without requiring prior preparation or pretreatment (Sudhakar et al. 2023). This attribute has led to the application of NMR for detecting adulteration in edible oils. When coupled with appropriate chemometrics, NMR facilitates enhanced visualization and rapid discrimination between pure and adulterated samples (Šmejkalová and Piccolo 2010; Ravaglia et al. 2019). For instance, Shi et al. (2018) explored the application of ^{1}H NMR combined with chemometrics to detect the adulteration of camellia oil with three distinct vegetable oils (Shi et al. 2018). The validated results demonstrated a rapid and accurate authentication of camellia oil, specifying the type of adulterant and the level of adulteration.

Figure 7.23a, shows the composition of extra virgin olive oil (EVOO) primarily comprises mono- and polyunsaturated fatty acids, with a very low content of bis-allylic fatty acids. In contrast, soybean oil contains 50–60% polyunsaturated fatty acids, predominantly linoleic acid (x-6). Therefore, the intensity of the bis-allylic

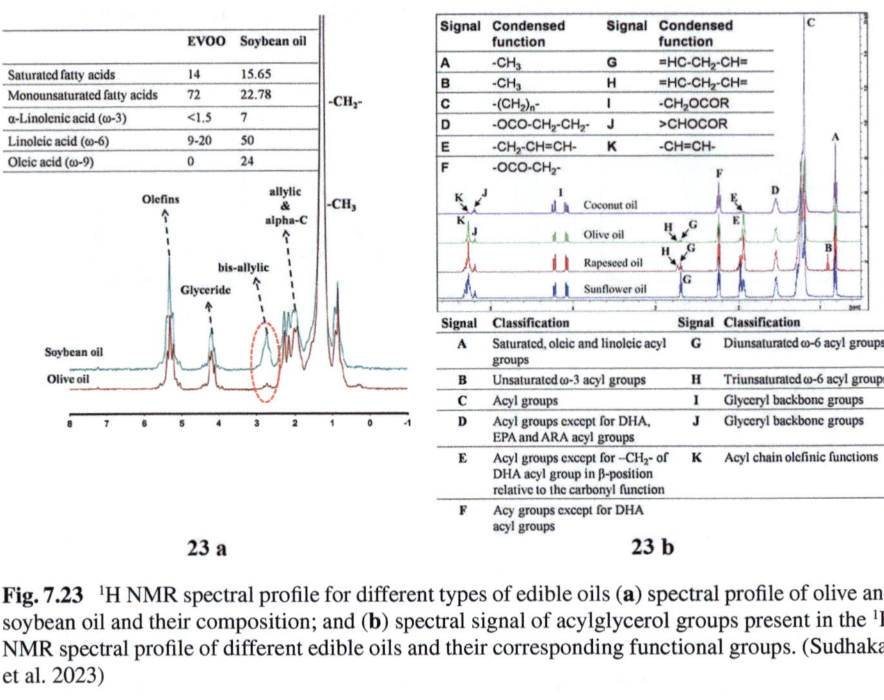

23 a 23 b

Fig. 7.23 ^{1}H NMR spectral profile for different types of edible oils (**a**) spectral profile of olive and soybean oil and their composition; and (**b**) spectral signal of acylglycerol groups present in the ^{1}H NMR spectral profile of different edible oils and their corresponding functional groups. (Sudhakar et al. 2023)

peak for soybean oil and adulterated EVOO is higher than that of pure EVOO (Fig. 7.23a). The comprehensive assessment of components corresponding to bis-allylic protons provides a clear indication of adulteration; if the intensity surpasses a predefined threshold, it implies the presence of polyunsaturated fatty acids (Riegel 2015).

The ^{1}H NMR spectral profile of different types of edible oils is contingent on the concentration of saturated and unsaturated fatty acids (Fig. 7.23b). The intensity of peaks for a specific compound differs among different oil types due to variations in concentration (Le Gresley et al. 2019).

A critical examination on NMR technology for edible oil adulteration suggests that this technology, coupled with suitable chemometric techniques, is effective for authenticating various edible oils even at low levels of adulteration. NMR technology offers abundant information regarding the compounds present in oils, emphasizing the importance of selecting specific spectra that clearly differentiate between edible and adulterant oils. ^{1}H NMR, ^{13}C NMR, and ^{31}P NMR have distinct applicabilities based on their advantages and limitations. ^{1}H NMR and ^{31}P NMR are suitable for cases where higher sensitivity and rapid analysis are required, whereas ^{13}C NMR, with lower sensitivity and lengthier quantitative analysis, serves a different purpose (Dais and Hatzakis 2013). NMR spectroscopy demands a profound understanding of the free fatty acids in oils due to their complex nature, necessitating the expertise of a technical operator. However, this challenge can be addressed through the application of big data and cloud technology, where NMR data from various researchers worldwide, focusing on different types of oils with diverse botanical and geographical origins, can be shared. Despite its initial high cost and time-consuming analysis, the application of NMR technology for authenticating edible oils is currently limited to laboratory settings (Sudhakar et al. 2023).

7.3.3.5 Fourier Transform Infrared Spectroscopy (FTIR)

FTIR spectroscopy involves the analysis of the interaction between infrared light and a sample. This technique offers insights into the functional groups and chemical bonds present in the sample. The identification of adulterants is achieved by comparing the infrared spectrum of the sample with reference spectra, focusing on characteristic absorption peaks.

7.3.4 Molecular Techniques

DNA testing is a robust tool for verifying the authenticity of edible flowers and detecting potential adulteration. Basically, the DNA testing procedure consists of four main steps including extraction, quantitation, amplification, and electrophoresis (Demeke and Jenkins 2010).The process involves extracting DNA from the sample and analysing specific regions or markers unique to different flower species.

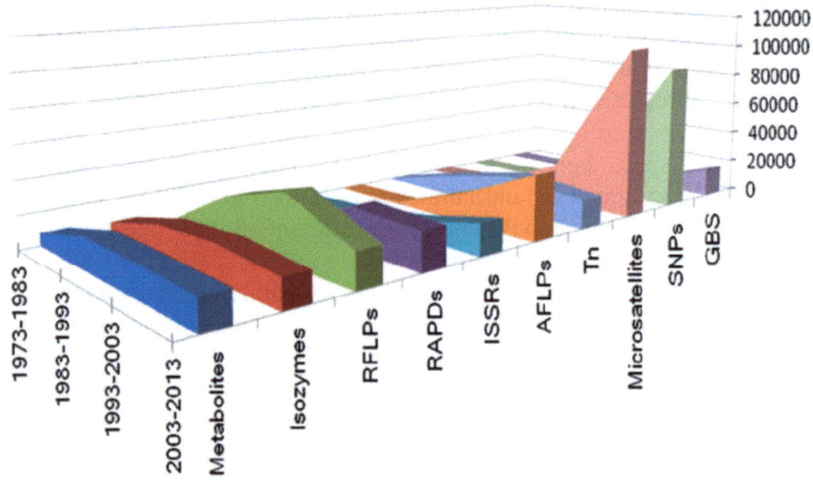

Fig. 7.24 Popularity of different molecular marker systems in last 40 years. Popularity has been interpreted based on total number of citations (right panel) for each marker system on google scholar. RFLP: restriction fragment length polymorphism, RAPD: random amplified polymorphic DNA, ISSRs: Inter-simple sequence repeats, rep-PCR or tn-PCR: Repetitive sequence based PCR, SNPs: single nucleotide polymorphisms, GBS: Genotyping by sequencing. (Grover and Sharma 2016)

In fact, advancements in molecular biology have accelerate the development of various molecular markers, especially in the past 40 years (Fig. 7.24).

Considerable improvement has been made in the resolution and usability of DNA marker systems due to advancements in three detection techniques: the hybridization-based technique, the polymerase chain reaction (PCR) based technique, and the DNA sequence-based technique. Table 7.3 represents the advantages and disadvantages of DNA-based methods used for the detection of adulteration (Grover and Sharma 2016; Amiteye 2021; Haji et al. 2023).

These fundamental marker methods can be divided into two groups: (1) non-PCR-based techniques or hybridization based techniques and, (2) PCR-based techniques (Agarwal et al. 2008) with its own specific applications and methodologies.

7.3.4.1 PCR Based Techniques

The use of PCR-based methods is cost-effective, accurate, and straightforward and they are sensitive and specific. These techniques have great potential for identifying adulterants and verifying the authenticity of goods (Haji et al. 2023), here we describe PCR and DNA barcoding as PCR Based techniques. Other PCR-based markers techniques (RAPD, AFLP, ISSR, SNP etc.) will not be discussed here.

Table 7.3 Advantages and disadvantages of DNA-based adulteration detection methods (El Sheikha 2019; Haji et al. 2023)

Technique	Advantages	Disadvantages
DNA-based methods (including PCR, sequencing and hybridization-based methods)	1. High specificity 2. High throughput 3. Highly stable DNA biomarkers 4. Highly sensitive 5. Robust, reproducible, and efficient 6. Adaptable	1. Technically challenging 2. Current high cost 3. Resource intensive 4. Large consumable requirement 5. Need to develop new bioinformatics algorithms to manage large amounts of data

Polymerase Chain Reaction (PCR) Method

Polymerase Chain Reaction (PCR) is a principal detection method that amplifies specific DNA regions of interest, enabling their detection and analysis. By designing primers that target known DNA sequences of authentic flowers, PCR confirms the presence or absence of specific flower species in the sample. For example, Marc El Beyrouthy and colleagues examined multiple samples reportedly containing sweet violet from various herbalist shops in Lebanon (El Beyrouthy et al. 2013). *Viola odorata* L. (Fig. 7.25) is a plant species belonging to the Viola genus, which is indigenous to Europe and Asia. It is also referred to as Sweet Violet, English Violet, Common Violet, or Garden Violet. Locally, it is known as Banafsaj in Lebanon and its flowers are commonly utilized for treating influenza, coughs, and are believed to possess diuretic, expectorant, and sedative properties (El Beyrouthy 2009).

In Lebanon and the Middle East, there exists an imbalance between the need for and availability of *Viola odorata*, largely due to challenges in harvesting that could result in the adulteration and replacement of authentic material. Therefore, the objective of their research was to examine various samples purported to be sweet violet obtained from various herbalist stores in Lebanon.

The analysis of 10 samples of purported *V. odorata* sold in various regions of Lebanon involved macroscopic, microscopic, and DNA examinations. The comparison of the samples with authentic *V. odorata L.* obtained from the wild confirmed the presence of adulteration. The study revealed that most of the Lebanese sweet violet was mixed with *Cercis siliquastrum* L. (Fig. 7.25).

In their research, the scientists utilized qPCR for DNA analysis, a widely employed technique in both research and molecular diagnostics. They used this method to identify instances of plant adulteration and inappropriate use of medicinal plants. In the DNA analysis, a Cercis primer and a Viola primer were used. According to Table 7.4, the natural Viola DNA tested positive when paired with the Viola primer, but it tested negative when paired with the Cercis primer. This indicates that there was hybridization between the Viola primer and the natural Viola DNA, which was followed by the amplification of the detected DNA.

Fig. 7.25 *Viola odorata* (**a**) By Fritz Geller-Grimm—Own work, CC BY-SA 2.5, https://commons.wikimedia.org/w/index.php?curid=1510692 and (**b**) By Kjetil Lenes—Own work, CC BY 2.5, https://commons.wikimedia.org/w/index.php?curid=2022147, (**c**) By Bouba at French Wikipedia—photo by Bouba, CC BY-SA 3.0, https://commons.wikimedia.org/w/index.php?curid=1658073 and (**d**) By Kurt Stüber- caliban.mpiz-koeln.mpg.de/mavica/index.html part of www.biolib.de, CC BY-SA 3.0, https://commons.wikimedia.org/w/index.php?curid=8033

The research conducted by Marc El Beyrouthy and colleagues found that "Banafsaj" samples contained a high concentration of *Cercis siliquastrum* and a low concentration of *Viola odorata* (El Beyrouthy et al. 2013).

DNA Barcoding

The utilization of PCR has been shown to be very effective in DNA barcoding methodologies (Haji et al. 2023).

DNA barcoding entails sequencing a short DNA region, such as the rbcL or matK genes commonly found in plants (Nehal et al. 2021). By comparing the obtained DNA sequence with a reference database, the identity of the flower species can be determined, aiding in the detection of any substitution or misidentification.

DNA barcoding is a relatively recent tool employed by taxonomists, researchers, and scientists across various fields. It serves as a method for identifying the species of plants, animals, or fungi and involves the use of genetic markers, techniques such as PCR for DNA amplification, and the construction of dense genetic maps. This

Table 7.4 Results of electrophoresis of primers with the analysed samples

DNA	*Viola* primers	*Cercis* primers
DNA natural *Viola*	+	–
DNA natural *Cercis*	–	+
DNA natural *Viola*	+	–
DNA natural *Cercis*	–	+
Sample1 (water)	–	–
Sample2 (Tripoli)	+	+
Sample3 (Tripoli)	+	+
Sample4 (Jbeil)	+	+
Sample5 (Jounieh)	+	+
Sample6 (Aley)	+	+
Sample7 (Rabieh)	+	+
Sample8 (Saida)	–	+
Sample9 (sour)	+	+

Note: *V. odorata* flowers were bought from different herbalist shops in 2011 and 2012: Tripoli, Jounieh, Aley, Jbeil, Rabieh, Saida and Sour regions (El Beyrouthy 2009)

approach essentially provides a veritable "identity card" for different species, utilizing elements such as microsatellite markers of DNA, which are highly polymorphic and uniformly distributed in the genome but lack a known function (Barcaccia et al. 2015). In recent decades, DNA barcoding has evolved from being confined to research laboratories in universities and institutions, mainly for research purposes, to finding real-life applications. It has spurred collaborations between laboratories and industries, becoming a key player in ensuring product quality and safeguarding consumer health (Galimberti et al. 2019). The fundamental principle underlying DNA barcoding involves the amplification of homologous genes by PCR, followed by DNA sequencing. Critical sequences serve as a unique "barcode" to validate the identity and authenticity of various products, including the identification of DNA from different plant and animal species (Kress et al. 2015). This method is applied to test and authenticate raw materials in both the food and non-food industry processes (Waugh 2007). DNA barcoding is recommended as an inexpensive, rapid, and objective method, with certain groups of investigators achieving success rates as high as 80%, allowing them to identify up to 75% of tested natural dietary supplements (Li and Dao 2011; Grazina et al. 2020; Nehal et al. 2021). As a modern tool, DNA barcoding is increasingly reliant on technological advances, and policymakers recognize its potential to enhance scientific and societal understanding of biodiversity. DNA barcoding plays a crucial role in generating consistent data for informing policy decisions, while simultaneously fostering societal awareness and garnering public attention for the greater good (Krishna Krishnamurthy and Francis 2012).

7.3.4.2 Non-PCR-Based Techniques or Hybridization Based Techniques

Hybridization Method

As per Carles et al. (2005), the hybridization technique is utilized to identify adulterants by noting slight alterations in nucleotide strands in comparison to known DNA sequences (Carles et al. 2005). It allows for the simultaneous detection of potential species. While molecular methods such as sequencing and hybridization-based comprehensive approaches are essential for identifying biological adulterants, prior knowledge of the sequence is necessary to create primers for amplifying specific sequences (Lockley and Bardsley 2000; Bansal et al. 2017; Haji et al. 2023).

Restriction Fragment Length Polymorphism (RFLP)

RFLP includes a hybridization stage. In RFLP, the presence of DNA polymorphisms is identified by binding a chemically labelled DNA probe to a Southern blot of DNA that has been digested by restriction endonucleases, resulting in a distinct DNA fragment profile. The unique profile comes about due to changes in nucleotides or rearrangements of DNA, like insertions, deletions, or single nucleotide polymorphisms. The RFLP markers exhibit high levels of polymorphism, are inherited codominantly, and are highly reproducible. RFLP markers are considered superior due to their high heritability, locus specificity, and presence across the entire plant genome. The method also allows for the simultaneous screening of numerous samples. DNA blots can undergo repeated analysis by stripping and reproving (typically eight to ten times) with different RFLP probes. This method is not commonly employed due to its time-consuming nature, the need for costly and radioactive/toxic reagents, and the requirement for a substantial amount of high-quality genomic DNA. Generating probes requires prior sequence information, which adds complexity to the methodology. The emergence of these constraints prompted the creation of a fresh array of simpler techniques, referred to as PCR-based methods (Agarwal et al. 2008).

In a brief description by Konzen et al. (2017), they outline a general process for RFLP analysis (Konzen et al. 2017). To start, DNA is extracted from the leaf tissues of the subjects to be analysed. A significant amount of DNA is necessary for RFLP analysis, which can be a time-consuming task. When needed, PCR is utilized to amplify a targeted DNA segment, taking 2–3 hours to generate the necessary DNA quantity for efficient RFLP analysis. When possible, implementing PCR reduces the time needed for sample analysis. The isolated DNA or PCR product is then digested using selected restriction enzymes to create restriction fragments. This digestion process results in numerous fragments of varying lengths in the reaction mixture. The restriction DNA fragments that were obtained are examined through agarose or polyacrylamide gel electrophoresis (PAGE) to detect variations in DNA fragment sizes. During electrophoresis, DNA fragments, which are negatively charged, separate according to their size and charge (Amiteye 2021).

The current research employed PCR-RFLP, a DNA-based method, to verify the authenticity of *Boerhavia diffusa*, a herbal medicine often contaminated with *Trianthema portulacastrum* and *Trianthema monogyna*. By subjecting the PCR-amplified ribosomal internal transcribed spacer (ITS) gene product to restriction digestion using five different enzymes (MspI, HinfI, MboI, EcoRI, and EcoRV), distinct banding patterns were observed for *Boerhavia diffusa* and the adulterants. Among the five enzymes, MspI, HinfI, and MboI demonstrated potential for authenticating *B. diffusa*, as they produced unique patterns for this plant that differed from those of the adulterants. The remaining enzymes either failed to digest the ITS product or exhibited insufficient polymorphism to identify a specific genotype (Biswas et al. 2013).

No differences in the size of the ITS region were detected among the three plants *Boerhavia diffusa*, *T. portulacastrum*, and *T. monogyna*. When experimental restriction enzymes were used for ITS-RFLP analysis, *T. portulacastrum* and *T. monogyna* exhibited identical patterns, which were distinct from those of *B. diffusa*. Hence, restriction enzymes MspI, HinfI, and MboI could potentially serve as tools for creating molecular markers to distinguish between these plants, verify their authenticity, and prevent adulteration (Biswas et al. 2013).

Sequencing Based Method

The process of nucleic acid sequencing involves identifying the precise sequence of nucleotides in a specific DNA or RNA molecule. Over the last 10 years, there has been a tremendous rise in the utilization of nucleic acid sequencing, largely due to the widespread availability of sequencing technology in research and clinical laboratories worldwide (Grada and Weinbrecht 2013).

The use of whole genome sequences (WGS) has changed the way DNA markers are identified, shifting from identifying polymorphisms based on fragments to identifying single nucleotide polymorphisms (SNPs) based on sequences, in order to speed up the marker identification process and increase the number of useful markers. However, WGS technologies based on Sanger sequencing are expensive, time-consuming, and only provide information about the specific individual being targeted, which has limited its application in specific gene discovery. The use of large breeding populations is constrained by time and cost factors. The advancement of next generation sequencing (NGS) technologies and robust computational pipelines has significantly decreased the expense of whole genome sequencing, enabling the discovery, sequencing, and genotyping of thousands of markers in a single step (Stapley et al. 2010; Ray and Satya 2014).

Next-Generation Sequencing (NGS)

Next-generation sequencing (NGS) is a recently developed technology utilized for the sequencing of DNA and RNA as well as detecting variants and mutations. In a brief timeframe, NGS can sequence hundreds to thousands of genes or even entire genomes (Qin 2019). NGS technologies facilitate the rapid and cost-effective sequencing of large amounts of DNA. This method can identify multiple flower species present in a sample simultaneously, providing comprehensive information about the sample's composition.

It's important to note that these chemical analysis and DNA testing methods require specialized equipment, expertise, and access to reference databases or standards for accurate identification of adulterants. Consulting with experts, food safety authorities, or accredited laboratories is recommended for conducting these tests and interpreting the results.

NGS has become an effective tool for identifying numerous DNA sequence polymorphism-based markers within a short duration (Fig. 7.26), making it a valuable tool for next generation plant breeding (Ray and Satya 2014).

7.3.5 *Biosensors*

Biosensors are analytical devices that integrate biological components closely with physiochemical transducers or transducing microsystems, encompassing optical, electrochemical, thermometric, piezoelectric, or magnetic modalities (Coulet 2019). Renowned for their efficacy in assessing the quality of food products, biosensors offer sensitivity, specificity, reproducibility, a linear response range, short response and recovery times, and operational stability (Ahmed et al. 2016). Contemporary advancements in biosensor technology include the integration of nanotechnology, with the goal of revolutionizing existing protocols for quality and safety detection (Sionek et al. 2020). Utilizing a variety of nanomaterials, including graphene, graphite, and carbon nanotubes, which possess a high surface area and numerous active sites, improves the biocompatibility of these sensors. In the foreseeable future, these sophisticated nano biosensors are anticipated to have a crucial function in promptly and precisely identifying chemical, physical, or microbiological threats in meat preservation systems (Sharifi et al. 2020; Sionek et al. 2021). A variety of techniques have been developed using different sensors, especially biosensors, for the identification of compounds within the food matrix. In the following discussion, we will review and delve into some of these innovative biosensors.

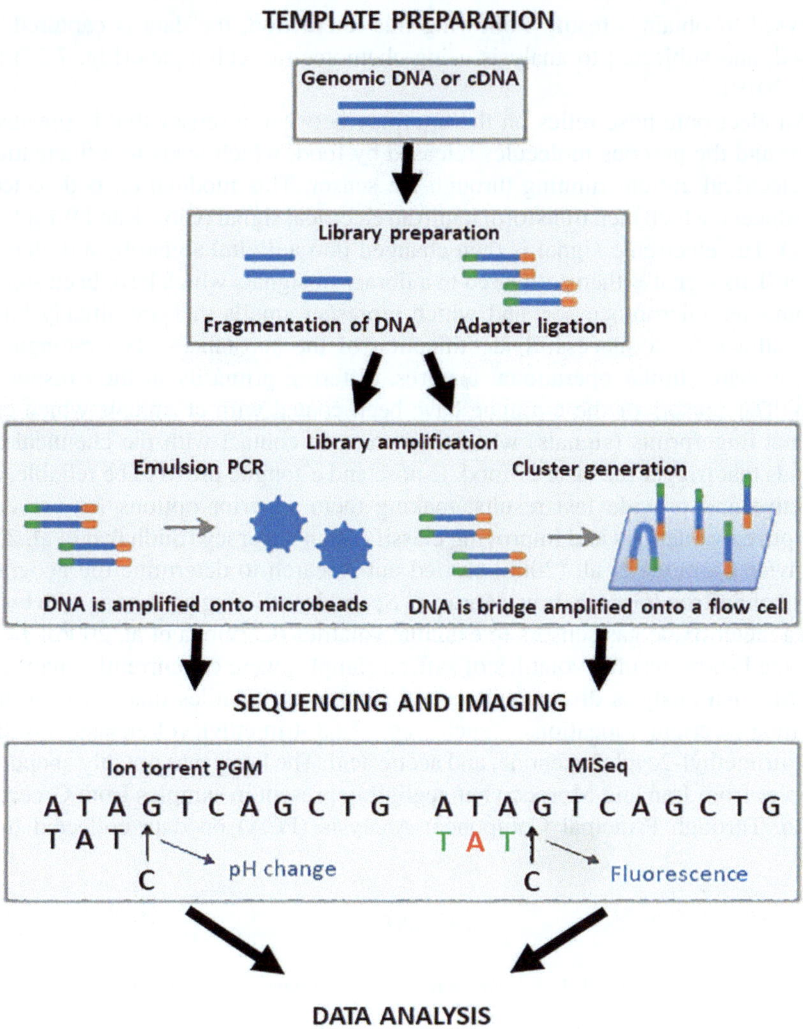

Fig. 7.26 Next-generation sequencing methodology. (Grada and Weinbrecht 2013)

7.3.5.1 E-nose and E-tongue

The operation of biosensors mimics the response of human sensory organs, particularly the nose and tongue, which offers a potential way to scientifically ascertain the genuineness and excellence of products. Two interesting biosensors designed to imitate human senses are the electronic tongue (e-tongue) and electronic nose (e-nose). These devices are equipped with a variety of sensors, such as metal oxide gas sensors, piezoelectric sensors, or electrochemical gas sensors.

By employing chemometric techniques, the signals (volatile compounds) from food materials can be sensed and processed by the sensors and the data can be analysed to obtain a result. Following this interaction, the data is captured, processed, and subjected to analysis using chemometric techniques (Fig. 7.27) (Roy et al. 2019).

An electronic nose relies on the interplay between a sensor that is sensitive to odour and the gaseous molecules released by food, which leads to a fluctuation in the electrical current running through the sensor. This modulation is detected by transducers, which then transform it into an electrical signal (Ciosek and Wróblewski 2011). The electronic signal is then changed into a digital signal by a digital converter. This signal is then compared to a library of signals which have been stored in a computer microprocessor, and which represent smells that are already known. This allows for a successful identification of the substances. The e-tongue and e-nose share similar operational features, differing primarily in their response to food. The sensors of the e-tongue have been coated with chemicals which create distinct fingerprints (signals) when they come in contact with the chemical compounds that trigger the taste of food. E-nose and e-tongue prove to be reliable, cost-effective and provide fast results, making them superior options for preventing deceptive evaluations and improving classification accuracy (Sudhakar et al. 2023).

Javier Carmona et al. (2006) carried out research to determine the geographic origin of Saffron (Greece, Iran, Morocco, Spain) by utilizing an E-nose with twenty-seven metal oxide gas sensors to examine volatiles (Carmona et al. 2006). To validate the E-nose results, volatiles of saffron samples were concurrently analysed by GC-MS. An analysis discovered several significant volatiles that aided in distinguishing certain locations, such as 2,6,6-trimethylcyclohexane-1,4-dione, 3,5,5-trimethyl-2cyclohexenone, and acetic acid. The latter was notably abundant in samples from Iran and Morocco but negligibly present in samples from Greece and Spain. Through Principal Component Analysis (PCA) on data collected by the

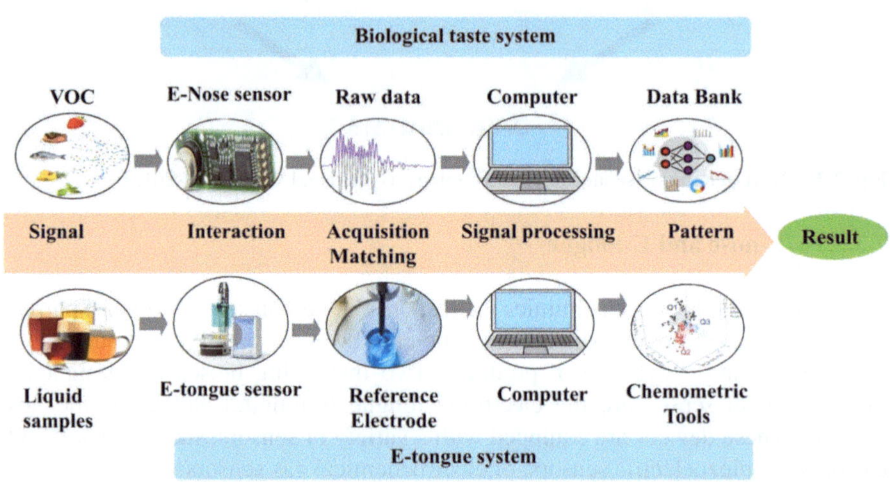

Fig. 7.27 Recognition process of data in electronic nose and electronic tongue. (Sudhakar et al. 2023)

E-nose, the researchers successfully differentiated the origin with a confidence level of 90%.

For the detection of adulterated Saffron samples from Iran, an E-tongue was employed. This device used voltammetry sensors based on screen-printed electrodes (SPE) in order to effectively analyse the relevant components of the Saffron and any adulterants present in solution form (Kumari et al. 2021).

Numerous research studies have explored the application of e-nose and e-tongue for detecting adulteration in edible oils (Dias et al. 2014; Haddi et al. 2011; Peris and Escuder-Gilabert 2016; Majchrzak et al. 2018; Semenov et al. 2019). Mildner-Szkudlarz and Jelen conducted a study to analyse the performance of solid-phase microextraction-fast gas chromatography with flame ionization detection (SPME-GCFID), solid-phase microextraction-mass spectrometry (SPME-MS), and metal oxide sensor-based electronic nose (HS-E nose) with PCA for recognizing adulteration of olive oil with hazelnut oil (Mildner-Szkudlarz and Jeleń 2008). The detection limits for these four methods ranged from 5% to 50% (v/v). Notably, the HS-E nose sensor showed an impressive repeatability score of 0.04–17.72% and an accuracy of ±2.85%.

Mildner-Szkudlarz and Jelen conducted research to identify extra virgin olive oil (EVOO) that had been adulterated with rapeseed oil and sunflower oil. This was done through the use of MOS e-nose and SPME-MS, which were then treated with PCA. The findings revealed that these two methods present a valid choice for a swift assessment of possible contamination in EVOO (Mildner-Szkudlarz and Jeleń 2010).

Kobra Heidarbeigi et al. (2015) developed an electronic nose containing six metal oxide semiconductor sensors. This electronic nose was utilized for identifying saffron, safflower, dyed corn stigma, and their combination as adulterants in saffron (Heidarbeigi et al. 2015).

7.3.5.2 Optical Based Biosensors

The selection and utilization of biorecognition elements are pivotal for conducting real-time analyses of target substrates or biologically active components of interest (Mildner-Szkudlarz and Jeleń 2010). As depicted in Fig. 7.28, biosensors are categorized in diverse manners, taking into account the biorecognition elements employed, which encompass enzymes, aptamers, whole cells, nanoparticles, immunosensors, and antibody-based sensors. The classification also extends to the type of transducers utilized, including electrochemical, optical, and mass-based/gravimetric transducers. Additionally, biosensors are differentiated based on the detection techniques applied, involving both label-free methods such as surface plasmon resonance (SPR), mass spectrometry, and acoustic wave, and label-based methods like fluorescence and chemiluminescence (Churi 2021).

Real-time monitoring of microbial activities in food can be done with optical biosensors. These sensors identify microbes in food by changing the level of the signal or refractive index, or by detecting the amount of microbial cells that attach to the biorecognition sensing region on the optical transducer (Nanda et al. 2022).

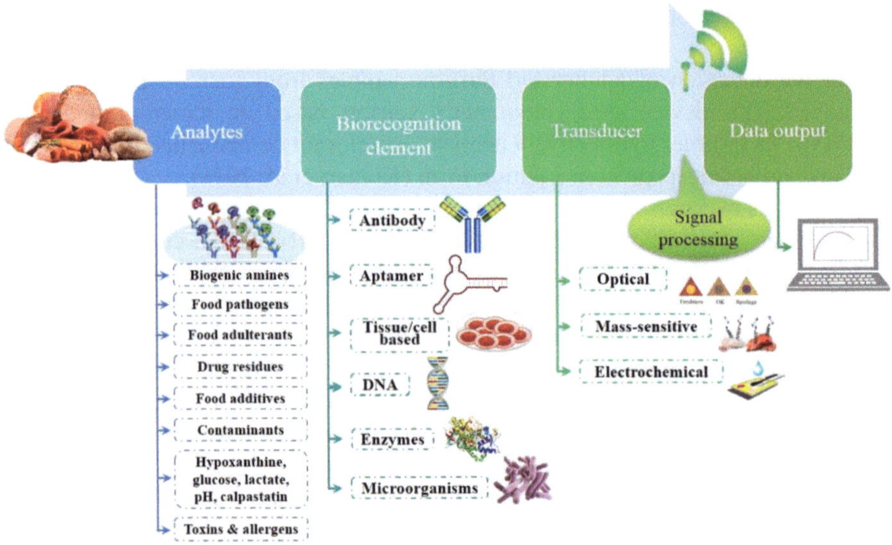

Fig. 7.28 Schematic diagram of the working principle of biosensors for the detection of different analytes. (Reproduced with permission from Nanda et al. 2022)

Among various optical analysis techniques, including colorimetric, fluorescence, localized surface plasmon resonance (LSPR), surface plasmon resonance (SPR) and chemiluminescence, the use of SPR as an optical biosensor is very widespread (Alamer et al. 2018).

SPR is a process wherein bioreceptors are fixed onto a metallic transmitter. The electromagnetic energy at a particular wavelength then interacts with the electron cloud of the transmitter, which results in a strong resonance. When bacterial cells come into contact with the metal surface, the refractive index is notably impacted (Alamer et al. 2018). Reflectance spectroscopy is commonly utilized for the detection of target pathogens. Various meat and meat products have been investigated using optical biosensors in a multitude of research projects to determine if any pathogenic organisms are present. For instance, a fiber optic immunosensor, incorporating robust immunomagnetic separation, detected *Listeria monocytogenes* in meat up to 3×10^2 CFU/mL (Darrel et al. 2003). For the purpose of distinguishing between pathogenic *L. monocytogenes* and other non-pathogenic or pathogenic species in artificially contaminated ready-to-eat (RTE) meat products, an aptamer-based fibre-optic biosensor was used (Gagaoua et al. 2020). Oh et al. (2017) found that *Salmonella Typhimurium* had levels of up to 4 log CFU/mL in pork meat within a half-hour timeframe when using localized SPR. A separate study by Zhang et al. (2017) established a multi-channel SPR biosensor for the specific recognition of three different foodborne pathogens (including *Escherichia coli* O157:H7, *Salmonella Enteritidis*, and *Listeria monocytogenes)* in food contaminated naturally. Nevertheless, in order to commercialize and expand the use of these optical biosensors, several problems must be addressed, such as elevated expenses, ensuring quality, disagreements on stability, matters of sensitivity, and configuration of the instruments. At various stages during processing, toxic materials like pesticides, antibiotics, veterinary drug residues, and hazardous additives may enter the food supply and contaminate entire batches. Techniques like HPLC, capillary electrophoresis, and mass spectrometry can be used to evaluate samples after certain steps of the processing. Nevertheless, these methods are costly, intricate, need high-level instruments, and the knowledge of a trained expert (Thakur and Ragavan 2013). Consequently, there has been an increasing demand for quicker, more dependable, and more delicate detection of impurities during processing stages using biosensor technology. Real-time testing and observation of food-borne contaminants via biosensors can substantially reduce the risk to people's health.

7.4 Conclusion

In past decade, edible flowers and their products have been mostly used due their remarkable role as a part of therapy, a recovery process or a healthy life style, and their significant value as an anticancer, anti-diabetic, cardio-protective, hepatoprotective, gastroprotective, and genoprotective properties. Furthermore, these flowers are highly prized as functional food components in the food and beverage sectors, playing a key role in enhancing visual appeal, providing distinctive fragrances and exotic tastes, offering beneficial nutrients, and contributing medicinal properties to a variety of food items and culinary preparations. In addition to their nutritional benefits and significant impact on health, edible flowers can also serve as a highly valued food option, contributing to the diversification of dietary choices in the food industry. Before recommending edible flowers as possible food sources and valuable providers of nutrients, it is crucial to carefully consider the fact that many of these types of flowers may contain adulterants, anti-nutrients, or toxins that could have negative effects on the health of consumers (Pires Jr et al. 2023).

In fact, the adulteration of food and edible flowers poses a significant public health risk, impacting millions of individuals annually. This leads to the development of cancer, lathyrism, liver disease, heart failure, kidney disease, and disorders of the nervous system. Producers and traders with economic motives engage in sophisticated adulteration of food products from time to time (Anagaw et al. 2024).

To overcome this sophisticated adulteration, and increase the effectiveness of their biological activity and values in medical and clinical application and food and drink industries, the control of adulteration strictly is necessary. The valuable input of scientists and researchers in developing technologies for detecting and measuring food adulteration is commendable in today's world. While various methods like physical, chemical, analytical, and immune-based techniques are employed for identifying adulterated products (Haji et al. 2023), further research is essential for creating new onsite testing analytical methods to detect adulteration in various edible flowers and food items. That means, the development, validation, and approval of additional methods are necessary, particularly those that are straightforward and uncomplicated to carry out. This point is more considerable for edible flowers as very little research studies have been done on their adulteration detection field (Anagaw et al. 2024).

There is a need to carry out education and awareness initiatives to educate people about the dangers of food adulteration and ways to safeguard themselves. Creating awareness in this area could potentially avert life-threatening illnesses and rescue the lives of unsuspecting individuals who fall prey to this danger (Haji et al. 2023).

The chapter of this book explores different methods employed to identify adulteration, along with their advantages and disadvantages, could be a good support for researchers in the field of edible flowers adulterants detection, and also could be a guidance, on the way to achieving the foods with higher quality, wholesome, and more safe.

References

Agarwal M, Shrivastava N, Padh H (2008) Advances in molecular marker techniques and their applications in plant sciences. Plant Cell Rep 27:617–631. https://doi.org/10.1007/s00299-008-0507-z

Agiomyrgianaki A, Petrakis PV, Dais P (2010) Detection of refined olive oil adulteration with refined hazelnut oil by employing NMR spectroscopy and multivariate statistical analysis. Talanta 80:2165–2171. https://doi.org/10.1016/j.talanta.2009.11.024

Ahmed MU, Zourob M, Tamiya E (2016) Food biosensors. Royal Society of Chemistry

Al Riza DF, Widodo S, Purwanto YA, Kondo N (2019) Combined fluorescence-transmittance imaging system for geographical authentication of patchouli oil. Spectrochim Acta A Mol Biomol Spectrosc 218:155–160. https://doi.org/10.1016/j.saa.2019.04.009

Alamer S, Eissa S, Chinnappan R, Herron P, Zourob M (2018) Rapid colorimetric lactoferrin-based sandwich immunoassay on cotton swabs for the detection of foodborne pathogenic bacteria. Talanta 185:275–280. https://doi.org/10.1016/j.talanta.2018.03.072

Ali AH (2022) High-performance liquid chromatography (HPLC): a review. Ann Adv Chem 6. https://doi.org/10.29328/journal.aac.1001026

Amiteye S (2021) Basic concepts and methodologies of DNA marker systems in plant molecular breeding. Heliyon 7. https://doi.org/10.1016/j.heliyon.2021.e08093

Anagaw YK, Ayenew W, Limenh LW, Geremew DT, Worku MC, Tessema TA, Simegn W, Mitku ML (2024) Food adulteration: causes, risks, and detection techniques. SAGE Open Med 12:20503121241250184. https://doi.org/10.1177/20503121241250184

Anastasaki EG, Kanakis CD, Pappas C, Maggi L, Zalacain A, Carmona M, Alonso GL, Polissiou MG (2010) Quantification of crocetin esters in saffron (Crocus sativus L.) using Raman spectroscopy and chemometrics. J Agric Food Chem 58:6011–6017. https://doi.org/10.1021/jf100143n

Azadmard-Damirchi S, Torbati M (2015) Adulterations in some edible oils and fats and their detection methods. J Food Qual Hazards Control 2:38–44

Bansal S, Singh A, Mangal M, Mangal AK, Kumar S (2017) Food adulteration: sources, health risks, and detection methods. Crit Rev Food Sci Nutr 57:1174–1189. https://doi.org/10.1080/10408398.2014.967834

Banti M (2020) Food adulteration and some methods of detection, review. Int J Food Sci Nutr 9:86–94. https://doi.org/10.11648/j.ijnfs.20200903.13

Barcaccia G, Lucchin M, Cassandro M (2015) DNA barcoding as a molecular tool to track down mislabeling and food piracy. Diversity 8:2. https://doi.org/10.3390/d8010002

Bharti A, Jain U, Chauhan N (2024) Progressive analytical techniques utilized for the detection of contaminants attributed to food safety and security. Talanta Open 10:100368. https://doi.org/10.1016/j.talo.2024.100368

Biswas K, Kapoor A, Biswas R (2013) Authentication of herbal medicinal plant-Boerhavia diffusa L. using PCR-RFLP. Curr Trends Biotechnol Pharm 7:725–731

Bounaas K, Bouzidi N, Daghbouche Y, Garrigues S, De la Guardia M, El Hattab M (2018) Essential oil counterfeit identification through middle infrared spectroscopy. Microchem J 139:347–356. https://doi.org/10.1016/j.microc.2018.03.008

Bubli SY, Haque F, Khan MS (2021) Gas chromatography and mass spectroscopy (GC-MS) technique for food analysis. In: Khan MS, Shafiur Rahman M (eds) Techniques to measure food safety and quality. Springer, Cham, pp 195–217. https://doi.org/10.1007/978-3-030-68636-9_9

Calvano CD, De Ceglie C, D'Accolti L, Zambonin CG (2012) MALDI-TOF mass spectrometry detection of extra-virgin olive oil adulteration with hazelnut oil by analysis of phospholipids using an ionic liquid as matrix and extraction solvent. Food Chem 134:1192–1198. https://doi.org/10.1016/j.foodchem.2012.02.154

Carles M, Cheung MKL, Moganti S, Dong TT, Tsim KW, Ip NY, Sucher NJ (2005) A DNA microarray for the authentication of toxic traditional Chinese medicinal plants. Planta Med 71:580–584. https://doi.org/10.1055/s-2005-864166

Carmona M, Martínez J, Zalacain A, Rodriguez-Mendez ML, de Saja JA, Alonso GL (2006) Analysis of saffron volatile fraction by TD–GC–MS and e-nose. Eur Food Res Technol 223:96–101. https://doi.org/10.1007/s00217-005-0144-5

Casado N, Fernández-Pintor B, Morante-Zarcero S, Sierra I (2022) Quick and green microextraction of pyrrolizidine alkaloids from infusions of mallow, calendula, and hibiscus flowers using ultrahigh-performance liquid chromatography coupled to tandem mass spectrometry analysis. J Agric Food Chem 70:7826–7841. https://doi.org/10.1021/acs.jafc.2c02186

Catharino RR, Haddad R, Cabrini LG, Cunha IB, Sawaya AC, Eberlin MN (2005) Characterization of vegetable oils by electrospray ionization mass spectrometry fingerprinting: classification, quality, adulteration, and aging. Anal Chem 77:7429–7433. https://doi.org/10.1021/ac0512507

Churi D (2021) Methods for detection of common adulterants in food. VIVA-Tech Int J Res Innov 1:4

Ciosek P, Wróblewski W (2011) Potentiometric electronic tongues for foodstuff and biosample recognition – an overview. Sensors 11:4688–4701. https://doi.org/10.3390/s110504688

Coulet PR (2019) Biosensor principles and applications. CRC Press

Dai H, Gao Q, He L (2020) Rapid determination of saffron grade and adulteration by thin-layer chromatography coupled with Raman spectroscopy. Food Anal Methods 13:2128–2137. https://doi.org/10.1007/s12161-020-01828-x

Dais P, Hatzakis E (2013) Quality assessment and authentication of virgin olive oil by NMR spectroscopy: a critical review. Anal Chim Acta 765:1–27. https://doi.org/10.1016/j.aca.2012.12.003

Darrel E, Goll D, Valery F, Thompson V, Li H (2003) The calpain system. Physiol Rev 83:731–801

Demeke T, Jenkins GR (2010) Influence of DNA extraction methods, PCR inhibitors and quantification methods on real-time PCR assay of biotechnology-derived traits. Anal Bioanal Chem 396:1977–1990. https://doi.org/10.1007/s00216-009-3150-9

Dhanya K, Sasikumar B (2010) Molecular marker based adulteration detection in traded food andAgricultural Commodities of Plant Origin with special reference to spices. Curr Trends Biotechnol Pharm 4:454–489

Dias LG, Fernandes A, Veloso AC, Machado AA, Pereira JA, Peres AM (2014) Single-cultivar extra virgin olive oil classification using a potentiometric electronic tongue. Food Chem 160:321–329. https://doi.org/10.1016/j.foodchem.2014.03.072

Do TKT, Hadji-Minaglou F, Antoniotti S, Fernandez X (2015) Authenticity of essential oils. Trends Analyt Chem 66:146–157. https://doi.org/10.1016/j.trac.2014.10.007

El Beyrouthy M (2009) Contribution à l'ethnopharmacologie libanaise et aux Lamiaceae du Liban. Acta Bot Gall 156:515–521. https://doi.org/10.1080/12538078.2009.10516175

El Beyrouthy M, Kafrouny M, Arnold N, Al-Hejin A, Siddig L, Hassan A (2013) Macroscopic, microscopic and DNA fingerprinting to fight adulteration of Banafsaj (Viola odorata L.) sold at the Lebanese herbal shops. Eur J Sci Res 104:642–651

El Sheikha AF (2019) DNAFoil: novel technology for the rapid detection of food adulteration. Trends Food Sci Technol 86:544–552. https://doi.org/10.1016/j.tifs.2018.11.012

Flanagan RJ, Taylor A, Watson ID, Whelpton R (2008) Thin-layer chromatography. In: Flanagan RJ, Taylor A, Watson ID, Whelpton R (eds) Fundamentals of analytical toxicology. Wiley, New York, pp 131–143

Foschi M, Tozzi L, Di Donato F, Biancolillo A, D'Archivio AA (2022) A novel FTIR-based chemometric solution for the assessment of saffron adulteration with non-fresh stigmas. Molecules 28:33. https://doi.org/10.3390/molecules28010033

Fritz E, Ölzant SM, Länger R (2008) Illicium verum Hook. f. and Illicium anisatum L.: anatomical characters and their value for differentiation. Sci Pharm 76:65–76. https://doi.org/10.3797/scipharm.0802-03

Gafner S, Bergeron C, Batcha LL, Angerhofer CK, Sudberg S, Sudberg ÉM, Guinaudeau H, Gauthier R (2003) Analysis of Scutellaria lateriflora and its adulterants Teucrium canadense and Teucrium chamaedrys by LC–UV/MS, TLC, and digital photomicroscopy. J AOAC Int 86:453–460. https://doi.org/10.1093/jaoac/86.3.453

Gagaoua M, Troy D, Mullen AM (2020) The extent and rate of the appearance of the major 110 and 30 kDa proteolytic fragments during post-mortem aging of beef depend on the glycolysing rate of the muscle and aging time: an LC–MS/MS approach to decipher their proteome and associated pathways. J Agric Food Chem 69:602–614. https://doi.org/10.1021/acs.jafc.0c06485

Galimberti A, Casiraghi M, Bruni I, Guzzetti L, Cortis P, Berterame NM, Labra M (2019) From DNA barcoding to personalized nutrition: the evolution of food traceability. Curr Opin Food Sci 28:41–48. https://doi.org/10.1016/j.cofs.2019.07.008

Gonzálvez A, Armenta S, De La Guardia M (2008) Trace elemental composition of curry by inductively coupled plasma optical emission spectrometry (ICP-OES). Food Addit Contam 1:114–121. https://doi.org/10.1080/02652030802520845

Goodner K, Rouseff R (2011) Practical analysis of flavor and fragrance materials, vol 219. Wiley Online Library. https://doi.org/10.1002/9781444343137

Grada A, Weinbrecht K (2013) Next-generation sequencing: methodology and application. J Invest Dermatol 133:1–4. https://doi.org/10.1038/jid.2013.248

Grazina L, Amaral JS, Mafra I (2020) Botanical origin authentication of dietary supplements by DNA-based approaches. Compr Rev Food Sci Food Saf 19:1080–1109. https://doi.org/10.1111/1541-4337.12551

Grover A, Sharma P (2016) Development and use of molecular markers: past and present. Crit Rev Biotechnol 36:290–302. https://doi.org/10.3109/07388551.2014.959891

Guyader S, Thomas F, Portaluri V, Jamin E, Akoka S, Silvestre V, Remaud G (2018) Authentication of edible fats and oils by non-targeted 13C INEPT NMR spectroscopy. Food Control 91:216–224. https://doi.org/10.1016/j.foodcont.2018.03.046

Haddi Z, Amari A, Ali AO, El Bari N, Barhoumi H, Maaref A, Jaffrezic-Renault N, Bouchikhi B (2011) Discrimination and identification of geographical origin virgin olive oil by an e-nose based on MOS sensors and pattern recognition techniques. Procedia Eng 25:1137–1140. https://doi.org/10.1016/j.proeng.2011.12.280

Haji A, Desalegn K, Hassen H (2023) Selected food items adulteration, their impacts on public health, and detection methods: a review. Food Sci Nutr 11:7534–7545. https://doi.org/10.1002/fsn3.3732

Heidarbeigi K, Mohtasebi SS, Foroughirad A, Ghasemi-Varnamkhasti M, Rafiee S, Rezaei K (2015) Detection of adulteration in saffron samples using electronic nose. Int J Food Prop 18:1391–1401. https://doi.org/10.1080/10942912.2014.915850

Howes M-JR, Simmonds MS, Kite GC (2004) Evaluation of the quality of sandalwood essential oils by gas chromatography–mass spectrometry. J Chromatogr A 1028:307–312. https://doi.org/10.1016/j.chroma.2003.11.093

Jakubczyk K, Łukomska A, Gutowska I, Kochman J, Janił J, Janda K (2021) Edible flowers extracts as a source of bioactive compounds with antioxidant properties – in vitro studies. Appl Sci 11:2120. https://doi.org/10.3390/app11052120

Joshi VC, Srinivas PV, Khan IA (2005) Rapid and easy identification of Illicium verum Hook. f. and its adulterant Illicium anisatum Linn. By fluorescent microscopy and gas chromatography. J AOAC Int 88(3):703–706

Konzen ER, Peron R, Ito MA, Brondani GE, Tsai SM (2017) Molecular identification of bamboo genera and species based on RAPD-RFLP markers. Silva Fenn (Hels) 51. https://doi.org/10.14214/sf.1691

Kress WJ, García-Robledo C, Uriarte M, Erickson DL (2015) DNA barcodes for ecology, evolution, and conservation. Trends Ecol Evol 30:25–35. https://doi.org/10.1016/j.tree.2014.10.008

Krishna Krishnamurthy P, Francis RA (2012) A critical review on the utility of DNA barcoding in biodiversity conservation. Biodivers Conserv 21:1901–1919. https://doi.org/10.1007/s10531-012-0306-2

Kumari L, Jaiswal P, Tripathy SS (2021) Various techniques useful for determination of adulterants in valuable saffron: a review. Trends Food Sci Technol 111:301–321. https://doi.org/10.1016/j.tifs.2021.02.061

Le Gresley A, Ampem G, Grootveld M, Percival BC, Naughton DP (2019) Characterisation of peroxidation products arising from culinary oils exposed to continuous and discontinuous thermal degradation processes. Food Funct 10:7952–7966. https://doi.org/10.1039/C9FO02065A

Li R, Dao Z (2011) Identification of Meconopsis species by a DNA barcode sequence: the nuclear internal transcribed spacer (ITS) region of ribosomal deoxyribonucleic acid (DNA). Afr J Biotechnol 10:15805–15807. https://doi.org/10.5897/AJB11.1877

Li S, Zhu X, Zhang J, Li G, Su D, Shan Y (2012) Authentication of pure camellia oil by using near infrared spectroscopy and pattern recognition techniques. J Food Sci 77:C374–C380. https://doi.org/10.1111/j.1750-3841.2012.02622.x

Li S, Xing B, Lin D, Yi H, Shao Q (2020) Rapid detection of saffron (Crocus sativus L.) adulterated with lotus stamens and corn stigmas by near-infrared spectroscopy and chemometrics. Ind crops. PRO 152:112539. https://doi.org/10.1016/j.indcrop.2020.112539

Lockley A, Bardsley R (2000) DNA-based methods for food authentication. Trends Food Sci Technol 11:67–77. https://doi.org/10.1016/S0924-2244(00)00049-2

Lozano-Sánchez J, Borrás-Linares I, Sass-Kiss A, Segura-Carretero A (2018) Chromatographic technique: high-performance liquid chromatography (HPLC). In: Modern techniques for food authentication. Elsevier, pp 459–526. https://doi.org/10.1016/B978-0-12-814264-6.00013-X

Majchrzak T, Wojnowski W, Dymerski T, Gębicki J, Namieśnik J (2018) Electronic noses in classification and quality control of edible oils: a review. Food Chem 246:192–201. https://doi.org/10.1016/j.foodchem.2017.11.013

Manasha S, Janani M (2016) Food adulteration and its problems (intentional, accidental and natural food adulteration). IJRFM 6:131–140

McGorrin RJ (2009) One hundred years of progress in food analysis. J Agric Food Chem 57:8076–8088. https://doi.org/10.1021/jf900189s

Meurer MC, Mees M, Mariano LNB, Boeing T, Somensi LB, Mariott M, Dos Santos AC, Longo B, França TCS, Klein-Júnior LC (2019) Hydroalcoholic extract of Tagetes erecta L. flowers, rich in the carotenoid lutein, attenuates inflammatory cytokine secretion and improves the oxidative stress in an animal model of ulcerative colitis. Nutr Res 66:95–106. https://doi.org/10.1016/j.nutres.2019.03.005

Mildner-Szkudlarz S, Jeleń HH (2008) The potential of different techniques for volatile compounds analysis coupled with PCA for the detection of the adulteration of olive oil with hazelnut oil. Food Chem 110:751–761. https://doi.org/10.1016/j.foodchem.2008.02.053

Mildner-Szkudlarz S, Jeleń HH (2010) Detection of olive oil adulteration with rapeseed and sunflower oils using Mos electronic nose and SMPE-MS. J Food Qual 33:21–41. https://doi.org/10.1111/j.1745-4557.2009.00286.x

Mosandl A (1995) Enantioselective capillary gas chromatography and stable isotope ratio mass spectrometry in the authenticity control of flavors and essential oils. Food Rev Int 11:597–664. https://doi.org/10.1080/87559129509541063

Nanda PK, Bhattacharya D, Das JK, Bandyopadhyay S, Ekhlas D, Lorenzo JM, Dandapat P, Alessandroni L, Das AK, Gagaoua M (2022) Emerging role of biosensors and chemical indicators to monitor the quality and safety of meat and meat products. Chem 10:322. https://doi.org/10.3390/chemosensors10080322

Negi A, Pare A, Meenatchi R (2021) Emerging techniques for adulterant authentication in spices and spice products. Food Control 127:108113. https://doi.org/10.1016/j.foodcont.2021.108113

Nehal N, Choudhary B, Nagpure A, Gupta RK (2021) DNA barcoding: a modern age tool for detection of adulteration in food. Crit Rev Biotechnol 41:767–791. https://doi.org/10.1080/07388551.2021.1874279

Nguyen C, Baskaran K, Pupulin A, Ruvinov I, Zaitoon O, Grewal S, Scaria B, Mehaidli A, Vegh C, Pandey S (2019) Hibiscus flower extract selectively induces apoptosis in breast cancer cells and positively interacts with common chemotherapeutics. BMC Complement Altern Med 19:1–14. https://doi.org/10.1186/s12906-019-2505-9

Nichols PL (2021) Automated and enabling technologies for medicinal chemistry. Prog Med Chem 60:191–272. https://doi.org/10.1016/bs.pmch.2021.01.003

Nováková L, Svoboda P, Pavlík J (2017) Ultra-high performance liquid chromatography. In: Liquid chromatography. Elsevier, pp 719–769. https://doi.org/10.1016/B978-0-12-805393-5.00029-4

Nowicka P, Wojdyło A (2019) Anti-hyperglycemic and anticholinergic effects of natural antioxidant contents in edible flowers. Antioxidants 8:308. https://doi.org/10.3390/antiox8080308

Oh SY, Heo NS, Shukla S, Cho H-J, Vilian AE, Kim J, Lee SY, Han Y-K, Yoo SM, Huh YS (2017) Development of gold nanoparticle-aptamer-based LSPR sensing chips for the rapid detection of salmonella typhimurium in pork meat. Sci Rep 7:10130. https://doi.org/10.1038/s41598-017-10188-2

Panda H (2003) The complete technology book on Herbal Perfumes & Cosmetics. National Institute of Industrial Research

Pariak Y, Güzeler N (2016) Nuclear magnetic resonance spectroscopy applications in foods. Curr Res Nutr Food Sci 4:161–168

Peña F, Cárdenas S, Gallego M, Valcárcel M (2005) Direct olive oil authentication: detection of adulteration of olive oil with hazelnut oil by direct coupling of headspace and mass spectrometry, and multivariate regression techniques. J Chromatogr A 1074:215–221. https://doi.org/10.1016/j.chroma.2005.03.081

Peris M, Escuder-Gilabert L (2016) Electronic noses and tongues to assess food authenticity and adulteration. Trends Food Sci Technol 58:40–54. https://doi.org/10.1016/j.tifs.2016.10.014

Pires Jr EO, Di Gioia F, Rouphael Y, García-Caparrós P, Tzortzakis N, Ferreira IC, Barros L, Petropoulos SA, Caleja C (2023) Edible flowers as an emerging horticultural product: a review on sensorial properties, mineral and aroma profile. Trends Food Sci Technol 137:31–54. https://doi.org/10.1016/j.tifs.2023.05.007

Poole C (2023) Instrumental thin-layer chromatography. Elsevier, Amsterdam

Pruthi J, Kulkarni B (1969) A simple technique for the rapid and easy detection of papaya seeds in black pepper berries. Indian Food Packer 23:51–52

Qin D (2019) Next-generation sequencing and its clinical application. Cancer Biol Med 16:4. https://doi.org/10.20892/j.issn.2095-3941.2018.0055

Ravaglia LM, Pizzotti ABC, Alcantara GB (2019) NMR-based and chemometric approaches applicable to adulteration studies for assessment of the botanical origin of edible oils. J Food Sci Technol 56:507–511. https://doi.org/10.1007/s13197-018-3485-3

Ray S, Satya P (2014) Next generation sequencing technologies for next generation plant breeding. Front Plant Sci 5. https://doi.org/10.3389/fpls.2014.00367

Reuhs BL (2017) High-performance liquid chromatography. In: Nielsen SS (ed) Food analysis. Food science text series. Springer, Cham, pp 213–226. https://doi.org/10.1007/978-3-319-45776-5_13

Revathy SS, Rathinamala R, Murugesan M (2012) Authentication methods for drugs used in Ayurveda, siddha and Unani systems of medicine: an overview. Int J Pharm Sci Res 3:2352–2361

Riegel SD (2015) Determination of olive oil adulteration with 60-MHz benchtop NMR spectrometry. Am Lab 47(2):16–19

Roy RB, Tudu B, Bandyopadhyay R, Bhattacharyya N (2019) Application of electronic nose and tongue for beverage quality evaluation. In: Engineering tools in the beverage industry. Elsevier, pp 229–254. https://doi.org/10.1016/B978-0-12-815258-4.00008-1

Rubiolo P, Sgorbini B, Liberto E, Cordero C, Bicchi C (2010) Essential oils and volatiles: sample preparation and analysis. A review. Flavour Fragr J 25:282–290. https://doi.org/10.1002/ffj.1984

Sadgrove NJ, Padilla-González GF, Phumthum M (2022) Fundamental chemistry of essential oils and volatile organic compounds, methods of analysis and authentication. Plants 11:789. https://doi.org/10.3390/plants11060789

Santos PM, Pereira-Filho ER, Colnago LA (2016) Detection and quantification of milk adulteration using time domain nuclear magnetic resonance (TD-NMR). Microchem J 124:15–19. https://doi.org/10.1016/j.microc.2015.07.013

Sciarrone D, Costa R, Ragonese C, Tranchida PQ, Tedone L, Santi L, Dugo P, Dugo G, Mondello L (2011) Application of a multidimensional gas chromatography system with simultaneous mass spectrometric and flame ionization detection to the analysis of sandalwood oil. J Chromatogr A 1218:137–142. https://doi.org/10.1016/j.chroma.2010.10.117

Semenov V, Volkov S, Khaydukova M, Fedorov A, Lisitsyna I, Kirsanov D, Legin A (2019) Determination of three quality parameters in vegetable oils using potentiometric e-tongue. J Food Compost Anal 75:75–80. https://doi.org/10.1016/j.jfca.2018.09.015

Sharifi S, Vahed SZ, Ahmadian E, Dizaj SM, Eftekhari A, Khalilov R, Ahmadi M, Hamidi-Asl E, Labib M (2020) Detection of pathogenic bacteria via nanomaterials-modified aptasensors. Biosens Bioelectron 150:111933. https://doi.org/10.1016/j.bios.2019.111933

Shi T, Zhu M, Chen Y, Yan X, Chen Q, Wu X, Lin J, Xie M (2018) 1H NMR combined with chemometrics for the rapid detection of adulteration in camellia oils. Food Chem 242:308–315. https://doi.org/10.1016/j.foodchem.2017.09.061

Shi T, Zhu M, Zhou X, Huo X, Long Y, Zeng X, Chen Y (2019) 1H NMR combined with PLS for the rapid determination of squalene and sterols in vegetable oils. Food Chem 287:46–54. https://doi.org/10.1016/j.foodchem.2019.02.072

Shukla S, Iqbal M (2015) Forensic analysis of the saffron: rapid authenticity testing. Int J Rese Appl Sci Eng Technol 3:139–143

Simmler C, Graham JG, Chen S-N, Pauli GF (2018) Integrated analytical assets aid botanical authenticity and adulteration management. Fitoterapia 129:401–414. https://doi.org/10.1016/j.fitote.2017.11.017

Sionek B, Przybylski W, Tambor K (2020) Biosensors in evaluation of quality of meat and meat products–a review. Ann Anim Sci 20:1151–1168. https://doi.org/10.2478/aoas-2020-0057

Sionek B, Przybylski W, Bańska A, Florowski T (2021) Applications of biosensors for meat quality evaluations. Sensors 21:7430. https://doi.org/10.3390/s21227430

Skaria BP (2007) Aromatic plants, vol 1. New India Publishing

Slichter CP (2013) Principles of magnetic resonance, vol 1. Springer Science & Business Media

Šmejkalová D, Piccolo A (2010) High-power gradient diffusion NMR spectroscopy for the rapid assessment of extra-virgin olive oil adulteration. Food Chem 118:153–158. https://doi.org/10.1016/j.foodchem.2009.04.088

Smillie T, Khan I (2010) A comprehensive approach to identifying and authenticating botanical products. Clin Pharmacol Ther 87:175–186. https://doi.org/10.1038/clpt.2009.287

Soffritti G, Busconi M, Sánchez RA, Thiercelin J-M, Polissiou M, Roldán M, Fernández JA (2016) Genetic and epigenetic approaches for the possible detection of adulteration and auto-adulteration in saffron (Crocus sativus L.) spice. Molecules 21:343. https://doi.org/10.3390/molecules21030343

Stapley J, Reger J, Feulner PG, Smadja C, Galindo J, Ekblom R, Bennison C, Ball AD, Beckerman AP, Slate J (2010) Adaptation genomics: the next generation. Trends Ecol Evol 25:705–712. https://doi.org/10.1016/j.tree.2010.09.002

Sudhakar A, Chakraborty SK, Mahanti NK, Varghese C (2023) Advanced techniques in edible oil authentication: a systematic review and critical analysis. Crit Rev Food Sci Nutr 63:873–901. https://doi.org/10.1080/10408398.2021.1956424

Takahashi JA, Rezende FAGG, Moura MAF, Dominguete LCB, Sande D (2020) Edible flowers: bioactive profile and its potential to be used in food development. Food Res Int 129:108868. https://doi.org/10.1016/j.foodres.2019.108868

Thakur M, Ragavan K (2013) Biosensors in food processing. J Food Sci Technol 50:625–641. https://doi.org/10.1007/s13197-012-0783-z

Thao NT, Satake A (2010) Enantiomeric and stable isotope analysis. Citrus Essential Oils, Flavor and Fragrance, pp 165–200. https://doi.org/10.1002/9780470613160

Thomas SN (2019) Chapter 10-mass spectrometry. In: Contemporary practice in clinical chemistry, pp 171–185

Todt H, Guthausen G, Burk W, Schmalbein D, Kamlowski A (2008) Time-domain NMR in quality control: standard applications in food. In: Webb GA (ed) Modern magnetic resonance. Springer, Dordrecht, pp 1739–1743. https://doi.org/10.1007/1-4020-3910-7_196

Upton R, David B, Gafner S, Glasl S (2020) Botanical ingredient identification and quality assessment: strengths and limitations of analytical techniques. Phytochem Rev 19:1157–1177. https://doi.org/10.1007/s11101-019-09625-z

Varliklioz Er S, Eksi-Kocak H, Yetim H, Boyaci IH (2017) Novel spectroscopic method for determination and quantification of saffron adulteration. Food Anal Methods 10:1547–1555. https://doi.org/10.1007/s12161-016-0710-4

Waksmundzka-Hajnos M, Sherma J (2010) High performance liquid chromatography in phytochemical analysis. CRC Press

Waugh J (2007) DNA barcoding in animal species: progress, potential and pitfalls. BioEssays 29:188–197. https://doi.org/10.1002/bies.20529

Whitsitt V, Beehner C, Welch C (2013) The role of good manufacturing practices for preventing dietary supplement adulteration. Anal Bioanal Chem 405:4353–4358. https://doi.org/10.1007/s00216-012-6663-6

Yang P-F, Yang Y-N, Feng Z-M, Jiang J-S, Zhang P-C (2019) Six new compounds from the flowers of Chrysanthemum morifolium and their biological activities. Bioorg Chem 82:139–144. https://doi.org/10.1016/j.bioorg.2018.10.007

Zhang L, Shuai Q, Li P, Zhang Q, Ma F, Zhang W, Ding X (2016) Ion mobility spectrometry fingerprints: a rapid detection technology for adulteration of sesame oil. Food Chem 192:60–66. https://doi.org/10.1016/j.foodchem.2015.06.096

Zhang X, Tsuji S, Kitaoka H, Kobayashi H, Tamai M, Ki H, Miyamoto T (2017) Simultaneous detection of Escherichia coli O157: H7, salmonella enteritidis, and listeria monocytogenes at a very low level using simultaneous enrichment broth and multichannel SPR biosensor. J Food Sci 82:2357–2363. https://doi.org/10.1111/1750-3841.13843

Zhu M, Shi T, Chen Y, Luo S, Leng T, Wang Y, Guo C, Xie M (2019) Prediction of fatty acid composition in camellia oil by 1H NMR combined with PLS regression. Food Chem 279:339–346. https://doi.org/10.1016/j.foodchem.2018.12.025

Chapter 8
Postharvest Handling and Shelf Life of Edible Flowers

Valentina Scariot, Matteo Caser, Nicole Mélanie Falla, and Laura Pistelli

Abstract After a few days from harvest, varying according to the species, edible flowers begin to wither, dehydrate, and discolour. Along with the decrease in visual appearance, a loss of nutraceutical contents occurs, causing an overall loss of quality. Therefore, it becomes necessary to adopt pre- and post-harvest strategies to prolong their shelf life, both preserving the organoleptic and nutraceutical characteristics. Before harvesting, good agricultural practices such as adequate mineral nutrition and irrigation, and pest and disease control need to be applied as they can greatly influence both produce quality and longevity. The harvesting procedure itself is also a fundamental aspect to be considered. After harvesting, external factors, among which mainly temperature, can induce changes in edible flowers' senescence. The main strategies used to delay the metabolism of edible flowers are preservation at low temperatures through refrigeration and freezing. Other opportunities to delay quality loss are represented by controlled and modified atmosphere packaging, and edible films and coatings, which act as a barrier against gas exchange and the growth of microorganisms. This chapter deals with the most recent knowledge that has been developed to maintain all the properties and quality of fresh edible flowers for a long time, extending their marketability.

Keywords Bioactive compounds · Edible coating · Modified atmosphere · Organoleptic alterations · Petal discolouration · Senescence · Visual quality

V. Scariot (✉) · M. Caser · N. M. Falla
Department of Agricultural, Forest and Food Sciences (DISAFA), University of Torino, Grugliasco, Italy
e-mail: valentina.scariot@unito.it

L. Pistelli
Department of Agriculture, Food, and Environment, University of Pisa, Pisa, Italy

M. M. Jowkar et al. (eds.), *Advances in Research on Edible Flowers*, Advances in Olericulture, https://doi.org/10.1007/978-3-032-04561-4_8

Abbreviations

CO_2	Carbon dioxide
LDPE	Low-density polyethylene
MRL	Maximum Residue Levels
1-MCP	1-methylcyclopropene
O_2	Oxygen
PP	Polypropylene
PVC	Polyvinylchloride
VOCs	Volatile organic compounds

8.1 Introduction

Edible flowers are widely used as fresh or minimally processed products in the preparation of refined and aesthetically beautiful dishes. Moreover, edible flowers are source of nutritional compounds and natural antioxidants that are beneficial to the consumer's health (Devecchi et al. 2021; Mlcek and Rop 2011). However, edible flowers represent a highly perishable product, with a rather short shelf life after harvesting (Fernandes et al. 2018a; Kelley et al. 2003). Unlike cut flowers which have long stems and are used for ornamental purposes, edible flowers are harvested with a very short stem and are packed without water, thus making them more fragile and more perishable (Fernandes et al. 2018a). Depending on the species, as early as 2–5 days after harvest, flowers undergo a series of physiological, biochemical, and ultrastructural changes that characterize the senescence process (Silva 2003): petals begin to wilt, dehydrate, lose their colour (browning) and fall off, limiting their marketability (Demasi et al. 2020; Fernandes et al. 2018a; Koike et al. 2015; Kou et al. 2012). The senescence process in edible flowers causes quality losses (Demasi et al. 2021b; Flores-López et al. 2016): changes occur in the visual appearance in terms of brightness (Landi et al. 2018; Marchioni et al. 2023), but also from a nutraceutical point of view, such as the alteration of their bioactive compounds, including numerous molecules with antioxidant activity (Mlcek and Rop 2011; Rogers 2012). Flower senescence is a complex metabolic process that leads to the degradation of nucleic acids, proteins, and membranes by the action of enzymes (Fernandes et al. 2018a; Macnish et al. 2010).

After harvesting, flowers, like fruits and vegetables, continue to survive despite the separation from their natural source of organic nutrients, minerals, and water, and the respiration pathway still occurs, to produce energy for the flowers. The transpiration remains, causing a loss of water from the plant tissues, with a decrease in weight and cellular turgor (Demasi et al. 2020; Flores-López et al. 2016) (Fig. 8.1). Therefore, the primary strategy for reducing respiration and thus preserving the quality is storing and handling edible flowers at low temperature, such as below 5 °C (Demasi et al. 2020).

Fig. 8.1 Visual quality loss during postharvest storage at low temperatures (4 °C) in petals of (**a**) *Lavandula angustifolia* Mill., (**b**) *Rosa canina* L., and (**c**) *Rosa pendulina* L. The purple bar corresponds to 10 mm, the orange bar corresponds to 2 cm, the pink bar corresponds to 2.5 cm

In addition, several flowers are very sensitive to ethylene, a gaseous phytohormone produced by the plant itself, and flowers. The presence of ethylene in the surrounding environment can greatly accelerate the senescence process (Edelman and Jones 2014; Marchioni et al. 2021) and many flowers can undergo senescence with an ethylene concentration of 0.5 μl L^{-1}. This concentration can be very often found in different steps of the distribution chain or in the packaging. Highly sensitive flowers must be protected from the ethylene action using specific inhibitors, such as 1-methylciclopropene (1-MCP) or reducing its biosynthesis (Scariot et al. 2014). The preservation of the quality of edible flowers must be obtained by the identification of their ethylene production level and sensitivity to this hormone.

Besides, the most used edible flowers are rich in antioxidant compounds that can help flowers to counteract senescence and thus transiently increase and ensure a longer shelf life (Cavaiuolo et al. 2013; Scariot et al. 2022). Overall, increased knowledge of the phytochemical characteristics and storability of various species may provide information to select species suitable for the emerging edible flower market.

Currently, there are a few legislations and legal requirements related to the cultivation, production, marketing, and storage of edible flowers, thus indicating the extremely important demand for strategic regulation and legislation to ensure the safe use of edible flowers (Demasi et al. 2021a; Purohit et al. 2021). Despite the current challenges, many projects have been striving to increase the production of edible flowers (Fernandes et al. 2020). A reference example could be the Interreg Alcotra project ANTEA (http://interregantea.eu) and the subsequent Interreg Alcotra project ANTES, which spread the knowledge about edible flower use and accelerated their production through the strategies for the improvement of several floral matrices (Demasi et al. 2021a).

8.2 Pre-harvest Strategies for Ensuring a Long Shelf Life

Edible flowers differ from ornamental flowers in terms of crop production practices and marketing, since their handling from planting to harvesting and processing needs extra care in order to avoid using agrochemicals that can be dangerous to human health. Postharvest resistance to pathogens is an important trait for selecting edible flowers. Different cultivars can show different sensitivity to *Botrytis cinerea* infection, as seen in the genus *Rosa* (Friedman et al. 2010). Good cultivation practices are fundamental to obtain high quality and durable products. Intensive care is required, as edible flowers are extremely sensitive and susceptible to various physical, chemical, and biological contaminants (Matyjaszczyk and Śmiechowska 2019; Pires Jr et al. 2023). In this context, various cultivation precautions can be adopted, such as avoiding wetting flowers and leaves to reduce the pathogens' attacks and keeping the substrate barely moist to avoid the formation of rot. Predatory insects, or parasitic organisms of pathogens, can be used to contain pathogenic insects (Copetta and Ruffoni 2022). Concerning the residues of authorized phytosanitary products, reference is made to the European database in which the Maximum Residue Levels (MRL) of the active substances permitted on the various crops are listed. Within this database, edible flowers are included in the category 'basil/edible flowers' (EU pesticide database; Tinivella et al. 2023). Besides the adoption of specific cultivation techniques, especially hydroponics, may increase their compound content and prolong their shelf life. Recently, Chrysargyris et al. (2019) obtained petunia flowers richer in antioxidants, applying a salinity of 50 mmol L^{-1} NaCl for 10 days before harvesting. Kopsell et al. (2016) enhanced carotenoids in petal tissue of two different cultivars of *Tagetes tenuifolia* modifying the blue and red LED compositions (Red:Blue in the range 10:90, 20:80, and 40:60). Marchioni et al. (2020a) assessed the possibility of increasing yield and nutraceutical properties of *Viola cornuta* blooms by cultivating the plants in pots under different greenhouse conditions. Additional LED lighting leads to increase production of *Viola cornuta* flowers, while the complementary presence of basal heating enhances the concentration of antioxidant molecules, as polyphenols, anthocyanins, and carotenoids

(Marchioni et al. 2020a). In other trials, pots containing *Begonia cucullata* flowers increased anthocyanins content after silicon application, while the non-thermal plasma technology did not improve any yield and quality of flowers (Traversari et al. 2021).

8.3 Harvesting and Handling Procedures to Maintain a High Quality

To preserve the freshness of the product, it is recommended that harvesting takes place in the morning, when the flowers are fully open, to ensure the best visual appearance of the final product (Pires Jr et al. 2023). However, flower tissues have a structural fragility, and improper handling may induce rapid degradation and contamination and reduction of visual quality caused by injuries and bruising.

Another issue of extreme relevance that should be taken into consideration at harvesting refers to the particularity of each species since a great variability between species or cultivars of the same species exists in terms of growing period and harvesting stage (e.g. blooming seasonally) (Fernandes et al. 2020). Furthermore, soil and climate conditions can also influence the senescence period of the flowers, including the harvesting stage, although few studies have reported this phenomenon for edible flowers (Cavaiuolo et al. 2013). Therefore, further knowledge about the specific requirements of each species is necessary to understand the harvesting period of flowers, to obtain the highest yield, as well as to define the best storage conditions (Shantamma et al. 2021).

Regarding harvesting, specific consideration should be given to wild edible flowers as it is not recommended to collect them by the roadsides or in places with pets access, due to residual organic contamination (Pires Jr et al. 2023). Moreover, the collection of flowers from wild plants should be performed by people experienced in the identification of the species or after a thorough search in literary tools such as guides and illustrated books, while the required protocols for wild plant collection should be followed to ensure that genetic erosion is avoided (Caser et al. 2023; Falla et al. 2021).

Immediately after harvest, the flowers should be sanitized, cooled, and stored at an average temperature between 1 and 5 °C to prolong shelf life, which can extend from 2 to 14 days under these conditions, depending on the product (Pires Jr et al. 2023).

After the proper sanitization process, the flowers should be stored in specific packages, namely polyethylene trays, or plastic collection boxes which should be kept under freezing conditions (Fernandes et al. 2018a; Pires Jr et al. 2023). In general, flower storage containers must be perforated to avoid humidity condensation and to reduce wilting and rotting of the petals (Fernandes et al. 2018a), while the low storage temperature should preserve the antioxidant activity of the floral tissues (Pires Jr et al. 2023).

8.4 Post-harvest Strategies for Preserving Fresh Edible Flowers

In order to improve distribution and marketing efficiency, the postharvest shelf life of edible flowers must be prolonged. Refrigeration, controlled and modified atmosphere, different packaging solutions, or edible films and coatings may be used to maintain all the properties and quality of edible flowers for a long time, and thus be able to meet consumer demand (Fernandes et al. 2018a). These techniques do not act directly on the flowers, but they do help slow down the metabolic and senescence processes that cause flowers to rapidly decline in quality (Demasi et al. 2020; Flores-López et al. 2016).

8.4.1 Low-Temperature Conservation

Temperature is one of the main environmental factors limiting the shelf life of flowers, as well as fruits and vegetables (Demasi et al. 2020; Liu et al. 2018). To prolong their lifetime, the supply chain mostly uses low-temperature storage by refrigeration (Fernandes et al. 2018a). The advantage is the delay of flower senescence and the avoidance of the subsequent deterioration of flower quality during storage. At low temperatures (2–8 °C), petal wilting and the process of cellular respiration are slowed down, the enzymes activity of the internal tissue breakdown is reduced, (Demasi et al. 2020; Liu et al. 2018), the growth of microorganisms is decreased and the production of ethylene is avoided. All these combined factors contribute to extending the shelf life of edible flowers (Fernandes et al. 2018a). During cool storage, the aesthetic appearance changes, and some organoleptic alterations may be noted, especially under prolonged storage, such as petal discolouration and wilting, dehydration, and darkening of the whole flowers (Pires Jr et al. 2023). The decrease in weight is due to the loss of cellular turgor, which is correlated to flower shape as evidenced in *Ageratum houstonianum*, with petals curling and crumpling (Marchioni et al. 2020b). In other species, the chromatic changes during cold storage are associated with the oxidation of phenolics producing dark brown pigmentations (e.g. *Tulbaghia simmleri* (Marchioni et al. 2023).

Floral scent is another important trait of edible flowers, which becomes relevant to consumers'choice and marketability, and it can be controlled during cold storage. The floral scent is due to the emission of volatile organic compounds (VOCs), a complex of compounds categorized into terpenoids, phenylpropanoids/benzenoids, and fatty acids derivatives (Fernandes et al. 2018b). Volatile terpenoids belong to the largest and most diversified class of floral VOCs, which in turn are subdivided into monoterpenes (C_{10}), sesquiterpenes (C_{15}), and diterpenes (C_{20}), according to their structure. The percentage of these compounds changes during cold storage and is peculiar to single species. The cold storage influenced mainly *Pelargonium odoratissimum* and *Salvia dorisiana* aroma, with an increase in total monoterpenes at

6 days of storage, with a predominance of limonene in *Pelargonium* (Marchioni et al. 2020b). While, *Ageratum houstonianum and Tagetes lemmonii* showed different behaviour, since they maintained the prevalence of the same class of constituents in all the analyzed conditions, even though the cold storage influenced the major compound abundance. For example, limonene increased in both species probably as a cold-induced response, especially in *A. houstonianum,* where it was the main compound after a 6-day storage; conversely, it did not persist in *T. lemmonii*. Moreover, monoterpenes increased in all the studied species during storage, probably as a response to cooling stress (Marchioni et al. 2020b).

In addition, the phytochemical content of flowers tends to vary according to the type of packaging, making the proper choice of packaging material an important aspect of ensuring the quality of the product until consumption (Demasi et al. 2020).

A study evaluated the shelf life of 17 wild edible flowers stored at 4 °C during 14 days, as well as assessing the maintenance of their bioactive compounds (Demasi et al. 2021b). Results showed how low temperatures increased the shelf life of most species, such as *Rosa pendulina* L. (up to 14 days) and *Rosa canina* L. (up to 10 days), except for *Taraxacum officinale* (L.) W.W.Weber ex F.H.Wigg., which did not reach the third day (Fig. 8.2). Moreover, during storage at 4 °C, anthocyanins remained almost unchanged for 14 days (Fig. 8.3), while total polyphenols increased in 8 species (*Borago officinalis* L., *Centaurea cyanus* L., *Lavandula angustifolia* Mill., *Leucanthemum vulgare* (Vaill.) Lam., *Primula veris* L., *Rosa canina, Taraxacum officinale*) and decreased in 4 species (*Allium ursinum* L., *Calendula officinalis* L., *Paeonia officinalis* L., *Tropaeolum majus* L.) (Fig. 8.4). The antioxidant activity also showed a different trend depending on the species: it increased in 4 species (*A. ursinum, Dianthus carthusianorum* L., *L. angustifolia, L. vulgare*), decreased in 4 species (C. *officinalis, P. officinalis, R. canina, R. pendulina*) and remained constant in the remaining 9 species during the 14 days of storage (Demasi et al. 2021b).

Several other studies assessed the effects of low-temperature storage on edible flowers, obtaining evidence of its effectiveness, i.e., squash blossoms were acceptable for up to 7 days (Villalta et al. 2004), violas, pansies, and nasturtiums up to 2 weeks (Kelley et al. 2003), as well as begonias (Demasi et al. 2020), and *Tagetes lemmonii* A. Gray up to 6 days (Marchioni et al. 2020b) (Table 8.1).

8.4.2 Edible Films and Coatings

Edible films and coatings are consumable materials commonly used by the food industry for fruits and vegetables (Bourtoom 2008). Films are thin coatings (layers) applied to the product, while coatings are a suspension or emulsion applied directly to the surface of the food. Both are formed by biological materials, such as proteins, lipids, and polysaccharides, and act as a barrier against gas exchange and moisture, and they prevent microbial growth or texture changes (Bourtoom 2008; Fernandes et al. 2018a; Zhao et al. 2019). They also constitute barriers against oxygen in order

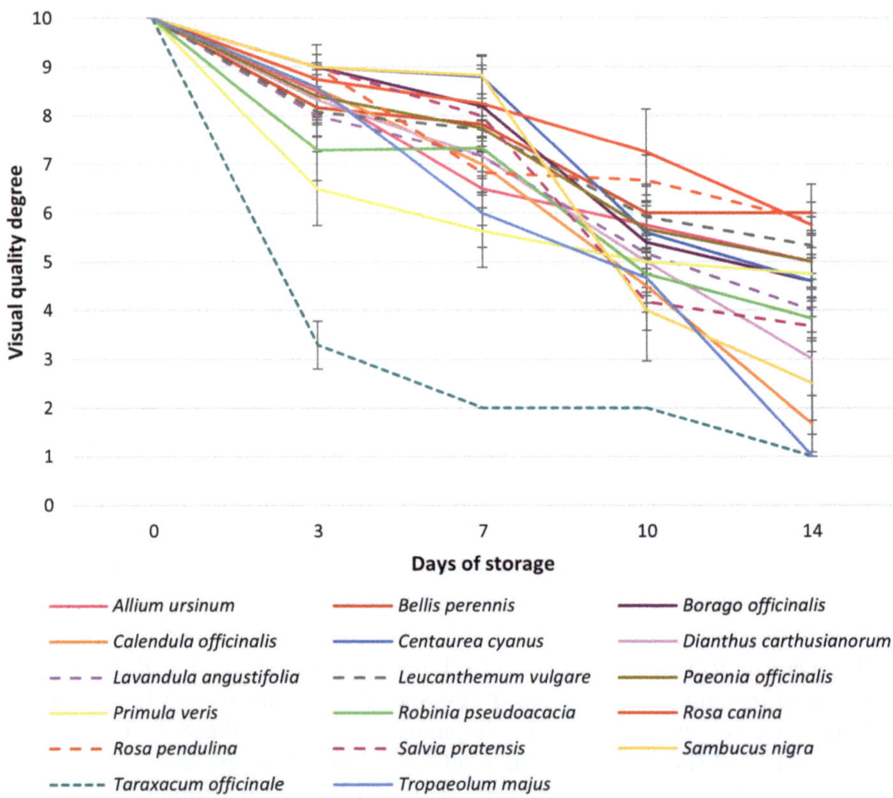

Fig. 8.2 The visual quality trend in edible flowers of 17 species during storage (day 0, 3, 7, 10, 14) at 4 °C, studied by Demasi et al. (2021b). Bars indicate the standard deviation

to reduce the respiration rate of food products, thus preserving their quality and extending their shelf life (Janjarasskul and Krochta 2010).

Edible films and coatings are an interesting alternative to synthetic packaging, since they are an integral part of the food consumed, avoiding to unpack the product, so that the pollution is reduced since there is no package to throw away (Bourtoom 2008). Moreover, they are usually made of renewable and edible components, thus degrading more rapidly than polymeric materials and being more environmentally friendly (Bourtoom 2008).

Polysaccharide-based films and coatings may be formed by cellulose, pectin derivatives, starch derivatives, exudate gums, and they are effective in delaying moisture loss from food products. However, being highly hydrophilic, they have scarce capacity as vapor barriers (Bourtoom 2008). Lipid-based films and coatings may be formed by natural wax, like paraffin wax or beeswax. Lipids act as a barrier against moisture loss due to their hydrophobic characteristics; however, because of this specific reason, they tend to form thick and fragile films. Therefore, they are often associated with a polysaccharide, which provides mechanical strength

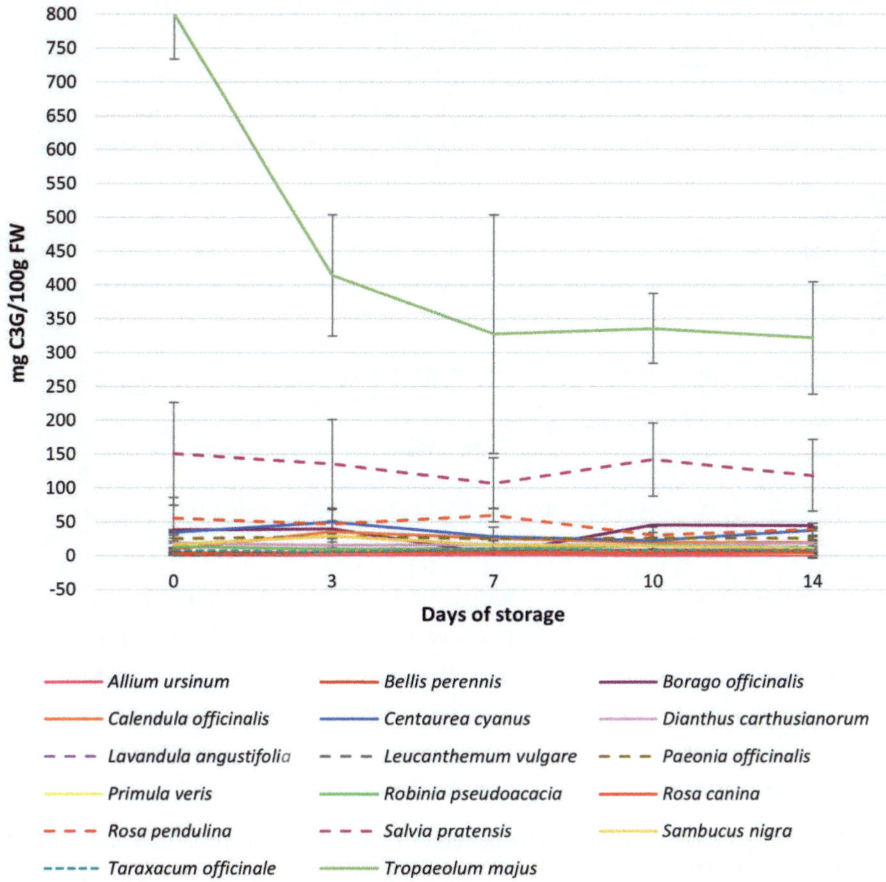

Fig. 8.3 Total anthocyanins trend in edible flowers of 17 species during storage (day 0, 3, 7, 10, 14) at 4 °C, studied by Demasi et al. (2021b). Bars indicate the standard deviation

(Bourtoom 2008). Protein-based films and coatings may be formed by gelatine, whey protein, casein, soy protein, wheat gluten, etc., and they constitute good oxygen barriers at low relative humidity (Bourtoom 2008). Edible films and coatings may also be formed by a heterogeneous combination of the previous categories, mixing polysaccharides, proteins and/or lipids (Bourtoom 2008).

Edible films and coatings are already used on fresh-cut fruits and vegetables, but they have rarely been applied to edible flowers and are under investigation. Fernandes and colleagues (Fernandes et al. 2018b) reported the effect of alginate coating on microbiological and physico-chemical quality of *Viola x wittrockiana* (pansy) edible flower stored at 4 °C. Their results showed a higher bioactive compound content and antioxidant activity than uncoated flowers, and a longer shelf life (14 days) than uncoated ones (7 days). Alginate coating was proven to be effective in preserving also the visual appearance of the studied flowers (Fernandes et al. 2018b).

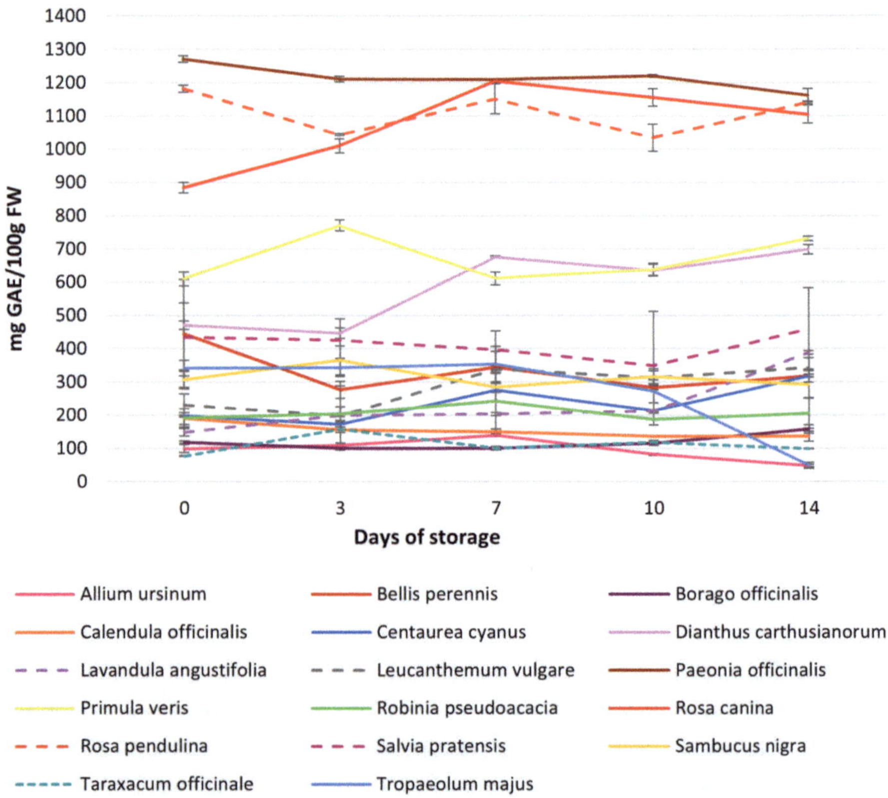

Fig. 8.4 Total polyphenols trend in edible flowers of 17 species during storage (day 0, 3, 7, 10, 14) at 4 °C, studied by Demasi et al. (2021b). Bars indicate the standard deviation

8.4.3 Packaging

Plants, through the process of cellular respiration, use carbon dioxide (CO_2) taken from the environment to produce sugars and oxygen (O_2). The high perishability of edible flowers is generally proportional to their respiration rate, which is usually quite high. After harvest, some external factors, such as temperature or packaging, may contribute to induce changes in the respiration rate of edible flowers (Fernandes et al. 2018a). Thus, packaging protects their fragile structure and isolates them from the external environment, reducing exposure to pathogens and contaminants. Currently, edible flowers are packed in rigid plastic containers; within these containers, the atmosphere surrounding the product can be unaltered or controlled (Fernandes et al. 2018a), as underlined in Table 8.1.

Table 8.1 Post-harvest conditions, shelf life, and bioactive compounds variation during storage of edible flowers reported in the literature

Edible flower	Packaging and Storage conditions	Acceptable storage days	Bioactive compounds variation during storage	References
Ageratum houstonianum Mill.	Polypropylene boxes (120 × 85 × 80 mm), 4 °C	2	Persistence of bioactive compounds.	Marchioni et al. (2020b)
Allium ursinum L.	Polypropylene boxes (123 × 114 × 50 mm), 4 °C	7	Reduction of polyphenols, persistence of anthocyanins, and increase in antioxidant activity.	Demasi et al. (2021b)
Antirrhinum majus L.	Polypropylene trays (13.5 × 17 × 3 cm^3) sealed with oxygen transmission film, 5 °C, 1-Methylcyclopropene (1-MCP) strip + passive modified atmosphere (MAP)	14	–	Kou et al. (2012)
Begonia semperflorens-cultorum Hort.	Polypropylene (PP) boxes (123 × 114 × 50 mm), 4 °C	9	Decrease in polyphenols, anthocyanins, and antioxidant activity.	Demasi et al. (2020)
Begonia semperflorens-cultorum Hort.	Polypropylene (PP) boxes (123 × 114 × 50 mm) in a bi-oriented polypropylene plastic film, 4 °C	9	Decrease in anthocyanins; persistence of polyphenols and antioxidant activity.	Demasi et al. (2020)
Borago officinalis L.	Polypropylene boxes (123 × 114 × 50 mm), 4 °C	7	Increased amount of polyphenols, the persistence of anthocyanins, and antioxidant activity.	Demasi et al. (2021b)
Borago officinalis L.	Polyethylene bags (50-μm thickness low-density) with 4 needle holes, from −2,5 °C to 5 °C	7	–	Kelley et al. (2003)
Calendula officinalis L.	Polypropylene boxes (123 × 114 × 50 mm), 4 °C	7	Reduction of polyphenols and antioxidant activity, persistence of anthocyanins.	Demasi et al. (2021b)

(continued)

Table 8.1 (continued)

Edible flower	Packaging and Storage conditions	Acceptable storage days	Bioactive compounds variation during storage	References
Calendula officinalis L.	Polypropylene trays (12 × 18 × 4 cm) with passive MAP by packing flowers in micro- and macro-perforated polypropylene film or continuous polyolefin film, 5 °C	10	Delayed decrease in polyphenols, sucrose and lycopene and loss of antioxidant activity, compared to control flowers.	Fadda et al. (2020)
Centaurea cyanus L.	Polypropylene boxes (123 × 114 × 50 mm), 4 °C	7	Increased amounts of polyphenols, persistence of anthocyanins, and antioxidant activity.	Demasi et al. (2021b)
Cucurbita pepo L.	Polystyrene containers (2 L), 2,5 or 5 °C	7	–	Villalta et al. (2004)
Dianthus carthusianorum L.	Polypropylene boxes (123 × 114 × 50 mm), 4 °C	7	Increased amount of polyphenols and antioxidant activity, persistence of anthocyanins.	Demasi et al. (2021b)
Dianthus caryophyllus L.	Polypropylene trays (13.5 × 17 × 3 cm^3) sealed with oxygen transmission film, 5 °C, 1-Methylcyclopropene (1-MCP) strip + passive modified atmosphere (MAP)	7	–	Kou et al. (2012)
Lavandula angustifolia Mill.	Polypropylene boxes (123 × 114 × 50 mm), 4 °C	7	Increased amount of polyphenols and antioxidant activity, persistence of anthocyanins.	Demasi et al. (2021b)
Leucanthemum vulgare (Vaill.) Lam.	Polypropylene boxes (123 × 114 × 50 mm), 4 °C	7	Increased amount of polyphenols and antioxidant activity, persistence of anthocyanins.	Demasi et al. (2021b)

(continued)

Table 8.1 (continued)

Edible flower	Packaging and Storage conditions	Acceptable storage days	Bioactive compounds variation during storage	References
Paeonia officinalis L.	Polypropylene boxes (123 × 114 × 50 mm), 4 °C	7	Reduction of polyphenols and antioxidant activity, persistence of anthocyanins.	Demasi et al. (2021b)
Pelargonium odoratissimum (L.) L'Hér 'Lemon'	Polypropylene boxes (120 × 85 × 80 mm), 4 °C	2	Increased amount of flavonoids, anthocyanins, and radical scavenging activity.	Marchioni et al. (2020b)
Primula veris L.	Polypropylene boxes (123 × 114 × 50 mm), 4 °C	3	Increased amount of polyphenols, persistence of anthocyanins, and antioxidant activity.	Demasi et al. (2021b)
Robinia pseudoacacia L.	Polypropylene boxes (123 × 114 × 50 mm), 4 °C	7	Persistence of polyphenols, anthocyanins, and antioxidant activity.	Demasi et al. (2021b)
Rosa canina L.	Polypropylene boxes (123 × 114 × 50 mm), 4 °C	10	Increased amount of polyphenols, persistence of anthocyanins, and reduction of antioxidant activity.	Demasi et al. (2021b)
Rosa pendulina L.	Polypropylene boxes (123 × 114 × 50 mm), 4 °C	10	Persistence of anthocyanins and polyphenols, reduction of antioxidant activity.	Demasi et al. (2021b)
Salvia discolor Kunth	Polyethylene terephthalate boxes (500 cm^3), 4 °C, under light	5	Reduction of polyphenols and anthocyanins. Persistence of flavonoids and antioxidant activity.	Landi et al. (2018)

(continued)

Table 8.1 (continued)

Edible flower	Packaging and Storage conditions	Acceptable storage days	Bioactive compounds variation during storage	References
Salvia dorisiana Standl	Polypropylene boxes (120 × 85 × 80 mm), 4 °C	6	Increased amount of flavonoids, anthocyanins, and radical scavenging activity.	Marchioni et al. (2020b)
Salvia pratensis L.	Polypropylene boxes (123 × 114 × 50 mm), 4 °C	7	Persistence of polyphenols, anthocyanins, and antioxidant activity.	Demasi et al. (2021b)
Sambucus nigra L.	Polypropylene boxes (123 × 114 × 50 mm), 4 °C	7	Persistence of polyphenols, anthocyanins, and antioxidant activity.	Demasi et al. (2021b)
Tagetes lemmonii A. Gray	Polypropylene boxes (120 × 85 × 80 mm), 4 °C	6	Increased amount of flavonoids, anthocyanins, and ascorbic acid.	Marchioni et al. (2020b)
Taraxacum officinale Weber	Polypropylene boxes (123 × 114 × 50 mm), 4 °C	1–2	Increased amount of polyphenols, persistence of anthocyanins, and antioxidant activity.	Demasi et al. (2021b)
Tropaeolum majus L.	Polypropylene boxes (123 × 114 × 50 mm), 4 °C	7	Persistence of polyphenols and antioxidant activity, reduction of anthocyanins.	Demasi et al. (2021b)
Tropaeolum majus L.	Polyethylene terephthalate boxes (500 cm^3), 4 °C, under light	5	Reduction of polyphenols. Persistence of flavonoids, and anthocyanins, increased antioxidant activity.	Landi et al. (2018)
Tropaeolum majus L.	Polypropylene trays (14 × 11 × 6 cm) covered with polypropylene (PP) film, 4–5 °C, active modified atmosphere (MAP: 5 or 10% CO_2; 10% O_2; 85 or 80% N_2)	7	Persistence of antioxidant activity.	Friedman et al. (2005)

(continued)

Table 8.1 (continued)

Edible flower	Packaging and Storage conditions	Acceptable storage days	Bioactive compounds variation during storage	References
Tulbaghia cominsii Vosa	Polyethylene terephthalate boxes (500 cm^3), 4 °C, under light	8	Reduction of polyphenols. Persistence of flavonoids, anthocyanins, and antioxidant activity.	Landi et al. (2018)
Tulbaghia simmleri Beauv.	Polyethylene boxes (120 × 85 × 80 mm), 4 °C, dark conditions	6	Persistence of polyphenols, flavonoids, anthocyanins, and antioxidant activity. Decrease in total monoterpenes	Marchioni et al. (2023)
Viola cornuta L.	Polypropylene (PP) boxes (123 × 114 × 50 mm), 4 °C	14	Increased amount of anthocyanins, and reduction of antioxidant activity.	Demasi et al. (2020)
Viola cornuta L.	Polypropylene (PP) boxes (123 × 114 × 50 mm) in a bi-oriented polypropylene plastic film, 4 °C	14	Reduction of polyphenols, anthocyanins, and antioxidant activity.	Demasi et al. (2020)
Viola tricolor L. 'Helen Mount'	Polyethylene bags (50-µm thickness low-density) with 4 needle holes, from −2,5 °C to 10 °C	7	–	Kelley et al. (2003)
Viola tricolor L. 'Helen Mount'	Polyethylene bags (50-µm thickness low-density) with 4 needle holes, from −2,5 °C to 2,5 °C	14	–	Kelley et al. (2003)
Viola x wittrockiana Gams ex Nauenb. & Buttler	Polyethylene bags (50-µm thickness low-density) with 4 needle holes, from −2,5 °C to 10 °C	7	–	Kelley et al. (2003)
Viola x wittrockiana Gams ex Nauenb. & Buttler	Polyethylene bags (50-µm thickness low-density) with 4 needle holes, from −2,5 °C to 2,5 °C	14	–	Kelley et al. (2003)
Viola x wittrockiana Gams ex Nauenb. & Buttler	Polyethylene bags (one flower per bag), 4 °C	4	Reduction of polyphenols, antioxidant activity, and monomeric anthocyanins.	Fernandes et al. (2017)

8.4.3.1 Normal Atmosphere Packaging

Uncontrolled atmosphere packaging is based on the use of a container, usually of various plastic materials, sealed with a film, itself of various materials. Few studies have focused on the influence of different plastic films on the properties of edible flowers without changing the atmosphere at the time of packaging.

Studies have looked at artichokes, broccoli, and cauliflower, analysing their post-harvest durability in different packages. Perforated polypropylene (PP), polyvinyl-chloride (PVC), low-density polyethylene (LDPE), and microperforated polypropylene films were evaluated (Fernandes et al. 2018a; Friedman et al. 2007). Demasi et al. (2020) investigated the shelf life of *Begonia x semperflorens-cultorum* Hort. and *Viola cornuta* L. edible flowers stored at 4 °C in plastic boxes (Fig. 8.5) inserted into a bi-oriented polypropylene plastic film (flowpack) or in plastic boxes closed with a plastic lid. Results showed that begonias' shelf life was positively influenced by flowpack, efficiently maintaining aesthetic quality and bioactive compounds for up to 9 days (Demasi et al. 2020). Conversely, violas' quality and bioactive compounds were better maintained in the plastic box with its lid, showing a 16-day shelf life.

In general, packaging with different films appears to improve physical protection and pathogen reduction, thereby prolonging shelf life and maintaining the bioactive components of the flowers, but some materials are found to be more suitable for certain species of edible flowers. Each species may react differently to the type of packaging, so further studies are needed to evaluate the most appropriate solution for each species.

Fig. 8.5 Edible flowers of *Viola cornuta* (**a**) and *Begonia semperflorens* (**b**) freshly harvested (day 0) and cold stored in plastic boxes. The yellow bar corresponds to 2 cm, the pink bar corresponds to 4 cm

8.4.3.2 Controlled and Modified Atmosphere Packaging

Controlled and modified atmosphere packaging technologies are already currently used by the food industry to extend the shelf life of fruits and vegetables (Aquino-Bolaños et al. 2013). By changing the gaseous environment surrounding the food, it is possible to reduce the respiration rate of the product, reduce microbial growth, and delay enzymatic deterioration (Fernandes et al. 2018a).

Controlled atmosphere packaging is a commercial system in which the gaseous concentrations of oxygen, carbon dioxide, and nitrogen are regulated, as well as the temperature and humidity of the storage environment. Modified atmosphere packaging, on the other hand, is a technique for sealing actively breathing food products in polymer film packages with modified levels of oxygen and carbon dioxide within the package (Fernandes et al. 2018a). In addition to the two packaging described above, 1-methylcyclopropene (1-MCP), a nontoxic ethylene antagonist that binds to ethylene receptors, blocking them, has also been considered in recent years. This protects the flower from the effects of ethylene, such as senescence and petal abscission (Fernandes et al. 2018a).

These three techniques have been used in combination with each other on edible flowers such as carnation or antirrhinum, proving to be effective in maintaining a fresh appearance and extending the shelf life of these flowers (Fernandes et al. 2018a). However, it is important to conduct further studies to investigate the reaction of each species to different conservation conditions, as different species have different behaviours when subjected to these treatments (Fernandes et al. 2018a).

8.5 Discussion and Conclusions

The global market of edible flowers was valued at USD 290.3 Million in 2020 and is expected to increase beyond USD 380.8 Million by the end of the year 2026, with an annual increase of about 4.24–4.85% (Pires Jr et al. 2023). This is mainly due to the increasing demand for edible flowers as natural and organic nutraceutical supplements, from the nutraceutical and food & beverages industries, as well as from the pharmaceutical industry. Harvested flowers need to be delivered within 3 or 4 days, and handling and packaging have to be done with great care in order to ensure a good quality product (Fernandes et al. 2020).

Edible flowers quality relies on their bioactive compound content, therefore it is important to study their properties and the best conservation methods to expand and extend their use. Some studies performed allowed the phytochemical characterization of several species both already commercially available and derived from wild harvesting, with the aim of strengthening and expanding an emerging supply chain. In addition, they helped increase knowledge of post-harvest techniques most suitable for the preservation of these fresh products.

Storage at low temperatures (4 °C) seems to be effective in extending the shelf life of packaged edible flowers, and is also suitable for maintaining their bioactive components, therefore it is a good strategy to adopt. The use of modified atmosphere seems promising for extending the shelf life of edible flowers, as well as new types of packaging and films. Reducing plastic packaging is certainly a goal to be pursued, and further studies will be needed to understand what the best solutions and materials are to use for storing edible flowers.

Flower species vary widely in their characteristics and phytochemical composition, so investigating different methods of preservation and processing is of great interest to improve knowledge on the most suitable technique to achieve maximum shelf life and maximum amount of bioactive compounds for each species.Funding InformationThe present chapter is the result of researches funded by the EU program Interreg V-A Francia Italia Alcotra, grant No. 1139: “ANTEA—Attività innovative per lo sviluppo della filiera transfrontaliera del fiore edule” and grant No. 8336: “ANTES – Fiori eduli e piante aromatiche: attività di capitalizzazione dei progetti ANTEA ed ESSICA”).

References

Aquino-Bolaños EN, Urrutia-Hernández TA, López Del Castillo-Lozano M, Chavéz-Servia JL, Verdalet-Guzmán I (2013) Physicochemical parameters and antioxidant compounds in edible squash (*Cucurbita pepo*) flower stored under controlled atmospheres. J Food Qual 36:302–308. https://doi.org/10.1111/jfq.12053

Bourtoom T (2008) Edible films and coatings: characteristics and properties. Int Food Res J 15:237–248

Caser M, Falla NM, Demasi S, Scariot V (2023) From fresh to dried lavender flower: changes in phytochemical profile according to drying method. Horticulturae 9:700. https://doi.org/10.3390/horticulturae9060700

Cavaiuolo M, Cocetta G, Ferrante A (2013) The antioxidants changes in ornamental flowers during development and senescence. Antioxidants 2:132–155. https://doi.org/10.3390/antiox2030132

Chrysargyris A, Tzionis A, Xylia P, Nicola S, Tzortzakis N (2019) Physiochemical properties of petunia edible flowers grown under saline conditions and their postharvest performance under modified atmosphere packaging and ethanol application. J Sci Food Agric 99:3644–3652. https://doi.org/10.1002/jsfa.9586

Copetta A, Ruffoni B (2022) Sostenibilità in pratica/3 – I fiori commestibili come nuovo modello di business sostenibile. Creafuturo

Demasi S, Falla NM, Caser M, Scariot V (2020) Postharvest aptitude of *Begonia semperflorens* and *Viola cornuta* edible flowers. Adv Hortic Sci 34:13–20. https://doi.org/10.13128/ahsc-7444

Demasi S, Caser M, Donno D, Enri SR, Lonati M, Scariot V (2021a) Exploring wild edible flowers as a source of bioactive compounds: new perspectives in horticulture. Folia Hortic 33:27–48. https://doi.org/10.2478/fhort-2021-0004

Demasi S, Mellano MG, Falla NM, Caser M, Scariot V (2021b) Sensory profile, shelf life, and dynamics of bioactive compounds during cold storage of 17 edible flowers. Horticulturae 7:166. https://doi.org/10.3390/horticulturae7070166

Devecchi A, Demasi S, Saba F, Rosato R, Gambino R, Ponzo V et al (2021) Compositional characteristics and antioxidant activity of edible rose flowers and their effect on phenolic urinary excretion. Pol J Food Nutr Sci 71:383–392. https://doi.org/10.31883/pjfns/142639

Edelman NF, Jones ML (2014) Evaluating ethylene sensitivity within the family Solanaceae at different developmental stages. HortScience 49:628–636

EU Pesticide Database (2024). https://ec.europa.eu/food/plant/pesticides/eu-pesticides-database/public/?event=homepage&language=EN. Accessed 20 Aug 2024

Fadda A, Palma A, Azara E, D'Aquino S (2020) Effect of modified atmosphere packaging on overall appearance and nutraceutical quality of pot marigold held at 5°C. Food Res Int 134:109248. https://doi.org/10.1016/j.foodres.2020.109248

Falla NM, Demasi S, Caser M, Scariot V (2021) Preliminary observations on *Viola calcarata* as a source of bioactive compounds: antioxidant activity and phytochemical profile of two alpine subspecies. Agronomy 11:2241. https://doi.org/10.3390/agronomy11112241

Fernandes L, Casal S, Pereira JA, Pereira EL, Ramalhosa E, Saraiva JA (2017) Effect of high hydrostatic pressure on the quality of four edible flowers: *Viola x wittrockiana*, *Centaurea cyanus*, *Borago officinalis* and *Camellia japonica*. Int J Food Sci Technol 52:2455–2462. https://doi.org/10.1111/ijfs.13530

Fernandes L, Saraiva JA, Pereira JA, Casal S, Ramalhosa E (2018a) Post-harvest technologies applied to edible flowers: a review. Food Rev Int 00:1–23. https://doi.org/10.1080/87559129.2018.1473422

Fernandes L, Pereira JA, Baptista P, Saraiva JA, Ramalhosa E, Casal S (2018b) Effect of application of edible coating and packaging on the quality of pansies (*Viola x wittrockiana*) of different colors and sizes. Food Sci Technol Int 24:321–329. https://doi.org/10.1177/1082013217753229

Fernandes L, Casal S, Pereira JA, Saraiva JA, Ramalhosa E (2020) An overview on the market of edible flowers. Food Rev Int 36:258–275. https://doi.org/10.1080/87559129.2019.1639727

Flores-López ML, Cerqueira MA, de Rodríguez DJ, Vicente AA (2016) Perspectives on utilization of edible coatings and nano-laminate coatings for extension of postharvest storage of fruits and vegetables. Food Eng Rev 8:292–305. https://doi.org/10.1007/s12393-015-9135-x

Friedman H, Vinokur Y, Rodov V, Goldman G, Resnick N, Umiel N (2005) *Tropaeolum majus* L. as edible flowers: growth and postharvest handling. Adv Hortic Sci 19:3–8

Friedman H, Rot I, Agami O, Vinokur Y, Rodov V, Reznick N et al (2007) Edible flowers: new crops with potential health benefits. Acta Hortic 755:283–289. https://doi.org/10.17660/actahortic.2007.755.36

Friedman H, Agami O, Vinokur Y, Droby S, Cohen L, Refaeli G et al (2010) Characterization of yield, sensitivity to *Botrytis cinerea* and antioxidant content of several rose species suitable for edible flowers. Sci Hortic 123:395–401. https://doi.org/10.1016/j.scienta.2009.09.019

Janjarasskul T, Krochta JM (2010) Edible packaging materials. Ann Rev Food Sci Technol 1:415–448. https://doi.org/10.1146/annurev.food.080708.100836

Kelley KM, Cameron AC, Biernbaum JA, Kenneth LP (2003) Effect of storage temperature on the quality of edible flowers. Postharvest Biol Technol 27:341–344

Koike A, Barreira JCM, Barros L, Santos-Buelga C, Villavicencio ALCH, Ferreira ICFR (2015) Edible flowers of *Viola tricolor* L. As a new functional food: antioxidant activity, individual phenolics and effects of gamma and electron-beam irradiation. Food Chem 179:6–14. https://doi.org/10.1016/j.foodchem.2015.01.123

Kopsell D, Belisle C, Whitlock C, Sams CE (2016) Genotype and lighting environment impact petal tissue pigmentation in *Tagetes tenuifolia*. Acta Hortic 1134:103–110

Kou L, Turner ER, Luo Y (2012) Extending the shelf life of edible flowers with controlled release of 1-Methylcyclopropene and modified atmosphere packaging. J Food Sci 77:188–193. https://doi.org/10.1111/j.1750-3841.2012.02683.x

Landi M, Ruffoni B, Combournac L, Guidi L (2018) Nutraceutical value of edible flowers upon cold storage. Ital J Food Sci 30:336–347

Liu W, Zhang J, Zhang Q, Shan Y (2018) Effects of postharvest chilling and heating treatments on the sensory quality and antioxidant system of daylily flowers. Hortic Environ Biotechnol 59:671–685. https://doi.org/10.1007/s13580-018-0087-y

Macnish AJ, Jiang C-Z, Negre-Zakharov F, Reid MS (2010) Physiological and molecular changes during opening and senescence of *Nicotiana mutabilis* flowers. Plant Sci 179:267–272. https://doi.org/10.1016/j.plantsci.2010.05.011

Marchioni I, Colla L, Pistelli L, Ruffoni B, Tinivella F, Minuto G (2020a) Different growing conditions can modulate metabolites content during post-harvest of *Viola cornuta* L. edible flowers. Adv Hort Sci 34:61–69. https://doi.org/10.13128/ahsc8395

Marchioni I, Pistelli L, Ferri B, Copetta A, Ruffoni B, Pistelli L, Najar B et al (2020b) Phytonutritional content and aroma profile changes during postharvest storage of edible flowers. Front Plant Sci 11:1–16. https://doi.org/10.3389/fpls.2020.590968

Marchioni I, Dimita R, Gioè G, Luisa P, Ruffoni B, Laura P et al (2021) The effects of post-harvest treatments on the quality of *Agastache aurantiaca* edible flowers. Horticulturae 7:1–13. https://doi.org/10.3390/horticulturae7040083

Marchioni I, Najar B, Copetta A, Ferri B, Ruffoni B, Pistelli L et al (2023) Phytonutritional and aromatic profiles of *Tulbaghia simmleri* Beauv. Edible flowers during cold storage. Adv Hort Sci 37:25–32. https://doi.org/10.36253/ahsc13857

Matyjaszczyk E, Śmiechowska M (2019) Edible flowers. Benefits and risks pertaining to their consumption. Trends Food Sci Technol 91:670–674. https://doi.org/10.1016/j.tifs.2019.07.017

Mlcek J, Rop O (2011) Fresh edible flowers of ornamental plants – a new source of nutraceutical foods. Trends Food Sci Technol 22:561–569. https://doi.org/10.1016/j.tifs.2011.04.006

Pires EDO Jr, Di Gioia F, Rouphael Y, García-Caparrós P, Tzortzakis N, Ferreira IC, Barros L, Petropoulos SA, Caleja C et al (2023) Edible flowers as an emerging horticultural product: a review on sensorial properties, mineral and aroma profile. Trends Food Sci Technol 137:31–54. https://doi.org/10.1016/j.tifs.2023.05.007

Purohit SR, Rana SS, Idrishi R, Sharma V, Ghosh P (2021) A review on nutritional, bioactive, toxicological properties and preservation of edible flowers. Future Foods 4:100078. https://doi.org/10.1016/j.fufo.2021.100078

Rogers HJ (2012) Is there an important role for reactive oxygen species and redox regulation during floral senescence? Plant Cell Environ 35:217–233. https://doi.org/10.1111/j.1365-3040.2011.02373.x

Scariot V, Paradiso R, Rogers H, De Pascale S (2014) Ethylene control in cut flowers: classical and innovative approaches. Postharvest Biol Technol 97:83–92. https://doi.org/10.1016/j.postharvbio.2014.06.010

Scariot V, Ferrante A, Romano D (2022) Editorial: edible flowers: understanding the effect of genotype, preharvest, and postharvest on quality, safety, and consumption. Front Plant Sci 13:1025196. https://doi.org/10.3390/antiox203

Shantamma S, Vasikaran ME, Waghmare R, Nimbkar S, Moses JA, Anandharamakrishnan C (2021) Emerging techniques for the processing and preservation of edible flowers. Future Foods 4:100094. https://doi.org/10.1016/j.fufo.2021.100094

Silva JA (2003) The cut flower: postharvest considerations. J Biol Sci 3(4):406–442

Tinivella F, Minuto A, Minuto G (2023) Protocolli di difesa: prodotti fitosanitari consentiti. In: Protocols. Interreg Alcotra ANTEA Project, WP N 3.8, protocol N. 34. https://www.interregantea.eu/docs/protocols/protocollo34.pdf. Accessed 20 Aug 2024

Traversari S, Pistelli L, Del Ministro B, Cacini S, Costamagna G, Ginepro M et al (2021) Combined effect of silicon and non-thermal plasma treatments on yield, mineral content, and nutraceutical proprieties of edible flowers of *Begonia cucullata*. Plant Physiol Biochem 166:1014–1021. https://doi.org/10.1016/j.plaphy.2021.07.012

Villalta AM, Ergun M, Berry AD, Shaw N, Sargent SA (2004) Quality changes of yellow summer squash blossoms (*Cucurbita pepo*) during storage. In: ISHS Acta Horticulturae 659: VII international symposium on protected cultivation in mild winter climates: production, Pest management and global competition, pp 831–834. https://doi.org/10.17660/ActaHortic.2004.659.108

Zhao L, Fan H, Zhang M, Chitrakar B, Bhandari B, Wang B (2019) Edible flowers: review of flower processing and extraction of bioactive compounds by novel technologies. Food Res Int 126:108660. https://doi.org/10.1016/j.foodres.2019.108660

Chapter 9
Postharvest Technologies Applied to Edible Flowers

Elka Fabiana Aparecida Almeida, Juliana Pinto de Lima, Ariane Castricini, and Gabriel Sthefano Lourenço

Abstract The sensitivity and short duration of edible flowers have led to the use of several technologies to increase their post-harvest life. These technologies enable consumers to have access to a quality product with characteristics similar to those at the time of harvest, and such methods also have a considerable impact on producers, who are able to benefit from longer flower shelf life. Edible flowers begin to lose their bioactive components soon after harvest, and technologies can be used to preserve them for fresh or dried consumption. This chapter covers the following main techniques for delaying the senescence of edible flowers: refrigeration, preservatives, irradiation, edible films, and drying methods. Some techniques, such as dehydration by freeze drying, fluidized bed drying, or spray drying, require sophisticated equipment and, as a result, are used on an industrial scale, mainly for extraction of bioactive compounds and dyes. Other post-harvest preservation methods are simpler, such as refrigeration, natural drying, or oven drying, and can be used by small farmers for marketing fresh products. These techniques can be applied separately or in combination, providing different responses according to the species, cultivar, and product use.

Keywords Dehydration · Drying · Edible films · Gourmet cooking · Postharvest life · Preservatives · Refrigeration

E. F. A. Almeida (✉) · J. P. de Lima
Universidade Federal de Minas Gerais, Montes Claros, MG, Brazil
e-mail: elka@ufmg.br

A. Castricini
Empresa de Pesquisa Agropecuária de Minas Gerais, Caldas, MG, Brazil

G. S. Lourenço
Universidade de Campinas, Campinas, SP, Brazil

M. M. Jowkar et al. (eds.), *Advances in Research on Edible Flowers*, Advances in Olericulture, https://doi.org/10.1007/978-3-032-04561-4_9

Abbreviations

FAO	Food and Agriculture Organization of the United Nations
FBD	Fluidized bed drying
FD	Freeze drying
FDA	Food and Drug Administration
HA	Hot air
IAEA	International Atomic Energy Agency
IR-HA	Infrared radiation combined with hot air
NDGA	Nordihydroguaiaretic acid
UV	Ultraviolet
WHO	World Health Organization
1-MCP	1-methylcyclopropene

9.1 Introduction

The beauty, aroma, flavour, nutritional properties, and low calorie content of edible flowers have led to a notable increase in their consumption around the world. However, these flowers are extremely sensitive and rapidly show signs of senescence, which is a natural and irreversible process involving a series of physiological, biochemical, and metabolic changes. This process is accompanied by changes in colour, flavour, nutritional content, texture, and other sensory properties, shortening shelf life (Pires et al. 2019).

The postharvest life of edible flowers depends on the species and on storage conditions; on average, these flowers have a shelf life ranging from 2 to 5 days, as a function of early petal abscission, discolouration, wilting, dehydration, and tissue darkening (Kou et al. 2012). Compared to cut flowers, used for decorative purposes, edible flowers are more perishable, as their stems are cut very short and they are stored without water reserves (Fig. 9.1) (Fernandes et al. 2019). Perishability is accelerated due to high water content, volatile compounds, and environmental conditions that favour microbial growth and biochemical breakdown. Therefore, it is necessary to use post-harvest techniques to extend flower shelf life and quality (Shantamma et al. 2021).

Maintaining the post-harvest quality of edible flowers depends on how they are handled at harvest, the type of packaging used, and storage conditions (temperature and relative humidity) appropriate for each species. In general, post-harvest procedures aim to slow or stop the changes that occur after harvest that result in flower senescence and loss of commercial value. The technologies available to preserve the quality and extend the shelf life of edible flowers after harvest are quite extensive and complex. To achieve these objectives, effective techniques can be applied, such as the use of edible films, refrigeration, preservative solutions, irradiation, heat drying, cold drying, freeze drying, fluidized bed drying, spray drying, and microwave drying.

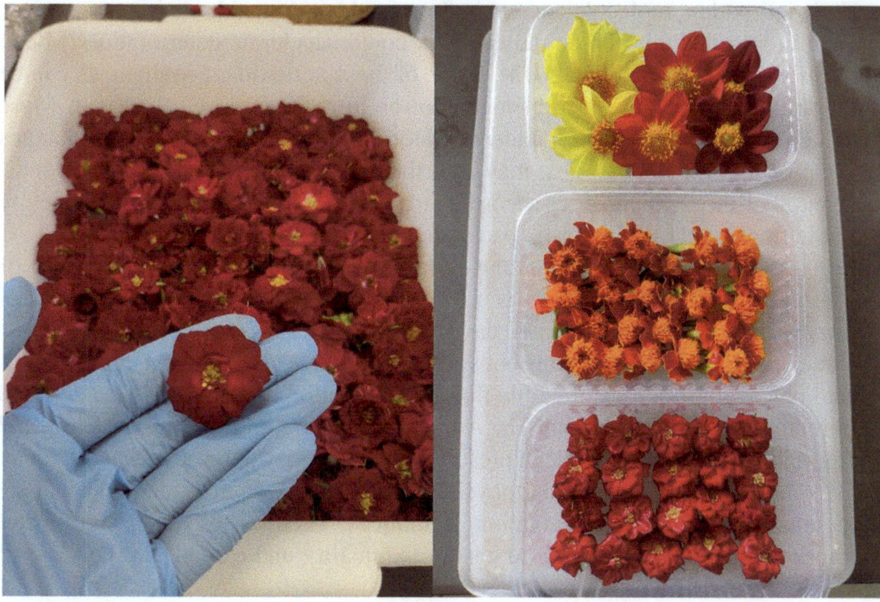

Fig. 9.1 Different edible flowers after harvest. Left: miniature rose. (Courtezy of Alice Cavalcante Costa); Right: top: Dahlia, middle: marigold and bottom: miniature rose. (Courtesy of Jacqueline Souza de Menezes)

9.2 Influence of Post-harvest Procedures on Flower Chemical and Physical Properties

Natural flower senescence results in physiological, biochemical, and metabolic changes that reduce the shelf life of flowers, affecting their colour, flavour, nutritional value, and other properties (Pires et al. 2019; Souza et al. 2020). One of the main physiological events that occur in the postharvest of flowers can be related to the action of ethylene. Flowers can be classified as climacteric (ethylene-sensitive) or non-climacteric (insensitive to exogenous ethylene exposure). Generally, a group of species from the same botanical family exhibits the same physiological behaviour in relation to this hormone, and knowledge about this sensitivity allows for the use of strategies to reduce the rate of petal senescence (Woltering and van Doorn 1998; Serek et al. 2006; Kou et al. 2012).

Generally, soluble solids content increases and titratable acidity decreases in flowers during storage, while pH undergoes specific changes depending on the species. Metabolism is generally faster in flowers than in other organs and is characterized by higher respiration rates. Flowers have high internal moisture, and loss of turgidity is one of the main causes of reduced aesthetic quality (Kou et al. 2012). Several flowers are also highly sensitive to ethylene, which causes wilting at extremely low concentrations, observable during packaging and shipping.

High antioxidant molecule content in flowers plays a crucial role in preservation, limiting the production of free radicals that could accelerate deterioration (Mlcek and Rop 2011; Cavaiulo et al. 2013). Some of these antioxidant compounds can be temporarily increased to ensure longer flower life (Cavaiulo et al. 2013). More information is essential to better understand longevity and physicochemical changes during post-harvest storage. That may contribute to improvements in storage conditions and shelf life, which varies according to species or cultivar. The total longevity of edible flowers placed in plastic packaging and kept in cold chamber at 5 ± 1 °C with 90–95% humidity varies among species—*Viola cornuta* has a longevity of only 4 days; *Antirrhinum majus* lasts up to 8 days; and *Tagetes patula* and *Dianthus chinensis* have greater longevity, 10 days (Pêgo et al. 2022).

Understanding the interactions between post-harvest procedures and the chemical and physical properties of edible flowers is essential to ensure their quality and viability in different applications. Harvesting, processing, storage, preservation, and transportation play a crucial role in preserving the integrity of these flowers (Li et al. 2023). Therefore, understanding physiological processes, maintaining adequate temperatures, and using preservation techniques, such as ethylene control and appropriate drying, are essential to ensure the quality and extend the shelf life of edible flowers.

9.3 Postharvest Techniques

9.3.1 Refrigeration

Several signs of senescence, such as loss of colour, wilting, dehydration, rapid browning, and petal abscission, are frequently observed during storage of plant products at room temperature, resulting from membrane damage and a decrease in bioactive compounds (Ikeura et al. 2023). Therefore, storage at low temperatures has proven to be an effective method to preserve the quality and extend the shelf life of edible flowers (Takahashi et al. 2020; Shantamma et al. 2021).

Temperature and humidity control during refrigeration extends the shelf life and maintains the characteristics of a product. Soon after harvest, edible flowers are cleaned, partially shaped, and cooled. They are generally stored at temperatures between 1 and 10 °C (Ikeura et al. 2023).

Various refrigeration techniques are used to maintain the quality of edible flowers during storage. Conventional cooling is one of the most used techniques, in which flowers are placed in refrigerated environments, ensuring precise temperature control. Water immersion is a technique in which flowers are dipped in cold water to reduce the temperature and preserve their appearance. A third option is a controlled atmosphere, which involves creating a specific environment with suitable levels of temperature, humidity, and gases to meet the individual needs of each species of edible flower. Choosing an appropriate technique is important for preserving the quality and maximizing the shelf life of the flowers, ensuring they remain fresh and attractive for an extended period (Menegaes et al. 2019).

Some studies have had promising results in investigating refrigeration and storage techniques for edible flowers. Refrigeration of nasturtium, marigold, and daisy at 4 °C with storage in a controlled atmosphere maintained microbiological contamination unchanged for 3 days (Wilczyńska et al. 2023). Another study analysed variation in vitamin C content in dahlia flowers during refrigerated storage at 8 °C, and the response in vitamin C levels differed according to flower colour. White and yellow dahlias showed a gradual decrease, whereas pink and multi-coloured dahlias maintained their content. Purple and cherry dahlias increased their ascorbic acid content as the days proceeded (Martínez-Damián et al. 2021). When stored at 4 °C, *Viola cornuta* had a shelf life of up to 16 days and better preserved compounds such as polyphenols, anthocyanins, and antioxidants (Demasi et al. 2020).

Associating low temperatures with other preservation methods has also been studied. Demasi et al. (2021) suggested application of combined technologies, such as modified atmosphere packaging, irradiation, or high hydrostatic pressure to improve efficient distribution of edible flowers, and thus contribute to market expansion. The physicochemical properties of the edible flowers *Viola cornuta*, *Viola tricolour*, *Antirrhinum majus*, *Dianthus chinensis,* and *Tagetes patula* were evaluated during the post-harvest period under refrigeration at 5 °C and relative humidity of 90–95%. A significant increase in total soluble solids and a reduction in titratable acidity were observed in the edible flowers studied, which showed commercial longevity, ranging from 6 to 14 days (Pêgo et al. 2022).

Refrigeration is also a possible preservation method in freezing flowers for maintaining quality for subsequent use in baked goods and ice cream, as well as their use in making ice cubes (Cohen and Fisher 2012). These applications provide attractive colours, elegance, and flavour to beverages such as cocktails, iced teas, water, juices, and alcoholic drinks. However, it is important to consider that storage in frozen foods may impact the levels of nutrients and bioactive compounds (Fernandes et al. 2019).

Borago officinalis, *Viola tricolour*, *Kalanchoe blossfeldiana*, and *Taraxacum officinale* preserved in ice cubes maintained a good visual appearance for up to 3 months, while flowers frozen in their natural form maintained appearance for only up to 1 month. Consistent changes in bioactive compounds or antioxidant activity were not observed in the one-month treatment period. As for microbial quality, both flowers frozen in their natural form and those preserved in ice cubes reduced or maintained the levels of microorganisms compared to fresh samples, except for *T. officinale*. Thus, preservation in ice cubes (Fig. 9.2) seems to be more suitable for maintaining flower quality for up to 3 months, while freezing alone is also a viable alternative for shorter periods, up to 1 month (Fernandes et al. 2020).

Although refrigeration has significant benefits, it is also a technique that presents challenges in preserving edible flowers. Some species are more sensitive to low temperatures and may suffer damage, such as browning and wilting. Additionally, exposure to inappropriate humidity levels can accelerate deterioration. However, solutions such as the use of insulating packaging, species-specific refrigeration

Fig. 9.2 *Rosa chinensis* L. (left) and *Coriandrum sativum* L. (right) preserved in ice cubes

techniques, and precise humidity control can be applied to overcome these challenges (Menegaes et al. 2019).

In conclusion, refrigeration plays an essential role in preserving the quality and maximizing the shelf life of edible flowers. Its benefits range from maintaining appearance and aroma to preserving texture and flavour. Despite specific challenges, such as sensitivity to low temperatures, solutions are being developed to overcome these obstacles. As research advances and more efficient and sustainable techniques emerge, refrigeration will continue to play a key role in meeting the growing demand for high-quality edible flowers.

9.3.2 Preservative Solutions

Edible flowers respond to treatments with preservative solutions (Fig. 9.3) to maintain their quality during storage. Solutions of ascorbic acid, salicylic acid, and others can be used (Souza et al. 2016). Ascorbic acid is an antioxidant, and when applied to flowers, it can increase their durability, as it protects tissues against oxidative damage. Salicylic acid is an elicitor that can also be used after harvest as it activates defence mechanisms, extending the shelf life of plant products.

Sucrose is an important sugar in the post-harvest of flowers, as high concentrations of carbohydrates are a prerequisite for long durability (Marissen 2001). This occurs because carbohydrates are the main source of energy necessary for conducting biochemical and physiological processes of the flower after separation from the mother plant (Nowak and Rudnicki 1990). However, unlike cut flowers, there is little scientific information on the use of sucrose in the preservation of edible flowers. For the edible *Acca sellowiana* flowers, the use of sucrose at 2%, 5%, or 10% does not affect durability, but the use of salicylic acid and ascorbic acid are effective in maintaining quality (Souza et al. 2016).

In ethylene-sensitive species, it is necessary to protect flowers using specific inhibitors, such as 1-methylcyclopropene (1-MCP) (Demasi et al. 2020; Scariot

Fig. 9.3 *Viola* edible flowers to which the preservative solution process was applied

et al. 2014; Scariot et al. 2022). The use of 1-MCP at 500 nL L^{-1} proved to be an efficient treatment to increase the durability of *Acca sellowiana* flowers (Souza et al. 2016). Edible carnation and snapdragon flowers that received 1-MCP at 5 μL L^{-1} together with storage in a modified atmosphere also had longer shelf life compared to the control (Kou et al. 2012).

In addition to the products mentioned, nordihydroguaiaretic acid (NDGA) has also been studied in the post-harvest preservation of edible flowers. This product proved to be effective in reducing yellowing, elongation, and disintegration of daylily (*Hemerocallis citrina*) flower buds during storage. NDGA also helped reduce the production of reactive oxygen species, slowing lipid peroxidation and preserving antioxidant capacity. Furthermore, NDGA affected the synthesis of different phytohormones, regulating physiological progress related to post-harvest senescence (Li et al. 2023).

9.3.3 Edible Films and Coatings

Films or coatings classified as edible, based on different materials and application techniques, can be an alternative in maintaining the quality of edible flowers and providing longer marketing time, thus enabling conservation and logistical requirements to markets far from the production site to be met. Films or coatings form a modified atmosphere around the flowers, reducing water loss through transpiration. Some coatings also act against the growth of microorganisms and are applied or formed directly onto the food surface.

Edible coatings are thin membranes, imperceptible to the naked eye and with diverse structural characteristics, and these coatings depend on the formulation of the precursor film-forming solution (Assis and Britto 2014). The coatings must be non-toxic and safe for use in food (FDA 2024), as they become part of the consumed product. The coatings produced from biological macromolecules and biopolymers; proteins and polysaccharides are most used. Polysaccharides such as cellulose, chitin, xanthan gum, guar gum, pectin, and starch (Fig. 9.4) and polyelectrolyte polysaccharides such as carboxymethylcellulose, chitosan, and alginate are

Fig. 9.4 On the left, cassava starch biofilm being applied to edible miniature roses. On the right, miniature rose flowers under conditions for biofilm drying. (Courtesy of Maria Cecília Afonso Fonseca)

hydrophilic materials. In the class of hydrophobic materials are, for example, hydrophobic proteins, oils and fatty acids, paraffins, and long-chain alcohols (Assis and Britto 2014).

Edible coatings have been widely studied for fruit preservation. However, the most suitable application of these coatings and their effectiveness in conserving the quality of edible flowers are still not well known, even though these coated flowers have a long history of consumption (Zhao et al. 2019). Based on promising results with artichoke and broccoli, edible coatings on edible flowers for traditional consumption, using chitosan and alginate coatings alone, enriched with bio-preservatives, or combined with other procedures as refrigeration, could be a promising post-harvest technology to increase the shelf life of other edible flowers (Fernandes et al. 2019).

The use of alginate coating helped to extend the shelf life of red, white, violet, and yellow pansies, combined with storage at 4 °C, and the appearance of the coated flowers improved compared to uncoated ones. Therefore, this treatment can be considered safe and effective for post-harvest preservation of the species, with an increase in flower durability ranging from 3 to 14 days in accordance with petal thickness and colour (Fernandes et al. 2018a, b).

The edible alginate coating showed promise in increasing the shelf life of *Borago officinalis*, as the coated edible flowers maintained the fresh appearance during refrigerated storage for 5 days, 4 days longer than the uncoated flowers. However, the flowers reportedly became fragile and easily damaged during handling (Fernandes et al. 2018c).

Due to the various alternatives of materials with potential for creating edible coatings and the various species of edible flowers already on the market, studies are necessary involving sensory evaluations, consumer preference and acceptance, cost-benefit analysis, assessment of quality and shelf life, ways of application, and the food safety of the materials.

9.3.4 *Irradiation*

Irradiation is a low-cost, non-chemical, and effective commercial alternative for extending the shelf life of foods in general. This technique uses radiation sources to modify the molecular structure, resulting in effects such as elimination of pests and pathogenic microorganisms, reduction in enzymatic activity, and preservation of nutritional and sensory characteristics. Effects depend on the treatment dose, moisture content, substance composition, and storage conditions (Hallman 2011; Chu et al. 2015). The Radura symbol (Fig. 9.5), is an international indicator that appears on packaging to inform consumers that the food has been irradiated.

This technique offers potential innovative benefits and effects to the industry, but it also requires attention to safety issues. Ionizing energy has been used as a postharvest method to delay ripening and senescence of plant products, yet it is important to follow strict legal controls. As established by the General Standard for Irradiated Foods of the Codex Alimentarius, ionizing radiation used in food processing is limited to high-energy photons, such as gamma rays from ^{60}Co or ^{137}Cs radionuclides, X-rays generated by machines with an energy level less than or equal to 5 MeV, and electrons generated by machines with an energy level less than or equal to 10 MeV (Codex 2003). This is recognized as a safe technology by several authorities, including the World Health Organization (WHO), the International Atomic Energy Agency (IAEA), and the Food and Agriculture Organization of the United Nations (FAO) (Farkas 2006).

In several countries, such as Argentina, Australia, Bangladesh, Thailand, the USA, India, and New Zealand, foods have undergone irradiation as a preservation

Fig. 9.5 Radura (international indicator that the food has been irradiated)

method for many years (IAEA 2009). Irradiation of edible flowers has become a prominent promising technology, involving controlled exposure to sources of ionizing (gamma rays, X-rays, or electron beams) and non-ionizing (ultraviolet) radiation. However, few studies have investigated the effect of irradiation on the quality and composition of edible flowers (Fernandes et al. 2018e). Therefore, more studies are required to evaluate the effects of irradiation on edible flowers, including their physicochemical, sensory, and nutritional characteristics, as well as their microbiological safety (Koike et al. 2021).

Some preliminary studies have shown promising results regarding the effect of irradiation on edible flowers. For example, a study carried out by Simoni et al. (2018) investigated the effects of gamma irradiation on the sensory and nutritional characteristics of *Bauhinia variegate* flowers. The results indicated that moderate doses of irradiation did not negatively affect the sensory properties of the flowers, such as colour, texture, and flavour.

Furthermore, Koike et al. (2021) studied the presence of carotenoids in *Dianthus chinensis* processed by ionizing radiation and observed that higher doses (1.0 KGy) increased the presence of lutein (type of carotenoid) in the flowers, indicating preservation of the nutritional quality of these edible flowers. Another study carried out by Koike et al. (2015) evaluated the tolerance of some edible flowers to gamma radiation. *Borago officinalis* does not tolerate doses greater than 0.3 kGy (dose necessary to eliminate pests and insects). In contrast, *Dianthus chinensis*, *Viola tricolour*, *Viola odorata*, and *Lobularia maritima* were tolerant to gamma radiation doses of up to 1 kGy. High doses of gamma irradiation can cause wilting, browning, and lesions in the petals of edible flowers (Koike et al. 2015).

In addition to the use of gamma irradiation, studies have been carried out on the use of electron beams in edible flowers. Villavicencio et al. (2018) used electron beams at different doses on *Bauhinia variegata* L. flowers and analysed possible changes in the nutritional and chemical properties, cytotoxicity, and antioxidant and anti-inflammatory activities of the flowers. The results indicated that irradiation doses of 0.5 kGy had a minimal effect on the nutritional profile and cytotoxicity, though there was an increase in antioxidant activity at 0.5 kGy, possibly related to greater presence of phenolic compounds in the edible flowers when exposed to this dose. A similar response was observed by Koike et al. (2015) in analysis of *V. tricolour* flowers, indicating that the antioxidant activity of this flower does not significantly decrease with irradiation. Furthermore, phenolic compounds are higher in irradiated samples compared to non-irradiated ones. Although the authors did not provide an explanation for this response, one might assume that irradiation enhanced the secondary metabolism of these flowers.

Moderate doses of ultraviolet (UV) light have been used in post-harvest research on edible flowers, based on the concept of hormesis. This approach suggests that the application of low or sub-lethal doses of an agent capable of inducing stress (such as UV radiation) can lead to beneficial adaptive responses and improvements in the quality of edible flowers. UV radiation can be categorized into three ranges: UV-A (320–400 nm), UV-B (280–320 nm), and UV-C (200–280 nm). Broccoli (*Brassica oleracea*) irradiated with UV-B showed notable effective inhibition of chlorophyll degradation during storage, suggesting suppression of the activities of chlorophyll-degrading enzymes (Aiamla-or et al. 2010). Similar results were observed in broccoli irradiated with UV-C, showing reduced chlorophyllase and chlorophyll peroxidase activity compared to the control group (Costa et al. 2006).

However, fungal infestation is common in edible flowers, and the irradiation dose required to prevent this infestation has a negative effect on quality. In this case, it is desirable to use a lower irradiation dose to preserve the functional properties of edible flowers, combining this treatment with another preservation method, such as refrigeration (Purohit et al. 2021). Cia et al. (2007) suggested the combination of UV-C treatment and gamma irradiation, which would allow the use of a lower dose of gamma rays and maintain flower quality. This type of treatment could be effective in inactivating fungi, ensuring greater longevity, a lower senescence rate, and a higher content of phenolic compounds.

In summary, irradiation is a promising technology for extending the shelf life and improving the post-harvest quality of edible flowers. Preliminary studies indicate that adequate doses of irradiation can preserve the sensory, nutritional, and functional characteristics of flowers while improving food safety. Challenges and safety concerns exist; but with continued research, appropriate regulations, and consumer awareness, this technique has the potential to expand the edible flower market. However, it is important to highlight that more research is necessary to more fully understand the effects of irradiation on different species of edible flowers, as well as to establish specific guidelines for their commercial application.

9.3.5 Dehydration of Edible Flowers

Edible flowers can be consumed fresh or dried. Therefore, in addition to the use of preservatives, edible films, irradiation, and other techniques already discussed, dehydration or drying is a very promising preservation alternative. Freshly harvested flowers exhibit both aesthetics and chemical composition that are always altered, regardless of the dehydration method, due to water loss, exposure to different light intensities, temperature variations, injuries, and other phenomena that occur, even with the best possible care. Therefore, it is necessary to use a dehydration technique that preserves the quality of the flowers as much as possible, so they can retain the properties desired by the consumer as it avoids the problem of loss of visual quality and valuable nutraceutical properties (Fig. 9.6).

Dehydration is a very traditional technique used to preserve plant products. Reducing the water content reduces the risk of microorganism development. This technique reduces biochemical processes and therefore maintains nutritional quality and increases the shelf life of the dehydrated product. Dehydration can be carried out by different methods: hot air or sun drying, cold drying, freeze drying, fluidized bed drying, spray drying, and microwave drying.

Fig. 9.6 Damascus roses with preserved colour after the dehydration process

9.3.5.1 Natural Dehydration

Natural dehydration in the sun or shade occurs without the need for sophisticated equipment; it therefore has reduced cost. However, this method is slow; it depends on climatic conditions; flower appearance is affected; there may be microbiological contamination; and the nutritional content and desirable antioxidant properties are modified and reduced (Zhao et al. 2019).

9.3.5.2 Hot Air Drying

Flowers can be dehydrated by hot air, using specific equipment for this purpose. This process consists of exposing edible flowers to a temperature above that of the environment, thereby removing the water in them (Afzal and Abe 1999). However, the constitution of edible flowers makes them sensitive to excess heat, so dehydrating them requires greater attention; it may be an unfeasible method because of loss of quality. Ornamental quality declines with reduced size; and darkening or discolouration and changes in biochemical aspects and in flavour may occur, which vary according to the species (Tanko et al. 2005; Fernandes et al. 2017).

The temperature and time for the hot air drying procedure vary according to the species. On average, according to Rani and Reddy (2015), the ideal temperature for drying flowers in a hot air oven is 45 °C to 65 °C. However, in *Begonia cucullata* 'Willd' flowers, the best preservation of antioxidant compounds (polyphenols, flavonoids, and anthocyanins) and sensory characteristics occurred when the flowers were dried at 60–70 °C (Marchioni et al. 2022). Nevertheless, although bioactive compounds were preserved, hot air drying strongly affected the visual quality of the flowers, and they presented a dark colour.

9.3.5.3 Cold Drying

Cold drying is based on the principle of air entering the drying cell to be cooled to below the dew point on a cold surface; the moisture is retained by the cold surface and the air is used in the drying process (Aykin-Dinçer and Erbas 2019). This process preserves foods, extending their shelf life and preventing deterioration of foods sensitive to heat. Although the processing period is long, drying at low temperatures has positive effects on the quality of biological materials (Kilic 2009). Despite the advantages of this method, research on its use in the dehydration of edible flowers is still required.

9.3.5.4 Freeze Drying

Freeze drying is a dehydration process by sublimation, where the ice in the food is directly transformed into water vapor, without going through the liquid water state. Freeze-dried foods are high value-added products that retain a large part of their original nutrients, since low temperatures are used in their processing (Vieira et al. 2012).

The high quality of the freeze-dried product is due to the absence of water. This absence stops microbiological activity and deterioration processes, in combination with the low temperatures that occur in this process (Ratti 2001). The disadvantages of freeze drying are the high initial cost of equipment and electricity consumption. It is also necessary to adequately calibrate the equipment according to the species that will be freeze dried. In research evaluating the effects of different freeze-drying processes on the moisture content, colour, and resistance of pink roses and carnations, the recommendation was to use the highest drying temperature (47 °C) and freezing time studied (4 h) (Chen et al. 2000).

9.3.5.5 Fluidized Bed Drying

Fluidized bed drying (FBD), which offers significant advantages, such as high heat and mass transfer, uniformity, moisture reduction with shorter time, and a high drying rate (Kassem et al. 2011) have high dehydration potential for edible flowers. This technique provides a uniform temperature of the equipment bed throughout the drying period. FBD dryers are capable of drying all types of foods, except those that are very cohesive and difficult to fluidize. FBD allows the conservation of products with high phytochemical content and antioxidant capacity. However, it is necessary to precisely set the temperature and air speed according to each species, as the thermal process can cause chemical degradation, especially of pigments, and reduce flower quality. *Moringa oleifera* flowers powder can be preserved through fluidized bed drying. This technique is able to increase the nutritional value and antioxidant activity of flowers for use in food composition and in the pharmaceutical industry (Manju et al. 2021).

9.3.5.6 Spray Drying

Spray drying is widely used to preserve powdered fruit and vegetable juices (Shishir and Chen 2017), but is also being studied for powdered edible flowers (Fuzetti et al. 2022; Lisiecka et al. 2023). The objective of this technique is to obtain high-quality food powders, which requires optimization of spray drying conditions. The most significant factors in spray drying are the inlet temperature and the carrier agent. A high drying temperature damages sensitive bioactive compounds, while the carrier agent protects such compounds. Other factors involved in spray drying can affect the physicochemical and microstructural properties of the product. When spray drying is successful, it improves the physical properties of powdered products while preserving bioactive compounds with the help of carrier agents or their combinations (Shishir and Chen 2017).

One of the main interests of the industry in spray-drying dehydration is the use of this method for subsequent extraction of pigments from edible flowers. Supercritical fluid extraction is used with different strategies for powdered flowers pigment extraction (López-Hortas et al. 2022). Natural dyes from edible flowers can be used in different types of foods. For example, natural dyes derived from dahlias and roses through spray drying have already been tested in yogurts. The use of these edible flowers in food not only provides colour, but also enriches the food with nutrients, antioxidant compounds, and other components that remain even after dehydration (Pires et al. 2018).

Anthocyanin-based dyes are most in demand, due to their antioxidant capacity; they can come from edible flowers processed through spray drying. Research on poppies, roses, and tulips showed an increase in phenolic compounds in tulips and higher anthocyanin content in poppies. However, all the flowers tested showed high activity of bioactive compounds (Bayram et al. 2015).

After spray drying, edible flowers can be microencapsulated, with considerable potential for diverse uses as a result of the antioxidant activity of these products, as already observed for *Malvae arboreae* and *Clitoria ternatea* flowers. Through this process, microencapsulated edible flowers have potential for use in beverages, various foods, desserts, and cosmetics and in the pharmaceutical industry (Fuzetti et al. 2022; Lisiecka et al. 2023; Pal and Bhattacharjee 2018).

9.3.5.7 Microwave Drying

Microwave drying is another method studied for drying edible flowers. During microwave heating, the external temperature is slightly lower than the internal temperature due to the evaporation of external water, greatly increasing the drying rate. It is important to highlight that it is necessary to consider the microwave power, the drying time, and the ideal temperature for each species. Even with reduction in drying time, some delicate species, such as roses, carnations, and dahlias may exhibit stains and petal fall, impairing the quality of the dehydrated flower (Safeena and Patil 2014). The use of microwaves has already been efficiently tested for

Hemerocalis (Ding et al. 2012), *Eucommina ulmoides* (Dong et al. 2011), and *Magnolia liliflora* (Li et al. 2015). The use of silica gel can effectively reduce damage through this drying process in more sensitive species, as roses for example (Safeena and Patil 2014).

All drying methods have advantages and disadvantages, and it is necessary to evaluate the cost-benefit of the process according to the species and intended use. Fernandes et al. 2018d evaluated the effect of convection drying with hot air, shade drying, and freeze drying on the bioactive compounds and antioxidant activity of cornflower petals (*Centaurea cyanus*), as well as on some physical and chemical properties. Drying in the shade had the highest values of monomeric anthocyanins, flavonoids, hydrolyzable tannins, total reducing capacity, and antioxidant activity. The greatest losses were observed with hot air. Shade drying is a very promising process that should be considered as a suitable method for dehydrating cornflower petals (Fernandes et al. (2018d). Marchioni et al. (2022) analysed wild *Begonia cucullata*, subjecting them to freeze-drying and drying in hot air at different temperatures. The authors found that the ideal temperature for preservation of antioxidant compounds in drying these flowers is 60–70 °C, and the freeze-dried samples showed a marked loss of pigmentation. Therefore, although freeze drying is one of the most sophisticated methods for dehydration, it needs to be tested to determine the effectiveness for each species.

Natural and vacuum drying were beneficial for increasing the phenolic content and antioxidant capacity of the chrysanthemum (*Chrysanthemum morifolium*) flower, and vacuum drying was able to retain the antioxidant components of this flower for a short period of time (Lu et al. 2020). Siriamornpun et al. (2012) evaluated the colour, carotenoid (lycopene, β-carotene, and lutein) content, and phenolic compound content in marigold flower powder obtained through freeze-drying (FD) for 48 h, through hot air (HA) at 60 °C for 4 h, and through infrared radiation combined with hot air (IR-HA). Colour changes were more limited when using IR-HA dehydration compared to FD and HA. Flowers dried by HA had the highest levels of β-carotene, while the use of IR-HA and FD provided the highest levels of lutein and lycopene. Infrared radiation drying combined with FIR-HA hot air convection is a suitable method for marigold with regard to preservation of colour, antioxidant properties, and bioactive compounds in the powdered flower.

9.4 Conclusions

There are several technologies already available to preserve the post-harvest quality of edible flowers, and the success of each one depends on the species, cultivar, and intended use. Furthermore, as explained above, technologies for preservation of edible flowers can be used alone or in combination. Therefore, further investigation is required, as each species of edible flower has its own particularities; and as there is great diversity in the use of these flowers, preservation techniques must be appropriately specified.

Acknowledgments The authors acknowledge the Fundação de Amparo à Pesquisa do Estado de Minas Gerais—FAPEMIG for financial support.

References

Afzal TM, Abe T (1999) Some fundamental attributes of far infrared radiation drying of potato. Dry Technol 17:138–155. https://doi.org/10.1080/07373939908917522

Aiamla-or S, Kaewsuksaeng S, Shigyo M et al (2010) Impact of UV-B irradiation on chlorophyll degradation and chlorophyll-degrading enzyme activities in stored broccoli (*Brassica oleracea* L. Italica group) florets. Food Chem 120(3):645–665. https://doi.org/10.1016/j.foodchem.2009.10.056

Assis OBG, Britto D (2014) Revisão: coberturas comestíveis protetoras em frutas: fundamentos e aplicações. Braz J Food Technol 17(2):87–97. https://doi.org/10.1590/bjft.2014.019

Aykin-Dinçer E, Erbas M (2019) Cold dryer as novel process for producing a minimally processed and dried meat. Innov Food Sci Emerg Technol 57:102–113. https://doi.org/10.1016/j.ifset.2019.01.006

Bayram O, Sagdic O, Ekici L (2015) Natural food colourants and bioactive extracts from some edible flowers. J Appl Bot Food Qual 88:170–176. https://doi.org/10.5073/JABFQ.2015.088.024

Cavaiulo M, Cocett G, Ferrante A (2013) The antioxidants changes in ornamental flowers during development and senescence. Antiox 2(3):132–155. https://doi.org/10.3390/antiox2030132

Chen W, Gast KLB, Smithey S (2000) The effects of different freeze-drying processes on the moisture content, colour, and physical strength of roses and carnations. Sci Hort 84(3/4):321–332. https://doi.org/10.1016/S0304-4238(99)00106-5

Chu EH, Shin EJ, Park HJ et al (2015) Effect of gamma irradiation and its convergent treatment for control of postharvest *Botrytis cinerea* of cut roses. Rad Phys and Chem 115:22–29. https://doi.org/10.1016/j.radphyschem.2015.05.042

Cia P, Pascholati SF, Benato EA et al (2007) Effects of gamma and UV-C irradiation on the postharvest control of papaya anthracnose. Postharvest Biol Technol 43(3):366–373. https://doi.org/10.1016/j.postharvbio.2006.10.004

Codex General Standard for Irradiated Foods. Codex Alimentarius Commission (2003) CODEX STAN:106–1983

Cohen W, Fisher J (2012) The book of gardening projects for kids: 101 ways to get kids outside dirty and having fun. Timber Press, London

Costa L, Vicente AR, Civello PM et al (2006) UV-C treatment delays postharvest senescence in broccoli florets. Postharvest Biol Technol 39(2):204–210. https://doi.org/10.1016/j.postharvbio.2005.10.012

Demasi S, Falla NM et al (2020) Postharvest aptitude of *Begonia semperflorens* and *Viola cornuta* edible flowers. Adv Hort Sci 34(1):13–20. https://doi.org/10.13128/ahsc7444

Demasi S, Mellano MG, Falla NM et al (2021) Sensory profile, shelf life, and dynamics of bioactive compounds during cold storage of 17 edible flowers. Horticulture 7(7):1–26. https://doi.org/10.3390/horticulturae7070166

Ding S, You J, An K et al (2012) Effective diffusivities and energy consumption of daylily in microwave drying. Int J Food Sci Technol 47:2648–2654. https://doi.org/10.1111/j.1365-2621.2012.03147.x

Dong J, Ma X, Fu Z et al (2011) Effects of microwave drying on the contents of functional constituents of *Eucommia ulmoides* flower tea. Indu Crops Prod 34:1102–1110. https://doi.org/10.1016/j.indcrop.2011.03.026

Farkas J (2006) Irradiation for better foods. Trends Food Sci Technol 17(4):148–152. https://doi.org/10.1016/j.tifs.2005.12.003

Fernandes L, Casal S, Pereira JA et al (2017) Effect of high hydrostatic pressure (HHP) treatment on edible flowers' properties. Food Biop Technol 10(5):799–807. https://doi.org/10.1007/s11947-017-1887-2

Fernandes L, Casal S, Pereira JA et al (2018a) Effect of alginate coating on the physico-chemical and microbial quality of pansies (*Viola ×wittrockiana*) during storage. Food Sci Biotechn 27:987–996. https://doi.org/10.1007/s10068-018-0326-0

Fernandes L, Pereira JA, Baptista P et al (2018b) Effect of application of edible coating and packaging on the quality of pansies (*Viola × wittrockiana*) of different colours and sizes. Food Sci Technol Intern 24(4):321–329. https://doi.org/10.1177/1082013217753229

Fernandes L, Pereira JA, Saraiva JA et al (2018c) The effect of different post-harvest treatments on the quality of borage (*Borago officinalis*) petals. Acta Sci Pol Technol Aliment 17(1):5–10. https://doi.org/10.17306/J.AFS.0533

Fernandes L, Casal S, Pereira JA et al (2018d) Effects of different drying methods on the bioactive compounds and antioxidant properties of edible Centaurea (*Centaurea cyanus*) petals. Braz Food Technol 21:1–10. https://doi.org/10.1590/1981-6723.21117

Fernandes L, Saraiva JA, Pereira JA et al (2018e) Post-harvest technologies applied to edible flowers: a review. Food Rev Int 35(2):132–134. https://doi.org/10.1080/87559129.2018.1473422

Fernandes L, Casal S, Pereira JA et al (2019) An overview on the market of edible flowers. Food Rev Int 36(3):258–275. https://doi.org/10.1080/87559129.2019.1639727

Fernandes L, Casal S, Pereira JA et al (2020) Freezing of edible flowers: effect on microbial and antioxidant quality during storage. Jour of Food Sci 85(4):1151–1159. https://doi.org/10.1111/1750-3841.15097

Food and Drug Administration (FDA) (2024) Generally recognized as safe (GRAS) (2024). Silver Spring. https://www.fda.gov/food/food-ingredients-packaging. Accessed on 25 Jan 2024

Fuzetti CG, Castilhos MB, Nicoletti VR (2022) Microencapsulation of natural blue dye from butterfly pea (*Clitoria ternatea* L.) flowers: the application of different carriers. J Food Proc Pres 46(1):e16420. https://doi.org/10.1111/jfpp.16420

Hallman GJ (2011) Phytosanitary applications of irradiation. Compr Rev Food Sc Food Saf 10(2):143–151. https://doi.org/10.1111/j.1541-4337.2010.00144.x

IAEA – International Atomic Energy Agency (2009) Food irradiation combating bacteria around the world. https://www.iaea.org/newscenter/news/food-irradiation-combating-bacteria-around-world

Ikeura H, Kobayashi F, Kai T et al (2023) Effects of different storage conditions on the colour, antioxidant activity, and volatile components of edible roses. Sci Hort 310:111707. https://doi.org/10.1016/j.scienta.2022.111707

Kassem AS, Shokr AZ, Mahdy ARE et al (2011) Comparison of drying characteristics of Thompson seedless grapes using combined microwave oven and hot air drying. J Saudi Soc Agric Sci 10:33–40. https://doi.org/10.1016/j.jssas.2010.05.001

Kilic A (2009) Low temperature and high velocity (LTHV) application in drying: Characteristics and effects on the fish quality. J Food Eng 91(1):173–182. https://doi.org/10.1016/j.jfoodeng.2008.08.023

Koike A, Barreira JCM, Barros L et al (2015) Edible flowers of *Viola tricolour* L. as a new functional food: antioxidant activity, individual phenolics and effects of gamma and electron-beam irradiation. Food Chem 179:6–14. https://doi.org/10.1016/j.foodchem.2015.01.123

Koike ACR, Araujo ES, Negrão BG et al (2021) Analysis of carotenoids in edible flowers of *Dianthus chinensis* processed by ionizing radiation. Braz J Rad Sci 9:1–11. https://doi.org/10.15392/bjrs.v9i1A.1577

Kou L, Turner ER, Luo Y (2012) Extending the shelf life of edible flowers with controlled release of 1-Methylcyclopropene and modified atmosphere packaging. J Food Sci 77:188–193. https://doi.org/10.1111/j.17503841.2012.02683.x

Li FY, Liu SW, Guo S (2015) Effects of drying methods on chemical composition and antioxidant activity of the flowers of *Magnolia liliflora*. Acta Hort Sin 42(6):1150–1156. https://doi.org/10.16420/j.issn.0513-353x.2014-1033

Li K, Xu R et al (2023) Nordihydroguaiaretic acid (NDGA) maintains the postharvest quality of fresh daylily flower buds (*Hemerocallis citrina*) by enhancing antioxidant metabolism and regulating phytohormone equilibrium. Postharvest Biol Technol 200:112326. https://doi.org/10.1016/j.postharvbio.2023.112326

Lisiecka K, Dziki D, Gawlik-Dziki U et al (2023) Influence of soluble fiber as a carrier on antioxidant and physical properties of powders produced based on the spray drying of *Malvae arboreae flos* aqueous extracts. Foods 12(18):1–18. https://doi.org/10.3390/foods12183363

López-Hortas L, Rodríguez P, Díaz-Reinoso B et al (2022) Supercritical fluid extraction as a suitable technology to recover bioactive compounds from flowers. J Supercr Fluids 188:1–21. https://doi.org/10.1016/j.supflu.2022.105652

Lu Q, Wang S, Xue S et al (2020) Effects of drying methods on phenolic components in different parts of *Chrysanthemum morifolium* flower. J Food Process Preserv 44(12):e14904. https://doi.org/10.1111/jfpp.14982

Manju KM, Rekha P, Kumar N (2021) Effect of fluidized-bed and freeze-drying techniques on physicochemical, nutritional, thermal, and structural properties of *Moringa oleifera* flowers leaves and seeds. J Food Process Preserv 45(9):e15719. https://doi.org/10.1111/jfpp.15719

Marchioni I, Taglieri I, Dimita R (2022) Postharvest treatments on sensorial and biochemical characteristics of *Begonia cucullata* Willd. Edible flowers. Foods 11(10):1–13. https://doi.org/10.3390/foods11101481

Marissen N (2001) Effects of pre-harvest light intensity and temperature on carbohydrate levels and vase life of cut roses. Acta Hort 543:331–335. https://doi.org/10.17660/ActaHortic.2001.543.40

Martínez-Damián MT, Mejía-Muñoz JM, Colinas-León MT et al (2021) Nutritional value, bioactive compounds and capacity antioxidant in edible flowers of dahlia. Acta Sci Pol Hort Cult 20(5):63–72. https://doi.org/10.24326/asphc.2021.5.6

Menegaes JF, Nunes UR, Bellé RA et al (2019) Post-harvesting of cut flowers and ornamental plants. Sci Agr Par 18(4):313–323. https://doi.org/10.18188/sap.v18i4.21261

Mlcek J, Rop O (2011) Fresh edible flowers of ornamental plants – a new source of nutraceutical foods. Trends Food Sci Technol 22(10):561–569. https://doi.org/10.1016/j.tifs.2011.04.006

Nowak J, Rudnicki RM (1990) Postharvest handling and storage of cut flowers, florist greens and potted plants. Timber Press, Portland, p 210p

Pal S, Bhattacharjee P (2018) Spray dried powder of lutein-rich supercritical carbon dioxide extract of gamma-irradiated marigold flowers: process optimization, characterization and food application. Powder Technol 327:512–523. https://doi.org/10.1016/j.powtec.2017.12.085

Pêgo RG, Fiorini CVA, Deco TAD et al (2022) Postharvest of edible flowers. Pesq Agrop Bras 57:1–7. https://doi.org/10.1590/S1678-3921.pab2022.v57.02953

Pires TCSP, Dias MI, Barros L et al (2018) Incorporation of natural colourants obtained from edible flowers in yogurts. LWT – Food Sci Technol 97:668–675

Pires TCSP, Barros L, Santos-Buelga C et al (2019) Edible flowers: emerging components in the diet. Trends Food Sci Technol 93:244–258. https://doi.org/10.1016/j.tifs.2019.09.020

Purohit SR, Rana S, Idrishi R et al (2021) A review on nutritional bioactive toxicological properties and preservation of edible flowers. Fut Foods 4:1–14. https://doi.org/10.1016/j.fufo.2021.100078

Rani P, Reddy M (2015) Dehydration techniques for flowers. Int J Appl Res 1(10):306–311

Ratti C (2001) Hot air and freeze-drying of high-value foods: a review. J Food Eng 49:311–319. https://doi.org/10.1016/S0260-8774(00)00228-4

Safeena SA, Patil VS (2014) Studies on the effect of microwave oven drying on flower quality of dried Dutch rose flowers. Natl Acad Sci Lett 37(1):19–24. https://doi.org/10.1007/s40009-013-0197-0

Scariot V, Paradiso R, Rogers H et al (2014) Ethylene control in cut flowers: classical and innovative approaches. Postharvest Biol Technol 97:83–92. https://doi.org/10.1016/j.postharvbio.2014.06.010

Scariot V, Ferrante A, Roma D (2022) Edible flowers: understanding the effect of genotype, preharvest, and postharvest on quality, safety, and consumption. Front Plant Sci 13:1–4. https://doi.org/10.3389/fpls.2022.1025196

Serek M, Woltering EJ, Sisler EC et al (2006) Controlling ethylene responses in flowers at the receptor level. Biotechnol Adv 24:368–381. https://doi.org/10.1016/j.biotechadv.2006.01.007

Shantamma S, Vasikaran EM, Waghmare R et al (2021) Emerging techniques for the processing and preservation of edible flowers. Fut Foods 4:1–12. https://doi.org/10.1016/j.fufo.2021.100094

Shishir MRI, Chen W (2017) Trends of spray drying: a critical review on drying of fruit and vegetable juices. Tren Food Sci Technol 65:49–57. https://doi.org/10.1016/j.tifs.2017.05.006

Simoni NK, Santos FF, Andrade TA et al (2018) The use of edible flowers in human food: sensory analysis of preparations. Int J Food Engin 4(2):140–143. https://doi.org/10.18178/ijfe.4.2.140-143

Siriamornpun S, Kaisoon O, Meeso N (2012) Changes in colour, antioxidant activities and carotenoids (lycopene, β-carotene, lutein) of marigold flower (*Tagetes erecta* L.) resulting from different drying processes. J Funct Foods 4(4):757–766. https://doi.org/10.1016/j.jff.2012.05.002

Souza AGD, Amarante CVTD, Steffens CA et al (2016) Postharvest quality of feijoa flowers treated with different preservative solutions and 1-Methylcyclopropene. Rev Bras de Frut 38(4):1–10. https://doi.org/10.1590/0100-29452016759

Souza AG, Carvalho J, Anami JM et al (2020) Refrigeração na conservação de flores de capuchinha. Agrotr 32:225–232. https://doi.org/10.21757/0103-3816-2020v32n3p225-232

Takahashi JA, Rezende FAGG, Moura MAF et al (2020) Edible flowers: bioactive profile and its potential to be used in food development. Food Res Int 129:1–14. https://doi.org/10.1016/j.foodres.2019.108868

Tanko H, Carrier DJ, Duan L et al (2005) Pre and post-harvest processing of medicinal plants. Plant Gen Resour: Charact Util 3(2):304–313. https://doi.org/10.1079/pgr200569

Vieira AP, Nicoleti JF, Telis VRN (2012) Liofilização de fatias de abacaxi: avaliação da cinética de secagem e da qualidade do produto. Braz J Food Technol 15(1):50–58

Villavicencio ALCH, Heleno SA, Calhelha RC et al (2018) The influence of electron beam radiation in the nutritional value, chemical composition and bioactivities of edible flowers of *Bauhinia variegata* L. var. candida alba Buch.-ham from Brazil. Food Chem 241:163–170. https://doi.org/10.1016/j.foodchem.2017.08.093

Wilczyńska A, Kukuiowicz A, Lewandowska A (2023) Effect of packaging on microbial quality of edible flowers during refrigerated storage. Pol J Food Nutr Sci 73(1):32–38. https://doi.org/10.31883/pjfns/159037

Woltering EJ, van Doorn WG (1998) Role of ethylene in senescence of petals: morphological and taxonomical relationships. J Exp Bot 39:1605–1616. https://doi.org/10.1093/jxb/39.11.1605

Zhao L, Fan H, Zhang M et al (2019) Edible flowers: review of flower processing and extraction of bioactive compounds by novel technologies. Food Res Int 126:1–14. https://doi.org/10.1016/j.foodres.2019.108660

Part II
Applications

Chapter 10
Edible Flowers for Infusion

Nursabrina Munawar and Norhayati Hussain

Abstract Edible flowers like marigold, chrysanthemum, yucca, gladiolus, and hibiscus are excellent sources of nutritional and phytochemical elements, beneficial for prevention of certain diseases. Commonly consume as infusion where water is used to provide a higher quality of extract, faster diffusion of phytochemical compounds, lower cost, safe and environmentally friendly process of extraction. Although there are several developed edible flower tisanes commercially, many other edible flowers with various health benefits remains unexplored. This work focused on identifying significant factors affecting edible flower infusion, nutritional content, and antioxidant method commonly used for edible flower infusion like 2,2- diphenyl-1-picrylhydrazyl (DPPH) radical scavenging, Ferric Reducing Antioxidant Power (FRAP), Total Phenolic Content (TPC), Total Flavonoid Content (TFC) and 2,2′-azino-bis (3-ethylbenzothiazoline-6-sulfonic acid) (ABTS). This review could enhance understanding of edible flower infusions by pinpointing key preparation parameters, thereby promoting the development and consumption of nutritious and health-boosting herbal infusions.

Keywords Antioxidant activities · Antioxidant power · Infusion preparation · Nutritional content · Phytochemicals · Sensory evaluation

N. Munawar
Faculty of Food Science and Technology, Department of Food Technology, Universiti Putra Malaysia, Selangor, Malaysia

Department of Food Technology, Faculty of Applied Sciences, Negeri Sembilan Branch, Universiti Teknologi MARA, Kuala Pilah, Malaysia

N. Hussain (✉)
Faculty of Food Science and Technology, Department of Food Technology, Universiti Putra Malaysia, Selangor, Malaysia
e-mail: aryatihussain@upm.edu.my

M. M. Jowkar et al. (eds.), *Advances in Research on Edible Flowers*, Advances in Olericulture, https://doi.org/10.1007/978-3-032-04561-4_10

Abbreviations

ABTS	2,2′-azino-bis 3-ethylbenzothiazoline-6-sulfonic acid
BSA E	Bovine Serum Albumin Equivalent
BC	Beta carotene
CAR	Carboxen
CE	Catechin equivalent
CUPRAC	Cupric Reducing Antioxidant Capacity
DMSO	Dimethylsulfoxide
DPPH	2,2-diphenyl-1-picrylhydrazyl
DVB	Divinylbenzene
DW	Dry Weight
EC	Epicatechin
EGC	Epigallocatechin
FC	Folin-Ciocalteu
FD	Flavour Dilution
FRAP	Ferric Reducing Antioxidant Power
FM	Fresh mass
GAE	Gallic Acid Equivalent
GCMS	Gas chromatography mass spectrometry
LPO	Lipid Peroxidation
PDMS	Polydimethylsiloxane
QDA	Quantitative Descriptive Analysis
RDA	Recommended Dietary Allowances
ROS	Reactive Oxygen Species
TEAC	Trolox Equivalent Antioxidant Capacity
TPC	Total Phenolic Content

10.1 Introduction

Flowers have played a significant role in human society. They are frequently reported as natural wonders and symbols of beauty. Other than being cultivated for aesthetic purposes, they also possess some nutritional and biological qualities (Kumari and Ujala Bhargava 2021). A flower must be harmless, nontoxic, and nutritive in order to be considered as edible (Nicknezhad et al. 2023). Although edible flowers have long been consumed but little information about their cultivation area, processing method and infusion parameters has been published in the literature. Few data are available regarding the types of cultivars which commonly consumed like chrysanthemum, rose saffron. Approximately 75% of these crops are grown worldwide, with Europe and Asia coming in second and third respectively. Edible flower eating is unique to the region in which it is grown and is mostly carried out by the local population (Jadhav et al. 2023).

According to (Mlcek et al. 2021), the number of edible flowers varies depending on region, but approximately there are about 97 families, 100 genera, and 180 species are scattered around the globe. Several well-known edible flowers are begonia, rose, tulip, hibiscus, borage, calendula, carnation, cornflower, daylily, nasturtium, pansy, and others. In addition to the ornamental flowers described above, edible flower includes several vegetables (broccoli and cauliflower), herbs (common sage, chives, marjoram, mint, thyme, and summer savoury), or the blossoms of some fruit trees (elderberry and apple). The inflorescences like broccoli, cauliflower and artichoke are incorporated as part of the daily diet in countries like India, China, Pakistan, Sri Lanka, and Nepal (Jadhav et al. 2023). Species like *Elettariopsis slahmong*, *Curcuma longa*, and *Kaempferia galanga* are important components in spicy and savoury meals from Peninsular Malaysia. Famous cuisine associated with Malaysia such as *masak lemak cili api*, (an aromatic dish cooked in coconut milk added with meat or seafood and flavoured with ingredients like chili peppers, turmeric, lemongrass, and other spices) and *Urap* (a traditional Javanese salad made from a mixture of any fresh vegetable and aromatic herb including gotu kola, winged bean, water dropwort, *Cosmos caudatus* with grated coconut dressing) (Rachkeeree et al. 2018). Figure 10.1 shows a few popular Malaysian cuisine incorporated with one or more edible flower plant to elevate the sensorial qualities of the dish and nutritional composition of the dish.

Edible flowers have been traditionally used as ingredients in salads, soups, entrees, and desserts in numerous countries including in Asia, Europe and the Middle Eastern (Kumari and Ujala Bhargava 2021; Ngoitaku et al. 2016). In some countries, the flower petal is incorporated into a cup of hot water to make a refreshing tea infusion (Nicknezhad et al. 2023). Despite the same preparation as black and green teas made from young *Camellia sinensis* leaves, many plants such as flowers are infused in hot water which is beneficial to health. There are various studies on edible flowers from the scientific community. Edible flowers are famous with their own functionality purposes to enhance health. Beyond their use as decorations and

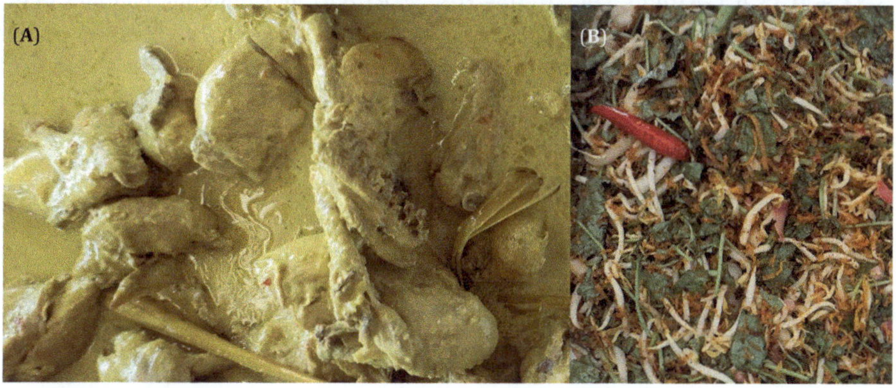

Fig. 10.1 Malaysian cuisines added with edible flowers known for its aromatic properties. (**a**) *Masak lemak cili api* with chicken as the main protein; (**b**) *Urap* with bean sprouts and winged bean as main vegetables

in culinary applications, certain edible flowers also offer as a nourishment and chemoprotective properties namely *Calendula officinalis*, *Begonia × tuberhybrida*, and *Tropaeolum majus* are studied in advance research across the globe to provide a scientific basis for the medicinal benefits (Mlcek et al. 2021). Lipid peroxidation and strong reactive oxygen species (ROS) scavenging activities were also reported by (Mlcek et al. 2021) in daylily (*Hemerocalli* cultivar methanolic extract with 1:10 plant to solvent ratio (w/v). *C. officinalis* extract has higher antioxidant activity to scavenge free radicals than the synthetic antioxidant, butyl-hydroxytoluene (BHT), which is frequently used as food additive. Commonly found in India, the edible flower of *Costus speciosus* Koen, also known as crepe ginger, is reported to have a number of health benefits, including those related to anti-inflammatory, anticarcinogenic, antidiabetic, antimicrobial, jaundice, skin disease, pneumonia, urinary disease, anaemia, dropsy, and rheumatism (Jadhav et al. 2023). In a study by (Nicknezhad et al. 2023), revealed that colour was found as the most significant component (63%) in decision making of influencing consumers tendency to buy and eat new food products apart from its nutritional and medicinal values. It was reported that the highest acceptance score from 82 panellist (in ascending order) was *Chrysanthemum morifloium* (purple spray chrysanthemum) (score 83.28), followed by *Tagetes erecta* L. (orange marigold) (74.38), *Hibiscus syriacus* (purple hisbiscus) (68.75), *Galdiolus* sp. (white gladiolus) (65.1) and *Yucca elephantipes* (yucca) (59.81).

In Malaysia, chrysanthemum is an edible flower easily available in the food market or pharmacy. Chrysanthemum is scientifically proven in lowering fevers and wind-heat syndrome, treats bruises, sprains, reduce inflammation, cholera and malaria. It also has antipyretic and antihypertensive features (Eisa et al. 2022). Mlcek et al. (2021) reported that the daylilies had the highest phenolic content (6.59 g GAE/kg) and flavonoid content (3.76 ± 0.23 g of RE/kg of fresh mass, FM) as compared to rose (TPC 4.45 ± 0.20 g GAE/kg; TFC2.43 ± 0.18 g RE/kg), French marigolds (TPC 4.78 ± 0.44 g GAE/kg; TFC 2.02 ± 0.17 g RE/kg), pot marigold (TPC 3.65 ± 0.19 g GAE/kg; TFC 2.05 ± 0.20 g RE/kg), nasturtium (TPC 3.23 ± 0.35 g GAE/kg; TFC 1.29 ± 0.32 g RE/kg) and tuberous begonia (TPC 4.82 ± 0.27 g GAE/kg; TFC 2.32 ± 0.21 g RE/kg). The daylilies, with yellow colour petals was reported to have the highest value in DPPH assay (7.94 ± 0.026 g of GAE/kg of FM) compared to other edible flowers. Farhat et al. (2022) studied on phytochemicals profile found in water extract of *Althaea officinalis* flowers, grown in Lebanon. These infusions were regularly consumed for cancer treatment in Lebanon population. The findings reported that the water extract of *A. officinalis* has a significant anti-hemolytic property with inhibition of red blood cell hemolysis at percentages ranging from 98% at 0.01 mg/ml to 87% with 25 mg/m. The water extract (6 mg/ml) also has 75% scavenging activity after analysed using DPPH assay, which indicates the high ability of the water extract to protect from free radicals and chelating metal ions.

Liu et al. (2020) reported that antioxidant activities, antiradical capacities, organic acid (ascorbic acid, citric acid, malic acid, glucose, and fructose), 18 phenolic compounds, total phenolics (from the range of 31.08 ± 0.19 to 52.94 ± 0.47 mg gallic acid/g dry weight), gallic acid, catechin, and 52 volatile compounds are found in tea infusion of rose hip species from Turkey. Phenolic compounds as well as flavonoids has been used as antioxidant for human health, curing and preventing many diseases like cancer (Tungmunnithum et al. 2018). *Osmanthus fragrans*, the edible flowering plant cultivated for its ornamental properties has been used traditionally to promote saliva secretion, deodorize and clear phlegm and toothaches treatment (Fu et al. 2022). Mao et al. (2017) reported on the potent synergistic effect of aceteoside from *O.fragrans* flower with gallic acid from green tea as compared to Pu′er, Black and Oolong Tea. The highest contribution to the antioxidant synergy was from the combination of acteoside and gallic acid (35.05%). Followed by other pairs, with acteoside and EC (19.17%), acteoside and EGC contributing 11.69%, and acteoside and catechin contributing 9.97%.

Demasi et al. (2021) investigated about 17 edible flowers (*Allium ursinum* L., *Borago officinalis* L., *Calendula officinalis* L., *Centaurea cyanus* L., *Cichorium intybus* L., *Dianthus carthusianorum* L., *Lavandula angustifolia* Mill., *Leucanthemum vulgare* (Vaill.) Lam., *Paeonia officinalis* L., *Primula veris* L., *Robinia pseudoacacia* L., *Rosa canina* L., *Rosa pendulina* L., *Salvia pratensis* L., *Sambucus nigra* L., *Taraxacum officinale* Weber, and *Tropaeolum majus* L. to assess the sensory profile during harvest, shelf life and bioactive compounds dynamics during cold storage. Each panellist evaluated each species independently using Quantitative Descriptive Analysis (QDA) test. The evaluated sensory attributes include four descriptors for taste (sweet, sour, bitter, and salt), three aroma descriptors (smell, specific flower aroma, and herbaceous aroma) and spiciness, chewiness, astringency. In *A. ursinum* the intensities of smell and garlic aroma were very high with 8.3 and 8.8 score, respectively, while the highest values for smell (9.0) and aroma intensity (8.2) was reported in *L. angustifolia*. *R. pendulina* has the longest shelf life of 14 days in terms of acceptable visual quality than the other 16 flower samples. During harvest, *P. officinalis*, *R. canina*, and *R. pendulina* exhibited the highest polyphenols in a range of 884–1271 mg compared to other 14 flower samples.

According to Kato-Noguchi and Kurniadie (2021), the plant *Lantana camara* L. which has a fragrant smell can be toxic and poisonous upon ingestion. *Lantana camara* L. could potentially injure the liver once ingested because it contains hepatotoxic pentacyclic triterpenoids namely lanthanides. It is important to learn that plant also has their own defensive mechanism to shield themself from injuries caused by animals. The plant species which has toxic and poison traits is avoided from being consumed by the herbivorous animal. Therefore, great caution must be taken where not all the flowers are edible as it could be detrimental to human and animal health.

10.2 Nutritional Content of Various Edible Flowers

The demand of attractive and tasty food is increasing globally where the quality of foodstuffs and aesthetic aspects contribute to the appearance of consumed meals. Demasi et al. (2021) stated that edible flowers could be utilised to improve the aesthetic appearance, colour and aroma of individual dishes. In Mexico, Dahlia flowers are frequently consumed in different type of dishes, for example in dried soups (Pires et al. 2017). Other than it contributes to the appealing appearance of food, edible flowers are also used during cooking preparation for flavouring and garnishing thus improved nutritional value of the cuisine (Mlcek et al. 2021). Some of the examples are dried hibiscus flowers incorporated into tea, while rose petals is used in sweets like rose-flavoured ice cream or rose water syrup (Nicknezhad et al. 2023). Edible flower is also related with biologically active substances like antioxidant capacity, flavonoids and total phenolics (Pires et al. 2018). In particular, edible flowers like *Rosa canina* L., *Rosa pendulina* L., *Salvia pratensis* L., *Sambucus nigra* L. have been studied on the bioactive compounds responsible for the antioxidant properties (Demasi et al. 2021).

Edible flowers can generally be eaten whole, yet occasionally, only parts of it should be eaten, depending on the variety of flower. For instance, only the petals of plants like rosa spp., calendula (*Calendula officinalis* L.), and roselle (*Hibiscus sabdariffa* L.), are eaten. Moreover, certain parts of some flowers like the white base of rose and chrysanthemum petals need to be removed because of their bitterness. The pollen part of flower is usually removed as it can detract flower's flavour and causing allergies in sensitive people (Fernandes et al. 2018). There are various segregated parts namely the petals, sepals, stamen, stigma, stem and the ovary.

The following are the list of parts available in a flower:

Petal	Sections of a flower that frequently have striking colours and collectively known as corolla
Stigma	Part of the pistil where pollen germinates
Stamen	Portion of a flower that produces pollen and typically has a thin thread supporting the anther
Anther	Section of the stamen that produces pollen
Pistil	Portion of a flower that produces ovules. Often, the ovary supports a lengthy style that is capped with a stigma. A fully developed ovule is a seed, and an adult ovary is a fruit
Sepal	Outermost, frequently green, leaf-like elements of the flower that envelop a budding flower, and collectively form a calyx as in roselle

Pollen, petals and nectars are the three important factors to determine the nutrients content and type of phytochemicals present in an edible flower. In pollen, there are proteins, carbohydrates, carotenoids, flavonoids, saturated and unsaturated lipids. Pollen has a very mild flavour and taste. A nectar usually produced by gland called nartaries. Nectar glands is found on a flower's petals, anthers, stamens, sepals, pistils, styles, ovaries, or other regions, depending on the species. A nectar usually consists of a mixture of amino acids, free sugars (fructose, glucose, and sucrose), proteins, inorganic ions, lipids, organic acids, phenolic compounds, alkaloids, terpenoids, and other substances. Petals are abundant in antioxidants, vitamins, and minerals such as potassium, phosphorus, calcium, and magnesium (Amrouche et al. 2022).

Majority of macronutrients in flowers are carbohydrates (86.12 g/100 g dry weight DW, whereas the amount of fibre varies greatly (ranging from 6.1 to 55.4 g/100 g DW). In general, flowers comprise of 70–95% water (Fernandes et al. 2017) protein and fat content are considered to be low in edible flower (Pires et al. 2017). Water makes up between 70% and 90% of the constituents of edible flowers. Carbohydrate is the most prevalent macronutrient in the edible flower, followed by water. The amount of carbohydrates in edible flowers is larger than in other types of flowers; Rosa spp. has one of the highest documented amounts, ranging from 80 to 90 g /100 g depending on dry weight. Edible flowers have free sugars like fructose, glucose, and sucrose; fructose makes up 50% of these sugars, with glucose and sucrose following closely after (Pires et al. 2017).

According to previous studies, the edible flower contains more than 60 g of carbohydrates, with a maximum of 90 g depending on the dry weight. These edible flowers can also be incorporated into food formulations for individuals on diets or formulations based on low carbohydrate diets. The bioactive ingredients found in edible flowers will also be incorporated into this formulation and increasing the food's nutritional value. The edible flowers also have substantial quantities dietary fibres but their content varies with the species of edible flowers (Jadhav et al. 2023). *Tagetes erecta* reported a high dietary fibre content at 55.4 g/100 g. Dietary fibre is essential to lower the cholesterol level in blood, lower the risk of cardiovascular diseases and other related health problems (Alzahrani 2017). Other edible flowers contain fibre which include *Antirrhinum majus* (50.9%) and *Viburnum inopinatum* (25.3%). An edible flower has a good source of fibre and, when consumed in a regular diet, can satisfy the daily need of 14 g of fibre per 1000 kcal (Schneeman 2018).

The edible flower contains a low percentage of protein (Suksathan et al. 2021) reported that the wild edible flower of *Bauhinia variegata* and *Shorea roxburghii* have higher protein content of 10.26 g and 10.20 g per 100 g, respectively than *Gmelina arborea* and *Viburnum inopinatum*, indicating that both edible flowers are great source of protein. (Rachkeeree et al. 2018) reported on the insight of edible flower of Zingiberaceae family found in Thailand. The authors reported that protein content of edible flower from Zingiberaceae family range from 0.06% to 1.41%. Even though some researches have indicated that edible flowers have a low protein content, these flowers can still be employed in food applications by including a source of protein in the dish. To compensate for their low protein level, the edible flowers used in Mexican recipes are combined with eggs to make soups or tacos (Jadhav et al. 2023). Vitamins are less explored micronutrients in edible flower. Vitamin C is an essential source in regulating human's immunity as a free radical scavenger in human's body system. As mentioned in Table 10.1, it has been found that *Viburnum inopinatum* has the highest vitamin C content of 3.47 mg per 100 g, followed by *Shorea roxburghii*, *Bauhinia variegata* and lastly, *Gmelina arborea*.

Table 10.1 Research on edible flowers which are commonly consumed in different regions of the world

Flower used	Country	Extraction/ infusion parameter	Main findings	Reference
Clitoria ternatea L.	Southeast Asia	Steeping in 100 ml water at different temperature and time; 25 °C for 12, 24 and 36 h 45 °C for 1, 2 and 3 h 60 °C for 15, 30, and 45 min	Steeping process at 60 °C for 45 min shows significantly higher antioxidant activity, total phenolic and monomeric anthocyanin content than other infusion parameters used	Soedirga et al. (2023)
Chrysanthemum iridium, *Clitoria ternatea* and *Jasminum sambac* L.	Indonesia	Infusion temperature: 30, 50 and 70 °C Infusion time: 1,3 and 5 min	Combination of brewing temperature at 50 °C for 3 min contributed to the strongest antioxidant inhibition power value (DPPH assay) of 34.61 ± 0.91% for the dried flowers	Nugroho et al. (2023)
Jasminum sambac	China	Optimal extraction conditions using a 50/30μm DVB/ CAR/PDMS fiber, tea/water ratio of 1:25 and extraction time of 5 min	80% of flowers is the key parameter affecting the jasmine fragrance intensity and comprehensive score in jasmine tea infusion	An et al. (2023)
Althaea officinalis L.	Lebanon	About 15 g of *Althaea officinalis* L. is added into 250 ml of water (3 h of maceration and stirring at room temperature, incubated for 10 min at 60 °C	Antioxidant properties; TFC value (33.78 ± 0.78 mg QE/g) is higher than TPC value (16.56 ± 0.58 mg GAE/g; the three most predominant phenolics are: Caffeic acid (302 mg/kg of dry extract); Apigenin (283 mg/kg of dry extract); Galangin (213 mg/kg of dry extract)	Farhat et al. (2022)

(continued)

Table 10.1 (continued)

Flower used	Country	Extraction/ infusion parameter	Main findings	Reference
Hibiscus sabdariffa	Turkey	Different infusions Hot infusion (98 °C for 16 and 40 min) Cold infusion (25 °C in 24 h)	Total phenolic and antioxidant activity (Trolox) and colour of the hot Roselle infusions presented >50% higher values than the cold sample; Total concentration of aroma was higher in the cold infusion obtained by (176544.9µg/L) than in the hot infusion (in 40 min-134831.3µg/L) and (in 16 min-89627.0µg/L)	Zannou et al. (2020)
Sambucus nigra L.	Europe	Infusion of elderberry flower, berries and leaves with 200 mL of boiling water (15 min) filtered (Whatman no. 541 filter paper)	Antioxidant compounds (phenolic, flavonoid and anthocyanin) from tea prepared using flowers higher than berries. Antioxidant activities; DPPH, FRAP and free-radical scavenging activity from flower tea higher than berries	Viapiana and Wesolowski (2017)
Helichrysum italicum	France	Standards and flower extracts in water infusions at 90 °C in 15 min and in 70% (v/v) acetone	Aqueous extraction by infusion led to higher antioxidant capacities as analysed using ABTS assay however DPPH assay and TPC was higher in ace-tone extract flower than water extract. Higher antifungal activities were obtained against *S. cerevisiae* and other fungi	Primitivo et al. (2022)
Taraxacum officinale Prunus domestica subsp. syriaca *Papaver rhoeas* L.	Poland	About 0.5 g of plant material is infused in 100 ml distilled water at 70 °C, 80 °C, and 90 °C for 10 min	The infusion temperature did not have an influence on fluoride content of common poppy and mirabelle plum. One glasses consumption of these infusions can contribute to 0.5–2.1% of RDA fluoride	Janda-Milczarek et al. (2019)

Pereira et al. (2017) studied on (*Crithmum maritimum* L.), the sea fennel flower revealed that the tisanes, especially those made from its leaves, contained abundant phenolic compounds precisely, p-hydroxybenzoic acid, ferulic acid, epicatechin, pyrocatechol, and 4-hydroxybenzaldehyde. The infusions from sea fennel flowers possess strong antioxidant properties, an intriguing mineral composition, and based on initial in vitro toxicological assessments, can be considered safe for consumption as non-toxic beverages. Nartea et al. (2023) reported *C. maritimum* can be considered a high source of lutein, vitamin A, and vitamin E up to 19.1, 1.85 and 52.81 mg/100 g of dried leaves, respectively.

Comparatively, the nutritional comparison between dried flowers and their infusions reveals significant differences. Dried flowers of *Dahlia mignon* (Dahlia), *Rosa gallica* 'Francesa' (French Rose), *Calendula officinalis* L. (Pot Marigold), and *Centaurea cyanus* L. (Cornflower) are rich in carbohydrates, energy, and soluble sugars. Cornflower has the highest carbohydrate content at 88.39 ± 0.13 g/100 g, Pot Marigold has the most energy of 421.58 ± 3.54 kcal/100 g, whereas French Rose has the highest soluble sugar concentration of 10.75 ± 1.05 g/100 g. Conversely, their infusions have much lower macronutrient values. Carbohydrate values in the infusions are low, with Dahlia and French Rose having the highest levels at 0.19 ± 0.02 g/100 ml. The energy content in infusions of Cornflower is also low, ranging from 0.56 ± 0.04 kcal/100 ml in Cornflower to 0.80 ± 0.08 kcal/100 ml for French Rose. Soluble sugars in the infusions are significantly reduced, with Dahlia and French Rose at 0.17 ± 0.01 g/100 ml and Cornflower and Pot Marigold at 0.14 ± 0.01 g/100 ml. This data highlights the substantial nutrient retention in dried flowers as compared to their infusions (Pires et al. 2017).

Other than that, Fernandes et al. (2018) reported *Moringa oleifera* flowers have a high protein content (21.3 ± 0.4 g/100 g), than other dried flowers like *Calendula officinalis* L. (6.43 g/100 g) (Pires et al. 2017), *Tagetes erecta* (7.9 g/100 g) (Jadhav et al. 2023) and *Gmelina arborea* (5.85 g/100 g) (Suksathan et al. 2021). In contrast, a study from (Isla et al. 2021) reveal that infusions of flowers such as *Larrea divaricata*, *Acacia caven*, and *Geoffroea decorticans* have significantly lower protein content ranging from 2.91 ± 0.23 to 3.13 ± 0.36 g BSA E/100 g. The results indicate that dried petals have a significantly higher amount of nutrients in comparison to their infusions. The process of drying has maintained the integrity of these nutrients, while the process of infusion only results in much lower nutrient extraction. This suggests that although infusions provide many health advantages, but it does not deliver the same level of nutritional content as dried petals directly.

10.3 Phytochemical Compounds in Edible Flower

Various studies reported on contribution of edible flowers towards human nutrition especially in term of immune system, cardiovascular disease, anti-inflammatory capacity, and anticancer ability. The benefits are mainly contributed by bioactive compounds like total phenolic content and antioxidant that can be found naturally in edible flowers (Kumari and Ujala Bhargava 2021). Caffeine and volatile compound are the important chemical constituents in tea brew which contribute to the taste and flavour. Polyphenols exert the antioxidant effects through free radical-scavenging abilities, helping to prevent cardiovascular disease, chronic gastritis, and some cancer (Bi et al. 2016). However, a benefit of edible flowers beverage is it does not contain caffeine, suitable for those who have a caffeine allergy (Amrouche et al. 2022). Killgore and Kamimori (2020) reported that daily caffeine consumption may help maintain performance, however not able to reduce subjective ratings of sleepiness, fatigue, or perceived performance during the entire period of wakefulness.

Total phenolic content, also known as phenols or phenolics, consist of hydroxyl group bonded specifically to an aromatic hydrocarbon group. Total phenolic compound is commonly found in plants sources (Nour et al. 2012). Examples of phenolic compounds are chlorogenic acid, cinnamates, flavonols, procyanidins, and carotenoids. Temperature, light and oxygen, chemical treatment, natural compounds and dissimilarity between fruit varieties and processing methods can cause variations of bioactive ingredients and content in fruits and vegetable (Debelo et al. 2020). Phenolic compounds in plant is linked with high antioxidant potential, and these phenolic antioxidants are applied in therapeutic application to delay the development of type 2 diabetes hence control hyperglycemia (Bastos et al. 2019).

Folin-Ciocalteu (FC) assay is commonly used to determine the total phenolic content because it is simple and effective. He et al. (2015) carried out research on the phenolic compounds and antioxidant activities of edible flower of *Pyrus pashia* which proved that high phenolics content (ranged from 28.1 ± 1.9 to 168.9 ± 4.7 mg GAE/g extract) and its topical antioxidant activities are present in *Pyrus pashia* flowers, collected from Kunming City in Yunnan Province, China. The total phenolic content is detected by FC method whereas 1,1-diphenyl-2-picrylhydrazyl (DPPH) assay, ferric reducing antioxidant power (FRAP) assay, β-carotene bleaching (BC) assay, and inhibition of lipid peroxidation (LPO) in mouse tissues were used for the detection of antioxidant activity in *Pyrus pashia* flower. Similar research also conducted by Li et al. (2014) where the total phenolic contents (ranged from 0.63 ± 0.03 to 35.84 ± 1.67 mg GAE/g wet weight) and antioxidant capacities of 51 edible and wild flowers from China were analysed. FRAP assay, trolox equivalent antioxidant capacity (TEAC) assay, FC and HPLC were deployed in the research of Li et al. (2014).

10.4 Edible Flower as Infusion

Edible flowers have been used across the globe for cooking at a very long time because of their aromatic scent, attractive hues, and beneficial qualities (Fig. 10.2). Owing to its unique qualities, the edible flower has a great deal of potential for use in the development of several processed food products intended for human consumption. A comparison of the bioactive profile and sensory characteristics of edible flower infusion using *Jasminum officinale* (Jasmine flower), *Rosa gallica* (France rose) bud, and *Osmanthus fragrans* (Osmanthus flower) was conducted by (Hussain et al. 2019). The study reported that Osmanthus flower contains the highest caffeine (4.96 ± 1.94µg/ml), total phenolic content (4.33 ± 0.03 mg GAE/g) with overall acceptability score of 6.16 ± 2.05 (out of 9-point hedonic scale) compared to

Fig. 10.2 Edible flower with attractive Colours: (**a**) *Taraxacum officinale* (dandelion); (**b**) *Rosa gallica* (French rose); (**c**) *Jasminum officinale* (common jasmine); (**d**) *Papaver rhoeas* L. (corn poppy); and (**e**) *Prunus domestica* (European plum). (Source: http://pixabay.com)

France Rose Buds and Jasmine Flower. Janda-Milczarek et al. (2019) studied an infusion from *Taraxacum officinale* (dandelion), *Prunus domestica* subsp. syriaca (mirabelle plum), and *Papaver rhoeas* L. (common poppy). The study shows dandelion infusions (0.5 g in 100 ml water), brew at temperature 70, 80 and 90 °C had higher fluoride content (0.0605 mg/L to 0.2529 mg/L) as compared to poppy and Mirabelle plum infusion. The consumption of 1–2 glasses of the dandelion infusion per day may function as a good source of fluoride.

According to Młynarczyk et al. (2018) who studied on elderberry flower infusion stated that the beneficial contents of edible flowers are dependent on the preparation of the infusion, which includes the temperature at which the infusion or extraction is made, the time spent infusing, the type of infusion used, the solvent used, and the process of drying the plant materials. Viapiana and Wesolowski (2017) reported that the elderberry infusion from *Sambucus nigra* L. flower had higher phenolic compounds than infusion from of the berry from the same plant. The total phenolic content (TPC) in elderflower infusion is in a range of 15.23 to 35.57 mg GAE/g DW while elderberry infusion is 19.81 to 23.90 mg of gallic acid equivalents/g dry weight of sample (GAE/g DW). The TPC of the elderberry flower infusion are strongly correlated with the antioxidant activities with DPPH (0.78) and FRAP assay (0.65) suggesting the elder beverages as important dietary sources for prevention of disease by oxidative stress. (Farhat et al. 2022) stated that consumption of infusions made of *Althaea officinalis* flowers might contain potential chemical compounds including polysaccharides, flavonoids, phenolic acids and coumarins. These major constituents of the plant contributed to anti-cancer effects and can be used in the future for pharmacological intents. Significantly, *Althaea officinalis* water extract analysed using DPPH assay found 75% free radical scavenging activity resulted at a concentration of 6 mg/ml.

Despite the genetic variation that caused the difference between physicochemical characteristics and taste of edible plant, edible flowers are often use in food preparation as natural food colouring, food decoration, and medicine (Nicknezhad et al. 2023). Many researches on edible flower infusion highlighted the optimum infusion parameters, extraction methods, antioxidant activities and others. Soedirga et al. (2023) reported that steeping time of *Clitoria ternatea* infusion at 60 °C for 45 min showed significantly higher antioxidant activity of IC50 of 251.89 ± 32.02 ppm, total phenolic and monomeric anthocyanin content of 138.42 ± 15.88 mg gallic acid equivalent (GAE)/g sample and 1.23 ± 0.07 mg/L, respectively compared to steeping at temperature 25 °C (for 12, 24, 36 h) and 45 °C (for 1, 2 and 3 h). However, the result reported by (Nugroho et al. 2023) stated that brewing temperature of 50 °C for 3 min has the strongest antioxidant inhibition power value (DPPH assay) of 34.61 ± 0.91% of the dried flowers (i.e. *Chrysanthemum indicum*, and *Jasminum sambac* L.) than brewing at 30 and 70 °C for 1, 3 and 5 min. There are many factors contributed to the differences in edible flower phytochemicals content since each edible flower have different genetic, environmental factors (light, climates, soil composition and developmental factors such as nutrient status, water availability, harvesting time and cultural practice (Lin et al. 2016). Most researchers reported on the comparison between hot and cold edible flower infusion where cold infusion is becoming popular and preferred by the youngsters. Zannou et al. (2020) reported that the hot infusion of *Hibiscus sabdariffa* calyx have 50% higher total phenolic, antioxidant activity (Trolox) and colour than its cold infusion. Apart from antioxidant properties, Primitivo et al. (2022) reported that higher antifungal activities were obtained in *Helichrysum italicum* against *Saccharomyces cerevisiae* and other fungi extracted in DMSO from the aerial parts of the plant (MIC of 6.325 μg/mL).

10.5 Factors Affecting Edible Flower Infusion

10.5.1 Temperature

The infusion preparation process, which includes water temperature, infusion duration, and number of extractions, affects the edible flower infusion's antioxidant capacity and polyphenol content. It also has an impact on the infusion's sensory qualities (Skolmowska et al. 2020). The ideal temperature settings to infuse edible flowers in water vary according to species and were discussed in many studies. The sensory attributes of edible flower infusions like aroma, flavour, colour and mouthfeel are also related to the diverse ingredients used in their preparation. The solubilization of tea components in the plant matrix and the diffusion of water into tea particles are both significantly affected by the temperature during extraction (Yadav et al. 2018).

The research on identification of optimum brewing temperature to extract the phenolics compounds in rose hip tea bag commonly found in Turkey market was conducted by (Akar et al. 2021). Different temperatures (0, 15, 30, 45, 60, 75, 90 °C) which include cold infusions temperatures were selected to determine the effect of temperature on the antioxidant activity of the rose hip infusion (Akar et al. 2021). Highest antioxidant activities were found in rose hip infusion at 90 °C with TPC (615 GAE g /mL) and FRAP (580 M TEAC) than infusion at other used temperatures while CUPRAC assay was reported to be the highest value in cold infusion (45 °C) (Akar et al. 2021) further reported that each antioxidant assay has different mechanism therefore has different sensitivity towards different phenolic substances. Similarly, Nugroho et al. (2023) reported that the antioxidant activity of purple chrysanthemum flower is the highest when brewed at higher 90 °C than *Jasminum sambac* L. (Jasmine) and *Clitoria ternatea* L. (Butterfly pea flower). Chang et al. (2021) indicates that black tea made of boiling water had greater total phenolic content (TPC), 50.4 ± 5.2 to 178.6 ± 16.4 mg gallic acid equivalent/g dry leaf whenever the black tea is boiled at 60 to 100 °C. In addition, the colour and flavour changes in black tea also are due to different brewing temperatures (Yadav et al. 2018).

According to Kushargina et al. (2022), water temperature at 85 and 95 °C will reduce the antioxidant effects and vitamin C content of butterfly pea flowers (*Clitoria ternatea* L.) to make functional drinks. They studied three brewing temperatures (75, 85 and 95 °C) and two infusion time (5 and 9 min) in butterfly pea flower infusion (1 g/250 ml water, w/v). The study showed that majority of panellist preferred infusion made at a temperature of 75 °C, for 9 min based on organoleptic testing (colour, aroma, taste). The flower infusion prepared at 75 °C for 9 min steeping time show a moderate antioxidant activity in DPPH assay (135.29 ppm) compared to infusion brewed at (85 and 95 °C) for 5 and 9 min. Nugroho et al. (2023) reported that vitamin C content and antioxidant activity of infused edible flower namely chrysanthemum, jasmine flowers, and butterfly pea flowers from Semarang City Indonesia were not affected by temperature or brewing time. The dry sample demonstrated optimum antioxidant activity at 50 °C for 3 min, which yield a 34.61% of free radical inhibition.

Although infusions are commonly enjoyed as hot tea, cold infusions can provide similar benefits for a specified flower. Bellachioma et al. (2022) mentioned that cold infusion exhibit is the best method to maximize the advantageous characteristics of *Crocus sativus* petal infusions. The approach produced a significant TPC of 186 ± 36 mg GAE/L and the greatest TFC of 91 ± 17 mg CE/L. In addition, cold infusions had a very high phenolic content, with a total phenolic content of 61.9 mg/g DM. Other than that, sensory analysis indicated that cold tepal infusions were favoured for their invigorating and delightful flavour, complemented by floral notes, making them well-suited for summer beverages. Hence, the research highlights that the cold infusion technique (2 h infusion at room temperature) is the most effective approach to extract the highest amount (61.9 mg/g DM) of phenolics compounds from *Crocus sativus* tepals, while simultaneously ensuring desirable sensory characteristics. This confirms the significance of cold infusion in the development of saffron-based beverages with functional features.

The study on Roselle infusions sought to identify the major odorants in infusions prepared using both hot and cold brewing procedures. The study revealed furans were the most prominent aroma group, followed by alcohols in cold infusion (Zannou et al. 2020). Next, cold infusion brew at 25 °C for 24 h had the highest overall aroma concentration (176,544.9μg/L), while the hot infusions had concentrations of 134,831.3μg/L at 98 °C for 16 min and 89,627.0μg/L at 98 °C for 40 min. The major odorants with the highest potency were furfural (FD factor of 4096) and 5-methyl-2-furfural (FD factor of 1024), which both contributed to caramel and bready aromas. In summary, optimal brewing conditions rely on the intended result. Moderate to high temperatures ranging from 70 °C to 90 °C are recommended for antioxidant activity and mineral extraction. Lower temperatures (70 °C) and shorter periods (5 min) are beneficial for preserving vitamin C. Cold infusion methods provide advantages for some flowers by yielding significant amounts of phenolic compounds, desirable sensory attributes and improved aromatic profiles.

10.5.2 Time

Preparation time is an important aspect in order to produce the highest bioactive compound with an acceptable sensorial property. Bellachioma et al. (2022) observed the extraction efficiency of *Crocus sativus* tepal infusion at different condition; 2 h at 4 °C and overnight at room temperature. There is no significant different in term of the extraction efficiency of total phenol in room temperature (186 ± 55 mg GAE/L) and chill (4 °C) 186 ± 36 mg GAE/L, respectively indicating that increasing infusion time at lower temperature does not affect the extraction yield from the *C.sativus* tepals. It is interesting to highlight that the longer infusion time in lower temperature is as efficient as short time infusion at high temperature due to comparable phytochemicals compounds extraction efficiency.

Study by Nikniaz et al. (2016) found that white tea antioxidant activity: (FRAP, total polyphenol and phenolic content was at its optimum temperature of 90 °C and the antioxidant capacity increased after 5–7 min of infusion compared to 15 and 30 min of extraction. Similarly, Pérez-Burillo et al. (2018) found the best water temperature and brewing time for white tea is at 98 °C for 7 min. Application of these parameters allow a greater number of bioactive compounds and antioxidant activity dissolved into the tea brew with tasty and slightly bitter-astringent drinks. There is a linear correlation (0.6020 to 0.8998) between the total polyphenols, antioxidant activity, catechin concentration and bitterness. Skolmowska et al. (2020) reported on the influence of different steeping conditions (temperature ranging from

75 °C to 95 °C and time ranging from 1 to 45 min) on catechins content in green tea. It was found that brewing tea at 85 °C for 3 min was optimal in terms of epigallocatechin gallate content in green tea infusion (50.69 mg/100 ml), which is in accordance with commonly recommended steeping conditions of green tea.

10.6 Sensory Evaluation of Edible Flower Infusion

Sensory analysis plays a crucial role in determining the best uses and combinations for edible flowers infusion in the food industry. Variation in taste and aroma among different flower species are attributed to their distinct chemical compositions, which consist of hundreds of compounds including organic acid, flavonoids and volatile compounds (Nicknezhad et al. 2023). Benvenuti et al. (2016) studied on the antioxidant power, anthocyanin content and organoleptic performance of 12 edible flowers provided by Carmazzi Farm, Tuscany, northwestern province of Lucca, Italy. These flowers include *Ageratum houstonianum*, *Antirrhinum majus*, *Begonia semperflorens*, *Borago officinalis*, *Calendula officinalis*, *Dianthus × barbatus*, *Fuchsia hybrida*, *Pelargonium peltatum*, *Petunia × hybrida*, *Tagetes erecta*, *Tropaeolum majus*, and *Viola × wittrockiana*. The sensory analysis of the fresh petals was conducted by 87 untrained panellists from Pisa University community. Benvenuti et al. (2016) concluded that *Tropaeolum majus* (8.2 score) has higher sensory score compared to other 11 edible flowers, demonstrating a strong spiciness, sweetness, softness and scent in its organoleptic characteristics. Its flavor was associated with radish. The high anthocyanin content was reported in *Viola × wittrockiana*, *Dianthus × barbatus*, *Pelargonium peltatum* with a value of 12.4, 13.3, 12.5 mg cyn-3-glu eq. 100 g^{-1} FW, respectively was associated to a high antioxidant activity. An et al. (2023) reported that the aroma quality of jasmine tea relied on combination of factors such as the origin of the jasmine flowers, scenting temperatures and even the weather during the production of jasmine tea. Approximately, 25 key volatiles compounds which were identified as the main compounds (i.e. grassy, dull odour and jasmine fragrance compounds) that responsible for the formation of aroma in jasmine tea infusion. In contrast with study by Hussain et al. (2019) which reported jasmine flower infusion was the least preferred by untrained panellist (4.96 ± 1.65) compared to France rose bud (5.90 ± 1.62) and Osmanthus flower infusion (6.16 ± 2.05) from 9-point hedonic scale. Despite having lower acceptability, jasmine flower infusion had the highest number of volatile compounds observed using GCMS with major detected compounds (31.98%) was pregna-5, 14-diene-3, 20-diol-18-carboxylic acid, 3-acetate-, lactone. In contrast, flower petals colour is one of the significant factors affecting overall acceptability of sensory analysis conducted by Mlcek et al. (2021) on six ornamental edible flower namely *Begonia × tuberhybrida*, *Tropaeolum majus*, *Calendula officinalis*, Rosa, Hemerocallis spp., and *Tagetes patula*. *Hemerocallis* spp. or daylily score the highest (8.2 out of 9 points Hedonic scale) in the sensory analysis and the most acceptable overall taste (8.4 point) compared to other species. The petal's colour is one of the significant factors to the panellist. The most consumed part of the flower is the flower bud. The daylily has a mild taste, with a mixture of sour, astringent, and spicy flavour. These flowers act as alternative to mustard due to their peppery, radishes, and spicy taste. It is the only flower that does not have the sweet flavour and is evaluated as slightly spicy. Mlcek and Rop (2011) reported the sweet taste of *Rosa damascena* is perceived as pleasant. The taste is resulted from the content of saccharose, which are a typical smell of individual flowers.

Nicknezhad et al. (2023) reported that purple chrysanthemum was more preferred compared to orange marigold, white gladiolus, purple hibiscus, and yucca by the 82 panellists in Rasht, one of the cities in north-central Iran. The researchers reported the contrast findings with the study by Mlcek et al. (2021) which identified visual criteria as the most important elements for consumers in Iran, rather than taste properties such as sweet, sour, bitter, astringent, and spicy. The scent of edible flower is highly effective in attracting consumers and stimulate their curiosity, causing motivation for the consumption of edible flower. In general, colour and fragrance are essential factors to select edible flower as a food source. Upon visual inspection, the flowers' attractiveness and desirability were more significant than their smell. It indicates that the aesthetic appearance such as colour, visual, and smelling sense stimulation are more important in food selection than oral senses, aroma, taste, and texture.

Although infusions are commonly enjoyed as hot tea, cold infusions may provide similar benefits for a certain flower. (Bellachioma et al. 2022) mentioned that saffron tepal cold infusion (2-h extraction at room temperature) exhibit the best method for maximizing the advantage characteristics of saffron tepal. The cold infusion of saffron tepal produced a significant TPC of 186 ± 36 mg GAE/L and TFC of 91 ± 17 mg CE/L compared to hot tepal infusion (TPC 83 ± 22 mg GAE/L and TFC of 34 ± 17 mg CE/L). Other than that, sensory analysis indicates that cold tepal infusions were favoured for their invigorating and delightful flavour, complemented by floral notes, making them well-suited for summer beverages. Hence, the research highlights that the cold infusion technique is the most effective approach to extract the highest number of advantageous compounds from *Crocus sativus* tepals, while simultaneously ensuring desirable sensory characteristics. This confirms the significance of cold infusion in the development of saffron-based beverages with functional features.

Both hot and cold Roselle infusions were studied for its major odorants that contributed to aroma characteristics. The study revealed furans were the most prominent aroma group, followed by alcohols in cold infusion (Zannou et al. 2020). The cold infusion brewed at 25 °C for 24 h had the highest overall aroma concentration, (176,544.9μg/L), compared to hot infusions (134,831.3μg/L) at for 98 °C for 16 min and 89,627.0μg/L at 98 °C for 40 min. The major odorants found with the highest potency were furfural (FD factor of 4096) and 5-methyl-2-furfural (FD factor of 1024), which both contributed to caramel and bready aromas.

In summary, optimal brewing conditions rely on the intended result. Moderate to high temperature ranging from 70 °C to 90 °C are recommended for hot infusion to achieve high antioxidant activity (i.e DPPH, FRAP and ABTS assay) and mineral extraction. Low temperature (70 °C) and short period (5 min) are beneficial to preserve vitamin C. Cold infusion methods provide advantages for many flowers with a significant amount of phenolic compounds, desirable sensory attributes, and improved aromatic profiles.

10.7 Conclusion and Future Prospective

This review presents a comprehensive analysis of steeping conditions, nutritional composition, antioxidant constituents, and preparation techniques for infusions made of edible flowers. It emphasizes the diverse macro and micronutrients in different edible flowers, highlighting their substantial dietary contributions and antioxidant properties. Although there has been a considerable amount of study conducted on edible flowers, there are still notable gaps, specifically for brewing parameters of flower infusion. Various brewing conditions which influence the characteristics of flower infusions in relation to cold infusion methods and effect on antioxidant qualities, including components that are sensitive to heat are not well studied. The nutritional content of dried flowers has been extensively studied but less study has been done on examining the nutritional compositions of edible flower infusions. This emphasizes the need to further investigate the essential area.

References

Akar Z, Orhan E, Durmuş S (2021) Effects of cold and hot infusions on antioxidant activities of rosehip tea bag. Turk J Agric Food Sci Technol 9:1455–1459. https://doi.org/10.24925/turjaf.v9i8.1455-1459.4276

Alzahrani QE (2017) Hydrodynamic characterisation of therapeutic glycan and glycan-like complexes PhD Thesis, University of Nottingham

Amrouche TA, Yang X, Capanoglu E et al (2022) Contribution of edible flowers to the mediterranean diet: phytonutrients, bioactivity evaluation and applications. Food Front 3:592–630. https://doi.org/10.1002/fft2.142

An H, Liu J, Chen Y et al (2023) Characterization of key volatile compounds in jasmine tea infusion with different amount of flowers. Food Chem X. https://doi.org/10.1016/j.fochx.2023.100750

Bastos RG, Salles BCC, Bini IF et al (2019) Phytochemical composition, antioxidant and in vivo antidiabetic activities of the hydroethanolic extract of Eugenia florida DC. (Myrtaceae) leaves. S Afr J Bot 123:317–332. https://doi.org/10.1016/j.sajb.2019.03.006

Bellachioma L, Rocchetti G, Morresi C et al (2022) Valorisation of Crocus sativus flower parts for herbal infusions: impact of brewing conditions on phenolic profiling, antioxidant capacity and sensory traits. Int J Food Sci Technol 57:3838–3849. https://doi.org/10.1111/ijfs.15713

Benvenuti S, Bortolotti E, Maggini R (2016) Antioxidant power, anthocyanin content and organoleptic performance of edible flowers. Sci Hortic 199:170–177. https://doi.org/10.1016/j.scienta.2015.12.052

Bi W, He C, Ma Y et al (2016) Investigation of free amino acid, total phenolics, antioxidant activity and purine alkaloids to assess the health properties of non-camellia tea. Acta Pharm Sin B 6:170–181. https://doi.org/10.1016/j.apsb.2015.11.003

Chang L, Lin S, Zou B et al (2021) Effect of frying conditions on self-heating fried spanish mackerel quality attributes and flavor characteristics. Foods 10. https://doi.org/10.3390/foods10010098

Debelo H, Li M, Ferruzzi MG et al (2020) Processing influences on food polyphenol profiles and biological activity. Curr Opin Food Sci 32:90–102. https://doi.org/10.1016/j.cofs.2020.03.001

Demasi S, Caser M, Donno D et al (2021) Exploring wild edible flowers as a source of bioactive compounds: new perspectives in horticulture. Folia Hortic 33:27–48. https://doi.org/10.2478/fhort-2021-0004

Eisa EA, Tilly-Mándy A, Honfi P et al (2022) Chrysanthemum: a comprehensive review on recent developments on in vitro regeneration. Biology. https://doi.org/10.3390/biology11121774

Farhat C, Younes H, Alyamani OA, Mrad M, Houran N, Khalifeh H, El-Makhour Y, Dbaibo G, Hage-Sleiman R (2022) Chemical characterization and in vitro biological evaluation of aqueous extract of Althaea officinalis l. flower grown in Lebanon. J Herb Med. https://doi.org/10.1016/j.hermed.2022.100575

Fernandes L, Casal S, Pereira JA et al (2017) Edible flowers: a review of the nutritional, antioxidant, antimicrobial properties and effects on human health. J Food Compos Anal 60:38–50. https://doi.org/10.1016/j.jfca.2017.03.017

Fernandes L, Ramalhosa E, Pereira JA et al (2018) The unexplored potential of edible flowers lipids. Agriculture

Fu CC, Xu FY, Qian YC, Koo HL, Duan YF et al (2022) Secondary metabolites of Osmanthus fragrans: metabolism and medicinal value. Front Pharmacol. https://doi.org/10.3389/fphar.2022.922204

He J, Yin T, Chen Y, Cai L, Tai Z, Li Z, Liu C, Wang Y, Ding Z (2015) Phenolic compounds and antioxidant activities of edible flowers of Pyrus pashia. J Funct Foods 17:371–379. https://doi.org/10.1016/j.jff.2015.05.045

Hussain N, Ishak I, Nurhazirah H et al (2019) Comparison of bioactive compounds and sensory evaluation on edible flowers tea infusion. Ital J Food Sci 31:2019–2265. https://doi.org/10.14674/IJFS-1071

Isla MI, Ezquer ME, Leal M et al (2021) Flower beverages of native medicinal plants from Argentina (Acacia caven, Geoffroea decorticans and Larrea divaricata) as antioxidant and anti-inflammatory. J Ethnopharmacol. https://doi.org/10.1016/j.jep.2021.114490

Jadhav HB, Badwaik LS, Annapure U et al (2023) A review on the journey of edible flowers from farm to consumer's plate. Appl Food Res. https://doi.org/10.1016/j.afres.2023.100312

Janda-Milczarek K, Jakubczyk K, Baranowska-Bosiacka I et al (2019) Infusions from edible flowers as a source of fluoride in diet. Fluoride 52:319–329

Kato-Noguchi H, Kurniadie D (2021) Allelopathy of Lantana camara as an invasive plant. Plants. https://doi.org/10.3390/plants10051028

Killgore WDS, Kamimori GH (2020) Multiple caffeine doses maintain vigilance, attention, complex motor sequence expression, and manual dexterity during 77 hours of total sleep deprivation. Neurobiol Sleep Circadian Rhythms. https://doi.org/10.1016/j.nbscr.2020.100051

Kumari P, Ujala Bhargava B (2021) Phytochemicals from edible flowers: opening a new arena for healthy lifestyle. J Funct Foods. https://doi.org/10.1016/j.jff.2021.104375

Kushargina R, Kusumaningati W, Yunianto AE (2022) Pengaruh bentuk, suhu, dan lama penyeduhan terhadap sifat organoleptik dan aktivitas antioksidan teh herbal bunga telang (*Clitoria ternatea* L.). J Gizi Indones 45:11–22. https://doi.org/10.36457/gizindo.v45i1.633

Li AN, Li S, Li HB, Xu DP, Xu XR, Chen F (2014) Total phenolic contents and antioxidant capacities of 51 edible and wild flowers. J Funct Foods 6:319–330. https://doi.org/10.1016/j.jff.2013.10.022

Lin D, Zhao J, Li Z et al (2016) An overview of plant phenolic compounds and their importance in human nutrition and management of type 2 diabetes. Molecules. https://doi.org/10.3390/molecules21101374

Liu W, Chen L, Huang Y et al (2020) Antioxidation and active constituents analysis of flower residue of *Rosa damascena*. Chin Herb Med 12. https://doi.org/10.1016/j.chmed.2020.05.005

Mao S, Wang K, Lei Y et al (2017) Antioxidant synergistic effects of Osmanthus fragrans flowers with green tea and their major contributed antioxidant compounds. Sci Rep. https://doi.org/10.1038/srep46501

Mlcek J, Rop O (2011) Fresh edible flowers of ornamental plants – a new source of nutraceutical foods. Trends Food Sci Technol 22:561–569. https://doi.org/10.1016/j.tifs.2011.04.006

Mlcek J, Plaskova A, Jurikova T et al (2021) Chemical, nutritional and sensory characteristics of six ornamental edible flowers species. Foods. https://doi.org/10.3390/foods10092053

Młynarczyk K, Walkowiak-Tomczak D, Łysiak GP (2018) Bioactive properties of Sambucus nigra L. as a functional ingredient for food and pharmaceutical industry. J Funct Foods 40:377–390. https://doi.org/10.1016/j.jff.2017.11.025

Nartea A, Orhotohwo OL, Fanesi B et al (2023) Sea fennel (Crithmum maritimum L.) leaves and flowers: bioactive compounds, antioxidant activity and hypoglycaemic potential. Food Biosci. https://doi.org/10.1016/j.fbio.2023.103417

Ngoitaku C, Kwannate P, Riangwong K (2016) Total phenolic content and antioxidant activities of edible flower tea products from Thailand. Int Food Res J 23:2286–2290

Nicknezhad S, Hashemabadi D, Allahyari MS et al (2023) Sensorial analysis of factors influencing consumers' perceptions toward the consumption of edible flowers in Iran. J Agric Food Res. https://doi.org/10.1016/j.jafr.2023.100580

Nikniaz Z, Mahdavi R, Lotfi Yagin N, Niknia L (2016) Effect of different brewing times on antioxidant activity and polyphenol content of loosely packed and bagged black teas (Camellia sinensis L.). Avicenna J Phytomed 6:313–321. https://doi.org/10.22038/ajp.2016.5914

Nour V, Trandafir I, Cosmulescu S (2012) HPLC determination of phenolic acids, flavonoids and juglone in walnut leaves. J Chromatogr Sci. https://doi.org/10.1093/chromsci/bms180

Nugroho P, Lewerissa K, Gentaarinda F (2023) Effect of brewing temperature and time on vitamin C content and antioxidant activity of edible flower infuses. J Pangan Agroind 11:167–177. https://doi.org/10.21776/ub.jpa.2023.011.04.1

Pereira CG, Barreira L, Bijttebier S et al (2017) Chemical profiling of infusions and decoctions of Helichrysum Italicum subsp. Picardii by UHPLC-PDA-MS and in vitro biological activities comparatively with green tea (Camellia sinensis) and rooibos tisane (Aspalathus Linearis). J Pharm Biomed Anal 145:593–603. https://doi.org/10.1016/j.jpba.2017.07.007

Pérez-Burillo S, Giménez R, Rufián-Henares JA, Pastoriza S (2018) Effect of brewing time and temperature on antioxidant capacity and phenols of white tea: relationship with sensory properties. Food Chem 248:111–118. https://doi.org/10.1016/j.foodchem.2017.12.056

Pires TCSP, Dias MI, Barros L et al (2017) Nutritional and chemical characterization of edible petals and corresponding infusions: valorization as new food ingredients. Food Chem 22:337–343. https://doi.org/10.1016/j.foodchem.2016.10.026

Pires TCSP, Dias MI, Barros L et al (2018) Edible flowers as sources of phenolic compounds with bioactive potential. Food Res Int 105:580–588. https://doi.org/10.1016/j.foodres.2017.11.014

Primitivo MJ, Neves M, Pires CL et al (2022) Edible flowers of Helichrysum Italicum: composition, nutritive value, and bioactivities. Food Res Int. https://doi.org/10.1016/j.foodres.2022.111399

Rachkeeree A, Kantadoung K, Suksathan R et al (2018) Nutritional compositions and phytochemical properties of the edible flowers from selected zingiberaceae found in Thailand. Front Nutr. https://doi.org/10.3389/fnut.2018.00003

Schneeman B (2018) Role of fiber in carbohydrate quality – meeting dietary fiber recommendations: what counts. Cereal Foods World 63:99–102. https://doi.org/10.1094/CFW-63-3-0099

Skolmowska D, Kołota A, Stachoń M, Lachowicz K (2020) Health-promoting properties of bioactive compounds of tea and the influence of method of infusion preparation and time on their content. Postepy Tech Przetw Spoz 30(57):131–141

Soedirga L, Matita IC, Sidharta J (2023) Physicochemical characteristics of butterfly pea flower petals steep obtained at different steeping temperature and time. Reaktor 1:9–15. https://doi.org/10.14710/reaktor.1.1.9-15

Suksathan R, Rachkeeree A, Puangpradab R, Kantadoung K, Sommano SR (2021) Phytochemical and nutritional compositions and antioxidants properties of wild edible flowers as sources of new tea formulations. NFS J 24:15–25. https://doi.org/10.1016/j.nfs.2021.06.001
Tungmunnithum D, Thongboonyou A, Pholboon A, Yangsabai A (2018) Flavonoids and other phenolic compounds from medicinal plants for pharmaceutical and medical aspects: an overview. Medicine 5. https://doi.org/10.3390/medicines5030093
Viapiana A, Wesolowski M (2017) The phenolic contents and antioxidant activities of infusions of Sambucus nigra L. Plant Foods Hum Nutr. https://doi.org/10.1007/s11130-016-0594-x
Yadav G, Farakte R, Patwardhan A (2018) Effect of brewing temperature, tea types and particle size on infusion of tea components. Int Food Res J 25:1228–1238
Zannou O, Kelebek H, Selli S (2020) Elucidation of key odorants in beninese roselle (Hibiscus sabdariffa l.) infusions prepared by hot and cold brewing. Food Res Int. https://doi.org/10.1016/j.foodres.2020.109133

Chapter 11
Edible Flowers for Non-alcoholic Beverages

Eleomar de O. Pires, Eliana Pereira, Lillian Barros, and Cristina Caleja

Abstract Edible flowers have been present in the human diet since antiquity, being used worldwide in the preparation of popular dishes and in gourmet cuisine. Its sensorial properties characteristics, combined with its functional properties, have been highly studied, which has attracted great interest from different industries. The food industry is particularly interested in exploring the potential application of edible flowers as natural ingredients with colouring, preservative, and functional characteristics. Within this sector, the non-alcoholic beverage market is vast and diversified, and has great potential for innovation, which is why this type of materials has been largely exploited. Additionally, and in view of the new demands of current consumers in the search for more natural products, it is important to create alternative drinks that are healthier, safer, and more functional. Edible flowers are assuming a significant and prominent role in the non-alcoholic beverage sector, and there is currently some offer on the market for consumers, namely floral waters, infusions, soft drinks, fermented drinks, among others. The range of edible flowers is very wide and with immense potential for exploitation, however, it is necessary to study in depth the characteristics of less known flower species in order to innovate in this food sector. In this sense, this chapter aims to describe the current aspects of edible flowers intended for the formulation of non-alcoholic beverages, in order to highlight the main discoveries and current challenges.

Keywords Beverage formulations · Consumer demands · Comestible flowers · Food innovation · Health benefits

---

E. de O. Pires
CIMO, LA SusTEC, Instituto Politécnico de Bragança, Campus de Santa Apolónia, Bragança, Portugal

Nutrition and Bromatology Group, Department of Food and Analytical Chemistry, Faculty of Food Science and Technology, University of Vigo, Ourense, Spain

E. Pereira · L. Barros · C. Caleja (✉)
CIMO, LA SusTEC, Instituto Politécnico de Bragança, Campus de Santa Apolónia, Bragança, Portugal
e-mail: ccaleja@ipb.pt

M. M. Jowkar et al. (eds.), *Advances in Research on Edible Flowers*, Advances in Olericulture, https://doi.org/10.1007/978-3-032-04561-4_11

Abbreviations

BCCA	Branched Chain Amino Acids
CAGR	Compound Annual Growth Rate
CBD	Cannabidiol
Co.	Company
CoQ10	Coenzyme Q10
DGAV	Direção-Geral de Alimentação e Veterinária
FAO	Food and Agriculture Organization of the United Nations
GABA	Gamma-AminoButyric Acid
Ltd.	Limited
SDGs	Sustainable Development Goals
UN	United Nations

11.1 Introduction

The flowers are complex plant structures that perform specific functions in terms of pollination and seed fertilization and are responsible for the sexual reproduction of angiosperms (Purohit et al. 2021). In general, the main constituents of a flower include its receptacles, the organ responsible for reproduction, where the stamens (masculine structures) and pistils (female structures) are found, while the petals and sepals are the external components responsible for protecting the reproductive organs and for attracting pollinators, due to their remarkable colours, characteristic scents, and diverse textures, and also (Benvenuti and Mazzoncini 2021).

These plant organs present a large diversity in terms of shape, size, colour, texture, and aroma, and can be classified according to their symmetry, number of sepals, petals, stamens, and pistils, as well as their type of fruiting (Benvenuti and Mazzoncini 2021). In addition, flowers can also play a crucial role in nature, contributing to the survival and evolution of plants (Wessinger and Hileman 2020), and in some aspects of human health, given their potential benefits linked to active and nutraceutical compounds, produced as secondary metabolites (Benvenuti and Mazzoncini 2021; Janarny et al. 2021; Faisal et al. 2022; Pensamiento-Niño et al. 2023). The flowers also affect the human social context through cultural, traditional, religious, and landscape interventions, and are present as components in the perfume, cosmetics, pharmaceutical, and food industries (Purohit et al. 2021).

Nowadays, edible flowers are conquering space in the world's gastronomy and food industry, either as ingredients in formulations or as more natural substitutes for food additives, such as colouring, preservatives, flavouring, and bioactive agents (Pensamiento-Niño et al. 2023). This movement also extends to the beverage industry, with a growing adherence of these plant structures in various products of the segment, as in the case of floral infusions, fermented beverages, floral concentrates, and powder preparations. Thus, this chapter aims to describe the current scenario of non-alcoholic beverages with added flowers and their derivatives to reveal their challenges and opportunities.

11.2 Edible Flowers: History and Consumer Culture

According to historical reports, it is known that flowers have been used in human food preparation since antiquity, being applied as main ingredients or together in various dishes, in order to enhance the appearance, flavour, colour, aroma and preservation aspects of the food products, as well as leveraging their nutritional properties (Xiong et al. 2014; Pires et al. 2019; Takahashi et al. 2020; Amrouche et al. 2022). The use of flowers in food preparation is a factor that has been attracting many curious consumers, awakening nostalgic and heritage feelings, while also generating countless discussions, due to a lack of knowledge, resistance, cultural conflict and neophobia (Alves et al. 2021; Pieterse et al. 2022). Otherwise, the branch of edible flowers is promising due to its enormous potential for innovation, with respect to formulations of foods and beverages, a trend that meets the current needs of consumers, due to their frequent demands for healthier, safer, and more environmentally friendly products (Zhao et al. 2019; Dujmović et al. 2022). According to recent literature, in addition to meeting the needs of human consumption, edible flowers are also considered potent foods for hunger relief, an aspect that corroborates with the Sustainable Development Goals (SDGs) of the United Nations (UN) (Syahadat et al. 2022).

In current terms, we can say that edible flowers are increasingly present in several countries around the world, and that they are becoming popular, day after day, in several places, either by encouraging their production, by spreading their culture of consumption, by being associated with several health benefits and their market expansion (Takahashi et al. 2020; Guiné et al. 2021; Faisal et al. 2022). Some examples of local applications that already exist today include the use of chrysanthemum and jasmine flowers in teas and desserts in the Chinese and Japanese cuisine (Dini 2019; Pérez-Gregorio 2021; Gayathry and John 2021), as well as chamomile, calendula in the western Himalayan region (Hegde et al. n.d.). In Europe, this practice dates back to ancient Greek and Roman times, when flowers such as roses, violets and lavender were used in sweets and desserts, as well as in soups and stews (dos Santos and Reis 2021).

Scientific publications also confirm the contribution of edible flowers as a component of the diet in Mediterranean regions, and this region is one of the most concerned about the subject in current times (Syahadat et al. 2022; Amrouche et al. 2022). In the Americas, pre-Columbian civilizations such as the Aztecs and Mayans already made use of flowers in their cooking, and the pumpkin flower is still used in Mexican cuisine (Figueredo-Urbina et al. 2022; Mulík et al. 2022), as well as in Brazil, flowers, and inflorescences such as artichoke, broccoli, cauliflower, were introduced by European and Asian immigrants (dos Santos and Reis 2021). While in African cuisine, flowers are commonly used for medicinal purposes, ritualistic, and in the preparation of refreshing drinks, as in the case of wild mallow, jasmine with many petals (Fernandes et al. 2019; Pieterse et al. 2022).

However, although the edible flower industry is constantly evolving, many challenges still have to be overcome given the lack of knowledge about this class of food, which is often considered a "new vegetable" in comparison to more popular and conventional food plant matrixes such as fruits and vegetables, thus triggering a scenario that is often neophobic (Ghosh and Rana 2021; Alves et al. 2021; Barani et al. 2022; Pensamiento-Niño et al. 2023). Given these various challenges we can list, the need to distinguish the purpose of cultivation of these plants by the consumers, since flowers for human consumption should follow the precepts of the horticultural techniques that aim to meet the food needs and not the floriculture that is the branch intended for landscaping and ornamentation (Benvenuti and Mazzoncini 2021). The implementation of legislation or quality standards for the cultivation, handling, and storage of flowers since these are currently scarce or even non-existent in several locations (Fernandes et al. 2019). The scientific and technological development of potential additives, food ingredients and foods themselves, from edible flowers for the food and beverage industry, aiming at their food safety, preservation of their bioactive properties and their maximum utilization in terms of yield and stability (Tundis et al. 2015; Rivas-García et al. 2022).

11.3 Current Aspects of the Beverage Industry

The beverage industry is characterized by its great dimension and diversity, in which innumerable elements are produced, distributed, and commercialized daily. This branch of industry is of notable and growing relevance on a global level since the beverage market is considered one of the most promising industrial sectors in the world, being responsible for the employment of huge human capital, and for presenting a remarkable financial turnover. According to a study conducted in 2023 by Research and Markets (2024), the beverage market is expected to grow at a CAGR (Compound Annual Growth Rate) of 3.47%, with financial turnover of US$ 1568.389 billion in 2021 and estimated to reach US$ 1991.845 billion by 2028. Another great example highlights the non-alcoholic beverage sector, which has been conquering its market and consumption space since the beginning of the twenty-first century, with a market growth forecast of around US$ 1 trillion (Bhagavathula et al. 2022). Whereas according to the Grand View Research (2025), The global market for non-alcoholic ready-to-drink (RTD) beverages was estimated at USD 931.18 billion in 2023 and is projected to exhibit a compound annual growth rate (CAGR) of 6.5% during the period from 2024 to 2030.

Besides the financial and economic aspects, this industry presents a constant evolution and innovation with the development of varieties of drinks such as soft drinks, juices, sports drinks, energy drinks and, especially, functional drinks, a phenomenon that is linked to frequent behavioural changes of consumers who have numerous demands and restrictions involving healthier alternatives (Luckow and Delahunty 2004; Lee et al. 2014; Carrad et al. 2015). Thus, the beverages produced in the industry range from simple drinks such as water, to more complex ones such as vegetable fermented and high-value-added alcoholic beverages. In terms of production, the beverage segment presents diversity given the wide production of various beverages, from soft drinks, juices, and water to alcoholic beverages such as beers, wines, and spirits.

The beverage supply chain distribution system is another vital area of the industry in which beverage products reach consumers'tables through a wide range of channels, namely retailers, supermarkets, bars, and restaurants. While the sale of beverages is promoted and influenced by a wide range of factors, including availability, price, quality, brand, and the advertising system adopted by large, medium, and small producers (Chinomona and Mahlatsi 2020).

One of the most relevant aspects today is the growing demand for healthy and functional beverages, given the abundant evidence about the health benefits of current beverage formulations. Consumers are looking for products that offer health benefits such as low-sugar, gluten-free, or rich in specific nutrients. The industry is therefore investing in research and development of new formulations and ingredients to meet these demands. In addition to this scenario, aspects of concern with the environmental impact of the beverage industry are also taken into consideration by consumers when purchasing their products (Dini 2019; Pérez-Gregorio 2021; Gayathry and John 2021).

Thus, it is a fact that consumers are increasingly aware of the ecological footprint of the products they consume and expect companies to implement sustainable practices throughout the production chain. To meet this demand, companies have been working on initiatives such as using recyclable materials, reducing greenhouse gas emissions, and implementing waste management practices. In addition, the beverage industry is currently undergoing a period of intense innovation and product diversification. Companies are launching new brands and product lines to meet different consumer preferences and to keep up with market trends.

Therefore, it is possible to state that the beverage industry is crucial to the global economy, since it covers a wide variety of consumer needs and motivations, ranging from health and hydration to socialization and entertainment (Chowdhury et al. 2022). In summary, the beverage industry is constantly adapting and evolving to meet the new demands and paradigms of the market and society. Companies that can anticipate these trends and innovate in their products and practices are more likely to stand out and maintain their competitiveness in the market.

11.4 Use of Edible Flowers in Beverage Formulations

Although the inclusion of flowers in beverages is not yet as popular in the industry as the use of fruits and herbs, this practice is gaining increasing space in the competitive and innovative scenario of the beverage sector. Some flowers, such as hops for beers and common daily infusions, such as hibiscus, chamomile, and lavender, are widely used in the formulation of beverages (Zannou et al. 2020; Ellis et al. 2022; Maleki et al. 2023; González-Salitre et al. 2023; Norhisham et al. 2023). However, to date, there are few scientific studies on non-alcoholic beverage formulations containing flowers, with most of them focused on floral infusions or the

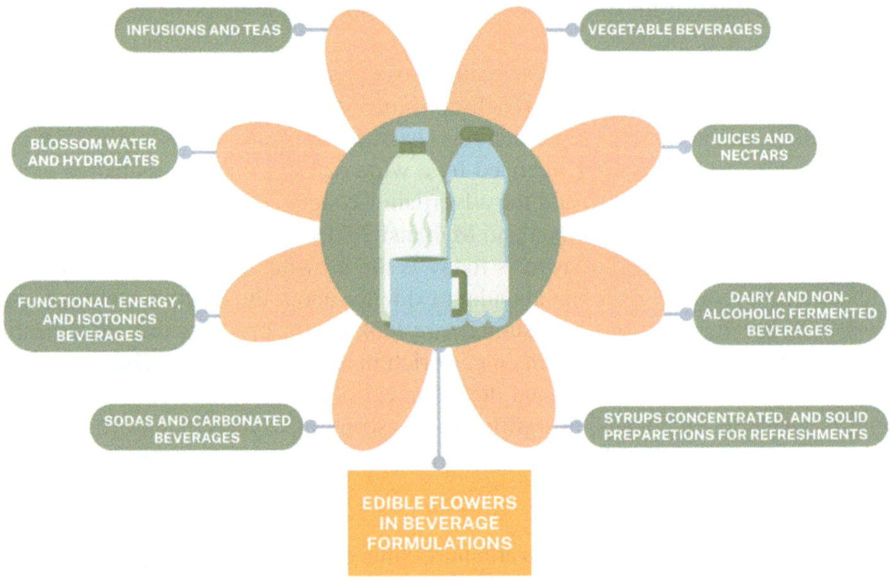

Fig. 11.1 Representative schemes for industrialized beverages consisted of edible flowers and their derivatives

development of flower-based fermented beverages. On the other hand, there are many products available in the market, both online and in physical stores, that feature flowers in their constituents in integral form, in parts (only petals) or as a percentage of extract. These products include botanical sodas, sugar syrups for drink preparation, blossom waters, and fermented beverages. Therefore, in the following topics, the industrial products found worldwide that use edible flowers or their derivatives as one of the components will be detailed (Fig. 11.1).

11.4.1 Teas and Infusions

Herbal teas and infusions are water-based beverages, consumed hot or cold, prepared through the process of infusing the plant material in contact with water in order to extract the phytochemical constituents (Bellachioma et al. 2022). By definition, tea is the beverage produced from the *Camellia sinensis* (L.) O. Kuntz plant and is currently one of the most widely consumed beverages worldwide (Wong et al. 2022). Additionally, infusions are prepared from various plant parts, such as roots, leaves, seeds, and especially flowers (Finimundy et al. 2020; Liu et al. 2023). These drinks can be found in supermarkets and tea shops and are also easily prepared at home with fresh or dried plants. Several scientific research have been conducted to investigate the actual effects of these drinks on the body, developed into interesting findings (Farzaneh and Carvalho 2015; Bellachioma et al. 2022; Casado et al. 2022; Primitivo et al. 2022; Monari et al. 2022; Wirz et al. 2022; Mahadi et al. 2022; Prihatningtias et al. 2022). These infusions are made from a wide variety of flowers, some of the most popular flowers being *Rosa* (Jiménez-López et al. 2017), *Lavandula* (Dobros et al. 2022; Batiha et al. 2023), *Matricaria* and *Chamaemelum* (Maleki et al. 2023), *Hibiscus* (Zannou et al. 2020; Choi et al. 2023; Norhisham et al. 2023), and *Jasminum* (Ferreres et al. 2014; An et al. 2023a) (Fig. 11.2). In addition to being refreshing and tasty, flower infusions can provide health benefits, functioned as therapeutic vehicles through the delivery of water-soluble phytochemicals (Bellachioma et al. 2022).

For example, chamomile is known for its calming properties and can help improve sleep quality (Adib-Hajbaghery and Mousavi 2017). Studies have shown that chamomile contains compounds with both relaxing and sedative properties, such as apigenin, which acts on the brain's GABA receptors, resulting in calming effects (Savage et al. 2018; Ebrahimi et al. 2022). *Hibiscus* contains anthocyanins and other flavonoids that act as antioxidants and help relax blood vessels, improving blood circulation and reducing blood pressure (Jalalyazdi et al. 2019; Ellis et al. 2022).

Lavender tea is also known for its calming and relaxing properties, as well as having positive effects on the digestive system (Aydinli and Karadağ 2023). Lavender contains compounds such as linalool, which act as natural anxiolytic, sedative, and analgesic (Hashimoto et al. 2023). Another hot flower-based beverage that has attracted attention is rose tea. Roses contain compounds such as flavonoids, carotenoids, fatty acids, and vitamins, which act as antioxidants and have positive effects on the immune system (Igual et al. 2022; Zhou et al. 2023). In addition, rose tea has anti-inflammatory and analgesic properties and can help relieve headaches and menstrual cramps (Singh et al. 2023).

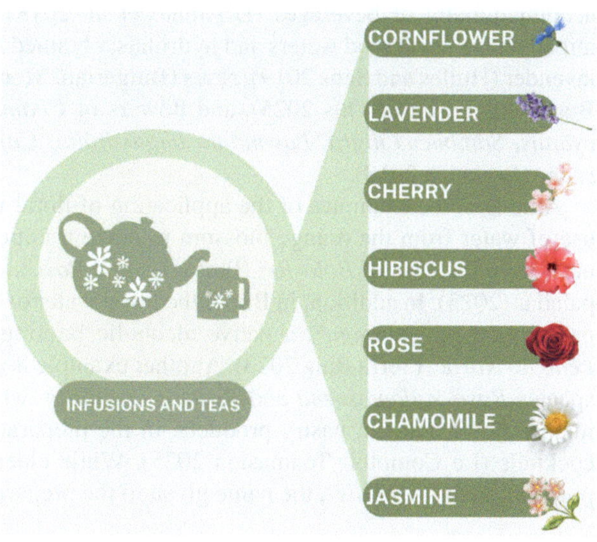

Fig. 11.2 Most popular edible flowers for infusion preparation

In summary, flower-based hot and cold drinks can offer health benefits such as calming, antioxidant, anti-inflammatory, and analgesic properties. However, it is important to remember that these beverages should not be used as single treatments for health problems, and it is always recommended to consult a health professional before starting any kind of regular consumption of teas and infusions. However, including flower infusions in the diet can be a natural and tasty way to increase antioxidant intake, promote heart health, improve sleep quality, and reduce stress and anxiety levels. Currently, edible flowers can be found together with herbal tea and infusions, either through a mix for preparing hot or cold drinks, or even products ready for immediate injection. Thus, Table 11.1 refers to commercial products and scientific research regarding infusions and teas from edible flowers.

11.4.2 Blossom Water, Hydrolats and Beverages with Essential Oils

The floral waters, or hydrolats, are obtained by the process of hydro distillation of the integral parts or petals of flowers, considered to be by-products of essential oils, such as, for example, in the process of obtaining the essential oils of chamomile, rose, orange and jasmine Fig. 11.3, they are mainly composed of water and present 1% of hydrophilic substances in their constitution (D'Amato et al. 2018; Popa et al. 2021). They are popularly used for medicinal purposes and in the manufacture of cosmetics and flavouring (Rodrigues et al. 2022), as an antimicrobial agents (Tornuk et al. 2011), but also in the preparation of confectionery dishes, sauces, desserts, drinks or cocktails (Soignons Nos Assi-ettes 2025), as well as in the preparation of fish, vegetables and salads (Aroma-zone.com 2022), as in the case of the hydrosol of the flowers of *Citrus medica* L. (Brigand and Nahon 2016), and also in the accompaniment of beverages (D'Amato et al. 2018). At an industrial level, we already have floral food waters and hydrolats, obtained from romain chamomile and lavender (Huiles and Sens 2019), roses (Bulgarian, Moroccan and Turkish) (Miracle Botanicals Essential Oils 2025), and flowers of *Chamaemelum nobile*, *Centaurea cyanus*, *Sambucus nigra*, *Lavandula angustifólia*, *Citrus aurantium*, *Rosa damascene* (Botanica 2018).

Among some examples of the application of floral waters, we can highlight the use of water from the orange blossom to make traditional European cuisine products, as in the case of *Bolo-Rei* (Portugal) and *Roscon de Reyes* (Spain) Directo al paladar (2023). In addition, in Brazil the floral water of orange flowers is used in the preparation of "*porongo*", a native alcoholic beverage prepared with "*Jurubeba Leão do Norte*" (Terra Raiz 2023). Another example is rose water obtained from the species *Rosa* × *damascena* and *Rosa* × *centifolia,* which is capable of imparting intense floral notes in pastry products, in the preparation of sorbet, yoghurts and cocktails (Le Comptoir Toamasina 2025). While elderflower water is used in the preparation of "*cordiais*", the name given to the preparation of fruit infusions in the

Table 11.1 Commercial products and scientific research on edible flower infusions

Factory	Product	Constituent	Origin	Reference
Japan GreenTea Co., Ltd. ®	Green tea	Cherry blossom (*Prunus* L.)	Japan	Japan GreenTea (2022)
The Tao of Tea	Infusion powder	Flowers of the tea plant (*Camellia sinensis* L.)	United States of America	The Tao of Tea (2025)
Vanalaya	Infusion mix	Lavender (*Lavandula* L.) Chamomile (*Matricaria* L.) Rose (*Rosa* L.) Hibiscus (*Hibiscus* L.) Butterfly bean (*Clitoria ternatea* L.)	India	Vital Global Ray Brands (2025)
YOGI TEA®	Infusion mix	Hibiscus (*Hibiscus* L.) Chamomile (*Matricaria* L.) Elderflower (*Sambucus nigra* L.) Linden (*Tilia* L.) Rose (*Rosa* L.) Lavender (*Lavandula* L.) Sunflower (*Helianthus* L.)	United States of America	Yogi (2023)
Blossoms	Infusion powder	Blossoms Lavender (*Lavandula* L.)	United States of America	Java Frost (2024)
Scientific research	Floral herbal infusions	Calendula (*Calendula* L.) Chamomile (*Matricaria* L.) Chrysanthemum (*Chrysanthemum* L.) Cornflower (*Centaurea cyanus* L.) Jasmine (*Jasminum* L.) Lavender (*Lavandula* L.) Neroli (*Citrus* L.) Rose (*Rosa* L.) Sweet Osmanthus (*O. fragrans* Lour.)	Taiwan	Ho et al. (2013)

Continued

Table 11.1 (continued)

Factory	Product	Constituent	Origin	Reference
	Flower infusion	Coffee flower (*Coffea* L.)	Germany	Wirz et al. (2022)
	Infusion and decoctions	Acacia (*Acacia caven* Molina) *Geoffroea decorticans* Burkart *Larrea divaricata* Cav.	Argentina	Isla et al. (2021)
	Flower teas	*Chrysanthemum* L.	Korea	Han et al. (2019)
	Infusions	*Sambucus nigra* L.	Poland	Viapiana and Wesolowski (2017)
	Infusions	Mallow (*Malva* Tourn. Ex L.) Calendula (*Calendula* L.) Hibiscus (*Hibiscus* L.)	Spain	Casado et al. (2022)
	Flower tea	*Xanthoceras sorbifolia* (*X. sorbifolium* Bunge)	China	Xiao et al. (2022)
	Flower tea	*Camellia sinensis* L.	China	Chen et al. (2020)
	Flower tea	Kunlun Chrysanthemum (*Coreopsis tinctoria* Nutt.)	China	Yu et al. (2021)
	Flower tea	*Eucommia ulmoides* Oliver	China	Cheng et al. (2022)
	Flower tea	*Coreopsis tinctoria* Nutt.	China	Jiang et al. (2015)
	Aqueous extract	*Althaea officinalis* L.	Lebanon	Farhat et al. (2022)
	Flower tea	*Syringa vulgaris* L.	Turkey	Yiğit et al. (2022)
	Tea infusions	Chrysanthemum (*Chrysanthemum* L.)	China	Peng et al. (2022)
	Flower teas	*Opuntia ficus-indica* L.	Algeria, Spain and Portugal	Amrane-Abider et al. (2022)

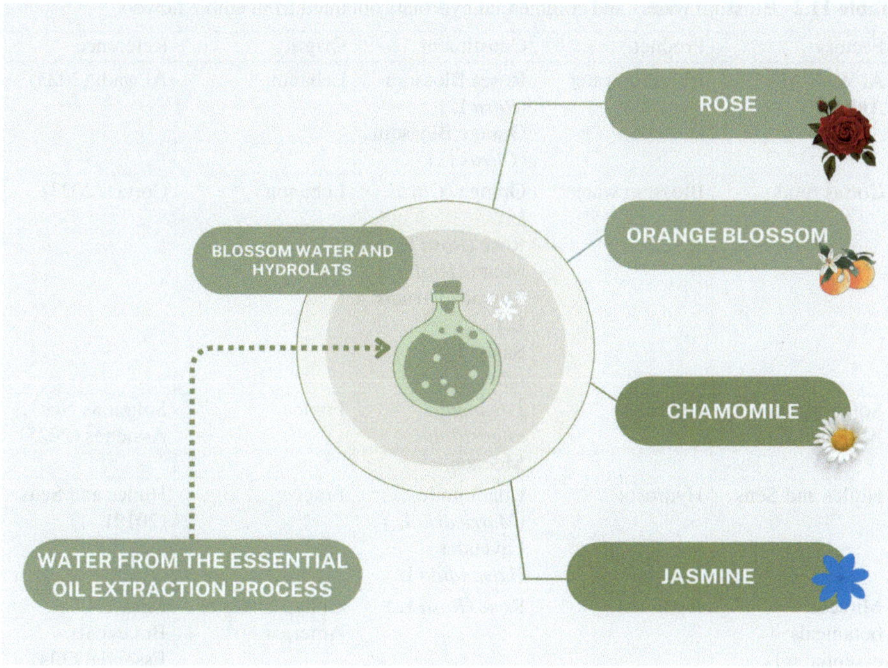

Fig. 11.3 Blossom waters from the essential oil extraction process

United Kingdom and Australia (Terri Salminen 2020). Table 11.2 reports some examples of floral waters and hydrosols that are currently being marketed.

The essential oils are used in the beverage industry because of their flavour, fragrance and antimicrobial, antioxidant and anti-infective properties (Preedy 2015; Adetuyi et al. 2024). Essential oils from flowers and plants, such as orange, lemon, lavender, rosemary and others, play a significant role in the food and beverage industry, with citrus fruits being the most commonly used in the production of soft drinks (Ameh and Obodozie-Ofoegbu 2016). Among the compounds present in these essential oils are limonene, citral, menthol and linalool, which are widely applied (Preedy 2015; Maleš et al. 2022). However, high concentrations of essential oils can cause negative organoleptic effects and limit their use due to instability, strong odor, potential toxicity and interaction with the food matrix.

To overcome these challenges, stabilization techniques such as encapsulation and emulsion have been employed to mitigate these limitations. These techniques are capable of increasing the feasibility of using essential oils, as well as improving the solubility, stability and efficacy of the oils, in addition to reducing their sensory impact in beverage formulations (Adetuyi et al. 2024). In addition to sensory aspects, essential oils can be used as food preservatives, integrated into barrier systems to protect and extend the shelf life of food (Chen et al. 2021). However, there are still challenges, such as determining the exact proportions given the complexity of the terpenes and terpenoids that make up these oils.

Table 11.2 Blossom waters and commercial hydrolats obtained from edible flowers

Factory	Product	Constituent	Origin	Reference
Al Wadi Al Akhdar	Blossom water	Roses Blossom (*Rosa* L.) Orange Blossom (*Citrus* L.)	Lebanon	Al wadi (2023)
Cortas food	Blossom water	Orange (*Citrus* L.) Rose (*Rosa* L.) Mint (*Mentha* L.) Thyme (*Thymus* L.) Sage (*Salvia officinalis* L.)	Lebanon	Cortas (2023)
Soignons nos Assiettes	Hydrolate	*Lavandula angustifolia* Moench.	France	Soignons Nos Assiettes (2025)
Huiles and Sens	Hydrosol	Chamomile (*Matricaria* L.) Lavender (*Lavandula* L.)	France	Huiles and Sens (2019)
Miracle botanicals essential oils	Hydrosol	Rose (*Rosa* L.)	United States of America	Miracle Botanicals Essential Oils (2025)
ProTec Botanica	Hydrolats	Chamomile (*Matricaria* L.) Cornflower (*Centaurea cyanus* L.) Elderflower (*Sambucus nigra* L.) Lavender (*Lavandula* L.) Orange Blossom (*Citrus* L.) Rose Blossom (*Rosa* L.)	United Kingdom	Botanica (2018)
Le Comptoir de Tomasina	Blossom water	Rose (*Rosa* L.)	France	Le Comptoir Toamasina (2025)

11.4.3 Functional, Energy, and Isotonic Beverages

Isotonic, energy and functional drinks serve different purposes and are formulated to meet specific body needs (Naumovski et al. 2023) (Fig. 11.4). Isotonic drinks are drinks formulated to provide extra hydration and nutrients for people who engage in intense or prolonged physical activity, such as long-distance running, endurance training, and team sports, these solutions can attenuate and prevent disturbances in

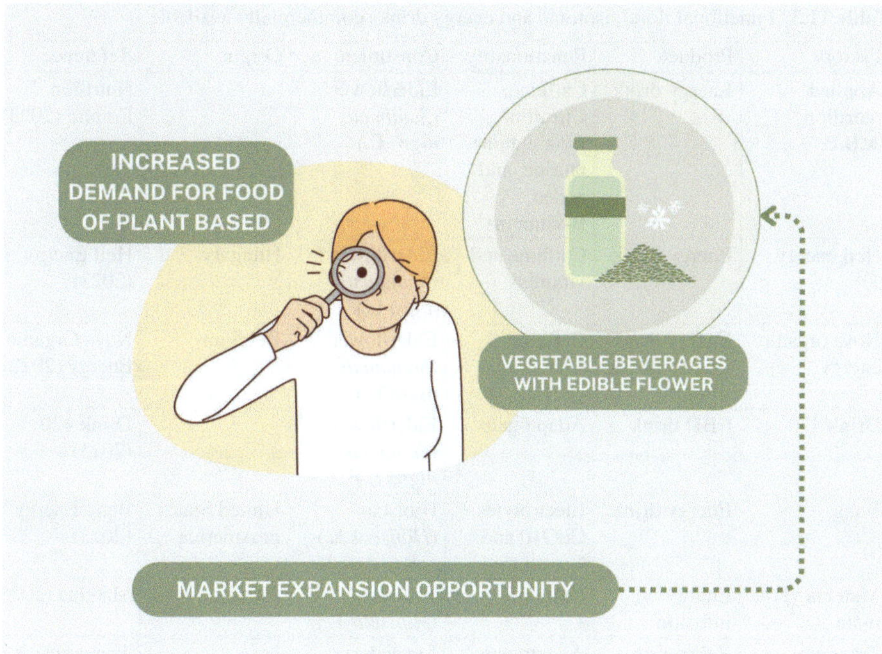

Fig. 11.4 Edible flowers as ingredients for functional, energy and isotonic drinks

the human body related to metabolism, cardiovascular aspects, thermoregulators and sports performance (Moreno et al. 2013).

These beverages usually contain water, electrolytes, carbohydrates, and sometimes vitamins and minerals (Fernandes et al. 2022). Electrolytes, such as sodium, potassium, and magnesium, help to maintain fluid balance in the body and to replenish electrolytes lost during intense exercise and sweating (Suh et al. 2023). Carbohydrates, which are an important source of energy for the body, are added to the drink to help maintain blood sugar levels and provide additional energy for the body during exercise (Naumovski et al. 2023). Energy drinks contain caffeine, taurine, B vitamins, and other ingredients that provide energy to the body (Jagim et al. 2023; Naumovski et al. 2023). These drinks are used to increase energy, improve performance and physical endurance, and reduce fatigue during intense activities, such as exercise and long-duration activities (Gutiérrez-Hellín and Varillas-Delgado 2021). While functional beverages are formulated to provide specific health benefits such as boosting the immune system, improving digestion, reducing stress, and increasing mental capacity (Dini 2019; Maleš et al. 2022). These drinks may contain ingredients such as probiotics, vitamins, minerals, herbs, and other nutrients (Corbo et al. 2014). Given the lack of scientific literature on functional, energy and isotonic drinks, Table 11.3 depicts the industrial formulations of this class of drinks with the addition of flowers and floral derivatives as food ingredients.

Table 11.3 Functional floral, isotonic and energy drinks commercially available

Factory	Product	Functionality	Constituent	Origin	Reference
Applied nutrition A.B.E.	Energy drink	Caffeine, Citrulline, Beta-alanine, taurine, and added B-vitamins	Elderflower (*Sambucus nigra* L.)		Nutrition Empire (2023)
Hell energy	Energy drink	Caffeine and vitamins	Elderflower (*Sambucus nigra* L.)	Hungary	Hell Energy (2023)
Nova organic energy	Energy drink	Caffeine	Elderflower (*Sambucus nigra* L.)	Belgium	Nova Organic Energy (2022)
Drink 420	CBD drink	Adaptogens	Elderflower (*Sambucus nigra* L.)		Drink 420 (2023)
Bang	Energy drink	Electrolytes, CoQ10 and Creatine	Hibiscus (*Hibiscus* L.)	United States of America	Bang Energy (2023)
Mateina yerba mate	Energy infusion	Caffeine	Hibiscus (*Hibiscus* L.)	Canada	Mateina (2025)
Tranquini	Relaxation drinks	Adaptogens	Lavender (*Lavandula* L.) Chamomile (*Matricaria* L.)		Tranquini (2025)
GIST	Sparkling water	Prebiotic	Hibiscus (*Hibiscus* L.) Calendula (*Calendula* L.) Chamomile (*Matricaria* L.)		Gist (2023)
Kin Euphorics	Suplemento dietético	Adaptogens and nootropics	Lavender (*Lavandula* L.) Vanilla (*Vanilla* Plum. ex Mill.) Passionflower (*Passiflora* L.)	United States of America	Kin euphorics (2023)

11.4.4 Soft Drinks, Sodas, and Carbonated Beverages

Soft drinks and carbonated beverages are types of carbonated beverages that contain dissolved carbon dioxide, resulting in an effervescent effect when they are opened or consumed (Abu-Reidah 2020; Tanguler and Sener 2022). In addition to soft drinks, which typically contain sugar, sweeteners, artificial colours and flavours, carbonated beverages include sparkling water, tonic water, sparkling mineral water and other flavoured beverages, which are often used as mixers in cocktails.

Currently, an excessive consumption of soft and sugary drinks is evident, which correlates with a significant increase in the risk of developing non-communicable chronic diseases (Kvalvik et al. 2022; Barajas-Torres et al. 2022; Saiki et al. 2022). These beverages are characterized by empty calories, which contribute to weight gain and obesity, conditions directly associated with problems such as diabetes, cardiovascular disease, and other illnesses (An et al. 2023b). In addition, soft drinks often have high levels of refined sugar, high fructose corn syrup, and artificial additives, which can compromise metabolic health and increase insulin resistance.

In light of this, the scientific community, governments of various nations, and health and food safety authorities are joining forces to investigate and mitigate the impacts of rampant consumption of these beverages (Beck et al. 2020; Anastasiou et al. 2023; Yan et al. 2023; Zhang et al. 2023; Inchingolo et al. 2023). Measures such as the implementation of taxes on these products and the control of marketing campaigns aimed at children and adolescents are being adopted (Wilking et al. 2023; Yan et al. 2023; Elliott et al. 2023). In fact, reducing consumption or choosing healthier beverage alternatives, such as organic and botanical soft drinks with low sugar and calorie content, along with a healthy diet and exercise routine, play an important role in contributing to and preventing chronic non-communicable diseases. From this perspective, edible flowers and their derivatives offer an answer to society's current demands, as they represent more natural alternatives to synthetic food additives, contributing to nutritional aspects and functionalities of these foods. However, to date, studies exploring the relationship between carbonated beverages and edible flowers are scarce, although there are already several products on the market that incorporate these flowers and their extracts, as evidenced in Table 11.4.

11.4.5 Vegetable Beverages

Plant-based beverages are beverages made from plant sources such as grains, nuts, seeds, and legumes (Boas et al. 2023). They are a popular alternative to cow's milk and can be used as a substitute in recipes, coffee, tea or consumed directly (Shkembi and Huppertz 2023). Plant-based beverages can be made from a variety of ingredients, such as soy, almonds, oats, rice, coconut, tree nuts, peanuts, among others (Drewnowski et al. 2021; Bocker and Silva 2022). These beverages are usually made by grinding or mixing the ingredients with water, and in some cases additional sweeteners and flavours are added. Plant-based drinks are popular and are gaining strength day by day among people who follow a vegan or vegetarian diet, or who have an intolerance or allergy to cow milk (Froiio et al. 2020; Drewnowski et al. 2021; Gustavo et al. 2023; da Silva et al. 2023; Slade 2023). They are also often consumed by people who is looking for healthier options or who want to reduce the consumption of animal-based food products (Pascual et al. 2018; Aydar et al. 2023) (Fig. 11.5).

Table 11.4 Carbonated beverages made from edible flowers

Factory	Product	Constituent	Origin	Reference
Belvoir farm	Organic cordial	Elderflower (*Sambucus nigra* L.)	United Kingdom	Belvoir Farm (2025)
Fentimans ltd	Botanical soft drinks	Elderflowers (*Sambucus nigra* L.)	United Kingdom	Fentimans (2025)
Drink a flower	Alcohol-free organic sparkling	Hibiscus (*Hibiscus* L.) and peppermint Black tea, ginger and chamomile (*Matricaria* L.) Elderberry (*Sambucus nigra* L.) and lemon	Belgium	Fruzzion (2022)
The London essence CO.	Craft soda	Jasmine (*Jasminum* L.) Elderflower (*Sambucus nigra* L.)		Wisehead Producitons (2025).
ST. STEVE'S	Craft soda	Chamomile flowers (*Matricaria* L.)		Mud Lake Farm, LLC (2020)
Iperos herbs	Carbonated soft drink	Elderflower (*Sambucus nigra* L.)	Greece	Iperos Herbs (2025)
Fever-tree	Floral soda	Orange Blossom (*Citrus* L.)	United Kingdom	Fever-Tree (2025)
Kimura Sakura cola	Cola soft drink	Cherry blossom (*Prunus* L.)	Japan	NikanKitchen (2023)
Thomas Henry cherry blossom tonic	Tonic	Cherry blossom (*Prunus* L.)	Germany	Bondston (2025)
Artisan drink company ltd.	Tonic	Eleder flower (*Sambucus nigra* L.) blossom Violet blossom (*Viola* L.)	United Kingdom	Artisan Drink Company Ltd (2025)
Petal	Botanical beverage	Rose (*Rosa* L.) Elderflower (*Sambucus nigra* L.) Marigold (*Calendula officinalis* L.) Dandelion (*Taraxacum* F.H.Wigg.)	United States of America	Petal Sparkling Botanicals (2025)

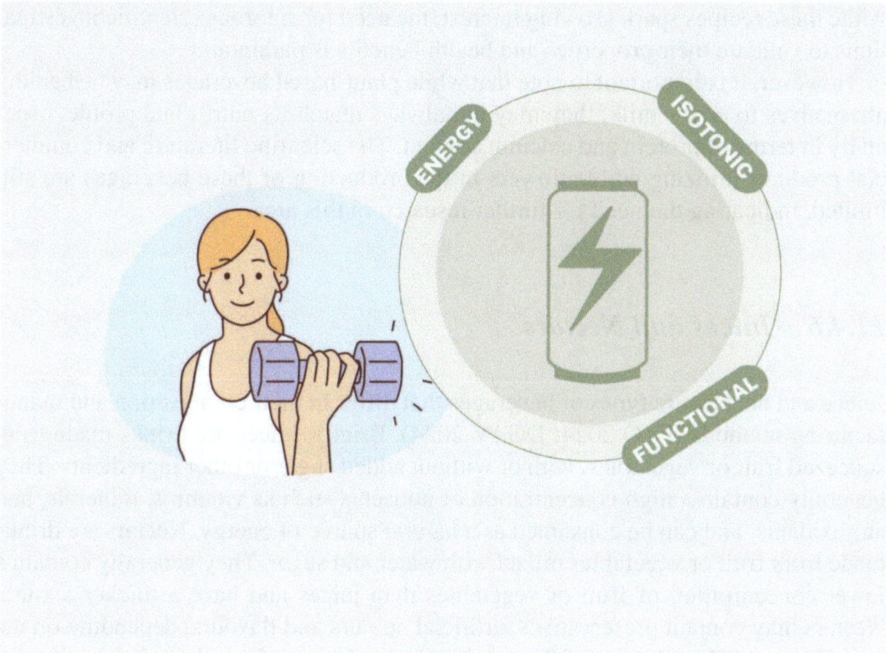

Fig. 11.5 Increased consumer demand for plant-based beverage

Incorporating edible flowers into plant-based beverage formulations represents a promising opportunity for innovation in the food industry. Flowers such as lavender, rose, and hibiscus offer unique sensory experiences and potential nutritional benefits (de Pires et al. 2023). For example, lavender, known for its relaxing properties, adds a subtle floral note to almond milk, enhancing its flavour profile. Similarly, rose petals contribute delicate aroma and distinct flavour to oat milk, potentially providing antioxidants and other bioactive compounds. Companies like "beber farm to bottle," located in northern California, USA, producing "Organic Almond-milk" in lavender and vanilla flavours (California 2025) and the Australian company Almo, with its "Chamomile Almond Milk" (Doorstep Organics 2021), are commercial examples embracing this trend.

The growing popularity of plant-based beverages and edible flower infusions is evident on various online platforms dedicated to cooking and health. Illustrated by websites like Elise Tries to Cook (Elise Tries to Cook 2023), Milky Plant (2023), Choosing Chia (2019), Plant Based Jess (2023), and Naturally Healthy Desserts (2017), these sources offer a diverse range of recipes, from chamomile and rose lattes to almond milk with lavender and hibiscus teas. Despite the abundance of information on the supposed nutritional and medicinal benefits of these beverages, the scarcity of scientific evidence supporting such claims is notable. The absence of systematic studies and controlled clinical trials limits the ability to accurately determine the effectiveness and potential health impacts of these preparations. Therefore, while these recipes spark growing interest, the need for rigorous scientific investigations to validate their properties and health benefits is paramount.

However, it is important to note that while plant-based beverages may be healthy alternatives to cow's milk, they may not always match its nutritional profile, especially in terms of protein and calcium content. The scientific literature and commercial products utilizing edible flowers in the production of these beverages are still limited, indicating the need for further research in this area.

11.4.6 Juices and Nectars

Juices and nectars are types of beverages that differ in their composition and manufacturing methods (FAO 2024; DGAV 2024). Briefly, juices are drinks made from squeezed fruit or vegetables, with or without added sugar or other ingredients. They generally contain a high concentration of nutrients such as vitamins, minerals, and antioxidants, and can be consumed as a natural source of energy. Nectars are drinks made from fruit or vegetables mixed with water and sugar. They generally contain a lower concentration of fruit or vegetables than juices and have a thicker texture. Nectars may contain preservatives, artificial colours and flavours, depending on the manufacturer. The addition of flowers in juice and nectar formulations is not a very widespread subject both in the scientific and industrial environment, given the scarce literature on the subject. In this way, The Table 11.5 reports some examples of available juice formulations with edible flowers.

11.4.7 Dairy and Non-alcoholic Fermented Beverages

Industrial non-alcoholic fermented beverages are drinks that have undergone a fermentation process but have little or no alcohol content. These drinks are often produced on an industrial scale and can be found in supermarkets and convenience stores. Fermented non-alcoholic beverages have been enjoyed by many cultures around the world for centuries and are known for their health benefits, particularly for the digestive system, as they are rich in probiotics and other beneficial bacteria (Marsh et al. 2014; Valero-Cases et al. 2020).

Kefir is a fermented drink made from milk or sweetened water (Guzel-Seydim et al. 2011, 2021; Azizi et al. 2021). It contains a variety of bacteria and yeasts that aid digestion and immunity and is rich in vitamins and minerals. Milk kefir has a consistency similar to yogurt, while water kefir is more like a carbonated drink. Kombucha is a fermented beverage made from sweetened tea and a culture of bacteria and yeast (Dufresne and Farnworth 2000; Villarreal-Soto et al. 2018; Laureys et al. 2020; Bortolomedi et al. 2022). It is known for its detoxifying and antioxidant properties, as well as its ability to improve digestion and the immune system (Dufresne and Farnworth 2000; Bortolomedi et al. 2022). Kombucha has a slightly

Table 11.5 Commercial juices and nectars formulations with edible flowers

Factory	Product	Constituent	Origin	Reference
Rabenhorst ®	Multifruit juice	Rose (*Rosa* L.)	Germany	Rabenhorst (2025)
TM CARPATY ENJOY	Flower drink	Acacia (*Acacia* Mill.) Elderflower (*Sambucus nigra* L.)	Ukraine	Carpaty Enjoy (2016)
Anni's vital shop	Juice	Elder blossom (*Sambucus nigra* L.)		Anni's (2020)
Naturae	Juice	Aloe Vera (*Aloe vera* L.) Chomomile (*Matricaria* L.)	Spain	Naturae (2025)
Vitamont	Smoothie detox	Hibiscus (*Hibiscus* L.)		Naturitas (2023)

sour taste and is typically consumed chilled (Nyiew et al. 2022). Yogurt and fermented milk drinks are fermented dairy-based foods and are rich in protein, calcium, and probiotics (Meydani and Ha 2000; Cheng 2010; Gahruie et al. 2015; Rashwan et al. 2023). They are versatile foods that can be eaten alone, mixed with fruits and cereals, or used in cooking recipes. There are many varieties of yogurt and dairy drinks available on the market, in natural, flavoured, and sweetened versions. These drinks are popular with people seeking healthier options because they contain probiotics, which are beneficial bacteria for the health of the digestive tract. Some of these drinks may also have anti-inflammatory and antioxidant properties, as well as aid digestion and immunity.

In fact, fermented beverages, as well as floral infusions, are the two classes of beverages with the most scientific literature present up to the present day, a fact that is reflected in the industry and in the market, given the immensity of products that can be found in stores. Physical and online on the internet. In the field of non-alcoholic fermented beverages, there has been a growing popularity of kombucha, a fact that can be demonstrated in Table 11.6.

11.4.8 Syrups, Concentrated Beverages, and Solid Preparations for Refreshments

Drink syrups, also known as cocktail syrups, are sweetened liquids used to impart flavour and texture to cocktails. They are made by combining sugar with other ingredients to create custom flavourings, such as fruit, herbs, spices, or flowers (Table 11.7). These syrups are used to sweeten and add flavour to cocktails and can be used as a main ingredient in drinks. In addition, syrups can be used to balance strong flavours, such as bitterness or acidity. Cocktail syrups can also be found in a

Table 11.6 Commercial dairy and non-alcoholic fermented beverages

Factory	Product	Constituent	Origin	Reference
B.effect	Kombucha	Jasmine Flower (*Jasminum* L.)	New Zealand	B.effect (2025)
One breath	Kombucha	Jasmine (*Jasminum* L.) Elderflower (*Sambucus nigra* L.)	United States of America	One-Breath-Beverage (2025)
Brew Dr. Kombucha	Kombucha	Hibiscus (*Hibiscus* L.)	United States of America	Brew Dr. Kombucha (2023)
Equinox	Kombucha	Elderflowers (*Sambucus nigra* L.)	United Kingdom	Abel and Cole (2023)
GT'S living foods	Kombucha	Lavender (*Lavandula* L.)	United States of America	GT's Living Foods (2025)
KTonic Kombucha	Kombucha	Chrysanthemum (*Chrysanthemum* L.) Hibiscus (*Hibiscus* L.)		KTonic Kombucha (2025)
Rise	Kombucha	Jasmine flower (*Jasminum* L.) Rose flower (*Rosa* L.)	Canada	RISE Kombucha Inc. (2025)
SantiveriBilbao	Kombucha	Hibiscus (*Hibiscus* L.)		Santiveri Bilbao (2021)
Seek north	Kombucha	Elderflower (*Sambucus nigra* L.)	United States of America	Seek North (2024)
PiQi	Kefir water	Butterfly pea (*Clitorea ternatea* L.)		PiQi Life (2023)
Prokey	Water kefir beverage	Hibiscus (*Hibiscus* L.)	Spain	Natura Center (2023)
Mlekarna Krepko	Milk kefir	Rose (*Rosa* L.)	Slovenia	Krepko (2023)
Botanitea	Kefir water	Apple Flower (*Malus* Mill.)	Singapore	Bontanitea (2025)
Scientific Researchs	Water kefir	*Clitorea ternatea* L.	Indonesia	Setiawati and Kusnadi (2021)
	Water kefir	*Hibiscus sabdariffa* L.	Indonesia	Hastuti (2015)
	Kefir	*Hibiscus sabdariffa* L.	Indonesia	Afifah (2019)
	Kefir	*Lavandula angustifolia* Mill.	Romania	Tiţa et al. (2022)
	Yogurt	*Hibiscus sabdariffa* L.	Brazil	de Moura et al. (2022)
	Kombucha	*Hibiscus sabdariffa* L	Brazil	Januário et al. (2020)
	Kombucha	*Clitorea ternatea* L.	Indonesia	Biotek et al. (2021)
	Kombucha	*Clitorea ternatea* L.	Indonesia	Pamungkas et al. (2022)
	Yogurt	*Hibiscus sabdariffa* L	Brazil	Arévalo-Pinedo et al. (2022)
	Yogurt	*Hibiscus sabdariffa* L	Brazil	Stella et al. (2023)

variety of flavours, from simple syrup, which is just sugar and water, to complex syrups that include flavours such as hibiscus, lavender, ginger, and more. They can be made at home or purchased at specialty beverage stores.

Flower concentrates are concentrated liquids obtained from flowers, usually by infusion or distillation. They are widely used in the culinary and beverage industry to add flavour and aroma to recipes. There are many different types of flowers concentrates available, including concentrates of rose, lavender, jasmine, hibiscus, and many others. Each type has its own distinctive flavour and aroma. Flower concentrates can be used in a wide variety of recipes, from desserts and confectionery to cocktails and soft drinks. They are also commonly used to flavour teas and infusions and can be added to marinades and sauces to add flavour and aroma. Powder drinks are powdered products that, when mixed with water or another liquid, create an instant drink. They are usually found in individual packets or large cans and can be used to make a variety of beverages such as iced teas, juices, soft drinks, sports, and energy drinks, and more.

Some powdered drinks contain added sugar and other artificial ingredients, while others are sweetened with natural sweeteners or contain no sugar at all. Some powdered drinks are also fortified with vitamins and minerals, making them a convenient option for those who wish to supplement their diet. Powdered drinks can be a practical option for those who are on the go or who don't have time to prepare fresh drinks at home. They can also be a more economical alternative to ready-to-drink beverages.

11.5 Development of Non-alcoholic Beverages Based on Edible Flowers: The Exploration of Future Applications

Flower-based beverages are becoming increasingly popular due to their healthier and more natural option compared to industrially made drinks with artificial additives. There are several possible future applications for flower-based beverages. One of these applications is in the functional beverage industry, where there is a growing demand for drinks that offer health benefits, such as stress reduction, immune system boost, and improved digestion (Tireki 2021). Flower-based beverages can be used to create functional beverages by adding other ingredients that offer these benefits.

A promising sector is floral waters and hydrosols, which are gaining space, both in their applications for the pastry segment and in the beverage industry, and because they are considered byproducts of the essential oil industry, which are often discarded or neglected (Jakubczyk et al. 2021; Tavares et al. 2022).

An important aspect for the future is the increase in efforts regarding plant-based beverages enriched with bioactive floral extracts. This is relevant to promote functionality and increase the benefits of these drinks, due to the growing demand from

Table 11.7 Commercial beverages with edible flowers

Factory	Product	Constituent	Origin	Reference
Matcha Oishii	Powder	Sakura blossom (*Prunus* L.)	Japan	Matcha Oishii (2025)
Japan cherry blossoms	Powder	Sakura flower (*Prunus* L.)	Japan and United States of America	Cherry Blossom Food (2025)
Torani	Puremade	Elderflower (*Sambucus nigra* L.) Rose (*Rosa* L.) Hibiscus (*Hibiscus* L.)	United States of America	R. Torre and Company (2023)
Bionutricia manufacturing Sdn Bhd	Standardized extract liquid concentrate	*Clitoria ternate* L.	Malaysia	Bionutricia Manufacturing (2024)
Amoretti	Syrup	Violet Lavender (*Lavandula* L.) Rose (*Rosa* L.) Orange blossom (*Citrus* L.)	United States of America	Amoretti (2025)
Le sirop de monin	Syrup	Cherry (*Prunus* L.) Elderflower (*Sambucus nigra* L.) Hibiscus (*Hibiscus* L.) Jasmine (*Jasminum* L.) Lavander (*Lavandula* L.) Rose (*Rosa* L.) Violet (*Viola* L.)	United States of America	Monin (2025)
Royal rose	Syrup	Rose (*Rosa* L.)	United States of America	Royal Rose Syrups (2025)
Soda farm	Syrup	Passionflower (*Passiflora* L.) Rose (*Rosa* L.) Chamomile (*Matricaria* L.) Lavender (*Lavandula* L.)	United States of America	Soda Farm (2023)
Blossoms syrup LTD	Syrup	Elderflower (*Sambucus nigra* L.)	United Kingdom	Blossoms Syrups (2025)
CHINRIU	Syrup	Sakura flower (*Prunus* L.)	Japan	Chinriu (2021)

(continued)

Table 11.7 (continued)

Factory	Product	Constituent	Origin	Reference
Jack Rudy cocktail CO.	Syrup	Elderflower (*Sambucus nigra* L.)		Jack Rudy Cocktail Co. (2025)
Split tree cocktail CO.	Syrup	Elderflower (*Sambucus nigra* L.)	Canada	Split Tree Cocktail Co. (2025)
Prosyro	Syrup	Lavender (*Lavandula* L.) Rose (*Rosa* L.) Elderflower (*Sambucus nigra* L.)	Canada	Prosyro (2025)
Raft	Syrup	Lavender (*Lavandula* L.) Hibiscus (*Hibiscus* L.) Vanilla (*Vanilla* Plum. ex Mill.)	United States of America	RAFT Cocktails (2025)
Dillon's	Syrup	Lavender (*Lavandula* L.)	Canada	Dillon's Small Batch Distillers (2025)
Kvas fine beverages Co.	Syrup	Lavender (*Lavandula* L.) Jasmine (*Jasminum* L.) Hibiscus (*Hibiscus* L.)	Canada	Kvas Fine Beverage Co (2025)
Charleston, So. - Bittermilk	Syrup	Elderflower (*Sambucus nigra* L.)		Bittermilk (2025)
Sauvie shrubs	Concentrated beverages	Rose (*Rosa* L.)		Sauvie Shrubs (2020)

the vegetarians, vegan and individuals with dietary restrictions related to milk consumption, such as lactose intolerance and milk allergies.

Beverages infused with CBD and adaptogens are also highly discussed topics and are gaining space in the industry and scientific community (Tireki 2021). Flower-based beverages can be used to create relaxing and therapeutic drinks infused with CBD, a compound found in cannabis, which can have positive effects on reducing anxiety and pain (Tireki 2021).

In the future, this trend is expected to continue to grow as more consumers are concerned about health and well-being, as well as environmental sustainability. The flower beverage industry can benefit from these trends by offering natural and sustainable products that meet these consumer needs. Additionally, technological innovations and growing interest in new flavours and gastronomic experiences can pave the way for new and innovative flower-based products and beverages. Companies that invest in research and development of new products in this direction can gain a competitive advantage in the market.

In summary, flower-based beverages have a variety of possible future applications. The important thing is to continue experimenting and creating new combinations to please consumers'taste buds. Finally, with the growing awareness of the importance of dietary diversity and the appreciation of local and seasonal products, there is an opportunity for local producers of edible flowers and herbs to work in collaboration with the beverage industry to create ecological and tasty options.

11.6 Concluding Remarks and Future Research Directions

The market for floral beverages has grown in recent years as the demand for natural, healthy products with sustainable ingredients has increased. The trend towards a more conscious and healthier lifestyle has led consumers to experiment with beverages with different ingredients, and the floral beverage industry has benefited from this movement. With technology advancing rapidly and interest in new flavours and dining experiences increasing, there is room for innovation in the floral beverage market. Companies that invest in the research and development of new products can gain a competitive edge in the market and stand out with differentiated and innovative products. In addition, awareness of dietary diversity and appreciation of local and seasonal produce provides an opportunity for local producers of edible flowers and herbs to work collaboratively with the beverage industry to create new tasty and eco-friendly options. However, it is worth noting that to date, there are few scientific studies related to the use of edible flowers in soft drink formulations when compared to other plant parts such as fruits and leaves. Moreover, it can be said that the existing literature is restricted and scarce, which is mostly centred on some specific types of beverages, an aspect that must be considered by the segment of research and innovation, given the wide range of products found in the market for beverages. Moreover, it can be said that the universe of edible flowers is plural, vast, and rich, which it is necessary for a scientific deepening about the best floral matrices, aiming at its best use, optimization of extraction of its components of interest for food applications, either as an ingredient or additive. In conclusion, the beverage market has a promising future, with opportunities for growth and innovation for companies that want to invest in natural products, sustainable and with differentiated ingredients.

Acknowledgments The authors are grateful to the Foundation for Science and Technology (FCT, Portugal) for financial support to CIMO (UIDB/00690/2020; DOI: 10.54499/UIDB/00690/2020) and (UIDP/00690/2020; DOI: 10.54499/UIDP/00690/2020) and SusTEC (LA/P/0007/2020; DOI: 10.54499/LA/P/0007/2020).

National funding by FCT, P.I., through the institutional and individual scientific employment program-contracts of L. Barros and E. Pereira's contracts, respectively, and C. Caleja thanks her contract through the project VIIAFoods (PRR Aviso n.° de agenda: C644929456-00000040), and E. Pires Jr. grateful to the Foundation for Science and Technology (FCT, Portugal) for financial support through national funds FCT/MCTES (2021. 05425.BD). This work is funded by the

European Regional Development Fund (ERDF) through the Regional Operational Program North 2020, within the scope of Project GreenHealth—Digital strategies in biological assets to improve well-being and promote Green Health, Norte-01-0145-FEDER-000042.

References

Abel & Cole (2023) Raspberry & Elderflower Kombucha, Organic, Equinox. https://www.abelandcole.co.uk/raspberry-elderflower-kombucha-organic-equinox-275ml. Accessed 05 May 2025

Abu-Reidah IM (2020) Carbonated beverages. In: Trends in non-alcoholic beverages, pp 1–36. https://doi.org/10.1016/B978-0-12-816938-4.00001-X

Adetuyi BO, Odelade KA, Olajide PA, Adetunji CO, Adetunji JB, Inobeme A, Godwin YD, Ajenifujah-Solebo O, Akinbo O, Popoola OA, Kolawole OM, Osemwegie OO, Yerima MB, Attanda ML (2024) Chapter 9 – the application of essential oil as an antimicrobial agent in dairy products. In: Adetunji CO, Sharifi-Rad J (eds) Applications of essential oils in the food industry. Academic, pp 99–105

Adib-Hajbaghery M, Mousavi SN (2017) The effects of chamomile extract on sleep quality among elderly people: a clinical trial. Complement Ther Med 35:109–114. https://doi.org/10.1016/J.CTIM.2017.09.010

Afifah AHR (2019) KUALITAS FISIKOKIMIA DAN ORGANOLEPTIK KEFIR SUSU KAMBING DENGAN EKSTRAK ROSELLA MERAH (Hibiscus sabdariffa L.) PADA PENYIMPANAN REFRIGERATOR

Al wadi (2023) Rose Water. Al Wadi Al Akhdar. https://www.alwadi.com/english/products/rose-water. Accessed 05 May 2025

Alves LC et al (2021) Consumers' perception of edible flowers using free word association. Res Soc Dev 10(4). https://doi.org/10.33448/RSD-V10I4.14011

Ameh SJ, Obodozie-Ofoegbu O (2016) Chapter 11 – essential oils as flavours in carbonated cola and citrus soft drinks. In: Preedy VR (ed) Essential oils in food preservation, flavour and safety. Academic, San Diego, pp 111–121

Amoretti (2025) Floral Syrups 3 Pack. https://amoretti.com/products/floral-syrups. Accessed 05 May 2025

Amrane-Abider M, Nerín C, Tamendjari A, Serralheiro MLM (2022) Phenolic composition, antioxidant and antiacetylcholinesterase activities of Opuntia ficus-indica peel and flower teas after in vitro gastrointestinal digestion. J Sci Food Agric 102(11):4401–4409. https://doi.org/10.1002/JSFA.11793

Amrouche TA, Yang X, Capanoglu E, Huang W, Chen Q, Wu L, Zhu Y, Liu Y, Wang Y, Lu B (2022) Contribution of edible flowers to the Mediterranean diet: phytonutrients, bioactivity evaluation and applications. Food Front 3(4):592–630. https://doi.org/10.1002/FFT2.142

An H, Liu J, Chen Y, Huang Y, Chen J, Liu Z et al (2023a) Characterization of key volatile compounds in jasmine tea infusion with different amount of flowers. Food Chem X 19:100750

An HJ, Kim Y, Seo YG (2023b) Relationship between coffee, tea, and carbonated beverages and cardiovascular risk factors. Nutrients 15(4):934. https://doi.org/10.3390/NU15040934/S1

Anastasiou K, Brooker PG, Cleanthous X, Tan R, Smith BPC, Riley M (2023) Oh so sweet: a comparative investigation of retail market composition of sweetened and flavoured beverages in Singapore and Australia. Nutrients 15(1):247. https://doi.org/10.3390/NU15010247/S1

Anni's (2020) Kyldeblomstdrik. In: Anni's Vital Shop. https://annisvitalshop.com/product/3023/. Accessed 05 May 2025

Arévalo-Pinedo R, Santos PZD, Rolin CR, Borges TD, Pizato S, Cortez-Vega WR (2022) PRODUÇÃO de IOGURTE DESNATADO TIPO sundae com GELEIA de HIBISCO (hibiscus sabdariffa). HOLOS 8

Aroma-zone.com (2022). Aroma-zone.com. In: Les hydrolats en cuisine. https://www.aroma-zone.com/info/dossier-thematique/les-hydrolats-en-cuisine

Artisan Drink Company Ltd (2025) Violet Blossom Tonic - Flavoured Tonic Water. In: Artisan Drinks Co. https://artisandrinks.com/products/violet-blossom-tonic. Accessed 05 May 2025

Aydar EF, Mertdinç Z, Demircan E, Çetinkaya SK, Özçelik B (2023) Kidney bean (Phaseolus vulgaris L.) milk substitute as a novel plant-based drink: fatty acid profile, antioxidant activity, in-vitro phenolic bio-accessibility and sensory characteristics. Innovative Food Sci Emerg Technol 83:103254. https://doi.org/10.1016/J.IFSET.2022.103254

Aydinli A, Karadağ S (2023) Effects of abdominal massage applied with ginger and lavender oil for elderly with constipation: a randomized controlled trial. Explore 19(1):115–120. https://doi.org/10.1016/J.EXPLORE.2022.08.010

Azizi NF, Kumar MR, Yeap SK, Abdullah JO, Khalid M, Omar AR, Osman MA, Mortadza SAS, Alitheen NB (2021) Kefir and its biological activities. Foods 10(6):1210. https://doi.org/10.3390/FOODS10061210

B.effect (2023) b.alive Jasmine Kombucha. https://beffect.nz/products/b-effect-kombucha-jasmine?_pos=1&_psq=jasmi&_ss=e&_v=1.0&variant=44340066910504. Accessed 05 May 2025

Bang Energy (2023) Bang Raging Raspberry Hibiscus. In: Pepsico Partners https://dev2.pepsicopartners.com/pepsico/en/USD/OTHER-BRANDS/BANG/RAGING-RASPBERRY-HIBISCUS/Bang-Raging-Raspberry-Hibiscus/p/1-136I6S-1. Accessed 05 May 2025

Barajas-Torres GC, Klünder-Klünder M, Garduño-Espinosa J, Parra-Ortega I, Franco-Hernández MI, Miranda-Lora AL (2022) Effects of carbonated beverage consumption on Oral pH and bacterial proliferation in adolescents: a randomized crossover clinical trial. Life 12(11):1776. https://doi.org/10.3390/LIFE12111776

Barani YH, Zhang M, Mujumdar AS, Chang L (2022) Preservation of colour and nutrients in anthocyanin-rich edible flowers: progress of new extraction and processing techniques. J Food Process Preserv 46(9):e16474. https://doi.org/10.1111/JFPP.16474

Batiha GES, Teibo JO, Wasef L, Shaheen HM, Akomolafe AP, TKA T, Al-kuraishy HM, Al-Garbeeb AI, Alexiou A, Papadakis M (2023) A review of the bioactive components and pharmacological properties of Lavandula species. Naunyn Schmiedeberg's Arch Pharmacol 396(5):877–900. https://doi.org/10.1007/S00210-023-02392-X

Beck AL, Martinez S, Patel AI, Fernandez A (2020) Trends in sugar-sweetened beverage consumption among California children. Public Health Nutr 23(16):2864–2869. https://doi.org/10.1017/S1368980020001147

Bellachioma L, Rocchetti G, Morresi C, Martinelli E, Lucini L, Ferretti G, Damiani E, Bacchetti T (2022) Valorisation of Crocus sativus flower parts for herbal infusions: impact of brewing conditions on phenolic profiling, antioxidant capacity and sensory traits. Int J Food Sci Technol 57(6):3838–3849. https://doi.org/10.1111/IJFS.15713

Belvoir Farm (2025) Organic Elderflower Cordia. In: Belvoir Farm. https://belvoirfarm.co.uk/product/organic-elderflower/ Accessed 05 May 2025

Benvenuti S, Mazzoncini M (2021) The biodiversity of edible flowers: discovering new tastes and new health benefits. Front Plant Sci 11:1812. https://doi.org/10.3389/FPLS.2020.569499/BIBTEX

Bhagavathula AS, Rahmani J, Vidyasagar K, Tesfaye W, Khubchandani J (2022) Sweetened beverage consumption and risk of cardiovascular mortality: a systematic review and meta-analysis. Diabetes Metab Syndr Clin Res Rev 16(4):102462. https://doi.org/10.1016/J.DSX.2022.102462

Bionutricia Extract (2024) Butterfly Pea Flower (*Clitoria ternatea* L.) Standardized Extract Liquid Concentrate, Fresh Natural Asian Gourmet Beverage or Bakery Ingredient, Fresh Natural Flavour & Colour l. https://bionutriciaextract.com/product/butterfly-pea-flower-clitoria-ternatea-l-extract-liquid/. Accessed 05 May 2025

Biotek J, Rezaldi F, Ningtias RY, Anggraenni SD, Ma A, Fatonah NS, Pertiwi FD, Dita LA, Mf F, Subekhi AI, Banten RSAHP, Pandeglang STB, History A (2021) PENGARUH METODE BIOTEKNOLOGI FERMENTASI KOMBUCHA BUNGA TELANG (Clitoria ternatea L)

SEBAGAI ANTIBAKTERI GRAM POSITIF DAN NEGATIF. Jurnal Biotek 9(2):169–185. https://doi.org/10.24252/JB.V9I2.25467

Bittermilk (2025). https://shop.bittermilk.com/products/no-2-tom-collins-with-elderflowers-hops. Accessed 05 May 2025

Blossoms Syrups (2025) Blossoms Elderflower Syrup. https://www.diffordsguide.com/beer-wine-spirits/6584/blossoms-elderflower-syrup. Accessed 05 May 2025

Boas MPPV et al (2023) Vegetable beverages prepared with brazilians nuts: a systematic review. Res Soc Dev 12(2):e5612239890–e5612239890. https://doi.org/10.33448/RSD-V12I2.39890

Bocker R, Silva EK (2022) Innovative technologies for manufacturing plant-based non-dairy alternative milk and their impact on nutritional, sensory and safety aspects. Future Foods 5:100098. https://doi.org/10.1016/J.FUFO.2021.100098

Bondston (2025) Thomas Henry. In: Bondston Premium Spirits Capital. https://www.bondston.com/thomas-henry-cherry-blossom-tonic-24x0-2l-p6023. Accessed 05 May 2025

Bontanitea (2025) Probiotea Kefir / Fruity Blossom. https://bontanitea.oddle.me/en_SG/. Accessed 05 May 2025

Bortolomedi BM, Paglarini CS, Brod FCA (2022) Bioactive compounds in kombucha: a review of substrate effect and fermentation conditions. Food Chem 385:132719. https://doi.org/10.1016/J.FOODCHEM.2022.132719

Botanica P (2018) ProTec Botanica. In: Protec Botanica. https://protecbotanica.com/products/hydrolats/

Brew Dr. Kombucha (2023) Jasmine green tea brewed with soothing chamomile and lavender. In: Love Lavender · Chamomile · Jasmine Green Tea. https://www.brewdrkombucha.com/raw-organic-kombucha/love#test1. Accessed 05 May 2025

Brigand JP, Nahon P (2016) Gastronomy and the citron tree (Citrus medica L.). Int J Gastron Food Sci 3:12–16. https://doi.org/10.1016/J.IJGFS.2016.03.001

California (2025) S of Beber Almondmilk. https://calosba.ca.gov/beber-almondmilk/. Accessed 10 Apr 2025

Carpaty Enjoy (2016) Acacia Flower Drink. https://carpaty-enjoy.ua/en/drinks/akaciya. Accessed 05 May 2025

Carrad AM, Louie JCY, Milosavljevic M, Kelly B, Flood VM (2015) Consumer support for healthy food and drink vending machines in public places. Aust N Z J Public Health 39(4):355–357. https://doi.org/10.1111/1753-6405.12386

Casado N, Fernández-Pintor B, Morante-Zarcero S, Sierra I (2022) Quick and green microextraction of pyrrolizidine alkaloids from infusions of mallow, calendula, and hibiscus flowers using ultrahigh-performance liquid chromatography coupled to tandem mass spectrometry analysis. J Agric Food Chem 70(25):7826–7841. https://doi.org/10.1021/ACS.JAFC.2C02186/ASSET/IMAGES/LARGE/JF2C02186_0005.JPEG

Chen D, Chen G, Sun Y, Zeng X, Ye H (2020) Physiological genetics, chemical composition, health benefits and toxicology of tea (Camellia sinensis L.) flower: a review. Food Res Int 137:109584. https://doi.org/10.1016/J.FOODRES.2020.109584

Chen K, Zhang M, Bhandari B, Mujumdar AS (2021) Edible flower essential oils: a review of chemical compositions, bioactivities, safety and applications in food preservation. Food Res Int 139:109809. https://doi.org/10.1016/j.foodres.2020.109809

Cheng H (2010) Volatile flavour compounds in yogurt: a review. Crit Rev Food Sci Nutr 50(10):938–950. https://doi.org/10.1080/10408390903044081

Cheng S, Jia H, Zhang Y, Zhou J, Chen X, Wu L, Wang J (2022) Geniposidic acid from Eucommia ulmoides Oliver Staminate flower tea mitigates cellular oxidative stress via activating AKT/NRF2 Signaling. Molecules 27(23):8568. https://doi.org/10.3390/MOLECULES27238568/S1

Cherry Blossom Food (2025) Japan Cherry Blossoms. In: Buy Cherry Blossom Powder Flavour and Sprinkles – Sakura Extract Concentrate. https://www.japancherryblossoms.com/powder/. Accessed 05 May 2025

Chinomona E, Mahlatsi MK (2020) Green supply chain practices and supply chain performance in the beverage industry of South Africa. J Contemp Manag 17(2):295–325. https://doi.org/10.35683/JCM20053.77

Choi Y, Park YH, Yang C, Kim DH, Lee KW, Lee MY (2023) Protocol for a randomized controlled trial evaluating the effect of Hibiscus syriacus L. flower extract on sleep quality. Front Nutr 10. https://doi.org/10.3389/FNUT.2023.1169193

Chowdhury MT, Sarkar A, Paul SK, Moktadir MA (2022) A case study on strategies to deal with the impacts of COVID-19 pandemic in the food and beverage industry. Oper Manag Res 15(1–2):166–178. https://doi.org/10.1007/S12063-020-00166-9/TABLES/4

Corbo MR, Bevilacqua A, Petruzzi L, Casanova FP, Sinigaglia M (2014) Functional beverages: the emerging side of functional foods. Compr Rev Food Sci Food Saf 13(6):1192–1206. https://doi.org/10.1111/1541-4337.12109

Cortas (2023) Cortas. https://www.cortasfood.com/category/flower-waters

D'Amato S, Serio A, López CC, Paparella A (2018) Hydrosols: biological activity and potential as antimicrobials for food applications. Food Control 86:126–137. https://doi.org/10.1016/J.FOODCONT.2017.10.030

da Silva LR, Velasco JI, Fakhouri FM (2023) Use of rice on the development of plant-based milk with antioxidant properties: from raw material to residue. LWT 173:114271. https://doi.org/10.1016/J.LWT.2022.114271

de Moura SCSR, Schettini GN, Gallina DA, Alvim ID, Hubinger MD (2022) Microencapsulation of hibiscus bioactives and its application in yogurt. J Food Process Preserv 46(4):e16468. https://doi.org/10.1111/JFPP.16468

de Pires EO et al (2023) Edible flowers as an emerging horticultural product: a review on sensorial properties, mineral and aroma profile. Trends Food Sci Technol 137:31–54. https://doi.org/10.1016/j.tifs.2023.05.007

de Rosso VV, Mercadante AZ (2007) Evaluation of colour and stability of anthocyanins from tropical fruits in an isotonic soft drink system. Innovative Food Sci Emerg Technol 8(3):347–352. https://doi.org/10.1016/J.IFSET.2007.03.008

DGAV (2024) Sumos e Néctares. In: Direção Geral de Alimentação e Veterinária (DGAV). https://www.dgav.pt/alimentos/conteudo/generos-alimenticios/regras-especificas-por-tipo-de-alimentos/sumos-e-nectares/ Accessed 05 May 2025

Dillon's Small Batch Distillers (2025) In: Dillon's Peach & Lavender Cocktail Syrup. https://www.dillons.ca/dillons-peach-lavender-cocktail-syrup.html. Accessed 05 May 2025

Dini I (2019) An overview of functional beverages. In: Functional and medicinal beverages: volume 11: the science of beverages, pp 1–40. https://doi.org/10.1016/B978-0-12-816397-9.00001-7

Directo al Paladar (2023) Qué es el agua de azahar, para qué sirve y dónde comprar el ingrediente fundamental del Roscón de Reyes para Navidad. In: Directo al paladar. https://www.directoalpaladar.com/ingredientes-y-alimentos/donde-comprar-agua-de-azahar. Accessed 05 May 2025

Dobros N, Zawada KD, Paradowska K (2022) Phytochemical profiling, antioxidant and anti-inflammatory activity of plants belonging to the Lavandula genus. Molecules 28(1):256. https://doi.org/10.3390/MOLECULES28010256

Doorstep Organics (2021) Almo - Chamomile Almond Milk. https://www.doorsteporganics.com.au/product/Chamomile-Almond-Milk-Almo-300ml?srsltid=AfmBOopimL_xlAhBAptHb-hLVqij6q3PAppmPjjqXRLSLczCKYXlBLwr6. Accessed 05 May 2025

dos Santos IC, Reis SN (2021) Edible flowers: traditional and current use. Ornament Hortic 27(4):438–445. https://doi.org/10.1590/2447-536X.V27I4.2392

Drewnowski A, Henry CJ, Dwyer JT (2021) Proposed nutrient standards for plant-based beverages intended as Milk alternatives. Front Nutr 8:796. https://doi.org/10.3389/FNUT.2021.761442/BIBTEX

Drink 420 (2023) Drink 420 Elderflower Lime. In: The CBD Drinks. https://www.thecbdstore.ie/shop/cbd-drinks/drink-420-elderflower-lime/ Accessed 05 May 2025

Dufresne C, Farnworth E (2000) Tea, Kombucha, and health: a review. Food Res Int 33(6):409–421. https://doi.org/10.1016/S0963-9969(00)00067-3

Dujmović M, Radman S, Opačić N, Uher SF, Mikuličin V, Voća S, Žlabur JŠ (2022) Edible flower species as a promising source of specialized metabolites. Plants 11(19):2529. https://doi.org/10.3390/PLANTS11192529/S1

Ebrahimi H, Mardani A, Basirinezhad MH, Hamidzadeh A, Eskandari F (2022) The effects of lavender and chamomile essential oil inhalation aromatherapy on depression, anxiety and stress in older community-dwelling people: a randomized controlled trial. Explore 18(3):272–278. https://doi.org/10.1016/J.EXPLORE.2020.12.012

Elise Tries to Cook (2023) Dairy Free Rose Latte In: Elise Tries To Cook. https://elisetriestocook.com/rose-latte/ Accessed 05 May 2025

Elliott C, Truman E, Black JE (2023) Tracking teen food marketing: participatory research to examine persuasive power and platforms of exposure. Appetite 186:106550. https://doi.org/10.1016/J.APPET.2023.106550

Ellis LR, Zulfiqar S, Holmes M, Marshall L, Dye L, Boesch C (2022) A systematic review and meta-analysis of the effects of Hibiscus sabdariffa on blood pressure and cardiometabolic markers. Nutr Rev 80(6):1723–1737. https://doi.org/10.1093/NUTRIT/NUAB104

Faisal Z, Saeed F, Afzaal M, Akram N, Shah YA, Islam F, Ateeq H (2022) Phytochemical profile and food applications of edible flowers: a comprehensive treatise. J Food Process Preserv 46(11):e17061. https://doi.org/10.1111/JFPP.17061

FAO (2024) General Standard For Fruit Juices And Nectars. In: Food and Agriculture Organization of the United Nations. https://www.fao.org/fao-who-codexalimentarius/sh-proxy/jp/?lnk=1&url=https%253A%252F%252Fworkspace.fao.org%252Fsites%252Fcodex%252FStandards%252FCXS%2B247-2005%252FCXS_247e.pdf. Accessed 05 May 2025

Farhat C, Younes H, Alyamani OA, Mrad M, Hourani N, Khalifeh H, El-Makhour Y, Dbaibo G, Hage-Sleiman R (2022) Chemical characterization and in vitro biological evaluation of aqueous extract of Althaea officinalis L. flower grown in Lebanon. J Herbal Med 34:100575. https://doi.org/10.1016/J.HERMED.2022.100575

Farzaneh V, Carvalho IS (2015) A review of the health benefit potentials of herbal plant infusions and their mechanism of actions. Ind Crop Prod 65:247–258. https://doi.org/10.1016/J.INDCROP.2014.10.057

Fentimans (2025) Gently Sparkling Elderflower. https://www.fentimans.com/products/gently-sparkling-elderflower?_pos=1&_psq=elder&_ss=e&_v=1.0 Accessed 05 May 2025

Fernandes L, Ramalhosa E, Baptista P, Pereira JA, Saraiva JA, Casal SIP (2019) Nutritional and nutraceutical composition of pansies (Viola × wittrockiana) during flowering. J Food Sci 84(3):490–498. https://doi.org/10.1111/1750-3841.14482

Fernandes MLP, de Souza AC, Júnior PSPC, Veríssimo LAA, Pylro VS, Dias DR, Schwan RF (2022) Sugary kefir grains as the inoculum for developing a low sodium isotonic beverage. Food Res Int 157:111257. https://doi.org/10.1016/J.FOODRES.2022.111257

Ferreres F, Grosso C, Gil-Izquierdo A, Valentão P, Andrade PB (2014) Assessing Jasminum grandiflorum L. authenticity by HPLC-DAD-ESI/MSn and effects on physiological enzymes and oxidative species. J Pharm Biomed Anal 88:157–161. https://doi.org/10.1016/J.JPBA.2013.08.040

Fever-Tree (2025) Raspberry & Orange Blossom Soda. In: Fever-Tree. https://fever-tree.com/en_GB/products/raspberry-orange-blossom-soda. Accessed 05 May 2025

Figueredo-Urbina CJ, Álvarez-Ríos GD, Zárraga LC, Figueredo-Urbina CJ, Álvarez-Ríos GD, Zárraga LC (2022) Edible flowers commercialized in local markets of Pachuca de Soto, Hidalgo, Mexico. Bot Sci 100(1):120–138. https://doi.org/10.17129/BOTSCI.2831

Finimundy TC, Pereira C, Dias MI, Caleja C, Calhelha RC, Sokovic M, Stojković D, Carvalho AM, Rosa E, Barros L, Ferreira ICFR (2020) Infusions of herbal blends as promising sources of phenolic compounds and bioactive properties. Molecules 25(9):2151. https://doi.org/10.3390/MOLECULES25092151

Froiio F, Cristiano MC, Mancuso A, Iannone M, Paolino D (2020) Vegetable-Milk-based yogurt-like structure: rheological properties influenced by gluten-free carob seed flour. Appl Sci 10(19):6963. https://doi.org/10.3390/APP10196963

Fruzzion (2022) Drink A Flower – Alcohol-free organic sparkling. In: Boissons saines à base de fleurs. https://www.drinkaflower.be/ Accessed 05 May 2025

Gahruie HH, Eskandari MH, Mesbahi G, Hanifpour MA (2015) Scientific and technical aspects of yogurt fortification: a review. Food Sci Human Wellness 4(1):1–8. https://doi.org/10.1016/J.FSHW.2015.03.002

Gayathry KS, John JA (2021) Functional beverages: special focus on anti-diabetic potential. J Process Preserv 45(11):e15974. https://doi.org/10.1111/JFPP.15974

Ghosh P, Rana SS (2021) Physicochemical, nutritional, bioactive compounds and fatty acid profiling of pumpkin flower (Cucurbita maxima), as a potential functional food. SN Appl Sci 3(2):1–14. https://doi.org/10.1007/S42452-020-04092-0/TABLES/5

Gist (2023) Prebiotic Herbal Soda. https://drinkgist.com/ Accessed 05 May 2025

González-Salitre L, González-Olivares LG, Basilio-Cortes UA (2023) Humulus lupulus L. a potential precursor to human health: high hops craft beer. Food Chem 405(Pt B). https://doi.org/10.1016/J.FOODCHEM.2022.134959

Grand View Research (2025) Non-alcoholic RTD Beverages Market Trends. In: Non-alcoholic RTD Beverages Market Size, Share & Trends Analysis Report By Product (Carbonated Soft Drinks, RTD Tea & Coffee, Functional Beverages, Juices), By Distribution Channel (Food Service, Retail), By Region, And Segment Forecasts, 2024–2030. https://www.grandviewresearch.com/industry-analysis/nonalcoholic-rtd-beverages-market-report. Accessed 05 May 2025

GT's Living Foods (2025) Synergy Raw Kombucha. https://gtslivingfoods.com/products/gts-synergy-kombucha-lavender-love. Accessed 05 May 2025

Guiné RPF, Florença SG, Ferrão AC, Bizjak MČ, Vombergar B, Simoni N, Vieira V (2021) Factors affecting eating habits and knowledge of edible flowers in different countries. Open Agric 6(1):67–81. https://doi.org/10.1515/OPAG-2020-0208/MACHINEREADABLECITATION/RIS

Gustavo GH, Bicas JL, Junior MRM (2023) Nutritional benefits of fruit and vegetable beverages obtained by lactic acid fermentation. In: Value-addition in beverages through enzyme technology, pp 177–198. https://doi.org/10.1016/B978-0-323-85683-6.00002-8

Gutiérrez-Hellín J, Varillas-Delgado D (2021) Energy drinks and sports performance, cardiovascular risk, and genetic associations; future prospects. Nutrients 13(3):715. https://doi.org/10.3390/NU13030715

Guzel-Seydim ZB, Kok-Tas T, Greene AK, Seydim AC (2011) Functional properties of kefir. Crit Rev Food Sci Nutr 51(3):261–268

Guzel-Seydim ZB, Gökırmaklı Ç, Greene AK (2021) A comparison of milk kefir and water kefir: physical, chemical, microbiological and functional properties. Trends Food Sci Technol 113:42–53. https://doi.org/10.1016/J.TIFS.2021.04.041

Han AR, Nam B, Kim BR, Lee KC, Song BS, Kim SH, Kim JB, Jin CH (2019) Phytochemical composition and antioxidant activities of two different colour chrysanthemum flower teas. Molecules 24(2):329. https://doi.org/10.3390/MOLECULES24020329

Hashimoto M, Takahashi K, Ohta T (2023) Inhibitory effects of linalool, an essential oil component of lavender, on nociceptive TRPA1 and voltage-gated Ca2+ channels in mouse sensory neurons. Biochem Biophys Rep 34:101468. https://doi.org/10.1016/J.BBREP.2023.101468

Hastuti AP (2015) Studi Aktivitas Antibakteri Kefir Teh Rosella Merah (Hibiscus sabdariffa L.) dari Berbagai Merk The Rosella Merah Yang Beredar di Pasaran

Hell Energy (2023) Hell Energy drink summer cool elderflower. In: Products. https://www.hellenergy.com/in/in-products/. Accessed 12 April 2023

Ho SC, Chang PW, Tong HT, Yu PY (2013) Inhibition of fluorescent advanced glycation end-products and N-Carboxymethyllysine formation by several floral herbal infusions. Int J Food Prop 17(3):617–628. https://doi.org/10.1080/10942912.2012.654566

Huiles & Sens (2019) Romain Chamomile Hydrosol Certified Organic. Hydrolats & Floral waters. https://www.huiles-et-sens.com/en/288-romain-chamomile-hydrosol-certified-organic.html. Accessed 05 May 2025

Igual M, García-Herrera P, Cámara RM, Martínez-Monzó J, García-Segovia P, Cámara M (2022) Bioactive compounds in rosehip (Rosa canina) powder with encapsulating agents. Molecules 27(15):4737. https://doi.org/10.3390/MOLECULES27154737

Inchingolo AM, Malcangi G, Ferrante L, Vecchio GD, Viapiano F, Mancini A, Inchingolo F, Inchingolo AD, Venere DD, Dipalma G, Patano A (2023) Damage from carbonated soft drinks on enamel: a systematic review. Nutrients 15(7):1785. https://doi.org/10.3390/NU15071785

Iperos Herbs (2025) Elderflower Herb Soda. In: Sodas. https://iperosherbs.com/product/elderflower-herb-soda/ Accessed 05 May 2025

Isla MI, Ezquer ME, Leal M, Moreno MA, Zampini IC (2021) Flower beverages of native medicinal plants from Argentina (Acacia caven, Geoffroea decorticans and Larrea divaricata) as antioxidant and anti-inflammatory. J Ethnopharmacol 281:114490. https://doi.org/10.1016/j.jep.2021.114490

Jack Rudy Cocktail Co (2025) Elderflower Tonic. https://jackrudycocktailco.com/collections/all/products/elderflower-tonic-2. Accessed 05 May 2025

Jagim AR et al (2023) International society of sports nutrition position stand: energy drinks and energy shots. J Int Soc Sports Nutr 20(1):2171314

Jakubczyk K, Tuchowska A, Janda-Milczarek K (2021) Plant hydrolates – antioxidant properties, chemical composition and potential applications. Biomed Pharmacother 142:112033. https://doi.org/10.1016/j.biopha.2021.112033

Jalalyazdi M, Ramezani J, Izadi-Moud A, Madani-Sani F, Shahlaei S, Ghiasi S (2019) Effect of hibiscus sabdariffa on blood pressure in patients with stage 1 hypertension. J Adv Pharm Technol Res 10(3):107. https://doi.org/10.4103/JAPTR.JAPTR_402_18

Janarny G, Gunathilake KDPP, Ranaweera KKDS (2021) Nutraceutical potential of dietary phytochemicals in edible flowers – a review. J Food Biochem 45(4):e13642. https://doi.org/10.1111/JFBC.13642

Januário JB, Moreira BR, Paraiso CM, Mizuta AG, Madrona GS (2020) Kombucha à base de Hibiscus sabdariffa L: avaliação tecnológica para produção de uma nova bebida/Kombucha based Hibiscus sabdariffa L: technological assessment for the production of a new beverage. Braz J Dev 6(1):3720–3732. https://doi.org/10.34117/BJDV6N1-264

Japan GreenTea (2022) Sweet Sakura Tea Green Tea. In: Japan GreenTea Co, Ltd. https://jp-greentea.co.jp/english/products/sweet-sakura-tea-green-tea-2/. Accessed 05 May 2025

Java Frost (2024) Blossoms Lavender Lemonade. Java Frost. https://javafrost.com/blossoms-lavender-lemonade-4-lb-bag.html. Accessed 05 May 2025

Jiang B, Le L, Wan W, Zhai W, Hu K, Xu L, Xiao P (2015) The flower tea Coreopsis tinctoria increases insulin sensitivity and regulates hepatic metabolism in rats fed a high-fat diet. Endocrinology 156(6):2006–2018. https://doi.org/10.1210/EN.2015-1015

Jiménez-López J, Ruiz-Medina A, Ortega-Barrales P, Llorent-Martínez EJ (2017) Rosa rubiginosa and Fraxinus oxycarpa herbal teas: characterization of phytochemical profiles by liquid chromatography-mass spectrometry, and evaluation of the antioxidant activity. New J Chem 41(15):7681–7688. https://doi.org/10.1039/C7NJ01743J

Kin euphorics (2023) Functional Beverage. https://www.kineuphorics.com/?srsltid=AfmBOoojSmkkKJhttt_rPR8i4uuhfb1p6ZQ_GjuDzW3sKJIfyw0k6E4O. Accessed 05 May 2025

Krepko (2023) Kefir collagen, vit. C and rose. In: In: Blossom In Full Splendour https://www.mlekarna-krepko.si/en/blog/blossom-in-full-splendour-kefir-krepki-suhec-with-collagen-vitamin-c-and-rose-extract/. Accessed 05 May 2025

KTonic Kombucha (2025) Cherry Blossom. Single Fermentation Kombucha (12, 16 oz bottles) — KTonic Kombucha. https://www.ktonickombucha.com/delivery/ktonic-kombucha-case-of-5-cherry-blossom. Accessed 05 May 2025

Kvalvik LG, Klungsøyr K, Igland J, Caspersen IH, Brantsæter AL, Solberg BS, Hartman C, Schweren LJS, Larsson H, Li L, Forthun I, Johansson S, Vasquez AA, Haavik J (2022) Association of sweetened carbonated beverage consumption during pregnancy and ADHD symptoms in the offspring: a study from the Norwegian mother, father and child cohort study (MoBa). Eur J Nutr 61(4):2153–2166. https://doi.org/10.1007/S00394-022-02798-Y/TABLES/4

Kvas Fine Beverage Co (2025) In: Cherry Hibiscus Simple Syrup. https://www.kvasfbco.com/products/cherry-hibiscus-simple-syrup-coming-soon. Accessed 05 May 2025

Laureys D, Britton SJ, De Clippeleer J (2020) Kombucha tea fermentation: a review. J Am Soc Brew Chem 78(3):165–174

Le Comptoir Toamasina (2025) Culinary Rose Water in Pastry (Bulgaria) https://lecomptoird-etoamasina.fr/en/natural-extract/13-culinary-rose-water-in-pastry-bulgaria.html. Accessed 05 May 2025

Lee PY, Lusk K, Mirosa M, Oey I (2014) The role of personal values in Chinese consumers' food consumption decisions. A case study of healthy drinks. Appetite 73:95–104. https://doi.org/10.1016/J.APPET.2013.11.001

Liu Y, Guo C, Zang E, Shi R, Liu Q, Zhang M, Zhang K, Li M (2023) Review on herbal tea as a functional food: classification, active compounds, biological activity, and industrial status. J Fut Foods 3(3):206–219. https://doi.org/10.1016/J.JFUTFO.2023.02.002

Luckow T, Delahunty C (2004) Which juice is 'healthier'? A consumer study of probiotic non-dairy juice drinks. Food Qual Prefer 15(7–8):751–759. https://doi.org/10.1016/J.FOODQUAL.2003.12.007

Mahadi JPW, Made N, Sanjiwani S, Sudiarsa W, Gayatri A (2022) Pembuatan infused water bunga [TEL]ang dan pemanfaatan bagi kesehatan. Jurnal Pengabdian Kepada Masyarakat Widya Mahadi 3(1):138–149. https://doi.org/10.5281/ZENODO.7447364

Maleki M, Mardani A, Manouchehri M, Farahani MA, Vaismoradi M, Glarcher M (2023) Effect of chamomile on the complications of cancer: a systematic review. Integr Cancer Ther 22

Maleš I, Pedisić S, Zorić Z, Elez-Garofulić I, Repajić M, You L, Vladimir-Knežević S, Butorac D, Dragović-Uzelac V (2022) The medicinal and aromatic plants as ingredients in functional beverage production. J Funct Foods 96:105210. https://doi.org/10.1016/J.JFF.2022.105210

Marsh AJ, Hill C, Ross RP, Cotter PD (2014) Fermented beverages with health-promoting potential: past and future perspectives. Trends Food Sci Technol 38(2):113–124. https://doi.org/10.1016/J.TIFS.2014.05.002

Matcha Oishii (2025) Cherry Blossom Powder Sakura. https://matchaoishii.com/products/cherry-blossom-powder-sakura. Accessed 05 May 2025

Mateina (2025) Mateina – Yerba Mate Infusion Hibiscus & Berries. https://mateina.ca/collections/zero-sugar-yerba-mate-drinks. Accessed 12 April 2023

Meydani SN, Ha WK (2000) Immunologic effects of yogurt. Am J Clin Nutr 71(4):861–872. https://doi.org/10.1093/AJCN/71.4.861

Miracle Botanicals Essential Oils (2025) What is a hydrosol? In: Hydrosols https://miraclebotanicals.com/collections/essential-oil-hydrosols-miracle-botanicals?utm_campaign=gs-2022-02-22&utm_source=google&utm_medium=smart_campaign&gclid=Cj0KCQjwpPKiBhDvARIsACn-gzCp4lB37R46P19hG6EvWJNpkJVu8wEVUGzsThncL7oQkscR4pF9rK8aAkFrEALw_wcB. Accessed 05 May 2025

Monari S, Ferri M, Salinitro M, Tassoni A (2022) Ethnobotanical review and dataset compiling on wild and cultivated plants traditionally used as medicinal remedies in Italy. Plants 11(15):2041. https://doi.org/10.3390/PLANTS11152041/S1

Monin (2025) Le Sirop. https://monin1912.com/collections/le-sirop?sort_by=title-ascending&filter.p.tag=Floral. Accessed 05 May 2025

Moreno IL, Pastre CM, Ferreira C, de Abreu LC, Valenti VE, Vanderlei LCM (2013) Effects of an isotonic beverage on autonomic regulation during and after exercise. J Int Soc Sports Nutr 10

Mud Lake Farm, LLC (2020) St. Steve's. In: Chamomile Soda – made with fresh chamomile flowers. https://www.ststeves.com/store/p83/chamomilesoda.html#/ Accessed 05 May 2025

Mulík S, Hernández-Carrión M, Pacheco-Pantoja SE, Aguilar-Ruiz N, Ozuna C (2022) Culinary uses of Mexican edible flowers: recipe analysis. Int J Gastron Food Sci 28:100539. https://doi.org/10.1016/J.IJGFS.2022.100539

Natura Center (2023) Prokey Hibisc. https://naturacenter.es/colesterol/61186-prokey-hibisco-500-ml-0643415972052.html. Accessed 05 May 2025

Naturae (2025) Organic aloe vera juice with chamomile. In: products. https://www.naturae.es/en/products/aloe-vera/healthy-aloe-vera-juices/organic-aloe-vera-juice-with-chamomile/. Accessed 05 May 2025

Naturitas (2023). https://www.naturitas.es/p/alimentacion/bebidas/zumos/smoothie-detox-de-hibisco-250-ml-vitamont?utm_source=awin&utm_medium=afiliacion&utm_campaign=904799&awc=30527_1684731708_beed6090faab4bd7e428b9f06867250e. Accessed 05 May 2025

Naumovski N, Mellor DD, Sugajski M, Buszewska-Forajta M, Buszewski B (2023) Functional beverages in the 21st century. Beverages 9(1):27. https://doi.org/10.3390/BEVERAGES9010027

NikanKitchen (2023) Kimura Sakura cola. In: NikanKitchen https://www.nikankitchen.com/en/products/3283/sandb-golden-curry-medium-hot-90g. Accessed 05 May 2025

Norhisham D et al (2023) Bioactivities of Kenaf biomass extracts: a review. PRO 11(4):1178. https://doi.org/10.3390/PR11041178

Nova Organic Energy (2022) Orange, Elderflower & Lime. https://www.novaorganicenergy.com/orange-elderflower-lime. Accessed 05 May 2025

Nutrition Empire (2023) Applied Nutrition A.B.E. (Energy Drink) Apple & Elderflower. In: Nutrition Empire. https://nutritionempire.net/en/product/applied-nutrition-abe-energy-drink-apple-elderflower-330ml. Accessed 05 May 2025

Nyiew KY, Kwong PJ, Yow YY (2022) An overview of antimicrobial properties of kombucha. Compr Rev Food Sci Food Saf 21(2):1024–1053. https://doi.org/10.1111/1541-4337.12892

One-Breath-Beverage (2025) One Breath Jasmine + Elderflower In: One Breath https://onebreath-beverage.com/products/one-breath-jasmine-elderflower. Accessed 05 May 2025

Pamungkas BT, Safitri A, Rezaldi F, Andry M, Agustiansyah LD, Fadillah MF, Hidayanto F, Hariadi H (2022) Antifungal Trycophyton Rubrum and Trycophyton Mentagrophytes in liquid Bath soap fermented probiotic Kombucha flower Telang (*Clitoria Ternatea* L) as a pharmaceutical biotechnology product. Biotik: Jurnal Ilmiah Biologi Teknologi dan Kependidikan 10(2):179–196. https://doi.org/10.22373/BIOTIK.V10I2.15160

Pascual L, Gras M, Vidal-Brotóns D, Alcañiz M, Martínez-Máñez R, Ros-Lis JV (2018) A voltammetric e-tongue tool for the emulation of the sensorial analysis and the discrimination of vegetal milks. Sensors Actuators B Chem 270:231–238. https://doi.org/10.1016/J.SNB.2018.04.151

Peng A, Lin L, Zhao M (2022) Discovery, characterization and stability evaluation of self-assembled submicroparticles in chrysanthemum tea infusions. Food Biosci 47:101642. https://doi.org/10.1016/J.FBIO.2022.101642

Pensamiento-Niño CA, Castañeda-Ovando A, Añorve-Morga J, Hernández-Fuentes AD, Aguilar-Arteaga K, Ojeda-Ramírez D (2023) Edible flowers and their relationship with human health. Biol Activities

Pérez-Gregorio R (2021) Phenolic compounds and functional beverages. Beverages 7(4):71. https://doi.org/10.3390/BEVERAGES7040071

Petal Sparkling Botanicals (2025) Petal. https://drinkpetal.com/products/original-rose?variant=36910609105055. Accessed 05 May 2025

Pieterse E, Millan E, Schönfeldt HC (2022) Consumption of edible flowers in South Africa: nutritional benefits, stakeholders' views, policy and practice implications. Br Food J. https://doi.org/10.1108/BFJ-10-2021-1091/FULL/PDF

PiQi Life (2023) Butterfly Pea Kefir. https://piqilife.com/products/butterfly-pea-kefir-750ml. Accessed 05 May 2025

Pires TCSP, Barros L, Santos-Buelga C, Ferreira ICFR (2019) Edible flowers: emerging components in the diet. Trends Food Sci Technol 93:244–258. https://doi.org/10.1016/J.TIFS.2019.09.020
Popa CL, Lupitu A, Mot MD, Copolovici L, Moisa C, Copolovici DM (2021) Chemical and biochemical characterization of essential oils and their corresponding Hydrolats from six species of the Lamiaceae Family. Plants 10(11):2489. https://doi.org/10.3390/PLANTS10112489
Preedy VR (2015) Essential oils in food preservation. Academic, Flavour and Safety
Prihatningtias R, Dewi PK, Armalina D, Istiadi H, Amelia RA, Fulyani F (2022) The effect of Roselle flower infusion (Hibiscus sabdariffa) on retinal ganglion cell apoptosis of Sprague Dawley rats exposed to cigarette smoke. J Biomed Transl Res 8(3):119–124. https://doi.org/10.14710/JBTR.V8I3.15442
Primitivo MJ, Neves M, Pires CL, Cruz PF, Brito C, Rodrigues AC, de Carvalho CCCR, Mortimer MM, Moreno MJ, Brito RMM, Taylor EJ, Millson SH, Reboredo F, Campos MJ, Vaz DC, Ribeiro VS (2022) Edible flowers of Helichrysum italicum: composition, nutritive value, and bioactivities. Food Res Int 157:111399. https://doi.org/10.1016/J.FOODRES.2022.111399
Prosyro (2025) In: Our Products. https://prosyro.com/our-syrups. Accessed 05 May 2025
Purohit SR, Rana SS, Idrishi R, Sharma V, Ghosh P (2021) A review on nutritional, bioactive, toxicological properties and preservation of edible flowers. Future Foods 4:100078. https://doi.org/10.1016/J.FUFO.2021.100078
R. Torre & Company (2023) Torani Syrups. https://www.torani.com/spring-sampler-50-ml-5-pack-variety-pack.html. Accessed 05 May 2025
Rabenhorst (2025) Inner Beauty With Biotin. https://www.rabenhorst.de/en/our-juices/inner-beauty-with-biotin/. Accessed 05 May 2025
RAFT Cocktails (2025) In: Raft Syrups. https://raftsyrups.com/collections/raft-syrups-1. Accessed 05 May 2025
Rashwan AK, Osman AI, Chen W (2023) Natural nutraceuticals for enhancing yogurt properties: a review. Environ Chem Lett 21(3):1907–1931. https://doi.org/10.1007/S10311-023-01588-0
Research and Markets (2024) Beverage Market - Global Forecasts from 2024 to 2029. In: Research and Markets. https://www.researchandmarkets.com/reports/5768689/beverage-market-global-forecasts. Accessed 05 May 2025
RISE Kombucha Inc. (2025) Lychee Jasmine Organic Kombucha. In: RISE https://risekombucha.com/kombucha/classic/lychee-jasmine/. Accessed 05 May 2025
Rivas-García L, Romero-Márquez JM, Navarro-Hortal MD, Esteban-Muñoz A, Giampieri F, Sumalla-Cano S, Battino M, Quiles JL, Llopis J, Sánchez-González C (2022) Unravelling potential biomedical applications of the edible flower Tulbaghia violacea. Food Chem 381:132096. https://doi.org/10.1016/J.FOODCHEM.2022.132096
Rodrigues L, Coelho E, Madeira R, Teixeira P, Henriques I, Coimbra MA (2022) Food ingredients derived from lemongrass Byproduct Hydrodistillation: essential oil, Hydrolate, and decoction. Molecules 27(8):2493. https://doi.org/10.3390/MOLECULES27082493
Royal Rose Syrups (2025) Royal Rose Syrups Organic. https://royalrosesyrups.com/products/rose-syrup?_pos=1&_sid=8ab0d3056&_ss=r. Accessed 05 May 2025
Saiki A, Yoshimi K, Nakagawa K, Nagasawa Y, Yoshizawa A, Yanagida R, Yamaguchi K, Nakane A, Maeda K, Tohara H (2022) Effects of thickened carbonated cola in older patients with dysphagia. Sci Rep 12(1):1–7. https://doi.org/10.1038/s41598-022-25926-4
Santiveri Bilbao (2021) Kombucha Hibiscus BIO. https://santiveri.net/producto/kombucha-hibiscus-bio/. Accessed 05 May 2025
Sauvie Shrubs (2020) Strawberry Rose Signature Shrub. https://www.sauvieshrubs.com/products/strawberry-rose-signature-shrub. Accessed 05 May 2025
Savage K, Firth J, Stough C, Sarris J (2018) GABA-modulating phytomedicines for anxiety: a systematic review of preclinical and clinical evidence. Phytother Res 32(1):3–18. https://doi.org/10.1002/PTR.5940
Seek North (2024) Baldor Specialty Foods. https://seeknorth.com/. Accessed 05 May 2025

Setiawati AE, Kusnadi J (2021) Optimization of fermentation time and grain concentration for water kefir production from butterfly pea flower (Clitoria ternatea). IOP Conf Ser Earth Environ Sci 924(1):012081. https://doi.org/10.1088/1755-1315/924/1/012081

Shkembi B, Huppertz T (2023) Glycemic responses of Milk and plant-based drinks: food matrix effects. Foods 12(3):453. https://doi.org/10.3390/FOODS12030453

Singh MK, Savita K, Singh S, Mishra D, Rani P, Chanda D, Verma RS (2023) Vasorelaxant property of 2-phenyl ethyl alcohol isolated from the spent floral distillate of damask rose (Rosa damascena mill.) and its possible mechanism. J Ethnopharmacol 313:116603. https://doi.org/10.1016/J.JEP.2023.116603

Slade P (2023) Does plant-based milk reduce sales of dairy milk? Evidence from the almond milk craze. Agric Resour Econ Rev 52(1):112–131. https://doi.org/10.1017/AGE.2022.22

Soda Farm (2023) Flower Power Drink Syrup. https://sodafarm.com/products/de-stress-herbal-drink-syrup. Accessed 12 April 2023

Soignons Nos Assiettes (2025) Lavande Fine Hydrolat Bio. In: Hydrolats Eaux florales Bio. https://soignons-nos-assiettes.fr/80-hydrolats-pour-la-cuisine. Accessed 05 May 2025

Split Tree Cocktail Co. (2025) Rhubarb & Elderflower Cordial. https://splittree.ca/products/cordial-drink-mix. Accessed 05 May 2025

Stella TR, Paraíso CM et al (2023) Hibiscus (hibiscus sabdariffa L.) extracts freeze-dried and encapsulated by ionic gelation: an approach for yogurt application. J Food Meas Charact:1–9. https://doi.org/10.1007/S11694-023-01818-7/TABLES/5

Suh SH, Waritanant T, Wattanavichean N, Jityen A, Osotchan T (2023) Electrolyte drink analysis by electronic tongue using carbon nanotube paste electrodes. J Phys Conf Ser 2431(1):012008. https://doi.org/10.1088/1742-6596/2431/1/012008

Syahadat RM, Saleh I, Christalista AAFA (2022) Tren riset pascapanen edible flower. SENTRI: Jurnal Riset Ilmiah 1(2):488–497. https://doi.org/10.55681/SENTRI.V1I2.241

Takahashi JA, Rezende FAGG, Moura MAF, Dominguete LCB, Sande D (2020) Edible flowers: bioactive profile and its potential to be used in food development. Food Res Int 129:108868. https://doi.org/10.1016/J.FOODRES.2019.108868

Tanguler H, Sener S (2022) Production of naturally flavoured and carbonated beverages using Williopsis saturnus yeast and cold fermentation process. Food Biosci 48:101750. https://doi.org/10.1016/J.FBIO.2022.101750

Tavares CS, Gameiro JA, Roseiro LB, Figueiredo AC (2022) Hydrolates: a review on their volatiles composition, biological properties and potential uses. Phytochem Rev 21(5):1661–1737. https://doi.org/10.1007/s11101-022-09803-6

Terra Raiz (2023) Água de flor de laranjeira. https://terraraiz.commercesuite.com.br/engarrafados/agua-de-flor-de-laranjeira Accessed 12 April 2023

Terri Salminen (2020) Elderflower water forever. In: Beautiful Food Stories. https://terrisalminen.com/?s=elder+flower&submit=Search. Accessed 05 May 2025

The Tao Of Tea (2025) Tea Flowers. In: The Tao Of Tea. https://taooftea.com/product/tea-flowers/. https://jp-greentea.co.jp/english/products/sweet-sakura-tea-green-tea-2/. Accessed 05 May 2025

Tireki S (2021) A review on packed non-alcoholic beverages: ingredients, production, trends and future opportunities for functional product development. Trends Food Sci Technol 112:442–454. https://doi.org/10.1016/j.tifs.2021.03.058

Tiţa O, Constantinescu MA, Tiţa MA, Opruţa TI, Dabija A, Georgescu C (2022) Valorization on the antioxidant potential of volatile oils of Lavandula angustifolia mill., Mentha piperita L. and Foeniculum vulgare L. in the production of kefir. Appl Sci 12(20):10287. https://doi.org/10.3390/APP122010287

Tornuk F, Cankurt H, Ozturk I, Sagdic O, Bayram O, Yetim H (2011) Efficacy of various plant hydrosols as natural food sanitizers in reducing Escherichia coli O157:H7 and salmonella Typhimurium on fresh cut carrots and apples. Int J Food Microbiol 148(1):30–35. https://doi.org/10.1016/J.IJFOODMICRO.2011.04.022

Tranquini (2025) Hibiscus. https://tranquini.com/shop/hibiscus/. Accessed 05 May 2025

Tundis R, Marrelli M, Conforti F, Tenuta MC, Bonesi M, Menichini F, Loizzo MR (2015) Trifolium pratense and T. repens (Leguminosae): edible flower extracts as functional ingredients. Foods 4(3):338–348. https://doi.org/10.3390/FOODS4030338

Valero-Cases E, Cerdá-Bernad D, Pastor JJ, Frutos MJ (2020) Non-dairy fermented beverages as potential carriers to ensure probiotics, prebiotics, and bioactive compounds arrival to the gut and their health benefits. Nutrients 12(6):1666. https://doi.org/10.3390/NU12061666

Viapiana A, Wesolowski M (2017) The phenolic contents and antioxidant activities of infusions of Sambucus nigra L. Plant Foods Hum Nutr 72(1):82–87. https://doi.org/10.1007/S11130-016-0594-X/TABLES/2

Villarreal-Soto SA, Beaufort S, Bouajila J, Souchard JP, Taillandier P (2018) Understanding Kombucha tea fermentation: a review. J Food Sci 83(3):580–588. https://doi.org/10.1111/1750-3841.14068

Vital Global Ray Brands (2025) Vanalaya Lavender Flower. In: Vital Global Ray Brands. https://vanalaya.com/lavender-tea/. Accessed 05 May 2025

Wessinger CA, Hileman LC (2020) Parallelism in flower evolution and development. Annu Rev Ecol Evol Syst 51(1):387–408

Wilking C, Moukalled S, Polacsek M (2023) Reducing student exposure to digital food and beverage marketing: policy and practice recommendations. J Sch Health. https://doi.org/10.1111/JOSH.13315

Wirz K, Schwarz S, Richling E, Walch SG, Lachenmeier DW (2022) Coffee flower as a promising novel food—chemical characterization and sensory evaluation. Biol Life Sci Forum 18(1):53. https://doi.org/10.3390/FOODS2022-12967

Wisehead Producitons (2025) The London Essence Co. https://www.londonessenceco.com/ae/. Accessed 05 May 2025

Wong M, Sirisena S, Ng K (2022) Phytochemical profile of differently processed tea: a review. J Food Sci 87(5):1925–1942. https://doi.org/10.1111/1750-3841.16137

Xiao J, Zou Y, Wen X, Guo Y, Hu F, Chen G, Wu Z, Lin Y, Wang Z, Sun L, Pan Y, Li N (2022) Functional contents and antioxidant potency of Chinese Wenguan flower tea. Food Control 138:109002. https://doi.org/10.1016/J.FOODCONT.2022.109002

Xiong L, Yang J, Jiang Y, Lu B, Hu Y, Zhou F, Mao S, Shen C (2014) Phenolic compounds and antioxidant capacities of 10 common edible flowers from China. J Food Sci 79(4):C517–C525. https://doi.org/10.1111/1750-3841.12404

Yan T, Shi L, Xu K, Bai J, Wen R, Liao X, Dai X, Wu Q, Zeng L, Peng W, Wang Y, Yan H, Dang S, Liu X (2023) Habitual intakes of sugar-sweetened beverages associated with gut microbiota-related metabolites and metabolic health outcomes in young Chinese adults. Nutr Metab Cardiovasc Dis 33(2):359–368. https://doi.org/10.1016/J.NUMECD.2022.10.016

Yiğit R, Çoklar H, Akbulut M (2022) Some physicochemical and phytochemical properties of Syringa vulgaris L. flower tea: influence of flower drying technique, brewing method and brewing time. J Food Meas Charact 16(5):4185–4197. https://doi.org/10.1007/S11694-022-01511-1/TABLES/5

Yogi (2023) Rose. In: Yogi®. https://www.yogitea.com/en/products/yogi-tea/rose/. Accessed 05 May 2025

Yu Q, Chen W, Zhong J, Qing D, Yan C (2021) Structural elucidation of three novel oligosaccharides from Kunlun chrysanthemum flower tea and their bioactivities. Food Chem Toxicol 149:112032. https://doi.org/10.1016/J.FCT.2021.112032

Zannou O, Kelebek H, Selli S (2020) Elucidation of key odorants in Beninese Roselle (Hibiscus sabdariffa L.) infusions prepared by hot and cold brewing. Food Res Int 133:109133. https://doi.org/10.1016/J.FOODRES.2020.109133

Zhang Y, Li R, Zhao Q, Fan S (2023) The impact of peer effect on students' consumption of sugar-sweetened beverages- instrumental variable evidence from North China. Food Policy 115:102413. https://doi.org/10.1016/J.FOODPOL.2023.102413

Zhao L, Fan H, Zhang M, Chitrakar B, Bhandari B, Wang B (2019) Edible flowers: review of flower processing and extraction of bioactive compounds by novel technologies. Food Res Int 126:108660. https://doi.org/10.1016/J.FOODRES.2019.108660

Zhou M, Sun Y, Luo L, Pan H, Zhang Q, Yu C (2023) Road to a bite of rosehip: a comprehensive review of bioactive compounds, biological activities, and industrial applications of fruits. Trends Food Sci Technol 136:76–91. https://doi.org/10.1016/J.TIFS.2023.04.006

Chapter 12
Edible Flowers for Liquors

Viviana Cavallaro, Davide Guffanti, Alice Petrini, Awais Ali, and Giacomo Cocetta

Abstract Aromatic plants have been used for the preparation of different kind of foods and beverages since ancient times. With this purpose, different plant organs were used, including flowers. The purpose of this chapter is to provide a series of information, as exhaustive as possible, relating to the flowering species most used worldwide to produce alcoholic beverages. The chapter has been structured based on the sensory properties conferred to beverages by the flowers of various species of agricultural interest. Following this approach, the identified species have been grouped into four categories: aromatic taste, bitter taste, bitter-aromatic taste, and sweet taste. For each of the identified species, information has been gathered and presented regarding the content and accumulation of metabolites responsible for sensory characteristics. Also, botanical characteristics, pedoclimatic requirements, and agronomic practices are presented, when available. Due to the availability of data and information, much of the content provided throughout the text is specific to the Northern Hemisphere, as indicated by references to certain months.

Keywords Alcoholic beverages · Beer · Botanicals · Brewery · Distilled · Spirits

Abbreviations

ADH	Alcohol dehydrogenase
CBD	Cannabidiol
HPS	High Pressure Sodium
IBU	International Bitterness Units

Authors Viviana Cavallaro and Davide Guffanti have equally contributed to this chapter.

V. Cavallaro · D. Guffanti · A. Petrini · A. Ali · G. Cocetta (✉)
Department of Agricultural and Environmental Sciences, Università Degli Studi Di Milano, Milan, Italy
e-mail: giacomo.cocetta@unimi.it

M. M. Jowkar et al. (eds.), *Advances in Research on Edible Flowers*, Advances in Olericulture, https://doi.org/10.1007/978-3-032-04561-4_12

LED	Light-Emitting Diode
LOX	Lipoxygenase
MH	Metal Halide
MVA	Mevalonate
Phe	Phenylalanine
THC	Tetrahydrocannabinol
Trp	Tryptophan
Tyr	Tyrosine
TyrA	Arogenate dehydrogenase
VOCs	Volatile compounds

12.1 Introduction

The use of aromatic plants flowers and spices to prepare beverages dates to ancient times, when different species, were used for the preparation of the so called 'Hippocratic wine' and other similar beverages, which were common in Greece, Italy and in other countries in the Mediterranean basin. These beverages were the precursors of famous liquors, which are still commercially relevant nowadays.

Flowers are plant organs that are particularly useful for these purposes. In fact, they provide a vast series of secondary metabolites, playing a key role in the interaction between plant and environment, and which, at the same time, give alcoholic beverages specific sensory traits, characterizing them in a significant way.

With the technological process of production of the alcoholic beverages, the extraction of pigments and aromatic compounds takes place, contributing to the definition of the key characteristics of each beverage.

Cultivation of edible flowers can thus be focused on maximizing the quality traits which are important for characterizing a specific alcoholic beverage.

The chapter has been organized as a classification of some of the edible flowers most used to produce alcoholic beverages. The chapter has been divided into subsections, corresponding to the main sensory properties which are conferred to the beverage by a specific flower species, including: aroma, bitter and sweetness.

Within these main categories, the most representative species have been described and all the available information about crop characteristics, crop needs, specific uses and cultivation techniques, have been reported, based on the current scientific literature.

12.2 Aromatic Taste

The VOCs are hydrophobic liquids produced in different parts of the plant, present in small quantities and following complex biogenetic pathways (Tonutti and Liddle 2010). Aromatic taste is related to VOCs concentration released from different floral tissues (Vallarino et al. 2018) such as petal, pistil, nectary, and calyx, depending on the species, cultivars, varieties and genotypes. The main compounds that are responsible for generating the aromatic taste in flowers are terpenoid, phenylpropanoid, benzenoid, fatty acid derivatives and amino acids (Mostafa et al. 2022). Terpenoids are the largest group of VOCs and their concentration in flowers depends on the growing phase (Kalua and Boss 2009). Acetyl-CoA and pyruvate, terpenoid precursors, are used to generate sesquiterpenes in the cytosol in the MVA pathway, or monoterpenes, diterpenes, and carotenoids in plastids in the non-MVA pathway. Sesquiterpenes are synthesized through the conversion of farnesyl diphosphate (Chen et al. 2011). Monoterpenes represent the main class of the aromatic compounds that confers flavour (Croteau et al. 2000); limonene, (E)-β-ocimene, myrcene, linalool, and α- and β-pinene are synthesized in plastids (Piechulla and Pott 2003). Carotenoids, photosynthesis accessory pigments, and chlorophyll-protective molecules, are presented at low concentration but are important volatile norisoprenoids' precursors that influence the aroma profile (Armstrong and Hearstt 1996; Baldermann et al. 2009). Phenylpropanoids and benzenoids are synthesized from Phe, an aromatic amino acid, through the shikimate acid pathway. Both biosynthetic pathways compete for phenylalanine in a conflicting process: if a large quantity of phenylpropanoids is generated, it will produce a small amount of benzenoids or not, and vice versa (Majetic and Sinka 2013). Fatty acids are the major precursors of volatile compounds in floral aromas, and they are catabolized through the LOX pathway. Phospholipids, triacylglycerols, and glycolipids are converted to linoleic and linolenic acids, and, at the end of biosynthetic pathway, ADH, esters and jasmonic acid (Muhlemann et al. 2014). Phe, Tyr, and Trp are the main aromatic amino acids precursors of different secondary metabolites. Phenylalanine is the precursor of some secondary metabolites, such as phenylpropanoids, cell wall lignin, flavonoids and anthocyanins (Tzin and Galili 2010). The last two are important to confer characteristic colour and sensory qualities to the flower (Taiz and Zeiger 2012). Tyrosine has the same biosynthetic pathway as Phenylalanine, converted by TyrA (Schenck et al. 2020). Tryptophan is produced starting from the transfer of an amino group of glutamines to chorismate to produce anthranilate and pyruvate, ending with the indole transportation to TSβ, which catalyses its condensation with serine (Parthasarathy et al. 2018).

The best way to conserve the aromatic taste is generally to use dried flowers, guaranteeing better storage without a modification in the content of aromatic compounds (Tonutti and Liddle 2010). The most important edible flowers that confer aroma to a lot of alcoholic beverages are summarized in Table 12.1.

Table 12.1 Flowers species and families used to confer aromatic taste to alcoholic beverages

Species	Family
Rosa damascena Herm.	Rosaceae
Rosa rugosa Thunb.	Rosaceae
Rosa centifolia L.	Rosaceae
Viola odorata L.	Violaceae
Lavandula angustifolia Mill.	Lamiaceae
Sambucus nigra L.	Adoxaceae
Cannabis sativa L.	Cannabaceae

12.2.1 Rosa

Rosa genus belongs to the Rosaceae family. These perennial plants are especially diffused in temperate regions in Europe, North America, northwest Africa and in different parts of Asia and Oceania (Chen et al. 2020). There exist a lot of species that can be bushy, sarmentose, climbing, creeping and shrubby that can reach up to 2 m in height. Originally rose grown wildly, but after a selection among hundreds of species, humans started to cultivate it in farms and private gardens. Generally, wild roses have at most five petals, while the cultivated ones are organized in multiple sets of petals. The flower size ranges from 1.25 cm to more than 17.5 cm in diameter. Roses are cultivated as an ornamental plant thanks to the colourful flowers, but also for the delightful fragrance and aromatic taste, depending on the variety and climatic conditions. Roses need temperatures in the range of 15–28 °C for optimal development, and at least 5–6 h of light during blooming stage. These plants can be cultivated in fields or in greenhouses, arranged, in the first case, in hedge rows to protect the flowers from wind damage, and in both cases to facilitate manual collection or with robotic systems. It is used in oxygenated clay-sandy soil rich in organic matter, with a pH of 6–7.5. Irrigation must be constant to keep the soil wet, avoiding contact with the blossoms. Granular fertilizers containing nitrogen, iron, magnesium, potassium, and phosphorus are used both in the field and in the greenhouse. At the harvest, the stems are cut to be sent to a location to be defoliated and to collect blossoms for steam distillation phase (Kovacheva et al. 2010). The main producers are Turkey with a blossom yield of 5000 t/ha, from which 2–2.5 tons of oil is obtained, and Bulgaria with a blossom yield of 3500 t/ha (Marzi and De Mastro 2008). Part of the essential oil is used in the alcohol production sector. The essential oils extraction is an expensive process, as 10 tons of rose petals are necessary to obtain 1 kg of rose essential oil. The most used species for alcoholic beverages production are *Rosa damascena* Herm., *Rosa rugosa* Thunb. and *Rosa centifolia* L. Especially in Europe and USA, many rose petal liquors are produced, as Crispin's Rose (added with apple), Bols Parfait Amour (added with violets, orange peel, almond and vanilla), Creme de Rose (made with the maceration of rose blossom and extracts), Rosolio (with the addition of different ingredients according to the region), Rose-Sauternes Cordial (dried rose petals are matched by a wine called Sauternes, made from fruit that has been affected by Botrytis cinerea), Culpeper's (macerated rose petals, sugar, and fruit in brandy) and Raspberry Rose (dried rose petals with raspberries) (Schloss 2013; Stewart 2013).

12.2.1.1 *Rosa damascena* Herm.

Rosa damascena Herm. (Fig. 12.1) is an artificial hybrid species obtained through crossing of *Rosa gallica* L. and *Rosa moschata* Herm. (Harkness 2003). It's a deciduous shrub that can reach up to 2 m in height. The leaves are pinnate and green in colour. Flowers have numerous thin overlapping white, pink or red petals with a

Fig. 12.1 Flower of *Rosa damascena* Herm. (pixabay.com)

globular form and have a diameter of 7 cm. This species is native to the Iran region and was introduced in Europe from Syria by crusaders in the fourteenth century (Bown 2001). In the past, damask rose was cultivated in Egypt, Greece, Roman Empire and nowadays is grown especially in Macedonia, Bulgaria, Turkey, and Morocco (Marzi and De Mastro 2008). Production starts after 3 years of development of the plant. The sowing phase takes place in spring or autumn and then blooming time depends on the temperatures; in warm regions the harvest is at the end of April, while in cool or temperate climates it is at the end of May or in June (Bown 2001). The concentration of aroma and compounds is highest once full flowering is reached. To preserve their properties, the flowers are harvested before dawn and processed fresh in the morning to obtain an optimal yield and avoid fermentation processes. The yield is about 4000 kg/ha of flowers. The petals are processed by steam distillation, obtaining a yellowish essential oil, with a pinkish and slightly spicy smell, and a sweetish, slightly bitter taste. The main compounds are monoterpenic alcohols (35%–45% citronellol, 10%–20% geraniol, nerol, phenylethyl, linalool, etc.), monoterpenic hydrocarbons (α-β-pinene, myrcene), esters, aldehydes and ketones. The yield is about 0.02%–0.05%. One of the most important liquors produced is with *Rosa damascena* Herm. essence, adding cucumber after the distillation process, is Hendrick's Gin (Kovacheva et al. 2010; Stewart 2013).

12.2.1.2 *Rosa rugosa* Thunb. and *Rosa centifolia* L.

Rosa rugosa (Fig. 12.2) is a suckering deciduous shrub plant that can reach up to 1.5 m in height. The leaves are pinnate and have a dark green colour, while in autumn they turn to yellow or orange red. Flowers are globular and have 5 dark pink to white petals, organized singly or in clusters, with a diameter of 6–9 cm. The blooming time depends on the temperatures but starts in May and the harvest is at the end of July. Its peculiarity is resistance to diseases and salinity; in fact, it's common to see this species on dunes and coastal grasslands This species is native to East

Fig. 12.2 Flower of *Rosa rugosa* Thunb. (pixabay.com)

Fig. 12.3 Flower of *Rosa centifolia* L. (pixabay.com)

China, Siberia, Korea, and Japan, and was introduced to western and central Europe as an ornamental plant (Bruun 2005).

Rosa centifolia (Fig. 12.3) is a deciduous shrub that can reach up to 1.5–2 m in height. The leaves are greyish-green pinnate. The flowers are globular and have numerous thin overlapping pink, white or dark red purple petals, with a diameter from 1.5 to 17.5 cm. The blooming time depends on the temperature but usually starts in June–July and the harvest in August. This species is native to India, but nowadays is diffused in Europe, especially cultivated in France (Grasse city) for the perfume industry and in Italy (Valle Scrivia, Genoa) to produce alcoholic beverages. Several important chemical compounds in flowers petals, as phenyl ethanol (43%), geranyl acetate (15.6%), geraniol (10.5%), linalool (6.9%), benzyl alcohol (3.3%), benzaldehyde (1.5%), nerol (5%–10%), citronellyl acetate (0.3%); other constituents are tannins, oloigomeric proantrocyanides, saccharine matter, mineral salts, mallic acid salt & tartaric acid, pectin (11%), riboflavin, sugars, purgative glycosides (multiflorin A & B). Genoa's farmers use both roses' species to produce an extract to use in alcoholic beverages too. Fresh flower petals are harvested before dawn and processed in dry form and macerate for 24 h with lemon and boiled water.

After that the essence is filtered and added with cane sugar and boiled for 20 min. This extract is used to flavour blond beer, wine, and cocktails (Jena et al. 2015).

12.2.2 Viola

The *Viola* genus belongs to the Violaceae family, with more than 600 species. Most of them are herbaceous perennials caulescent or acaulescent, few annual or shrubby plants. These are wildly diffused all over the world but are especially cultivated in temperate regions in North America, Europe, and Asia; these are also present in Oceania, Hawaii islands and Andes Mountain. Most of the species are cultivated especially as ornamental flowers, but these can also be found in the wild. Viola plants can reach up to 20 cm in height. The leaves are usually cordate and opposite or alternate, while the flowers are zygomorphic, with bilateral symmetry and solitary, and usually a diameter from 3 to 5 cm. The flower consists of five white, yellow, orange, red, pink, blue or purple petals. The flowering phase starts from winter to summer, earlier in warm areas (Connole 2022). Industrial cultivation takes place in greenhouses for extra-seasonal production: it is sown in trays and transplanted into pots after 8–12 weeks, in soil rich in organic matter and well oxygenated with a pH of 5.5–7. Constant watering is required to keep the soil humid, even if the viola is tolerant of dry conditions. Liquid fertilizers containing nitrogen, iron, magnesium, potassium, and phosphorus occur with the beginning of flowering phase. The plant needs high exposure to light to ensure a good blooming. Harvesting is done manually once the flower is ripe, by cutting the plant from the ground by separating the stem from the blossom (Ravenscroft 2015; Connole 2022). The flower fragrance and flavour are complicated due to ionone, a compound that interferes with scent receptors after a few mouthfuls; moreover, there's also a genetic factor how people can taste ionone: someone cannot smell or taste it at all, while others feel an unpleasant taste of soap rather than the flower aroma. Viola essence is extracted to produce liquors as Creme Yvette, Parfait d'Amour and Hpnotiq Harmonie. The Creme Yvette version, made by Cooper Spirits International, is made by a blend of cassis, berries, orange peel, honey, and an infusion of violet petals, while Parfait d'Amour is a liquor made with citrus, vanilla, spices, and roses or violets. Hpnotiq Harmonie is a liquor made by a mixture of infused violets, berries, and lavender (Stewart 2013; Tang et al. 2017).

12.2.2.1 *Viola odorata* L.

Viola odorata L. (Fig. 12.4) is a perennial evergreen herbaceous plant that can reach up to 20–30 cm in height. It is native to Europe and Asia and was introduced in North America. The leaves are bright green heart-shaped while the flowers are highly aromatic and white, pink, or purple in colour. The blooming time is from February to April. Aromatic extracts are used to flavour food and alcoholic

Fig. 12.4 Flowers of *Viola odorata* L. (actaplantarum.org)

beverages. In Austria, two varieties (Queen Charlotte and March) are used from Destillerie Purkhart for their Crème de violet liquor, obtained from an infusion of flowers, sugar and alcohol (Stewart 2013; Tang et al. 2017).

12.2.3 Lavandula

Lavandula genus belongs to the Lamiaceae family. It is an evergreen and perennial bush with stems with numerous ramifications. These plants are native to West Europe and North Africa mountains, but nowadays cultivated in all Europe, especially in France, Greece, and Turkey. They grow spontaneously in a Mediterranean climate. Mostly these species are used as ornamental flowers or for herbal purposes. Lavander plants can reach up to 40–100 cm in height. The leaves are coriaceous, linear, opposite and tomentose, greyish green in colour, while the flowers are purplish and small, gathered in vertexes forming spicastre 7 cm long. In temperate environments, flowering occurs from June to August. Planting the crop can last up to 9 years for lavendin and 12 years for lavander. The cultivation takes place mainly in the field. Clayey soils with pH 6.5–8 are used. Propagation is performed both through seeds and cuttings, although the latter solution is preferred because higher yields are obtained compared to the gamic production. In case of production in seedbeds, 600–700 plants are obtained from a quantity of seeds of 2 g/m^2. Due to the dormancy of the seeds, gibberellic acid, or treatments at 2–3 °C are used for 7–8 days. Once the plant has reached 10 cm in height, it is transplanted to have a planting density of 300–350 plants/m^2. The most modern cultivation techniques involve the use of semi-woody cuttings of 12–15 cm, taken from young mother plants in autumn or spring. They are first placed in the nursery for better rooting using rhizogenic products; after 1 year, in spring, the mechanized transplant takes place in the field with a distance between the rows of 150–200 cm, with a planting density of 1.5–2 plants/m^2. In the hottest periods, it is necessary to carry out irrigation with small volumes avoiding water stagnation which could damage the cutting rooting. Weeding should be done regularly for weed control. Fertilizations are carried out with a supply of nitrogen, phosphorus, and potassium equal to 50–60 kg/ha, with subsequent annual fertilizations in spring with a supply of 60–70 kg/ha by means of inter-row milling. The harvest takes place through mowing-binding or mowing-loading harvesting machines for which they cut the inflorescences and a small part of the herbaceous stem without damaging the woody part and the roots. Usually, the harvest takes place from the second year, with an inflorescence yield that can reach 5–7 t/ha for lavander and 12–15 t/ha for lavendin, while the shelled flowers yield is 1–1.5 t/ha for lavendin and a little less for lavender. It is better to harvest when at least 70% of the flowers are opened or when withering begins. It is necessary to harvest on dry days to avoid fermentation processes that could damage the product quality. To obtain the extract, the flowers are dried before and the product must be cleaned from leaves which would confirm a sour aroma. The whole plant emanates a strong smell due to the high essential oil content, which yields 3 times higher on dry than on fresh. The oil content varies according to the species, ranging from 0.5%–1.5% in lavender to 0.9%–3% in lavendin. The essence is a yellow green colour, mainly consisting of linalyl acetate (~50%), linalool, cineol, borneol and camphor. A quality oil has a high content of linalyl acetate and a low camphor content (maximum 10%–12%). The ratio of linalyl acetate to linalool must be greater than 1. Environmental and agronomic factors influence the quality of the oil, such as the cultivation altitude and fertilization: in fact, for the first case, for each additional meter of height there is a decrease in the oil content with the consequent improvement in quality thanks to the linalyl acetate increase at the expense of camphor; for the second case, an excess of potassium increases the level of camphor, while with nitrogen and phosphorus the opposite effect is obtained. Some lavender species may contain compounds which are toxic to human health (Basso 2009; Catizone et al. 2013; Stewart 2013).

12.2.3.1 *Lavandula angustifolia* Mill.

Lavandula angustifolia Mill. (Fig. 12.5) is an evergreen perennial suffruticose species, with erected and woody stems that can reach up to 20–40 cm in height. It is native to the Mediterranean area with dry climates and resistant to temperatures down to −20 °C in winter. As well as growing wildly up to high altitudes, it is cultivated at lower altitudes throughout Mediterranean Europe. The leaves are coriaceous, linear or linear-lanceolate with a revolute edge, opposite and more tomentose than lavendin, greyish green in colour. The inflorescence is consisting of whorls less dense than lavendin. The flowers are purplish and small. The main essential oil compounds are terpenes, tannins, flavonoids, rosmarinic acid and phytosterols. It is obtained a more valuable essential oil than lavendin. Lavender astringent essence is used to produce some alcoholic beverages, as Hpnotiq Harmonie liquor, made by a mixture of infused violets, berries and lavender, Levanduľový likér, a slovakian

Fig. 12.5 Flowers of *Lavandula angustifolia* Mill. (pixabay.com)

liquor made by fresh lavanders flowers and KuSshh liquor, made by flower maceration in spirit. Fresh flowers are used also to prepare cocktails as Lavander Martini (Basso 2009; Catizone et al. 2013; Stewart 2013).

12.2.4 Sambucus

Sambucus genus belongs to the Adoxaceae family. Shrub or small tree-shaped perennials species belong to this genus. These plants are native to temperate and subtropical regions in the world, especially diffused in Europe, Asia, North Africa, and North America. It grows well in full sun wildly at high altitudes, but it can be cultivated as an ornamental plant or in food, pharmaceutical, cosmetic and alcoholic beverage industry. Elder plants can reach up to 2–15 m in height. The leaves are opposite, pinnate, light green in colour, while the flowers are white or cream in colour, collected in bunches, which bloom in late spring, followed by small red, bluish, or black berries. The elder grows preferably in humid areas, from clay to sandy soils with pH 5.5–6.5. The propagation takes place by seed or by semi-woody cuttings of 20–30 cm rooted in the nursery to be transplanted in spring. Wide planting layouts are used with distance between rows of 2–3 m to guarantee space for the passage of harvesting machinery. The flowers harvest takes place preferably in hot afternoons when the fragrance is strongest. Matured flowers are collected, cutting the stem about 10–15 cm, under the blossom cluster (Vedel and Lange 1960; Marzi and De Mastro 2008).

12.2.4.1 *Sambucus nigra* L.

Sambucus nigra L. (Fig. 12.6) is a perennial shrub species, with highly branched stem that can reach up to 2–6 m in height. It is widespread both as a spontaneous species and as cultivated in vast Mediterranean areas. The leaves are opposite and

Fig. 12.6 Flowers of *Sambucus nigra* L. (actaplantarum.org)

composed of 5–7 ovate and acuminate light green leaves. The flowers are white, or cream collected in corymbs of 20–30 cm in diameter, while the berries are black in colour. Blooming occurs from May to July and the harvest takes place before the berries ripen in September. The flowers must mature to be dried, becoming yellow brown in colour and with a strong aroma. A semi-fat essential oil is obtained (0.03%–0.014%), through the dried flowers distillation, due to a high content of fatty acids which give it a very sweet and honeyed smell. Furthermore, there is a content of isoquercitrin, rutin and, to in minor quantity, astragaline. The extract, through benzene, is yellow-brown or dark green in colour. Sambucus essential oil is used to produce some alcoholic beverages, as St-Germain, a pale yellow liquor made by a secret process (not maceration) where the extract is combined with grape eau-de-vie, sugar, and some citrus fruit; flowers are used also to produce various elderflower wines (as Elderflower Champagne), characterized by a robust, fruity flavour, or cocktails, as Elderflower Cordial, made with water, sugar, elderflowers, lemons, oranges and citric acid, added with Champagne, Martini or Soda (Marzi and De Mastro 2008; Stewart 2013).

12.2.5 Cannabis

Cannabis genus belong to the Cannabaceae family. Hemp is an annual herb. This plant is native to Central Asia and has been introduced in Africa, Europe, and America. The stem is conical or cylindrical and the height can range according to the variety from 1 to 6 meters. Leaves are cotyledonary, small, deciduous, and fleshy, with a sustenance function in the emergence phase, while the others are palmate-seven, composed of three or more, serrated, dark green in colour. The vegetative cycle in a Western European climate ranges from 80 to 150 days, depending on the sum of temperatures. Hemp can be cultivated in the field, as a monoculture or in crop rotation, or indoors in greenhouses or in sheds. As a monoculture it is not common to use due to uncontrollable self-propagation weeds development as wild mustard and hemp broomrape. In rotation system it is recommended to cultivate sugar beet, maize or a leguminous crop before hemp, after wheat and before spring barley. Hemp is useful to leave the ground cleaned from weeds in autumn and to improve the soil structure. The best soil to grow hemp is loamy, with a pH of 6–8. The optimum temperature is in the range 19–25 °C and the irrigation must be more frequent in the flowering phase than in the seedling or vegetative phase as water stress leads to a yield loss interrupting the blooming phase. Fertilization (nitrogen, phosphorus, potassium, calcium, and magnesium) takes place based on climatic conditions and soil characteristics but usually takes place in the period of greatest growth from May to June. For sowing phase, the soil must have a temperature of 8–10 °C and usually takes place from mid-March to the end of May, sowing in a depth of 2–3 cm in soil, with a density of 35–80 kg/ha (80–100 plants/m^2). In the greenhouse, production and management are efficient and optimal as it is a controlled environment (temperature, humidity, photoperiod, weather conditions) and allows extra-seasonal production. Indoors, LED, HPS and MH light treatments are used to favour crop development. Among these, the one most used is the LED system, which thanks to the peaks in the spectrum with red and blue wavelengths, favours rapid plant development and greater biomass, leading to earlier flowering and seed production compared to other systems. Seedlings are used for sowing, and then, once the plants are vigorous, transplanted into pots one or more times to ensure maximum development. The soil must not be too wet and fertilization with granular fertilizer (nitrogen, phosphorus, potassium, calcium and magnesium) is carried out (Wellhoffer 2020). The harvest in fields normally is in late July or early August thanks to special machines that cut the part of the plant with the inflorescences, or manually in small fields or indoors. To harvest other parts of the plant a mower, swather cutter and combine harvester (Paun et al. 2020). There are dioecious varieties, with individuals with only male or female flowers, and monoecious, with individuals with flowers of both sexes. The male inflorescence is a cob, while the female one is a thick and tufted false ear. The male flowers are small buds (pollen sacs) without fluff, while the female ones have the pistil from which seeds develop. Hemp inflorescences must be harvest at full female flowers blooming. Hemp bloom from 10 to 12 weeks after sowing stage. Hemp is cultivated for several main uses in agroecosystem improvement crops, cosmetics, seeds for human nutrition, animal diets integration, textile and paper production, applications in green building and in bioenergy. Low-THC inflorescent varieties are extracted dry to obtain an essential oil through hydro distillation. *Cannabis sativa* L. (Fig. 12.7) varieties are the most used for this aim. The main essential oil compounds are terpenes, as monoterpenes (α-pinene, β-myrcene and trans-β-ocimene) sesquiterpenes (caryophillene and humulene, as the most abundant) that are responsible for the distinctive flavour and smell, different in each variety. Other important compounds are phyto-cannabinoids, typical of Cannabis plants, that are odourless, and flavonoids that confer a little bitter taste. The production ranges between 3.4–4.9 L/ha, The essential oil can be used to flavour alcoholic beverages without THC and CBD content, as Cannabis Sativa Vodka (Vodka flavoured with hemp flowers), Seshat Liquor (flowers and leaves infusion), Cannabis Spritz THL (citrus fruits, herbs and dried fruit infusion),

Fig. 12.7 *Cannabis sativa* L. (pixabay.com)

Canapito (inflorescences infusion), Cannabis Gin THL (gin with mixed berries and flowers), Ganja Lemon THL (lemon with hemp flowers) and several flavoured beers as Turn Hemp, Cannabis Lager and X-Mark Cannabis (Meier and Mediavilla 1998; Bouloc et al. 2013; Welling et al. 2016; Calzolari et al. 2017; Tang et al. 2017; Ranalli 2020).

12.3 Bitter Taste

Among all basic flavours—sweet, sour, salty, bitter, and umami—bitter is the most complex and the least understood. Perception of bitter taste depends on different factors, among them the specific individual receptors, the component of bitterness, and the interaction between them. Even if the perception of this taste is considered primarily as a defence mechanism which regulate intake of food and can be considered as an unpleasant characteristic by consumers, in recent decades bitterness is common accepted as one of the most important flavours in alcoholic beverages like beer, wine, and other liquors. Differently from beer and wine, for liquors and spirits several kinds of raw material (grains, fruits, flowers, or other carbohydrate-rich ingredients) can be used, conferring a bitter taste profile highly variable. Moreover, the processes of distillation, fermentation, and aging, together with the interaction with microbes, and quantity of higher alcohols have a strong impact on the final product. Bitter compounds in food and beverages are mainly peptides, ethers, lactones, phenols, terpenes, and flavonoids which mostly contain groups of -NO_2, =N-, -SH, -S-, -SO_3H, -S-S-, and =C=S. However, phenols and terpenes are considered as major responsible in generating bitter taste. The use of aromatic plants to this purpose has been already known from ancient history (specially for medical treatments), however changes in consumer preferences lead to an evolution in the type of flavoured alcoholic beverages consumed, but nevertheless aromatic plants and essential oils continue to have an important role (Drewnowski 1997; Tonutti and Liddle 2010; Luo et al. 2020). Different portions of the plant are utilized to confer bitter taste to alcoholic beverages but speaking of edible flowers the most important are summarized in Table 12.2. The biggest family is represented by the *Asteraceae*, by the group *Artemisia*.

12.3.1 Artemisia

Artemisia genus is mostly spread across the North Hemisphere, and it comprehends numerous herbaceous annual or perennial shrubby plants, largely utilized in liquor industry for their aromatic property. The different species are commonly divided into absinthe and *genepì*. The former group includes species famous for their bitter property, like *A. absinthium* L. (Wormwood, grand wormwood, absinthe, absinthium), *A. pontica* L. (Roman wormwood or small absinthe), and *A. vallesiaca* All. (Mountain wormwood); the latter includes different species, among which *A. genepi* Weber—black *genepì*—*A. mutellina* Will.—white genepì—and *A. glacialis* L.—major *genepi*, most utilized as aromatic compound of the Italian famous liquor *genepì*, which name might derive from the Arpitan dialect *zhènépi* or *jnépi*, used for several Artemisia species in Savoy, France. Species which have an important commercial impact are the ones included in the absinthe group (Lachenmeier 2010; Bora and Sharma 2011; Abad et al. 2012; Vouillamoz et al. 2015). They grow better in clayey-limestone drained soil, with good exposure to sunlight. The effect of temperature and light intensity on growth and production of the active ingredients of *Artemisia absinthium* L., for example, were studied using different accessions from different geographical areas. Studies have shown that higher temperatures and light

Table 12.2 Flowers species and families used to confer bitter taste to alcoholic beverages

Specie	Family
Artemisa absinthium L.	Asteraceae
Artemisia pontica L.	Asteraceae
Artemisia valesiaca all.	Asteraceae
Helichrysum italicum (Roth)	Asteraceae
Jasminum officinale L.	Oleaceae
Crocus sativus L.	Iridaceae
Hibiscus rosa-senensis Linn	Malvaceae

intensities favour an increase in the biomass produced, positively influencing both the morphological characteristics of the plant and the composition of the essential oil. Although it prefers temperatures between 23 and 25 °C, it can also resist the cold, surviving up to −18 °C (Nguyen and Németh 2016; Nguyen et al. 2018; Saunoriūtė et al. 2020).

Propagation of *A. pontica* and *A. absinthium* is performed through rhizome or bush division because plants are generally sterile. Generally, from a "mother" bush of good quality it is possible to obtain 30 "daughters" plants. Implant of rhizomes is performed in autumn (October–November), if the propagation is made from bush division implants can be made at the end of winter (February–March) (Judžentienė 2015a; Koul and Taak 2018). In the case of rhizomes propagation, 2–3 t/ha are used. In the case of seedlings, around 120,000–130,000 plants/ha. Plants are transplanted approximately 40–50 cm between rows and 30-40 cm across the row. It is possible to obtain plants in a seedbed, sowing with a density of 2 g/m^2. Organic fertilization is advisable after implant (500 q/ha) and, if the climatic trend is very dry, it is well to provide irrigation treatments in spring during plant establishment. Moreover, during vegetative stage, it is a good practice to perform a 2–3 steps chemical fertilization supplied with nitrogen, phosphorus, and potassium. In general, the harvest period starts at the second year from the implant, and it is performed during late spring-summer (between the end of May and August), when the flowers are in full blossom and they have the maximum quantity of essential oil, by mowing the plant at the base. The extraction is performed through alcoholic distillation of fresh flowers or maceration of dried flowers (method utilized for vermouth preparation). The yield of dry mass is about 100 q/ha (corresponding to 40–45 q/ha of commercialized product) (Marzi and De Mastro 2008; Basso 2009; Catizone et al. 2013). The yield of essential oil is 0.5%–0.8% and it is used both in traditional medicine and in modern pharmacology, for its tonic, febrifugal, choleretic and narcotic properties. The essential oil also has anthelmintic, anti-flu, anti-inflammatory and antimicrobial properties. It is used for the preparation of antiseptic, antidepressant, digestive, carminative and stimulant products. The characteristic bitter taste is due to the presence in the essential oil of tannins, monoterpenes and sesquiterpenes, absinthine, anabsinthine, and α-β-thujone. Harvest time must be carefully observed, to avoid yield losses of the essential oil or high accumulation in the flowers of toxic compounds such as thujone (Putievsky 1993; Lachenmeier 2010).

12.3.1.1 *Artemisia absinthium* L.

A. absinthium L. is an herbaceous perennial plant that can reach up to 1 m in height. It is a whitish-looking plant with alternate green and septate leaves. Yellow flowers are gathered in flower heads (Fig. 12.8), and they are generally sterile, rarely producing achenes. The shoot is woody and branched in the upper part. It is wildly spread in dry and rocky places all over Europe, Asia, and Africa and it has been introduced in North America for cultivation. Wormwood was already well known during the XIII century for medical treatment: in fact, this plant has tonic and

Fig. 12.8 Inflorescences of *Artemisia absinthium* L. (de Santis E. 2007; actaplantarum.org)

anti-inflammatory property. In popular tradition absinthe flowers were put in wine maceration for 4–5 days and mountaineers used to drink it every morning and after meal as digestive and vermifuge. All the organs of *A. abstinhum* are rich in bitter principles, and the principal components of the essential oil (0.2%–0.5% of the dry matter) obtained through distillation of flower (called *Absinthii herba*) or leaves (called *Absinthii folium*) are absinthine, anabsinthine, myrcene, pinene, limonene, eucalyptol, linalool, thujone, and phellandrene (Batiha et al. 2020).

Dried flowers are utilized to aromatize vermouth, bitter and other alcoholic/non-alcoholic beverages. During XIX-XX centuries this plant was extremely cultivated in France to make the famous absinthe liquor, obtained through the hydro-distillation of different parts of the plants together with other aromatic species such as fennel, coriander, hyssopus, and anise. However, due to the high consumption and the presence of thujone, which has toxic effect in high concentration, selling of absinthe was then forbidden in different Countries (Marzi and De Mastro 2008; Basso 2009; Catizone et al. 2013). Nowadays the maximum levels of thujone in beverages are 0.5 mg/kg with some exceptions, 35 mg/kg in bitter drinks. Many brands (around 200) of absinthe are produced in the USA, Australia, various European countries (Switzerland, France, Germany, Spain, Portugal, Belgium, Netherlands, England, Italy, Czech Republic). The major types are distinguished depending on to the colour of absinthe: la Verte, (green) and la Bleue (blue), but also are produced absinthe yellow (mead base) or red (an extremely rare kind has a cranberry red colour because of the hibiscus flowers used for its production) (Judžentienė 2015a).

12.3.1.2 *Artemisia pontica* L.

Artemisia pontica L. is originally from east Europe and Caucasus region. It is a perennial herbaceous rhizomatous species (60–80 cm in height), greyish, with numerous thin and ramified stems, alternate and septate leaves covered with hair (Fig. 12.9), yellow and tubular flowers reunited in flower heads and almost always

Fig. 12.9 Leaves of *Artemisia pontica* L. (Faggi G. 2012; actaplantarum.org)

sterile, so the propagation of this species is done by rhizome or bush division. Roman wormwood is nowadays highly cultivated in Italy, especially in Piedmont and Val d'Aosta regions (Marzi and De Mastro 2008; Basso 2009; Catizone et al. 2013).

The oil essence, composition and yield are like the one of *A. absinthum* L., presents a green-blue fluorescence due to chamazulene, a molecule belonging to sesquiterpenes. The yellow flowers are commonly used, together with leaves, to confer a bitter taste to vermouth (Talzhanov et al. 2005). In Europe, the legal definition of "vermouth" requires the obligation to use plants belonging to *Artemisia* group (generally *A. pontica* or *A. absinthium*). Vermouth is a particular aromatized wine developed in 1768 by Antonio and Benedetto Carpano. Nowadays Carpano, together with Martini and Cinzano, are the well-known vermouth 'brand (Tonutti and Liddle 2010).

12.3.1.3 *Artemisia vallesiaca* All.

Commonly known as mountain wormwood, this specie is wildly spread in different Italian regions such as Trentino and Val d'Aosta. Like *A. pontica* and *A. absinthium*, *A. vallesiaca* can be cultivated with a lifespan of 3–4 years, it is directly sown at the beginning of the spring in rows 60–70 cm (about 2.3 ft) apart. The subsequent cultivation operations are like those of the other species. Harvest is performed at the end of summer before the complete blossom. The liquor industry requires flowering tops which have a characteristic aroma (production 60–70 q/ha) (Marzi and De Mastro 2008).

12.3.2 Helichrysum italicum *(Roth)*

Helichrysum italicum belongs to the Asteraceae family and it is a perennial woody and branchy suffrutex wildly spreads in dry-rocky places of the Mediterranean area, up to a height of 600–800 m. The plant is compact, with woody stems of 25–40 cm and dense, linear, silver-green leaves. Light-yellow flowers are tubular and gathered in heads. Flower heads are in turn gathered in corymbs of 25–35 inflorescences. This plant belongs to the habitat of *Lauretum*, so it is thermophilic, typical of arid environments well protected from the cold (Marzi and De Mastro 2008; Basso 2009; Catizone et al. 2013; Ninčević et al. 2019).

Helichrysum italicum can be again divided in several subspecies of which two are the most utilized for essential oil extraction (Abbate et al. 2004; Cesaraccio and Pellizzaro 2004):

- *H. italicum* ssp. *italicum*, most common (Fig. 12.10) (Bianchini et al. 2001).
- *H. italicum* ssp. *microphillum*, found in Sardinia region, Corsica, and Balearic Island (Taglialatela-Scafati et al. 2013).

H. serotinum was previously considered as a subspecie of *H. italicum*. Nowadays it is a separate specie, found in the Iberic peninsula and recognizable from its larger leaves and late flowering. *H. serotinum* is also important for essential oil extraction (Galbany-Casals et al. 2011).

Even if in the Mediterranean area is a spontaneous plant, *Helichrysum* can also be cultivated, with a lifespan of 6 years, through direct sown, cutting propagation or bush division. Transplant in field is done in autumn and, depending on the availability of water and the exploitation of the crop, it is possible to recognize two kinds of cultivation: "thick" implant, 12–18 plants/m^2 (essential oil production); "sparse" implant, 3–5 plants/m^2 (production of dry matter). Helichrysum is a rustic and undemanding plant and for this reason it does not require high levels of nutrients during fertilization. Before preparation of implants, phosphorus and potassium must be distributed (80–100 kg/ha) and then must be then applied every 2–3 years. Nitrogen

Fig. 12.10 Flowers of *Helichrysum italicum* (Roth) ssp. *Italicum* (pixabay.com)

(30–50 kg/ha) is administered during sowing or transplant, after 50 days. Moreover, N application is required every year, at the start of the vegetative phase. Even if this plant is adapted to survive in dry climate, irrigation is an important factor to reach higher yield. However, to avoid water stagnation, irrigation is necessary at the beginning (sowing, transplant, first development stage) to favour nutrients absorption. The highest production is reached during the third and the fourth year. It is possible to obtain 8–12 t/ha of biomass, corresponding to 2–4 t/ha of dry weight, and equal to 5–15 kg/ha of essential oil. All the aerial parts of the plant are harvested to obtain essential oil, but it is possible to obtain a more sophisticated product by selecting only flowering tops (yield of dry flowers: 1.4–7 t/ha) (Marzi and De Mastro 2008; Basso 2009; Catizone et al. 2013). The harvested parts must be quickly dried or distilled, to avoid fermentation processes. Through the hydro-alcoholic distillation of flowers a bright-yellow essential oil rich in α-β- pinene, flavonoids, caffeic acid, and sesquiterpenes is obtained which confer the bitter taste. This oil has cough-expectorant, anti-inflammatory, analgesic, anti-rheumatic, anti-arthritic, purifying, healing, antibacterial and anti-allergic properties (Tomas-Barberan et al. 1990; Pietta et al. 1991). A Sardinian typical liquor (*Caru Elicrisio*) is obtained through a hydro-alcoholic infusion of *Helicrysum* blooms. A great demand for the H. *italicum* essential oil resulted in the huge expansion of its plantation cultivation in several European Countries (Ninčević et al. 2019).

12.3.3 Jasminum officinale

The genus *Jasminum* belongs to the family of Oleaceae, and it has its origin in different regions of Persia and Kashmir. *Jasminum officinale* L. (Fig. 12.11), commonly known as jasmine, is spread in mild-winter climate part of Mediterranean area, such as Egypt, Syria, Cyprus, South France, and Italy, but this flower is

Fig. 12.11 Flowers of *Jasminum officinale* L. (pixabay.com)

commonly cultivated also in India and China (Balkrishna et al. 2022). This specie has an ancient history: jasmine flowers were offered to the god Vishnu during celebration. European diffusion was around XVI century as cosmetic product and, during XX century the province of Reggio Calabria (Italy) became one of the largest producers, thanks to the positive climate conditions (flower yield of 4500 kg/ha, almost doubled compared to other countries in the same period, and 1500 kg of essential oil). Nowadays the market competition with India and North Africa, together with the high-cost production, led to the disappearance of this cultivation in Italy (today yield of essential oil is 200 kg) (Marzi and De Mastro 2008; Basso 2009; Catizone et al. 2013).

Jasmine cultivation is practiced in mild climate, free from frosts. Generally, the propagation is performed by cutting plants sown and grown up in a nursery and then transplanted during the following spring. Plant density is variable due to soil fertility and sun exposure, but it is around 5000–10,000 root cuttings. *J. officinale* lifespan of cultivation is up to 10–12 years, and the large production of flowers begins in the fourth year. Propagation is performed by cutting. Plant establishment must be advantaged through several irrigations, weed control, and pruning when necessary. Flowers have a scalar blossom from June, with their apex during August, but it can vary because it is strongly influenced by high temperature. In fact, in Egypt or Morocco flower blossom begins in June and finishes in December. Harvest is performed by hand in the early morning and then flowers are rapidly processed through a two-distillation phase to obtain the essential oil (from 1000 kg of flowers 2.8–3.4 kg of essential oil is obtained). The fragrance of jasmine is due to different terpene compounds (especially linalool and terpinol), phenyl-acetic acid and chitons (mostly jasmonate) (Bera et al. 2015; Rassem et al. 2018). Jasmine is especially relevant for perfume and cosmetic industry, but the flower can be utilized also for the preparation or aromatization of liquors and spirits: during XVIII-XIX centuries jasmine has been used in different alcoholic recipes. Nowadays jasmine is an uncommon ingredient in alcoholic beverages, due to the great cost of production (it occurs at least 6000 flowers to obtain 1 kg of product). However, it is possible to find a jasmine-infused Cognac (produced by Jacques Cardin) and different liquors made by two American distillers (Koval and GreenBar Collective) (Stewart 2013).

12.3.4 Crocus sativus *L.*

Saffron (*C. sativus* L.) (Fig. 12.12) is a perennial herbaceous plant belonging to the Iridaceae family. The adult plant consists of a bulb-tuber (globular corm, 3–4 cm in diameter) which contains about 20 undifferentiated gems, even if flowers and leaves originate from only three of them. During vegetative stage a sprout develops for each of the main buds (2–3 per corm). Sprouts, containing flowers and leaves, emerge from the ground wrapped in a whitish sheath, reaching up to 20–35 cm in height. The flower is a perigonium with six petals of an intense violet colour (Marzi and De Mastro 2008; Basso 2009; Catizone et al. 2013).

Fig. 12.12 Flowers of Crocus sativus L. (pixabay.com)

In the *Crocus* genus it is possible to find numerous species, some of them wildly spread, but saffron, originated in Minor Asia, is cultivated in all the Mediterranean area (from west to east) of Europe and North Africa (Mucci and Macchia 2001). The great importance of this plant for food, cosmetic, and medicinal uses has been already recognized in ancient age: in the frescoes of Knossos (1660 B.C.) it is possible to observe women employed in the collection of saffron flowers and Plinius claimed that best quality saffron originated in Cilicia. Phoenicians and Persians already knew the cultivation of this plant. In the Roman empire, Galicia Region was famous for saffron production, and during Middle Age saffron cultivated in Tuscany, Abruzzo, and Umbria was considered as one of the most valuable species (Khan et al. 2021). Nowadays, even if its cultivation has lost some interest due to the high cost of production, saffron is still utilized for its flavouring and colouring propriety (Jafari-Ghanavati and Saket 2020). The most precious varieties are:

- Italian saffron: red colour and large stigmas. The one cultivated in Aquila province is the most known and important.
- Spanish saffron: smaller stigmas but redder compared to Italian variety.
- French saffron: long and large stigmas, good colour (Marzi and De Mastro 2008; Basso 2009; Catizone et al. 2013).

The lifespan of this cultivation is 2–3 years, even if annual cultivation is preferred because it leads to higher production. At the end of the life cycle, it is recommended to perform a long crop rotation, with a return of saffron in the same field after 7–9 years to avoid fungal disease. Before implant soil must be fertilized with an organic fertilizer (supplemented with 50–60 kg/ha of P and K). In some Region like Sardinia, saffron cultivation is associated to almond trees one. During vegetative phase, *C. sativus* prefers Mediterranean climate (12–18 °C) even if it can adapt to cold temperatures. However, a mild and dry climate is required during flowering time because excessive rainfall during spring can increase the risk of fungal disease (especially *Fusarium* and *Rhizoctonia*). To overcome these problems, in the last decades different indoor systems have been proposed and between them also hydroponic and aeroponic cultivations (Souret and Weathers 2000; Mollafilabi et al. 2013; Schroeder et al. 2020).

Saffron is triploid and sterile, so its propagation is agamic: "daughter" corms originate from the "mother" corm during the vegetative phase. In the annual cultivation, corms selected for propagation must have at least 2.5 cm of diameter to generate flowers, but best results are obtained with corms which has a diameter > 4 cm. Density in field is about 500,000–600,000 corms and each corm produces 2–2.5 purple-violet flowers in a two-week period in autumn (mid-October, early November, potential yield of 1.2–1.5 million flowers/ha) (Rashed-Mohassel 2020). Studies have demonstrated that treatment of the corms with GA3 10 ppm or 2.4D at 50 ppm before transplants can enhance the number of flowers/corms (Mzabri et al. 2021). Inside the flower it is possible to recognize the most important part, the three-part red stigma which contains the active principles of saffron.

Flowers are manually harvested at sunrise and then they are immediately processed to separate and to dry stigmas. Dried stigmas must be intact and preserve their red coloration and aroma. One-hundred-forty thousand flowers are necessary to obtain 1 kg of dried stigmas. After the harvest and the desiccation of aerial parts, new corms are ready to be utilized in the next implants. The compounds which confer to saffron spice the typical bitter taste and the red orange colour are three: the crocin (colour), the picrocrocin (taste), and the safranal (aroma) (Mzabri and Addi 2019). Moreover, saffron is rich in carotenoids, like zeaxanthin and lycopene, and it contains more than 150 volatile compounds (Ríos et al. 1996; Kumar et al. 2009). The quality of saffron depends on the concentration of these three main metabolites providing the unique colour and flavour of stigmas. Their contents depend on the environment and cultural practices (Gresta et al. 2009). In liquor industry, stigmas (ore the pulverized spice) are utilized to confer the yellow colour to the Italian liquor "Strega" or Chartreuse-like liquors produced in Spain, France, and Italy. Also, the Benedictine producers admitted that saffron is one of secret ingredients. Again, the famous bitter Fernet Branca also uses this spice to confer the bitter taste to the beverage (Stewart 2013).

12.3.5 Hibiscus rosa-sinensis *L.*

Chinese hibiscus (*H. rosa-sinensis*) (Fig. 12.13) is a glabrous perennial shrub belonging to the family of Malvaceae. The petals may be smooth or scalloped, single or double. The leaves of the *Hibiscus* are ovate, and they grow from 5 to 15 cm. The fruit of the red Hibiscus is dry, and contains up to three seeds, each of which is kidney shaped and 2.5 cm long. This plant originated in China, but it is cultivated for its ornamental values in several place with moderate temperature and

Fig. 12.13 Flower of *Hibiscus ros-sinensis* L. (Rossi F. 2007; actaplantarum.org)

high humidity conditions. Plants are usually propagated by cuttings and there are several forms with different colours, however red flowered is preferred for its use in traditional medicine and as flavouring agent in food and beverages. Juices extracted from flowers and leaves have been used since ancient times as a natural remedy for disease and other symptoms: *Hibiscus* has in fact antitumor, antioxidant, analgesic, antipyretic, and anti-asthmatic proprieties. The edible flowers contain moisture, nitrogen, calcium, iron, and phosphorus. It has been reported that Chinese hibiscus contains a variety of compounds such as phlobatannins, saponin, glycosides, flavonoids, terpenoid, thiamine, riboflavin, and niacin (Jadhav et al. 2009; Alabade et al. 2018). Hibiscus flowers can be used to confer red colour to a particular and rare red absinthe (Judžentienė 2015a) or to obtain a liquor produced through an artisanal method, leaving dried flower in alcoholic maceration. An analysis of the main class of volatile compounds reveals that the aroma of the liquor is mainly determined by limonene, α-pinene, p-cymene, myrcene, and β-pinene. Flavonoids represent the main components involved in conferment of astringency and bitterness, while anthocyanins are responsible for the red colour of the beverage (Pieracci et al. 2021).

12.4 Aromatic Bitter Taste

Several edible flowers used for alcoholic beverage cannot be grouped only in aromatic or bitter categories (Tonutti and Liddle 2010). Some species (Table 12.3), which will be further developed, are characterized not only by bitter taste, but also contain important metabolites which contribute to aromatic notes.

Table 12.3 Species and family of the flowers utilized to conferring aromatic bitter taste to alcoholic beverages

Specie	Family
Hyssopus officinalis L.	Lamiaceae
Taraxacum officinalis L.	Asteraceae
Lonicera japonica Thunb.	Caprifoliaceae
Citrus × *aurantium* L.	Rutaceae
Humulus lupulus L.	Cannabaceae

Fig. 12.14 Flowers of *Hyssopus officinalis* L. (pixabay.com)

12.4.1 Hyssopus officinalis *L.*

The genus *Hyssopus* belongs to the family of Lamiaceae and includes 11 species. *Hyssopus officinalis* L. (Fig. 12.14) is commonly known as hyssop, is native to central and south Europe, west Asia, and north Africa (Judžentienė 2015b). *H. officinalis* has the highest essential oil content of the genus and the most economically important (Mutu et al. 2014).

It is a perennial, semi-shrubby herb that grows on dry, rocky, and calcareous soil, with xerophytic characteristics and tolerant to drought, cold and frost (Judžentienė 2015b; Guerrini et al. 2021; Nair 2022). The plant can reach a height of 0.2–0.6 m and have small, linear, lanceolate dark-green leaves. The strongly aromatic flowers bloom during the summer, from June to August, are grouped in whorls forming spike-like inflorescence, arranged unilaterally on terminal branches. Corollas are small in dimension, from 7 to 11 mm, two-lipped and erect. Flowers' colour could be blue, pink, pale violet, white or purple (Judžentienė 2015b; Guerrini et al. 2021). Indeed, *H. officinalis* is characterized by a great heterogeneity due to many subspecies and varieties, which results in morphological and biochemical differences within the species (Mutu et al. 2014). Thus, the characteristic aroma and the composition of essential oil can be influenced by genetic variations. The main components are the terpenes pinocamphene, isopinocamphene, β-pinene, 1,8-cineole, myrtenol, linalool, and pinocarvone but the metabolites ratio and nature can be influenced also by growing condition and agronomic practices (Khazaie et al. 2008; Kizil et al. 2016; Wesolowska and Jadczak 2018; Aćimović et al. 2021). Phenolic compounds and flavones are the most important phenolic compound in hyssop. All the components contribute to the bitter and spicy aroma of hyssop (Fathiazad and Hamedeyazdan 2011; Ravindran et al. 2012; Wesolowska and Jadczak 2018; Babich et al. 2021). The herb is present in the wild flora but is mainly cultivated to produce essential oil in Europe (France, Spain, Italy, Germany, Slovenia, Moldova, Bulgaria) and in Iran, USA, China, and India (Mutu et al. 2014; Judžentienė 2015b). It is use also for ornamental purposes. Hyssop is a self-fertile plant which can be cultivated for about 10 years. Propagation can be done from cutting, by division of adult plants or from seed (Judžentienė 2015b; Nair 2022). It grows well in light and rich soil, but can be adapted also to poor, dry, sandy soil as well. The ideal pH range of 6.6–8.5. The plant is well suited to dry period; thus, it represents a valuable alternative to marginal and difficult area or as alternative crop in dry years. Hyssop can be also grown in pots, which reduce water demand. Several studies also reported growing *H. officinalis* in hydroponic systems, where the manipulation of mineral composition and salinity of mineral solution can have positive effects on the amount of essential oil and vary the composition (Moro et al. 2011; Jahantigh et al. 2016; Skrypnik et al. 2022).

Hyssop flower are important ingredients in two important French liquors, Bénédictine and Chartreuse (Tonutti and Liddle 2010; Wesolowska and Jadczak 2018). The first one is a medicinal beverage with more than 20 plants and spices, where hyssop is responsible of giving a minty aroma. Chartreuse is a liquor made up from more than 130 plants, it exists in several forms according to the ratio of the plants. Hyssop flower confers a bitter but warm taste, so they are used in the preparation of several liquors and bitters, such as Italian Amaro Ramazzotti. It is an essential ingredient for liquors wine, such as vermouth, in which is used for both its bitter and aromatic notes (Butnariu and Butu 2019; Morata et al. 2019).

12.4.2 **Taraxacum officinale** ***F.H.Wigg.***

Taraxacum officinale F.H.Wigg. (Fig. 12.15) belongs to the Asteraceae family and is also known as dandelion. It can be found throughout the temperate zone of the northern hemisphere in Europe, North America, and Asia (Grauso et al. 2019; Lis and Olas 2019; Motti et al. 2022). *T. officinale* is a perennial weed adapted to

Fig. 12.15 Flower of Taraxacum officinale F.H.Wigg. (pixabay.com)

every soil type, presents a taproot and rosette-clustered dark green leaves, 5–40 cm long (Schütz et al. 2006; Grauso et al. 2019; Lis and Olas 2019) (Schütz et al. 2006; Grauso et al. 2019; Lis and Olas 2019). Shapes of leaves can be different, from lobe less to tooth-edged to highly incised. Each plant can develop flowering stalks 5–40 cm long with a terminal inflorescence, composed from 140 to 400 of yellow, ligulate flowers. Every dandelion can develop up to 10 flowers (Schütz et al. 2006). Usually, two blooming period can be observed, the first during April–May and the second in September (Lis and Olas 2019). Flowers of dandelion are rich in carotenoids, flavonoids, phenolic acids, and terpenes (Schütz et al. 2006; Martinez et al. 2015; Grauso et al. 2019; Lis and Olas 2019; Bashir and Peer 2022; Motti et al. 2022; Olas 2022; Zhang et al. 2022). Among the last, sesquiterpene lactones and caffeoylquinic acid derivatives are the most abundant, which contribute to the characteristic bitter taste (Lis and Olas 2019; Zhang et al. 2022). Dandelion is a rustic plant, which grows spontaneously on meadows, lawns, roadsides, gardens, orchards, and wasteland (Alexopoulos et al. 2021a). It can tolerate a wide range of conditions, from drought to frost. The ideal soil for cultivation should be rich in humus, calcium and water and protected by the wind. Plants can be cultivated also in greenhouses and in hydroponic system, to obtain standardized high-quality material independently from external condition (Martinez et al. 2015; Gill 2016; Alexopoulos et al. 2021a, b). Dandelion flowers have been used since Celts to produce wine: the fermentation process of ligulate flowers produced a rich in flavour wine, with an alcoholic percentage of 12%–13% (Grauso et al. 2019; Olas 2022). *T. officinale* is also used to make liquors or to add aroma to Belgian *saison* style beer. In this case, flowers are added after the boiling process to add flavour and not bitterness (Bashir and Peer 2022; Olas 2022).

Fig. 12.16 Flowers of *Lonicera japonica* Thunb. (pixabay.com)

12.4.3 Lonicera japonica *Thunb.*

Lonicera japonica Thunb. (Fig. 12.16) is a species native to eastern Asia, commonly used as ornamental plant thanks to its pleasant and fragrant flower and its rapid growth (Starr et al. 2003; Schierenbeck 2004; Li and Lu 2015; Zhou et al. 2020) (Starr et al. 2003; Schierenbeck 2004; Li and Lu 2015; Zhou et al. 2020). *Lonicera* is the largest genus of the family of Caprifoliaceae, which counts approximately 120 species, most of them known as honeysuckle (Ikeda et al. 1994; Schierenbeck 2004; Qiu et al. 2021; Ge et al. 2022). *L. japonica*, or Japanese or chines honeysuckle, is an evergreen climbing wine, with ovate leaves. For each axillary cymes two flowers are present, with a white corolla which can turn yellow or pink (Starr et al. 2003; Shang et al. 2011; Yuan et al. 2014; Li and Lu 2015). The flowering occurs from May to September. According to the different stage of flowering, from the bud to the complete gold flowering stage, the composition in metabolite can change. Hydrocarbons, alcohols, aldehydes, ketones, and esters are the most representative ones. Linalool and jasmone and their derivatives are the most important. Regarding the fragrance, the key odorants of Japanese honeysuckle flowers are benzaldehyde, 4-ethylphenol, decanoic acid, vanillin, 3-methyl-2-butenal, and β-ionone (Ikeda et al. 1994; Su et al. 2020).

Besides spontaneous growth, *L. japonica* can be cultivated from cuttings and seeds very easily, not only for ornamental purposes but also to produce flowers and berries, both traditionally used in Chinese traditional medicine (Starr et al. 2003; Hikosaka et al. 2017; Ge et al. 2022). Greenhouse cultivation can contribute to the standardization of the metabolites content (Yan et al. 2016). In addition, the use of supplemental lighting allows to obtain higher yield and a greater number of floral buds and make possible the growth of honeysuckles around the year. At the same time, the cultivation in controlled environment of *L. japonica* can help the control of the widespread of the plant. The ancient Chinese pharmacopeia suggests the preparation of wine brew with honeysuckle flower, to cure several diseases. Traditional recipes soak 45–55 g of honeysuckle flowers in 500 mL of wine for

25–50 days (Starr et al. 2003; Shang et al. 2011; Li and Lu 2015; Zhou et al. 2020). Some studies suggest alternative fermentation methods in order to perform not only a simple extraction of the active principles but also to produce new products. Honeysuckle can be used also to aromatize other alcoholic product, such as vodka (Zhou et al. 2020).

12.4.4 Citrus × aurantium *L.*

Citrus × aurantium L. (Fig. 12.17) is commonly known as bitter or sour orange and it is part of the genus *Citrus* (Agustí et al. 2014; Santis and Frangipane 2020). It belongs to Rutaceae family, which includes important worldwide fruits, such as oranges, tangerines, limes, lemons, and grapefruits (Agustí et al. 2014; Santis and Frangipane 2020). *C. aurantium* fruits have some different characteristics from oranges, like acid pulp and a more bitter albedo (Farag et al. 2020). Bitter orange is native of tropical region of Asia, and it can be found in tropics and subtropics, especially along the Mediterranean coast (Suryawanshi 2011; Agustí et al. 2014; Farag et al. 2020; Santis and Frangipane 2020). Tree of *C. aurantium* present long, ovate leaves and pure white flowers. The flowers are used for the extraction of essential oil, named neroli (Dugo et al. 2011; Karthikeyan 2014). Linalool is the most important and characterizing compound, followed by linalyl acetate and limonene (Karthikeyan 2014; Farag et al. 2020). Deep, loose, well-aerated and well-drained soil are the ideal condition for citrus growing, with pH range from 5.5 to 7.5

Fig. 12.17 Flowers of *Citrus × aurantium* L. (pixabay.com)

(Karthikeyan 2014; Santis and Frangipane 2020). Optimal growing temperature are around 29–35 °C, *C. aurantium* has not specific need of chilling (Ammar et al. 2012; Karthikeyan 2014). Citrus are cultivated in most of the tropical and subtropical regions, for both fruits and flowers production and commercialization (Karthikeyan 2014). Flowers not only used for essential oil production but also for aromatization of beverage. Flowers of *C. aurantium* can be used for aromatization of special beers but also to produce liquors, with other flowers such as bergamot and jasmine, other herbs or by own.

12.4.5 Humulus lupulus *L.*

Humulus lupulus L. (Fig. 12.18) is commonly known as hop and is part of Cannabaceae family (Basso 2009; Zangrando and Marconi 2018). It is native to Europe, western Asia and North America and it is appreciated for its secondary metabolite, due to its essential role in beer (Basso 2009), making hop the most important edible flower to produce alcoholic beverages (Václav et al. 2020). The most known varieties are five (Bocquet et al. 2018):

- *H. lupulus* var. Neomexicanus
- *H. lupulus* var. Lupuloides
- *H. lupulus* var. Pubescens
- *H. lupulus* var. Cordifolius
- *H. lupulus* var. Lupulus

Fig. 12.18 Flowers of Humulus lupulus L. (pixabay.com)

H. lupulus var. Lupulus is the most economic relevant. Varieties used for beer are very similar to their wild relative, excepted for some characteristics useful for production, such as shape, structure, and dimension of flowers (Acosta-Rangel et al. 2021; Afonso et al. 2022). Hop is a dioecious perennial herbaceous climbing plant (Marzi and De Mastro 2008; Basso 2009), which can reach 6–10 m height, growing very fast, up to 30 cm/day (Zangrando and Marconi 2018; Mpanga and Schalau 2020). The radical apparatus is about 3 m deep, roots and rhizomes made up a structure which is call rhizome crown. From the rhizomes new bines develop during spring. The bines, firstly herbaceous than woody, need structure on which grown on (Zangrando and Marconi 2018). Leaves are opposite, lobed with 3–5-7 lobes. From the axils of the leaves, secondary bines can develop, which will differentiate inflorescences. Male inflorescence are inconspicuous green-white grapes, while female are flower cones, known as hops (Marzi and De Mastro 2008). The herbaceous flower cones are made up of bract and bracteole on a strig and measure up to 2.5 cm length, with up to 60 flowers inside each cone. Attached to the base of the bracteoles the lupulin gland are located (Almaguer et al. 2014; Zangrando and Marconi 2018). Lupulin is a resinous substance made up of essential oils and hop acids. Hop acids are divided into alpha-acids, which include cohumulone, adhumulone and humulone, and beta-acids, which are colupolone, lupolone and adlupulone (Almaguer et al. 2014; Bocquet et al. 2018; Zangrando and Marconi 2018). Alpha acids are the most important metabolite in brewing process and different hop can be classified according to the alpha-acids present: if the content of cohumulone is less than 30% of the total of alpha acid, hop will be mostly aromatic; if the content of cohumulone is more than 30%, the bitterness will be the most characterizing aspects (Bocquet et al. 2018; Zangrando and Marconi 2018). Essential oils are important in giving aroma to hops (Marzi and De Mastro 2008); about 400 compounds have been identified by chromatographic analysis, but only 70 have been recognized. The most important are myrcene, humulene and caryophyllene, all terpenic hydrocarbons (Aberl and Coelhan 2012; Almaguer et al. 2014; Bocquet et al. 2018). Other important metabolites are polyphenols, such as xanthohumol (with a very strong bitter taste), and tannins, which are helpful in the sedimentation process of malt, but if present in too high concentration may result in defects (Zangrando and Marconi 2018).

The cultivation region of hop is strictly delimited to light availability and temperature, so the range is from 30–35° to 50–55° latitude. In the northern hemisphere it is commonly cultivated in Europe, Asia, and North America (Zangrando and Marconi 2018); in the southern hemisphere, cultivation zone is less extent and concentrate in South America (Argentina and Uruguay), a very little part of South Africa, Victoria and Tasmania state in Australia and in the meridional island of New Zealand. Usually hop plants are propagated using root shoots or rhizomes, which are planted during early spring in rows. Special structures are needed, usually made up of vertical wires reaching 5–7 m, supported by overhead trellis (Zangrando and Marconi 2018). In recent years, some studies have tried to employ espaliers with limited height (3 m) to favour mechanization of harvesting operation (Blackman and Wilson 2002; Zangrando and Marconi 2018). Hops are harvested at the end of the summer, usually from half August to half September; mature cones are light-green colour and bracts still close. After harvest, cones are separated from bines and dried, reducing water content from 80% to 6%–8% (Marzi and De Mastro 2008; Basso 2009; Zangrando and Marconi 2018). Hop is highly sensitive to variations in the number of hours of lighting. Indeed, photoperiod is a parameter to stop the dormancy, which is interrupted when 16–18 h of light are reached, and to promote the flowering, which occurs only when the hours of light decrease to a certain amount cultivar dependent. Together with the climate needs, photoperiod contribute to determine the "hop belt", the latitude range in which the cultivation of *H. lupulus* is possible (Acosta-Rangel et al. 2021). Hop plants need a vernalization period to have an optimal growth and production, usually intended as temperature from 4.4 to 6 °C for a period of 30–60 days (Bauerle 2019). Hops is adapted to growth in sandy and clay soil, with ideal condition in medium-textured soil with a good humidity and well drained (Basso 2009); optimum pH value is around 6. Some studies are focusing on the optimization of hop cultivation in soilless system (Tadevosyan et al. 2005, 2009). The adoption of such method presents several advantages: the protection and the independence from external factors, the possibilities of artificial illumination and temperature management, the reduction of plants' distance and the better water use efficiency (Sullivan et al. 2020).

As is well known, hop is used for beer production. It has an essential role in giving it its characteristic taste, in the formation of beer foam, and in the extension of its shelf life (Almaguer et al. 2014). As previously say, different hop can provide different flavour (Almaguer et al. 2015): hops which contain less alpha-acids have a more flavours and aroma, while if alpha-acids content is high, the hop will give more bitterness. To measure bitterness of different hops, the IBU scale is used. Highest IBU value equals to strongly bitter beer. Furthermore, the moment in which hop is added during the brewing process is pivotal. Usually, hop is used during boiling process, to give beer the bitter note (Marzi and De Mastro 2008); more long will be the boiling period since hop add, more the alpha acids will be isomerized in form enhancing bitterness. If hop is used after batches, the role of essential oil will be predominant, and the aromatic properties of the flower will enrich the final product (Bocquet et al. 2018; Zangrando and Marconi 2018; Acosta-Rangel et al. 2021). Hop is also used to produce other alcoholic beverage, such as liquor and bitter digestive (Vázquez-Araújo et al. 2013).

12.5 Sweet Taste

Alcoholic drinks produce both unpleasant and enjoyable flavours (Scinska et al. 2000). Bitterness is often seen as aversive (Kampov-Polevoy et al. 1999) and is something that humans are born with a distaste for. Humans, on the other hand, have a natural predisposition for sweet flavours. A pleasurable reaction can be induced by alcoholic beverages either directly through the ethanol (Kranzler et al. 2001) or indirectly through residual sugars or sweet mixers. Some human and animal

Table 12.4 Species and family of the flowers utilized to conferring sweet taste to alcoholic beverages

Specie	Family
Madhuca longifolia	Sapotaceae
Primula vulgaris Huds.	Primulaceae
Nelumbo nucifera	Nelumbaceae

evidence (Bogucka-Bonikowska et al. 2001) show connections between a liking for sweet foods and alcohol consumption, but other (Glendinning 1994; Steiner et al. 2001) do not. Some species, summarized in Table 12.4, are utilized to confer sweet taste to alcoholic beverages.

12.5.1 Madhuca longifolia

A tropical Indian tree, *Madhuca longifolia* (Sangeetha and Nagappan 2016) can be found mostly in Nepal, Myanmar, Sri Lanka, and the central, southern, and northern plains and forests of India. Madhka, Madkam, Mahuwa, Butter Tree, Mahua, Mahwa, Mee, or Vippa Chettu are some of its more popular names (Suryawanshi and Mokat 2021). It is a fast-growing tree with an average height of 20 meters, evergreen or semi-evergreen foliage, and it is a member of the Sapotaceae family (Sangeetha and Nagappan 2016). It is a significant tree in tropical mixed deciduous forests in the Indian states of Odisha, Chhattisgarh, Jharkhand, Uttar Pradesh, Bihar, Maharashtra, Andhra Pradesh, Madhya Pradesh, Kerala, Gujarat, West Bengal, and Tamil Nadu, demonstrating its adaptability to arid conditions. Flowers (Fig. 12.19) are numerous, near the ends of branches, drooping on pedicels. The corolla is yellowish-white while anthers have stiff hairs and there are 20–30 stamens. Fruits are ovoid, fleshy, green berries with 14 seeds. It can withstand minor frost and thrives in environments with average yearly daytime temperatures between 2 and 46 °C. It thrives best in regions with mean annual precipitation of between 550 and 1500 mm. Moreover, it prefers a deep, sandy-loam or loam soil with excellent drainage. It can also be found in clayey, calcareous, and shallow soils. Established plants can withstand drought. It begins bearing when the tree is about 10 years old and has a lengthy lifespan. Up to 90 kilogram of blooms can be produced annually by a fully established tree. If trees are cut down during the hot season when they are dormant, they produce dense branches which can provide a mean annual increment of 3–5 cubic meters/ha using a 25–30-year growing cycle.

Due to the high sugar content, dried Mahua flowers are a desirable source of fermented goods. Mahua wine can also be made with fresh flowers, according to reports (Yadav et al. 2009) and is consume as a country liquor by the rural and tribal people in India (Mishra and Poonia 2019). Dry mahua flowers have been used to make a variety of products, including alcohol, brandy, acetone, ethanol, lactic acid, and other fermented goods (Patel and Naik 2010). Mahua alcohol, which is derived from the blossoms of the *Madhuca longifolia*, is prized for its floral nuances, sweetness, and smoky undertones. After being manually harvested, these bulbous, pale

Fig. 12.19 Flowers of *Madhuca longifolia* J.F.Macbr. (wikipedia.org)

yellow, and sap-drenched blooms are sorted, steeped, and then fermented. Over burned fermenting fluids are distilled in pots and pans. Furthermore, mahua is utilized as a local alcoholic beverage and has medical benefits due to its anti-inflammatory and anti-microbial characteristics (Mishra and Poonia 2019). Essential minerals including Ca, P, Fe, and K are present in mahua flowers (Kumar and Pradhan 2013; Saif et al. 2020).

Mahua's final test involves splashing it upon an open flame to determine its potency and purity. The spirit is deemed to be at its ideal distillation stage if the fire produces a raging fire. Alcohol by volume in traditional mahua ranges from 10% to 25%. Mahua is typically sold by distilleries with a lower alcohol content—between 5%–7.5%. According to a survey by the Chhattisgarh State of Rural Development, up to 90% of mahua flowers in India are being utilized to produce alcohol.

12.5.2 Primula vulgaris *Huds.*

The common primrose, *Primula vulgaris* (Fig. 12.20), is a species of flowering plant in the Primulaceae family that is indigenous to parts of southwest Asia, northwest Africa, and western and southern Europe (Ramsey 2019). A perennial with a height range of 10–30 cm (4–12 inches), *Primula vulgaris* has a basal rosette of leaves that, in favourable circumstances, are essentially evergreen. On slopes and in meadows, it blooms in the northern hemisphere's early spring (February to April).

The leaves have an irregularly crenate to dentate edge and measure 5–25 cm long by 2–6 cm wide. They are frequently highly wrinkled. The toothing on the leaf blade is irregular and gradually attenuates toward the base. The leaf rosette's core conceals the one, incredibly short stalk. The 2–4 cm wide, softly perfumed flowers are produced singly on short, slender stems (Fisher et al. 2008).

Fig. 12.20 Flowers of *Primula vulgaris* Huds. (wikipedia.org)

Although white or pink varieties are frequently found in nature, the flowers are normally a pale-yellow colour. The actinomorphic flowers have a superior ovary that eventually develops into a capsule with valves that opens to release the tiny black seeds. The flowers are hermaphrodite but heterostylous; each plant produces either pin flowers or thrum flowers, which are longuistylous flowers with a conspicuous style capita (brevistylous flower: with the stamens prominent). Only the area between pin and thrum blooms is fertile. Pollination from pin to pin and thrum to thrum is ineffectual.

P. vulgaris can completely cover the ground in open forests and shady hedgerows under the correct conditions. Stream banks, shrubs, orchards, and clear, damp deciduous forests are where you'll mostly find it. It can also occasionally be seen in meadows. Plants grow best in Central Europe's moderate winter climates on nutrient-rich but lime-poor, humus-rich, loose, and frequently rocky loam soils.

A common plant in cottage gardens, wild primrose is easily obtained as seeds or young plants. It thrives in light shade and moist yet well-drained soil. By seed and division, it grows. The throat typically has a deep yellow hue. The nineteenth and early twentieth centuries saw a rise in the practice of growing designated colourful varieties of primrose. Both the blooms and the leaves, which range in flavour from mild lettuce to more bitter salad greens, are edible. The leaves can be used to make soup; however, it is best to combine them with other plants because they can occasionally be too powerful. The young flowers can be used to make primrose wine, while the leaves can be used to make tea (Fisher et al. 2008).

12.5.3 Nelumbo nucifera

Nelumbo nucifera (Fig. 12.21) is one of two existing species of aquatic plants in the family Nelumbonaceae, sometimes known as sacred lotus, Laxmi lotus, Indian lotus, or simply lotus. Its native range spans a very large area, from India and Sri

Fig. 12.21 *Nelumbo nucifera* Gaertn. (wikipedia.org)

Lanka (at altitudes up to 1400 m or 4600 ft. in the southern Himalayas (Mukherjee et al. 2009), through northern Indo-china and East Asia (north to the Amur region; the Russian populations have occasionally been referred to as "Nelumbo komarovii"), with isolated locations at the Caspian Sea, to nearly all of Island Southeast Asia, New Guinea, and northern and eastern Australia. It has been grown for its edible seeds for a very long time roughly 3000 years (Zhang et al. 2015) and is frequently grown in water gardens (Mukherjee et al. 2009).

It serves as both Vietnam's and India's national flower. While the leaves of the lotus are supported high above the water's surface or float on top of it, the lotus roots are buried in the dirt of the pond or river bottom. Typically, the flowers are located on stout stems that extend several centimetres above the foliage. The plant can grow in water up to 2.0 m (6 ft. 7 in) deep thanks to the long leaf stalks (petioles), which can reach that length. The horizontal breadth of the peltate leaf blade or lamina can reach 1 m (3 ft. 3 in) (Cook et al. 1974; Tungmunnithum et al. 2020). The magnificent flowers can reach up to 0.30 m (12 in) in diameter, while the leaves can be up to 0.80 m (31 in) in circumference. However, reports of 14 inches (35 cm) have been common.

Like humans and other warm-blooded animals, lotuses have the extraordinary capacity to regulate the temperature of their blossoms within a small range (Yoon 1996). After extended dormancy, lotus seeds can still germinate. A sacred lotus seed with an estimated age of 1300 years and 270 years was successfully germinated in 1994 (Shen-Miller et al. 1995). A 2.5 m (8 ft) depth of water is necessary for the sacred lotus to grow. A 0.30 m depth is the bare minimum in the water (12 in). This low water level, which warms up more quickly, is beneficial for greater development and flowering in chilly areas. A loamy, nutrient-rich soil is necessary for the sacred lotus.

A tiny piece of rhizome with at least one eye is either planted in ponds or directly into a field that has been flooded during the start of the summer season (from March through May in the northern hemisphere) (Lim 2016). There are numerous additional methods of propagation using seeds or buds (Tian Daike 2008). Additionally, tissue culture is a viable propagation technique for the creation of large quantities of consistent, true-to-type, disease-free materials in the future (Trang et al. 2019).

Three months after planting, the first leaves and blooms can be gathered. In the summer, flowers can be picked every 2 days, and every 3 days in the winter. The flower production reaches its peak 4 months after planting. For 3–4 months, flower harvesting is often done by hand (La-Ongsri et al. 2009).

The process of making sweet wines from lotuses is well documented, and research has investigated the amounts of tartaric acid, acetic acid, and alcohol extracted from the flowers of it (Suresh 2019). The tartaric acid percentage was recorded at 0.015, acetic acid percentage was 0.024, and alcohol percentage was 0.75, respectively. Lotus liquor demonstrated significant antioxidant effects in addition to wine, Although the precise makeup of active antioxidants is still unknown, polyphenol chemicals however, appear to be important (Lee et al. 2005). Additionally, a significant amount of total phenolic components has been found in the wine made from lotus buds (Yuwa-Amornpitak et al. 2012).

12.6 Conclusions

To date, there are numerous plant species of interest that can be utilized in the production process of alcoholic beverages. The sensory properties conferred to alcoholic beverages are often attributable to the vast array of secondary metabolites produced by various flower species. These metabolites could also contribute to enhancing health-promoting and antioxidant properties of beverages; however, it is important to point out that in the case of alcoholic beverages, consumption should be limited, to mitigate potential harm caused by alcohol consumption on human health.

Literature information regarding identified species remains limited, often concerning species underutilized from an agronomic and productive standpoint. Hence, there is a growing need to further promote the study and research not only concerning the identification of species of interest but also regarding the adaptation and improvement of agronomic and cultivation practices.

Acknowledgments Authors are thankful to Enzo de Santis, Giorgio Faggi, and Franco Rossi for the permission to use their images in this chapter. All the original photos can be found on https://www.actaplantarum.org/.

References

Abad MJ, Bedoya LM, Apaza L, Bermejo P (2012) The Artemisia L. genus: a review of bioactive essential oils. Molecules 17:2542–2566. https://doi.org/10.3390/molecules17032542

Abbate GM, Ruffoni B, Bisio A et al (2004) Analisi delle popolazione siciliana di Helicrysum: propagazione in vitro e indagini fitochimiche. Italus Hortus 11(4):206–209

Aberl A, Coelhan M (2012) Determination of volatile compounds in different hop varieties by headspace-trap GC/MS-in comparison with conventional hop essential oil analysis. J Agric Food Chem 60:2785–2792. https://doi.org/10.1021/jf205002p

Aćimović M, Pezo L, Zeremski T et al (2021) Weather conditions influence on hyssop essential oil quality. Processes 9(7):1152. https://doi.org/10.3390/pr9071152

Acosta-Rangel A, Rechcigl J, Bollin S et al (2021) Hop (Humulus lupulus L.) phenology, growth, and yield under subtropical climatic conditions: effects of cultivars and crop management. Aust J Crop Sci 15:764–772. https://doi.org/10.21475/ajcs.21.15.05.p3192

Afonso S, Dias MI, Ferreira ICFR et al (2022) The phenolic composition of hops (Humulus lupulus L.) was highly influenced by cultivar and year and little by soil liming or foliar spray rich in nutrients or algae. Horticulturae 8(5):385. https://doi.org/10.3390/horticulturae8050385

Agustí M, Mesejo C, Reig C, Martínez-Fuentes A (2014) Citrus production. In: Dixon GR, Aldous DE (eds) Horticulture: plants for people and places, Volume 1. Springer Netherlands, Dordrecht, pp 159–195

Alabade P, Jagtap K, Ade N (2018) An update review on Hibiscus rosa sinensis phytochemistry and medicinal uses. J Ayurvedic Herb Med 4:135–146. https://doi.org/10.31254/jahm.2018.4308

Alexopoulos AA, Assimakopoulou A, Panagopoulos P et al (2021a) Impact of salinity on the growth and chemical composition of two underutilized wild edible greens: Taraxacum officinale and reichardia picroides. Horticulturae 7:1–19. https://doi.org/10.3390/horticulturae7070160

Alexopoulos AA, Marandos E, Assimakopoulou A et al (2021b) Effect of nutrient solution ph on the growth, yield and quality of taraxacum officinale and reichardia picroides in a floating hydroponic system. Agronomy 11(6):1118. https://doi.org/10.3390/agronomy11061118

Almaguer C, Schönberger C, Gastl M et al (2014) Humulus lupulus - a story that begs to be told. A review. J Inst Brew 120:289–314. https://doi.org/10.1002/jib.160

Almaguer C, Gastl M, Arendt EK, Becker T (2015) Comparative study of the contribution of hop (Humulus lupulus L.) hard resins extracted from different hop varieties to beer quality parameters. J Am Soc Brew Chem 73:115–123. https://doi.org/10.1094/ASBCJ-2015-0327-01

Ammar AH, Lebrihi A, Mathieu F et al (2012) Chemical composition and in vitro antimicrobial and antioxidant activities of Citrus aurantium L. flowers essential oil (Neroli oil). Pak J Biol Sci 15:1034–1040. https://doi.org/10.3923/pjbs.2012.1034.1040

Armstrong GA, Hearstt JE (1996) Genetics and molecular biology of carotenoid pigment biosynthesis. FASEB J 10:228–237. https://doi.org/10.1096/fasebj.10.2.8641556

Babich O, Sukhikh S, Pungin A et al (2021) Evaluation of the conditions for the cultivation of callus cultures of hyssopus officinalis regarding the yield of polyphenolic compounds. Plants 10(5):915. https://doi.org/10.3390/plants10050915

Baldermann S, Yang Z, Sakai M et al (2009) Volatile constituents in the scent of roses. Floric Ornam Biotechnol 3:89–97

Balkrishna A, Sharma G, Sharma N et al (2022) Transition of Indian agriculture from glorious past to challenging future: a serious concern Acharya. Indian J Ecol 49(3):977–986. https://doi.org/10.55362/IJE/2022/3625

Bashir S, Peer LA (2022) Phytochemistry, biological properties, economic and ecological importance of Taraxacum officinale, A review. Int J Bot Stud 7:574–582

Basso F (2009) Piante officinali aromatiche e medicinali - Aspetti bioagronomici aromatici e fitoterapeutici. Pitagora Editrice, Bologna

Batiha GES, Olatunde A, El-mleeh A et al (2020) Bioactive compounds, pharmacological actions, and pharmacokinetics of wormwood (Artemisia absinthium). Antibiotics 9:1–25. https://doi.org/10.3390/antibiotics9060353

Bauerle WL (2019) Disentangling photoperiod from hop vernalization and dormancy for global production and speed breeding. Sci Rep 9:1–8. https://doi.org/10.1038/s41598-019-52548-0

Bera P, Kotamreddy JNR, Samanta T et al (2015) Inter-specific variation in headspace scent volatiles composition of four commercially cultivated jasmine flowers. Nat Prod Res 29:1328–1335. https://doi.org/10.1080/14786419.2014.1000319

Bianchini A, Tomi P, Costa J, Bernardini AF (2001) Composition of Helichrysum italicum (Roth) G. Don fil. subsp. italicum essential oils from Corsica (France). Flavour Fragr J 16:30–34. https://doi.org/10.1002/1099-1026(200101/02)16:1<30::AID-FFJ941>3.0.CO;2-F

Blackman JD, Wilson DW (2002) Support systems and training methods for dwarf hops (Humulus lupulus L.). J Hortic Sci Biotechnol 77:310–313. https://doi.org/10.1080/14620316.2002.11511498

Bocquet L, Sahpaz S, Hilbert JL et al (2018) Humulus lupulus L., a very popular beer ingredient and medicinal plant: overview of its phytochemistry, its bioactivity, and its biotechnology. Phytochem Rev 17:1047–1090. https://doi.org/10.1007/s11101-018-9584-y

Bogucka-Bonikowska A, Scinska A, Koros E et al (2001) Taste responses in alcohol-dependent men. Alcohol Alcohol 36:516–519. https://doi.org/10.1093/alcalc/36.6.516

Bora KS, Sharma A (2011) The genus Artemisia: a comprehensive review. Pharm Biol 49:101–109. https://doi.org/10.3109/13880209.2010.497815

Bouloc P, Allegret S, Arnaud L (2013) Hemp: industrial production and uses, Illustrate. CABI, Wallingford

Bown D (2001) New encyclopedia of herbs & their uses. DK ADULT, London

Bruun HH (2005) Biological Flora of the British Isles: Rosa Thunb. ex Murray. J Ecol 93:441–470. https://doi.org/10.1111/j.1365-2745.2005.01002.x

Butnariu M, Butu A (2019) Biotechnology of flavored or special wines. Elsevier Inc., New York

Calzolari D, Magagnini G, Lucini L et al (2017) High added-value compounds from Cannabis threshing residues. Ind Crop Prod 108:558–563. https://doi.org/10.1016/j.indcrop.2017.06.063

Catizone P, Barbanti L, Marotti I, Dinelli G (2013) Produzione ed impiego delle piante officinali. Patron Editore, Bologna

Cesaraccio C, Pellizzaro G (2004) Osservazioni fenologiche e ciclo stagionale della specie Helicrysum italicum (Roth.) Don subsp. Mycrophyllum (Willid.) Nyman. Italus Hortus 11:135–137

Chen F, Tholl D, Bohlmann J, Pichersky E (2011) The family of terpene synthases in plants: a mid-size family of genes for specialized metabolism that is highly diversified throughout the kingdom. Plant J 66:212–229. https://doi.org/10.1111/j.1365-313X.2011.04520.x

Chen X, Li J, Cheng T et al (2020) Molecular systematics of Rosoideae (Rosaceae). Plant Syst Evol 306(1):9. https://doi.org/10.1007/s00606-020-01629-z

Connole K (2022) Viola: herb of the year (TM) 2022. International Herb Association, Callahan

Cook CDK, Gut BJ, Rix EM, Schneller J (1974) Water plants of the world: a manual for the identification of the genera of freshwater macrophytes, 1st edn. Springer Science, Dordrecht

Croteau R, Kutchan TM, Lewis NG (2000) Natural products (Secondary Metabolites). In: Buchanan B, Gruissem W, Jones R (eds) Biochemistry and molecular biology of plants. Wiley, Rockville, pp 1250–1318

De Santis D, Frangipane MT (2020) Citrus aurantium L.: cultivar impact on sensory profile. Int J Gastron Food Sci 20:100203. https://doi.org/10.1016/j.ijgfs.2020.100203

Drewnowski A (1997) The science and complexity of bitter taste. Nutr Rev 55:31. https://doi.org/10.1111/j.1753-4887.2001.tb07007.x

Dugo G, Bonaccorsi I, Sciarrone D et al (2011) Characterization of oils from the fruits, leaves and flowers of the bitter orange tree. J Essent Oil Res 23:45–59. https://doi.org/10.1080/10412905.2011.9700446

Farag MA, Abib B, Ayad L, Khattab AR (2020) Sweet and bitter oranges: an updated comparative review of their bioactives, nutrition, food quality, therapeutic merits and biowaste valorization practices. Food Chem 331:127306. https://doi.org/10.1016/j.foodchem.2020.127306

Fathiazad F, Hamedeyazdan S (2011) A review on Hyssopus officinalis L: composition and biological activities. Afr J Pharm Pharmacol 5:1959–1966

Fisher MA, Oswald K, Adler W (2008) Exkursionsflora für Österreich, Liechtenstein und Südtirol (Excursion Flora for Austria, Lichtenstein and the South Tyrol). Biologiezentrum der Oberosterreichischen Landesmuseen, Berlin

Galbany-Casals M, Blanco-Moreno JM, Garcia-Jacas N et al (2011) Genetic variation in Mediterranean Helichrysum italicum (Asteraceae; Gnaphalieae): do disjunct populations of subsp. microphyllum have a common origin? Plant Biol 13:678–687. https://doi.org/10.1111/j.1438-8677.2010.00411.x

Ge L, Xie Q, Jiang Y et al (2022) Genus Lonicera: new drug discovery from traditional usage to modern chemical and pharmacological research. Phytomedicine 96:153889. https://doi.org/10.1016/j.phymed.2021.153889

Gill RK (2016) Nutrient management for growing dandelion (Taraxacum officinale L.) in nutrient film and deep flow hydroponics. ProQuest Dissertations and Theses 61

Glendinning JI (1994) Is the bitter rejection response always adaptive? Physiol Behav 56:1217–1227. https://doi.org/10.1016/0031-9384(94)90369-7

Grauso L, Emrick S, De Falco B, Lanzotti V (2019) Common dandelion: a review of its botanical, phytochemical and pharmacological pro fi les. Phytochem Rev 18:1115–1132. https://doi.org/10.1007/s11101-019-09622-2

Gresta F, Avola G, Lombardo GM et al (2009) Analysis of flowering, stigmas yield and qualitative traits of saffron (Crocus sativus L.) as affected by environmental conditions. Sci Hortic 119:320–324. https://doi.org/10.1016/j.scienta.2008.08.008

Guerrini A, Sacchetti G, Echeverria Guevara MP et al (2021) Wild Italian Hyssopus officinalis subsp. aristatus (Godr.) Nyman: from morphological and phytochemical evidences to biological activities. Plants 10:631. https://doi.org/10.3390/plants10040631

Harkness P (2003) The rose: an illustrated hystory. Firefly books, Richmond Hill

Hikosaka S, Iwamoto N, Goto E, Ching-Hui C (2017) Effects of supplemental lighting on growth & medicinal compounds of Japanese honeysuckle (Lonicera japonica Thunb.). Environ Control Biol 55:71–76. https://doi.org/10.2525/ecb.55.71

Ikeda N, Ishihara M, Tsuneya T et al (1994) Volatile components of honeysuckle (lonicera japonica thunb.) flowers. Flavour Fragr J 9:325–331. https://doi.org/10.1002/ffj.2730090609

Jadhav VM, Thorat RM, Sathe NS, Kadam VJ (2009) Hibiscus rosa sinensis Linn—"‘Rudrapuspa’": A Review. J Pharm Res 2:1168–1173

Jafari-Ghanavati M, Saket S (2020) Saffron and folklore. In: Koocheki A, Khajeh-Hosseini M (eds) Saffron - science, technology and health, 1st edn. Elsevier, Chennai, pp 3–13

Jahantigh O, Najafi F, Badi HN et al (2016) Essential oil composition of Hyssop (Hyssopus officinalis L.) under salt stress at flowering stage. J Essent Oil Res 28:458–464. https://doi.org/10.1080/10412905.2016.1153001

Jena J, Jitendra J, Vineeta T et al (2015) Rosa centifolia: plant review. Int J Res Pharm Chem 2(3):794–796

Judžentienė A (2015a) Wormwood (Artemisia absinthium L.) oils. In: Essential oils in food preservation, flavor and safety. Academic, pp 849–856. https://doi.org/10.1016/B978-0-12-416641-7.00097-3

Judžentienė A (2015b) Hyssop (Hyssopus officinalis L.) oils. In: Essential oils in food preservation, flavor and safety. Academic, pp 471–479

Kalua CM, Boss PK (2009) Evolution of volatile compounds during the development of cabernet sauvignon grapes (vitis vinifera l.). J Agric Food Chem 57:3818–3830. https://doi.org/10.1021/jf803471n

Kampov-Polevoy AB, Garbutt JC, Janowsky DS (1999) Association between preference for sweets and excessive alcohol intake: a review of animal and human studies. Alcohol Alcohol 34:386–395. https://doi.org/10.1093/alcalc/34.3.386

Karthikeyan V (2014) Citrus aurantium (bitter orange): a review of its traditional uses, phytochemistry and pharmacology. Int J Drug Discov Herb Res 4:766–772

Khan MH, Alie BA, Qureshi AMI et al (2021) Saffron: a therapeutic and prophylactic nutrition for human population. In: Aftab T, Hakeem KR (eds) Medicinal and aromatic plants. Springer, Cham, pp 303–324

Khazaie HR, Nadjafi F, Bannayan M (2008) Effect of irrigation frequency and planting density on herbage biomass and oil production of thyme (Thymus vulgaris) and hyssop (Hyssopus officinalis). Ind Crop Prod 27:315–321. https://doi.org/10.1016/j.indcrop.2007.11.007

Kizil S, Güler V, Kirici S, Turk M (2016) Some agronomic characteristics and essential oil composition of Hyssop (Hyssopus officinalis L.) under cultivation conditions. Acta Sci Pol Hortorum Cultus 15(6):193–207

Koul B, Taak P (2018) The Artemisia genus: a review on traditional uses, phytochemical constituents, pharmacological properties and germplasm conservation. J Glycomics Lipidomics 07:1–7. https://doi.org/10.4172/2153-0637.1000142

Kovacheva N, Rusanov K, Atanassov I (2010) Industrial cultivation of oil bearing rose and rose oil production in Bulgaria during 21ST century, directions and challenges. Biotechnol Biotechnol Equip 24:1793–1798. https://doi.org/10.2478/V10133-010-0032-4

Kranzler HR, Sandstrom KA, Van Kirk J (2001) Sweet taste preference as a risk factor for alcohol dependence. Am J Psychiatry 158:813–815. https://doi.org/10.1176/appi.ajp.158.5.813

Kumar S, Pradhan S (2013) Madhuca Lonigfolia (Sapotaceae): a review of its traditional uses, phytochemistry and pharmacology. Int J Biomed Res 2:30–36

Kumar R, Singh V, Devi K et al (2009) State of art of saffron (Crocus sativus L.) agronomy: a comprehensive review. Food Rev Int 25(1):44–85

Lachenmeier DW (2010) Wormwood (Artemisia absinthium L.)-A curious plant with both neurotoxic and neuroprotective properties? J Ethnopharmacol 131:224–227. https://doi.org/10.1016/j.jep.2010.05.062

La-Ongsri W, Trisonthi C, Balslev H (2009) Management and use of Nelumbo nucifera Gaertn. in Thai wetlands. Wetl Ecol Manag 17:279–289. https://doi.org/10.1007/s11273-008-9106-6

Lee HK, Choi YM, Noh DO, Suh HJ (2005) Antioxidant effect of Korean traditional lotus liquor (Yunyupju). Int J Food Sci Technol 40:709–715. https://doi.org/10.1111/j.1365-2621.2005.00990.x

Li H, Lu C (2015) Lonicera japonica Thunb 金银花 (Jinyinhua, Honey Suckle). In: Liu Y, Wang Z, Zhang J (eds) Dietary Chinese Herbs. Springer Vienna, Vienna, pp 693–702

Lim TK (2016) Edible medicinal and non-medicinal plants. Springer, Dordrecht

Lis B, Olas B (2019) Pro-health activity of dandelion (Taraxacum officinale L.) and its food products—history and present. J Funct Foods 59:40–48. https://doi.org/10.1016/j.jff.2019.05.012

Luo Y, Kong L, Xue R et al (2020) Bitterness in alcoholic beverages: the profiles of perception, constituents, and contributors. Trends Food Sci Technol 96:222–232. https://doi.org/10.1016/j.tifs.2019.12.026

Majetic CJ, Sinka BN (2013) Diverging pathways: differential benzenoid and phenylpropanoid volatile production in Phlox subulata L. cultivars. Biochem Syst Ecol 50:75–81. https://doi.org/10.1016/j.bse.2013.03.047

Martinez M, Poirrier P, Chamy R et al (2015) Taraxacum of fi cinale and related species — an ethnopharmacological review and its potential as a commercial medicinal plant. J Ethnopharmacol 169:244–262. https://doi.org/10.1016/j.jep.2015.03.067

Marzi V, De Mastro G (2008) Piante Officinali - Coltivazione, trattamenti di post-raccolta, contenuti di principi attivi, impieghi in vari settori industriali ed erboristici, 2018th edn. Adda, Bari

Meier C, Mediavilla V (1998) Factors influencing the yield and the quality of hemp (Cannabis sativa L.) essential oil. J Int Hemp Assoc 5:16–20

Mishra A, Poonia A (2019) Mahua (Madhuca longifolia) flowers: review on processing and biological properties. Nutr Food Sci 49:1153–1163. https://doi.org/10.1108/NFS-12-2018-0358

Mollafilabi A, Koocheki A, Moeinerad H, Kooshki L (2013) Effect of plant density and weight of corm on yield and yield components of saffron (crocus sativus l.) under soil, hydroponic and plastic tunnel cultivation. Acta Hortic 997:51–58. https://doi.org/10.17660/ActaHortic.2013.997.5

Morata A, Vaquero C, Palomero F et al (2019) Technology of vermouth wines. In: Grumezescu AM, Holban AM (eds) Alcoholic beverages volume 7: the science of beverages. Elsevier, pp 35–64

Moro A, Zalacain A, De Mendoza JH, Carmona M (2011) Effects of agronomic practices on volatile composition of hyssopus officinalis L. Essential oils. Molecules 16:4131–4139. https://doi.org/10.3390/molecules16054131

Mostafa S, Wang Y, Zeng W, Jin B (2022) Floral scents and fruit aromas: functions, compositions, biosynthesis, and regulation. Front Plant Sci 13:1–23. https://doi.org/10.3389/fpls.2022.860157

Motti R, Paura B, Cozzolino A, de Falco B (2022) Edible flowers used in some countries of the Mediterranean Basin: an ethnobotanical overview. Plants 11:1–22

Mpanga I, Schalau J (2020) Hop production in northern Arizona: opportunity and challenges for small-scale growers? The use of bio-effectors for crop nutrition view project hop production in northern Arizona: opportunity and challenges for small-scale growers? The University of Arizona Cooperative Extension

Mucci F, Macchia M (2001) Zafferano (Crocus sativus L.). Baldoni e Giaroni. Patron Editore. In: Baldoni R, Giardini L (eds) Coltivazioni erbacee, vol 2. Patron Editore, Bologna

Muhlemann JK, Klempien A, Dudareva N (2014) Floral volatiles: from biosynthesis to function. Plant Cell Environ 37:1936–1949. https://doi.org/10.1111/pce.12314

Mukherjee PK, Mukherjee D, Maji AK et al (2009) The sacred lotus (Nelumbo nucifera) - phytochemical and therapeutic profile. J Pharm Pharmacol 61:407–422. https://doi.org/10.1211/jpp/61.04.0001

Mutu A, Clapco S, Martea R et al (2014) Intraspecific genetic variability of Hyssopus officinalis L. Iaşi, România. J Exp Mol Biol 15(2):1

Mzabri I, Addi M (2019) Traditional and modern uses of saffron (Crocus sativus). Cosmet 6:1–11. https://doi.org/10.3390/cosmetics6040063

Mzabri I, Rimani M, Kouddane N, Berrichi A (2021) Study of the effect of pretreatment of corms by different concentrations of gibberellic acid and at different periods on the growth, flowering, and quality of saffron in eastern Morocco. Adv Agric 2021(1):9979827. https://doi.org/10.1155/2021/9979827

Nair KP (2022) Hyssop. In: Nair KP (ed) Herbal and Acidulant Tree Spices. Springer, Cham, pp 65–68

Nguyen HT, Németh ZÉ (2016) Sources of variability of wormwood (Artemisia absinthium L.) essential oil. J Appl Res Med Aromat Plants 3:143–150. https://doi.org/10.1016/j.jarmap.2016.07.005

Nguyen HT, Radácsi P, Gosztola B, Németh ÉZ (2018) Effects of temperature and light intensity on morphological and phytochemical characters and antioxidant potential of wormwood (Artemisia absinthium L.). Biochem Syst Ecol 79:1–7. https://doi.org/10.1016/j.bse.2018.03.005

Ninčević T, Grdiša M, Šatović Z, Jug-Dujaković M (2019) Helichrysum italicum (Roth) G. Don: taxonomy, biological activity, biochemical and genetic diversity. Ind Crop Prod 138:111487. https://doi.org/10.1016/j.indcrop.2019.111487

Olas B (2022) New perspectives on the effect of dandelion, its food products and other preparations on the cardiovascular system and its diseases. Nutrients 14(7):1350

Parthasarathy A, Cross PJ, Dobson RCJ et al (2018) A three-ring circus: metabolism of the three proteogenic aromatic amino acids and their role in the health of plants and animals. Front Mol Biosci 5:1–30. https://doi.org/10.3389/fmolb.2018.00029

Patel M, Naik SN (2010) Flowers of Madhuca indica J. F. Gmel.: present status and future perspectives. Indian J Nat Prod Resour 1:438–443

Paun A, Stroescu G, Zaica A et al (2020) Analysis of the process of green hemp stalks sequential harvesting. E3S Web Conf 180:1–8. https://doi.org/10.1051/e3sconf/202018003026

Piechulla B, Pott MB (2003) Plant scents - mediators of inter-and intraorganismic communication. Planta 217:687–689. https://doi.org/10.1007/s00425-003-1047-y

Pieracci Y, Pistelli L, Lari M et al (2021) Hibiscus rosa-sinensis as flavoring agent for alcoholic beverages. Appl Sci 11(21):9864. https://doi.org/10.3390/app11219864

Pietta P, Mauri P, Gardana C (1991) High-performance liquid chromatographic determination of flavonoid glucosides from Helichrysum italicurn. J Chromatogr 537:449–452. https://doi.org/10.1016/S0021-9673(01)88917-2
Putievsky E (1993) Cultivation and processing of medicinal plants. Wiley, Chichester
Qiu S, Bai M, Zhao P et al (2021) Phytochemical and network-based chemotaxonomic study of Lonicera japonica Thunb. Biochem Syst Ecol 94:104210. https://doi.org/10.1016/j.bse.2020.104210
Ramsey AJ (2019) Considering cytonuclear interactions in the face of heteroplasmy: evidence from Daucus carota (Apiaceae), a gynodioecious plant species. Dissertation
Ranalli P (2020) La canapa - miglioramento genetico, sostenibilità, utilizzi, normativa di riferimento. Edagricole Calderini, Bologna
Rashed-Mohassel MH (2020) Evolution and botany of saffron (Crocus sativus L.) and allied species. In: Koocheki A, Khajeh-Hosseini M (eds) Saffron - science, technology and health, 1st edn. Woodhead Publishing, Chennai, pp 37–57
Rassem HH, Nour AH, Mohammed Yunus R (2018) GC-MS analysis of bioactive constituents of jasmine flower. J Chem Eng Ind Biotechnol 4:52–59. https://doi.org/10.15282/jceib.v4i1.3883
Ravenscroft, D (2015) Growing annuals & perennials in your garden: The gardnersHQ plant growing guide. GardenersHQ. https://www.gardenershq.com/
Ravindran PN, Divakaran M, Pillai GS (2012) Other herbs and spices: Achiote to Szechuan pepper, Second Edition. Woodhead Publishing Limited, Cambridge, UK
Ríos JL, Recio MC, Giner RM, Meñez S (1996) An update review of saffron and its active constituents. Phytother Res 10:189–193. https://doi.org/10.1002/(SICI)1099-1573(199605)10:3<189::AID-PTR754>3.0.CO;2-C
Saif M, Varma R, Kant R, Gupta RK (2020) Madhuca longifolia (Mahua): a comprehensive ethno pharmacological review. Int J Chem Stud 8:172–175. https://doi.org/10.22271/chemi.2020.v8.i2c.8762
Sangeetha R, Nagappan D (2016) Madhuca Longifolia (Sapotaceae): a review of its phytochemical and pharmacological profile. Int J Pharm Bio Sci 7(4):106–114. https://doi.org/10.22376/ijpbs.2016.7.4.b106-114
Saunoriūtė S, Ragažinskienė O, Ivanauskas L et al (2020) The influence of meteorological factors on phytochemical composition of Artemisia pontica L. Chemija 31:278–283. https://doi.org/10.6001/CHEMIJA.V31I4.4325
Schenck CA, Westphal J, Jayaraman D et al (2020) Role of cytosolic, tyrosine-insensitive prephenate dehydrogenase in Medicago truncatula. Plant Direct 4:1–15. https://doi.org/10.1002/pld3.218
Schierenbeck KA (2004) Japanese honeysuckle (Lonicera japonica) as an invasive species; history, ecology, and context. Crit Rev Plant Sci 23:391–400. https://doi.org/10.1080/07352680490505141
Schloss A (2013) Homemade liqueurs and infused spirits: innovative flavor combinations, plus homemade versions of Kahlúa, Cointreau, and other popular liqueurs. Storey Publishing, LLC, North Adams
Schroeder FG, Lozoya DR, Ruser P (2020) Hydroponic forcing of saffron (Crocus sativus L.). Acta Hortic 1273:281–287. https://doi.org/10.17660/ActaHortic.2020.1273.37
Schütz K, Carle R, Schieber A (2006) Taraxacum-a review on its phytochemical and pharmacological profile. J Ethnopharmacol 107:313–323
Scinska A, Koros E, Habrat B et al (2000) Bitter and sweet components of ethanol taste in humans. Drug Alcohol Depend 60:199–206. https://doi.org/10.1016/S0376-8716(99)00149-0
Shang X, Pan H, Li M et al (2011) Lonicera japonica Thunb.: Ethnopharmacology, phytochemistry and pharmacology of an important traditional Chinese medicine. J Ethnopharmacol 138:1–21. https://doi.org/10.1016/j.jep.2011.08.016
Shen-Miller J, Mudgett MB, Schopf JW et al (1995) Exceptional seed longevity and robust growth: ancient sacred lotus from China. Am J Bot 82:1367–1380. https://doi.org/10.2307/2445863
Skrypnik L, Feduraev P, Styran T et al (2022) Biomass, phenolic compounds, essential oil content, and antioxidant properties of hyssop (Hyssopus officinalis L.) grown in hydroponics

as affected by treatment type and selenium concentration. Horticulturae 8:1037. https://doi.org/10.3390/horticulturae8111037
Souret FF, Weathers PJ (2000) The growth os saffron (Crocus sativus L.) in aeroponics and hydroponics. J Herbs Spices Med Plants 7:25–35. https://doi.org/10.1300/J044v07n03_04
Starr F, Starr K, Loope L (2003) Lonicera japonica. Maui: United States Geological Survey-Biological Resources Division
Steiner JE, Glaser D, Hawilo ME, Berridge KC (2001) Comparative expression of hedonic impact: affective reactions to taste by human infants and other primates. Neurosci Biobehav Rev 25:53–74. https://doi.org/10.1016/S0149-7634(00)00051-8
Stewart A (2013) The drunken botanist. Algonquin Books, New York
Su K, Zhang X, Liu SQ et al (2020) Identification of key odorants in honeysuckle by headspace-solid phase microextraction and solvent-assisted flavour evaporation with gas chromatography-mass spectrometry and gas chromatograph-olfactometry in combination with chemometrics. PLoS One 15:1–14. https://doi.org/10.1371/journal.pone.0237881
Sullivan CAO, Mcintyre CL, Dry IB et al (2020) Reading amino acids in a nanopore. Nat Biotechnol 38:7–9. https://doi.org/10.1038/s41587-019-0401-y
Suresh A (2019) Biochemical analysis and production of wine from chrysanthemum, Ixora, lotus, hibiscus and Nerium flowers. International Journal of Microbiology Research, ISSN 975–5276
Suryawanshi JAS (2011) An overview of Citrus aurantium used in treatment of various diseases. Afr J Plant Sci 5:390–395
Suryawanshi YC, Mokat DN (2021) Morphophysiological seed variability in Mahua trees from Western Ghats and its impact on tribal life. Proc Natl Acad Sci India Sect B Biol Sci 91:227–239. https://doi.org/10.1007/s40011-020-01223-w
Tadevosyan AH, Mairapetyan SK, Buniatyan RJ et al (2005) Humulus hydroponics in Armenia. Acta Hortic 668:235–240
Tadevosyan AH, Mairapetyan SK, Alexanyan JS et al (2009) Introduction, hydroponic growing possibility and productivity of some hop varieties in the Ararat Valley. Acta Hortic 848:141–148
Taglialatela-Scafati O, Pollastro F, Chianese G et al (2013) Antimicrobial phenolics and unusual glycerides from Helichrysum italicum subsp. microphyllum. J Nat Prod 76:346–353. https://doi.org/10.1021/np3007149
Taiz L, Zeiger E (2012) Metaboliti secondari e difese delle piante. In: Fisiologia Vegetale, 4th edn. Piccin, Padova
Talzhanov NA, Sadyrbekov DT, Smagulova FM et al (2005) Components of Artemisia pontica. Chem Nat Compd 41:178–181. https://doi.org/10.1007/s10600-005-0107-x
Tang K, Struik PC, Yin X et al (2017) A comprehensive study of planting density and nitrogen fertilization effect on dual-purpose hemp (Cannabis sativa L.) cultivation. Ind Crop Prod 107:427–438. https://doi.org/10.1016/j.indcrop.2017.06.033
Tian D (2008) Container production and post-harvest handling of lotus (Nelumbo) and micropropagation of herbaceous peony (Paeonia). Auburn University, Auburn
Tomas-Barberan F, Iniesta-Sanmartin E, Tomas-Llorente F, Rumbero A (1990) Antimicrobial phenolic compounds from three Spanish Helichrysum species. Phytochemistry 29(4):1093–1095. https://doi.org/10.1016/0031-9422(90)85410-H
Tonutti I, Liddle P (2010) Aromatic plants in alcoholic beverages. A review. Flavour Fragr J 25:341–350. https://doi.org/10.1002/ffj.2001
Trang NTQ, Hieuthao TT, Hong HTK (2019) Study on the anatomical morphology of lotus varieties (Nelumbo nucifera Gaertn.) in Vietnam. Plant Cell Biotechnol Mol Biol 20:95–105
Tungmunnithum D, Renouard S, Drouet S et al (2020) A critical cross-species comparison of pollen from Nelumbo nucifera gaertn. vs. Nymphaea lotus L. for authentication of thai medicinal herbal tea. Plants 9:1–14. https://doi.org/10.3390/plants9070921
Tzin V, Galili G (2010) New insights into the shikimate and aromatic amino acids biosynthesis pathways in plants. Mol Plant 3:956–972. https://doi.org/10.1093/mp/ssq048
Václav B, Karel K, Milan K et al (2020) Distribution of root system of hop plants in hop gardens with regular rows cultivation. Plant Soil Environ 66:317–326. https://doi.org/10.17221/672/2019-PSE

Vallarino JG, Erban A, Fehrle I et al (2018) Acquisition of volatile compounds by gas chromatography–mass spectrometry (GC-MS). Methods Mol Biol 1778:225–239. https://doi.org/10.1007/978-1-4939-7819-9_16

Vázquez-Araújo L, Rodríguez-Solana R, Cortés-Diéguez SM, Domínguez JM (2013) Study of the suitability of two hop cultivars for making herb liqueurs: volatile composition, sensory analysis, and consumer study. Eur Food Res Technol 237:775–786. https://doi.org/10.1007/s00217-013-2050-6

Vedel H, Lange J (1960) *Trees and bushes in wood and hedgerow*. Eyre Methuen, London, UK

Vouillamoz JF, Carlen C, Taglialatela-Scafati O et al (2015) The génépi Artemisia species. Ethnopharmacology, cultivation, phytochemistry, and bioactivity. Fitoterapia 106:231–241. https://doi.org/10.1016/j.fitote.2015.09.007

Wellhoffer J (2020) The effects of light spectra on the growth and development of greenhouse CBD Hemp. Master's Thesis, Cornell University

Welling MT, Liu L, Shapter T et al (2016) Characterisation of cannabinoid composition in a diverse Cannabis sativa L. germplasm collection. Euphytica 208:463–475. https://doi.org/10.1007/s10681-015-1585-y

Wesolowska A, Jadczak D (2018) Comparison of the chemical composition of essential oils isolated from hyssop (Hyssopus officinalis L.) with blue, pink and white flowers. J Essent Oil Bear Plants 21:938–949. https://doi.org/10.1080/0972060X.2018.1530613

Yadav P, Garg N, Diwedi DH (2009) Effect of location of cultivar, fermentation temperature and additives on the physico-chemical and sensory qualities on mahua (Madhuca indica J. F. gmel.) wine preparation. Indian J Nat Prod Resour 8:406–418

Yan K, Cui M, Zhao S et al (2016) Salinity stress is beneficial to the accumulation of chlorogenic acids in honeysuckle (Lonicera Japonica Thunb.). Front Plant Sci 7:1–10. https://doi.org/10.3389/fpls.2016.01563

Yoon CK (1996) Heat of lotus attracts insects and scientists

Yuan Y, Yang J, Yu X et al (2014) Anthocyanins from buds of Lonicera japonica Thunb. var. chinensis (Wats.) Bak. Food Res Int 62:812–818. https://doi.org/10.1016/j.foodres.2014.03.026

Yuwa-Amornpitak T, Koguchi M, Teramoto Y (2012) Antioxidant activity of herbal wine made from cassava starch. World Appl Sci J 16:874–878

Zangrando T, Marconi M (2018) Birra: Materie prime, tecnologia, stili 2, 1st edn. Edagricole, Bologna

Zhang Y, Lu X, Zeng S et al (2015) Nutritional composition, physiological functions and processing of lotus (*Nelumbo* nucifera Gaertn.) seeds: a review. Phytochem Rev 14:321–334. https://doi.org/10.1007/s11101-015-9401-9

Zhang Y, Hu Y, Li W et al (2022) Updates and advances on pharmacological properties of Taraxacum mongolicum Hand. -Mazz and its potential applications. Food Chem 373:131380. https://doi.org/10.1016/j.foodchem.2021.131380

Zhou X, Borras-Hidalgo O, Ruan W et al (2020) Analysis of the flavor and active compounds in *Lonicera japonica* wine produced by a new extraction and fermentation method. Am J Food Sci Technol 8:185–195. https://doi.org/10.12691/ajfst-8-5-3

Chapter 13
Edible Flowers for Different Food Preparations

César Ozuna and Rey Castañeda-Rodríguez

Abstract Worldwide, edible flowers are a topic of interest in gastronomy and scientific research. However, their consumption, known as floriphagia, is as old as the beginning of the great civilizations of humanity. To date, floriphagia has gained significant relevance due to new gastronomic trends and potential health benefits. This chapter aims to introduce the different culinary preparations for edible flowers and the trends to generate added value in them. Edible flowers are mainly classified by their need for heat treatment before culinary preparation. The type of cooking method to which edible flowers are subjected depends on the texture and chemical composition of the flower. The main dishes that include flowers are salads, soups, creams, stews, desserts, pickled flowers, or as spices. Among the trends to generate added value, drying processes, modified atmospheres, and bioactive compound and pigment extraction stand out. In conclusion, despite the great progress that has been made on the properties of fresh edible flowers, information regarding the nutritional properties of cooked flowers still needs to be made available. Moreover, it is essential to continue developing processes that allow the incorporation of edible flowers into processed foods and thus generate novel, nutritious, and attractive functional foods for the consumer.

Keywords Cooking methods · Gastronomy · Floriphagia · Functional foods · Pigments · Spice

C. Ozuna (✉)
Departmento de Alimentos, Divisi6n de Ciencias de la Vida, Campus Irapuato-Salamanca, Universidad de Guanajuato, Irapuato, Mexico
e-mail: cesar.ozuna@ugto.mx

R. Castañeda-Rodríguez
Instituto Universitario de Ingeniería de Alimentos-Food UPV, Universitat Politècnica de València, Valencia, Spain
e-mail: lrcasrod@doctor.upv.es

M. M. Jowkar et al. (eds.), *Advances in Research on Edible Flowers*, Advances in Olericulture, https://doi.org/10.1007/978-3-032-04561-4_13

13.1 Introduction

> The petals of the roses are very carefully detached, trying not to prick your fingers, because apart from the fact that it is very painful (the sting), the petals can be impregnated with blood, and this, besides altering the flavour of the dish, can cause chemical reactions, quite dangerous.... (Recipe: Quail in rose petals, in Like Water For Chocolate, Esquivel 1992)

Floriphagia, or flower consumption, is not new or the result of a gastronomic trend in haute cuisine. Flowers have been part of our diet since the beginning of the great ancient civilizations, such as the Greek, Roman, Chinese, and Mesoamerican (Mulík and Ozuna 2020; Rodrigues and Spence 2023). Flowers and their consumption are deeply rooted in various cultures worldwide. In ancient cultures, there were deities related to flowers and the different expressions associated with them. For example, in pre-Hispanic Mexico, Xochipilli, which means prince of flowers, was considered the god of games, spring, flowers, love, music, and dance (Mulík and Ozuna 2020).

Edible flowers are harmless and non-toxic parts of a plant that have health benefits when included in the human diet (Chitrakar et al. 2019). Therefore, the consumption of flowers depends on them being microbiologically innocuous and free of pesticides and non-organic fertilizers (Lara-Cortés et al. 2013). Importantly, floriphagia can cause adverse health effects without the proper information. Not all flowers are edible, as toxic varieties exist (Matyjaszczyk and Śmiechowska 2019). Some of the toxic compounds of flowers are derived from the secondary metabolism of the plants, such as the case of alkaloids, which are hydrosoluble. Thus, these compounds can be efficiently removed by blanching the flowers before cooking (Sotelo et al. 2007). Some studies have reported that the toxic effects of plants are related to their location. For example, the leaves of *Allium schoenoprasum* contain sulphurous substances. Compounds such as hydrogen cyanide, coumarin, and erucic acid must follow the acceptable daily intake established by the Joint FAO/WHO Expert Committee on Food Additives (Pires et al. 2019). Among the group of flowers not suitable for consumption due to its high content of toxic compounds are anthurium (*Anthurium* spp.) (calcium oxalate and cyanogenic glycosides), lily (*Lilium* spp.) (calcium oxalate), and begonia (*Begonia* spp.) (oxalic acid), among others (Lara-Cortés et al. 2013).

Today, in an ever-evolving world, the population's nutritional needs have changed toward healthier lifestyles. Edible flowers have high nutritional value and contain bioactive compounds, which modulate the oxidative stress caused by free radicals (Kritsi et al. 2022). For this reason, various research projects have focused on extracting pigments and compounds with antioxidant activity to add them to other types of products (Bragueto Escher et al. 2019; Hnin et al. 2019). Additionally, the popularity of edible flowers can be analysed from a culinary perspective since more and more chefs and cooks use them as ingredients in numerous recipes. This is because, due to their various physicochemical properties, flowers provide flavours and aromas to the dishes that include them as an ingredient (Mulík and Ozuna 2020).

This chapter introduces the different culinary preparations for edible flowers and the trends to generate added value. First, an overview of edible flowers in international gastronomy through time is given. Then, how different cooking methods can modify the properties of flower products are described. Next, the main edible flowers consumed raw and those that require cooking to be incorporated into dishes, emphasizing their nutritional and functional properties and main culinary preparations are introduced. Finally, innovative technologies and strategies for adding value to edible flowers are discussed.

13.2 Edible Flowers in World Gastronomy Through Time

The consumption of edible flowers evokes colourful, fresh, and aromatic dishes. However, there is more to them and different ways to prepare them (Fig. 13.1).

For example, edible flowers can be the main ingredient in several drinks and dishes, such as salads, soups, creams, main dishes, desserts, and even pickled products (Table 13.1). Globally, edible flower consumption originates in the traditional cuisine of different countries where floriphagia is part of their culture.

In ancient Rome, roses (*Rosa* spp.) were used to prepare drinks, salads, purees, and omelettes (Takahashi et al. 2020). During medieval times in France, common

Fig. 13.1 Some dishes made edible flowers: (**a**) Salads (*Tagetes erecta* L.); (**b**) Bakery products (*Lavandula angustifolia*); (**c**) Soups (*Cucurbita* spp.); (**d**) Stewed with fish (*Capparis spinosa* L.); (**e**) Boiled and then stir-fried (*Agave* spp.); (**f**) Dishes with egg (*Cucurbita* spp.); (**g**) Stuffed peppers with Mexican marigold (*Tagetes erecta* L.). (Photos: A, B, C, F, and G: ceased by Stanislav Mulík. Photo E: ceased by Gabriela Castañeda Rodríguez. Photo D: created by César Ozuna)

Table 13.1 Edible flowers and their gastronomic uses around the world

Common name	Scientific name	Gastronomic uses
Agave	*Agave* spp.	Meat dishes, batter-fried, filled with fresh cheese, scrambled with eggs.
Cornflower, bachelor's button	*Centaurea cyanus*	Salads and infusions
Common stock	*Matthiola incana*	Desserts
Artichoke thistle	*Cynara cardunculus* L.	Soups, creams, steamed, garnish
Caper berries	*Capparis spinosa* L.	Prickled for salads and stews
Common poppy	*Papaver rhoeas*	Aroma for wines and oils
Orange daylily	*Hemerocallis fulva*	Salads, soups, compotes, garnish, and desserts
Banana flower	*Musa* spp.	Salads, soups, curries, fried or stir-fried dishes, and processed products
Begonia	*Begonia x tuberhybrida, B. semperflorens*	Candied and salads
Borage	*Borago officinalis*	Salads and colouring agents in vinegars
Broccoli	*Brassica oleracea var. italica*	Salads, soups, creams, and side dishes
Marigold	*Tagetes erecta* L.	Salads, soups, creams, desserts, and corn-based drinks (*atole*)
Balloon flower	*Platycodon grandiflorus*	Salads
Nasturtium	*Tropaeolum majus*	Salads
Pumpkin or squash	*Cucurbita* spp.	Stews, soups, battered, stuffed, tacos, quesadillas, and desserts
Carnations	*Dianthus caryophyllus, D. barbatus, D. plumarius*	Salads
Cauliflower	*Brassica oleracea var. botrytis* L.	Salads, soups, creams, side dish
Coral	*Erythrina* spp.	Dishes accompanied by egg, beans, onion, and tomato
Chrysanthemum	*Chrysanthemum* spp.	Salads, soups, sauces, and vinaigrettes
Dandelion	*Taraxacum officinale*	Salads and soups
Dahlia	*Dahlia* spp.	Salads and desserts
Geranium	*Pelargonium* spp.	Drinks, salads, and desserts
Sunflower	*Helianthus annuus*	Salads
Gladiolus	*Gladiolus* spp.	Salads
Roselle	*Hibiscus sabdariffa* L.	Drinks, infusions, stuffing in some dishes, and desserts
Lavender	*Lavandula angustifolia*	Salads, creams, soups, infusions, and desserts
Common mallow	*Malva sylvestris*	Salads
Chamomile	*Matricaria recutita*	Infusions, desserts, and to macerate oils and vinaigrettes
Plumeria	*Plumeria rubra*	Infusions and drinks

(continued)

Table 13.1 (continued)

Common name	Scientific name	Gastronomic uses
Yarrow	*Achillea millefolium*	Infusions and drinks
Wild pansy	*Viola tricolor*	Salads and desserts
Violet	*Viola odorata*	Infusions, salads, soups, and desserts
Yucca	*Yucca* spp.	Salads, dishes with egg, pork, shrimp, or vegetables.

marigold flowers were used as salad ingredients (Mlcek and Rop 2011). In the seventeenth century, violets (*Viola odorata* L.) were used as a source of sugar and as a colouring agent for syrups (Mlcek and Rop 2011). Furthermore, edible flowers have been used in Europe as ingredients for preparing herbal infusions (herbal teas) due to their health benefits. Currently, the Mediterranean region is deeply rooted in the consumption of flowers in different preparations. For example, artichokes (*Cynara cardunculus* L.), capers (*Capparis spinosa* L.), and saffron (*Crocus sativus*) are an important part of both the Mediterranean cuisine and diet. In Portugal and Spain, traditional cheese is produced with sheep's milk and the rennet obtained using thistle flowers (*Cynara cardunculus*) as a coagulating agent (Gomes et al. 2019).

As for the Asian continent, China has a long tradition of consuming edible flowers. Since ancient times, Chinese culture has used edible flowers as ingredients in herbal cuisine to enhance flavours and garnish (Zheng et al. 2018). Additionally, due to their beneficial properties, edible flowers have been used to make herbal medicines and infusions. The main edible flowers used in Chinese culture are *Paeonia suffruticosa*, *Lilium brownii* var. *viridulum*, *Flos lonicerae*, *Rosa chinensis*, *Lavandula pedunculata*, *Prunus persica*, *Hibiscus sabdariffa*, *Flos carthami*, *Chrysanthemum morifolium*, and *Flos rosae rugosae* (Xiong et al. 2014). Moreover, in Thailand and India, flowers are deeply rooted in their culture (Panyayong and Srikaeo 2022). The use of these flowers is primarily related to traditional medicine. However, because of their texture, some edible flowers, either in raw or cooked, have been incorporated into various dishes, such as salads, soups, and curries. The main species consumed in these countries are *Nelumbo nucifera*, *Tagetes* spp., *Musa* spp., and *Plumeria rubra*.

As for Latin American countries, edible flowers are deeply rooted in Mexican and Central American countries since edible flowers were used by pre-Hispanic civilizations (Sotelo et al. 2007; Mulík and Ozuna 2020). Traditional Mexican cuisine combines a great diversity of endemic ingredients, as well as others introduced into the country following historical events (Mulík et al. 2022). Mexican edible flowers include more than 100 plant species from approximately 50 genera (Mapes and Basurto 2016), which grow under specific environmental conditions. The main Mexican endemic edible flowers include *Agavaceae* spp., *Tagetes* spp., *Cucurbita* spp., *Yucca* spp., *Erythrina* spp., *Opuntia ficus-indica* L., *Dahlia* spp., *Quararibea funebris*, among others (Mulík and Ozuna 2020). Unlike other parts of the world, in Mexico, edible flowers are not eaten raw but are consumed after being cooked following specific recipes (Mulík et al. 2022).

At a global level, new gastronomic trends and "haute cuisine" have resumed using edible flowers in their dishes. These flowers are mainly used as a garnish or included as ingredients in salads, sauces, and desserts. For this reason, edible flowers are trendy and in fashion in the gastronomic sector. Thus, chefs, researchers, and food technologists are increasingly interested in the nutritional and functional properties of edible flowers. They are also interested in new flower species that could be used to develop new dishes.

13.3 Cooking Methods and Their Effects on the Nutritional Content of Plant Products

There are different ways to consume edible flowers based on different cultures and traditions. However, due to their chemical composition, some edible flowers require cooking processes to eliminate specific toxic anti-nutritious components. Heat cooking is a process that involves changes in the sensory, physicochemical, and nutritional properties of food. In a recent study, Mulík et al. (2022) reported that the main culinary techniques used to cook Mexican endemic edible flowers are boiling and stir-frying, which take between 11 and 20 min.

Boiling is probably the most widely used cooking method in preparing different plant products. The contact time between food and boiling water depends on the recipe and the processes to which plant matrices are subjected. Boiling modifies the physicochemical (Monalisa et al. 2020), bioactive (Yang et al. 2019), and nutritional properties (Moorthy et al. 2020) of plant products. Some of the physical changes include loss of turgor and loss of cellular adhesion forces. Both phenomena lead to the loss of firmness or softening of the physical structure of the plant matrix (Ng and Waldron 1997).

Regarding the changes in the nutritional properties of plant products, boiling, a process carried out in water, causes the leaching of hydrosoluble compounds. However, the non-polar components are not efficiently extracted. Therefore, boiling and steaming are culinary techniques that, if used at moderate cooking times, can contribute to preserving certain pigments. Recently, it has been demonstrated that squash and agave flowers steamed for short times, around 3 min, preserved their chlorophyll and carotenoid contents (Castañeda-Rodríguez et al. 2023, 2024a).

Steaming is a culinary technique in which the plant product is not in direct contact with the boiling water but with the steam it produces. Steaming plant products has advantages in heat transfer. However, energy loss during the process can increase cooking times compared to other methods (Putriani et al. 2022). Some researchers have demonstrated that steaming increases the diffusion of polar soluble compounds and preserves hydrophobic compounds. Thus, unlike boiling, steaming can preserve hydrophobic compounds and reduce their leaching. These findings are relevant for sensitive plant products, such as edible flowers.

Stir-frying is a culinary technique that uses hot oil in a pan. During this process, the oil at high temperatures and the constant movement generated during the process intensify heat and mass transfer (Adler-Nissen 2007; Zhou et al. 2019), consequently, cause the cooking of the food. One of the critical points of stir-frying is the oil composition and the temperature that the cooking oil can reach. During stir-frying, heat is transferred to the oil, and it is heated depending on the process temperature; if the temperature is high enough, the oil will reach its melting point. Thus, when the temperature exceeds 200 °C, some oils (extra virgin olive oil, sunflower, canola, among others) can degrade and form oxidation products with harmful effects on health (Santos et al. 2013; Zribi et al. 2014; Hosseini et al. 2016). For this reason, it is important to use oils with high melting points. The composition of the oil is also of great relevance since it can affect the composition of the phenolic compounds of the plant products (Moreno et al. 2007). Furthermore, heating and continuous stirring accelerate the lipid oxidation of the oil (Song et al. 2015).

13.4 Edible Flowers and Their Culinary Preparations

13.4.1 Flowers That Are Consumed Without Being Subjected to Cooking

There is conflicting information about the benefits of eating raw or cooked food. For example, the proteins and certain complex carbohydrates of cooked foods are easier to assimilate by the human body than those from their raw counterparts. On the other hand, thermo-sensitive compounds, such as polyphenols, carotenoids, and some vitamins, are in higher amounts in raw foods. However, exposure to medium temperatures for short times has been shown to have beneficial effects on the bioactive compounds of other foods. Thus, it was shown that stir-frying (90 °C, 4 min) kale (*Brassica oleracea* L. var. *acephala*) and red cabbage (*Brassica oleracea* L. var. *capitata*) increases their total phenolic compound content by 5% (Murador et al. 2016). As for edible flowers (agave and squash), it has been shown that during the first minutes (1–3 min) of cooking by boiling, steaming, and stir-frying, their main bioactive compounds (chlorophyll, carotenoids, and phenolic compounds) are lost (Castañeda-Rodríguez et al. 2023, 2024a). However, the solubility of each bioactive compound seems to have an important influence on this phenomenon. Thus, non-polar compounds (carotenoids and chlorophyll) are mainly lost during stir-frying and polar compounds (phenolic compounds) during boiling (Castañeda-Rodríguez et al. 2023, 2024a).

Most flowers that are consumed in their raw form are used as fresh ingredients for salads or to decorate or infuse hot preparations. One of the great advantages of consuming fresh flowers is that, as in most plant products, since they are not cooked, their appearance, colour, texture, and nutritional composition remain the same.

In fresh flowers, the colour in their raw form can be directly related to their nutritional value. To date, dietary approaches suggest people incorporate fruits and vegetables of various colours into their daily diet. Therefore, based on these approaches, the consumption of flowers of different colours could have a similar effect. According to their colour, the flowers that can be consumed raw can be separated into four large groups by following their predominant colour. First, we have white flowers, like elder flowers (*Sambucuss* spp.) (Fig. 13.2a), jasmine (*Jasminum* spp.) (Fig. 13.2b), and capers (*Capparis spinosa* L.) (Fig. 13.2c). Then, we have yellow and orange flowers, the most characteristic being dandelion (*Taraxacum officinale*) (Fig. 13.2d), sunflower (*Helianthus annuus*) (Fig. 13.2e), and common marigold (*Calendula officinalis*) (Fig. 13.2f). There are also the pink or red flowers, like the carnations (*Dianthus caryophyllus*) (Fig. 13.2g), begonia (*Begonia* spp.) (Fig. 13.2h),

Fig. 13.2 Edible flowers that are consumed raw. (**a**) Elder (*Sambucuss* spp.); (**b**) Jasmine (*Jasminum* spp.); (**c**) Capers (*Capparis spinosa* L.); (**d**) Dandelion (*Taraxacum officinale*); (**e**) Sunflower (*Helianthus annuus*); (**f**) Common marigold (*Calendula officinalis*); (**g**) Carnation (*Dianthus caryophyllus*); (**h**) Begonia (*Begonia* spp.); (**i**) Common poppy (*Papaver rhoeas*); (**j**) Wild pansy (*Viola tricolor*); (**k**) Lavender (*Lavandula* spp.); (**l**) Saffron (*Crocus sativus*). (Images: Creative Commons license (CC BY-SA 2.0); original photo available at https://www.pxfuel.com/)

and common poppy (*Papaver rhoeas*) (Fig. 13.2i). Finally, we have the blue, violet, and purple flowers, such as wild pansy (*Viola tricolor*) (Fig. 13.2j), lavender (Lavandula spp.) (Fig. 13.2k), and saffron (*Crocus sativus*) (Fig. 13.2l). The yellow, orange, and red colors of flowers may be mainly due to the presence of carotenoids. Although the colour red, like blue, violet, and purple, can also be related to the presence of anthocyanins.

The consumption of the compounds responsible for the colour of foods has been associated with different beneficial effects on consumers' health. For example, the yellow and orange pigments in fruit and vegetables have been associated with a decrease in the risk of cardiovascular diseases and different types of cancer (cervical, gastric, pancreatic, prostate, and ovarian, among others) (Blumfield et al. 2022). As for blue pigments, these have been shown to decrease the risk of hypertension and esophageal cancer (Blumfield et al. 2022). On the other hand, red pigments decrease the risk of hypertension and contracting cervical, pancreatic, laryngeal, pharyngeal, and lung cancer (Blumfield et al. 2022). In addition to causing a lower fracture risk (Blumfield et al. 2022). Although the previously mentioned flowers can be used as ingredients for salads, some of them, such as saffron, can be lightly toasted to enhance their flavour and aroma. Violet flowers, for example, can be added to teas and even caramelized to garnish cakes and other desserts.

A specific case of flowers consumed raw is that of capers (*Capparis spinosa* L.) (Fig. 13.2c). Capers are the flower buds produced by the caper or caper tree. Capers are usually eaten pickled; that is, they have previously been marinated in a solution of vinegar and salt. The typical commercial processing of pickled capers begins with pre-treatment in a high-salt brine (approximately 16–20% NaCl w/v). After pre-treatment, capers are desalted and packed in containers with a covering liquid that contains approximately 6% NaCl and 1% acetic acid (Inocencio et al. 2000). The main production of pickled capers occurs in Mediterranean countries such as Greece, Italy, Turkey, Morocco, and Spain. Thus, in these countries, capers are used in salads and several pasta, fish, and meat dishes (Fig. 13.1d). As for their health benefits, it has been reported that capers are rich in flavonoids such as quercetin and kaempferol. (Inocencio et al. 2000) reported that based on a hypothetical daily intake of capers, a serving of salad (20 capers, 100 g) would provide 20–80 mg of quercetin (30–160 mg of flavonol glycosides). These values indicate that a serving of capers provides at least the same amount of quercetin as 100 g of onions.

13.4.2 Flowers That Require Cooked

There are several flowers that look more like a common vegetable than a flower. Several vegetables from the cruciferous family, like broccoli (*Brassica oleracea* var. *italica*) (Fig. 13.3a) and cauliflower (*Brassica oleracea* var. *botrytis* L.) (Fig. 13.3b), are flower heads widely used in the gastronomy of most countries in the world (Ahmed and Ali 2013; Hwang 2019; López-Hernández et al. 2022; Czarnowska-Kujawska et al. 2022). In Southeast Asian countries and India, banana inflorescences (*Musa* spp.) (Fig. 13.3c) are commonly used in their gastronomy. The culinary treatment given to this product is very similar to that of any other vegetable (Panyayong and Srikaeo 2022). Finally, artichoke (*Cynara cardunculus* L.) (Fig. 13.3d) is another flower that is widely used in Mediterranean countries and, due to its composition and enzyme activity, is used as a plant rennet in cheese production (Gostin and Waisundara 2019).

Broccoli (*Brassica oleracea* var. *italica*) (Fig. 13.3a) is known for having high contents of dietary fibre, minerals, vitamins, antioxidants (ascorbic acid and polyphenols) (López-Hernández et al. 2022), and anticancer compounds (glucosinolates

Fig. 13.3 Edible flowers that require cooking for consumption. (**a**) Broccoli (*Brassica oleracea* var. *italica*); (**b**) Cauliflower (*Brassica oleracea* var. *botrytis* L.); (**c**) Banana (*Musa* spp.); (**d**) Artichoke (*Cynara cardunculus* L.); (**e**) Squash (*Cucurbita pepo* L.); (**f**) Agave (*Agave* spp.); (**g**) Mexican marigold (*Tagetes Erecta* L.); (**h**) Yucca (*Yucca* spp.); (**i**) Coral (*Erythrina* spp.); (**j**) Roselle (*Hibiscus sabdariffa* L.). (Figures a, b, c, d, and h: Creative Commons license (CC BY-SA 2.0); original photo available at https://www.pxfuel.com/; https://commons.wikimedia.org/. Photos e, f, i, and j: created by Rey Castañeda-Rodríguez. Photo C: created by Edith Ozuna. Photo G: ceased by Stanislav Mulík)

and their derivatives) (Soundararajan and Kim 2018). Worldwide, broccoli is consumed raw, steamed, and stir-fried in soups, creams, and other preparations. Several studies have demonstrated that the best cooking methods for this vegetable are steaming and microwaving. These methods avoid the loss of hydrosoluble compounds with antioxidant capacity, such as polyphenols and glucosinolates (López-Hernández et al. 2022; Czarnowska-Kujawska et al. 2022). However, in some preparations, such as soups and creams, these compounds would remain in the cooking liquid media.

Like broccoli, cauliflower (*Brassica oleracea* var. *botrytis* L.) (Fig. 13.3b) is rich in vitamins, polyphenols, and glucosinolates (Ahmed and Ali 2013). Glucosinolates are one of the most important bioactive compounds in cruciferous vegetables (Cartea and Velasco 2008), specifically for their chemoprotective, anti-inflammatory, and anticancer properties (Miękus et al. 2020). However, the availability of these functional compounds depends on the cooking process (Ahmed and Ali 2013). Hwang (2019) demonstrated that steaming is still the best method to maintain the concentration of glucosinolates and the antioxidant capacity of cauliflower. Additionally, (Ahmed and Ali 2013) reported that steaming, stir-frying, or microwaving preserves the minerals, pigments, and bioactive compounds of cauliflower.

In Thai cuisine, banana inflorescences, also known as banana blossom, banana flower, or banana heart (*Musa* spp.) (Fig. 13.3c), are consumed fresh, in salads, fried or stir-fried dishes, soups, curries, or processed products (Panyayong and Srikaeo 2022). Banana inflorescences have beneficial health effects due to their high contents of high-quality protein, dietary fibre, vitamins, minerals, and bioactive compounds with functional effects. As for the dietary protein content in banana inflorescences, these flowers have a 0.54 ratio (essential/non-essential amino acids), which is higher than the requirements recommended for adults by the World Health Organization (Ramu et al. 2017). In Asian and African traditional medicine, banana inflorescences are used as a remedy for dysentery, hypertension, diabetes, ulcers, and bronchitis (Panyayong and Srikaeo 2022).

Artichokes (Fig. 13.3d) belong to the *Asteraceae* (Compositae) family. This flower is mainly produced in Italy, Egypt, and Spain (Gostin and Waisundara 2019; Kayahan and Saloglu 2021). This flower is important in the Mediterranean diet due to its nutritional and bioactive properties. The antimicrobial, anti-inflammatory, antioxidant, anti-obesity, and anti-genotoxic effects of artichoke have been previously demonstrated. This functionality is mainly due to its high content of polyphenols and inulin, which can be sensitive to heat treatments (Kayahan and Saloglu 2021). Boiling, steaming, and stir-frying have detrimental effects on the concentration of flavonoids in the edible parts of the artichoke. However, steaming and stir-frying artichokes increase the concentration of their main polyphenols: 5-O-caffeoylquinic acid and 1,5-O-dicaffeoylquinic acid. The increase in these polyphenols could be attributed to the formation of different dicaffeoylquinic acid isomers during cooking (Ferracane et al. 2008; Gostin and Waisundara 2019).

Moreover, other edible flowers require heat treatment for consumption. However, their production is not extensive, making their consumption seasonal. In this sense, squash blossoms (*Cucurbita pepo* L.) (Fig. 13.3e), *Agave* spp. flowers (Fig. 13.3f),

Tagetes spp. flowers (Fig. 13.3g), yucca flowers (*Yucca* spp.) (Fig. 13.3h) and *Erythrina* spp. flowers (Fig. 13.3i) are consumed in some regions of North and Central America (Mulík and Ozuna 2020). Except for the squash blossom, the consumption of other edible flowers is limited to certain culinary preparations deeply rooted in the traditional cuisine of the communities that consume them. Finally, within this section of edible flowers, we have incorporated the Roselle flower (*Hibiscus sabdariffa* L.) (Fig. 13.3j), which is widely consumed in different tropical and subtropical regions around the world.

Pumpkin was one of the first domesticated plants in America; therefore, it is considered endemic to the continent. Thus, in America, there is a great diversity of the genus *Cucurbita*, of which four species are considered Mesoamerican: *Cucurbita argyrosperma*, *Cucurbita ficifolia*, *Cucurbita moschata,* and *Cucurbita pepo* (Mapes and Basurto 2016). In the *Cucurbitaceae* family, plants usually have unisexual flowers; moreover, the same plant can produce both male and female flowers. Thus, its sexual expression is regulated by genetic, hormonal, and environmental factors. In winter, the plants have a high number of female flowers, while in summer conditions, the production of male flowers is favoured (Manzano et al. 2011). Male blossoms of *Cucurbita pepo* L. are long and pedicellate, with a bell-shaped calyx and a tubular corolla. On the other hand, female flowers have thick peduncles that reach 5 cm in length, an ovoid ovary, and a larger corolla than males. In addition, it has been demonstrated that female blossoms have more phenolic compounds and antioxidant capacity than the male ones (Tarhan et al. 2007). Also, a study has found that the squash blossom contains two new glycosylated flavonols, rhamnazin 3-rutinoside and isorhamnetin-3-rutinoside-4-rhamnoside (Ghosh and Rana 2021). Additionally, this flower has a higher protein and carbohydrate content than other edible flowers. (Sotelo et al. 2007). The squash blossom is one of the most used flowers in the gastronomy of various American, European, and Asian countries due to its colour, texture, flavour, and annual availability (Ghosh and Rana 2021). Squash blossoms, with or without the fruit, are sold in markets around the world. Thus, within floriphagia, this flower is one of the most used when cooking since it can be incorporated into various preparations. This flower can be stuffed or battered and is commonly used in stews, soups, and desserts (Fig. 13.1c and f). In Mexico, it is a main ingredient for making quesadillas (Mulík and Ozuna 2020).

Yucca flowers (*Yucca* spp.) (Fig. 13.3i) can be found in arid regions and are widely collected in different countries in North and Central America. These flowers are rich in protein and carbohydrates (Castañeda-Rodríguez et al. 2024b). Additionally, they are an important source of ascorbic acid and calcium. The petals of yucca flowers are consumed without the stamens because of their bitter and unpleasant taste. In addition, before their consumption, yucca flowers must be boiled, and the cooking water removed to eliminate their characteristic bitterness. In traditional cuisine, yucca flowers are combined with tomato, lime juice, and salt in a salad that can be used as a side dish. Furthermore, as part of the fusion of Spanish and Mesoamerican cuisines, in various recipes, this yucca flower salad can be seasoned with olive oil. Moreover, these flowers can be fried or stuffed and combined with eggs, pork, shrimp, or vegetables.

Tagetes erecta L. flowers (Fig. 13.3f) have a characteristic intense yellow-orange colour. Chemically, the colouration of these flowers is due to two pigments: lutein and zeaxanthin. Due to the presence of these pigments, marigold or cempasúchil flowers (*Tagetes erecta* L.) have been incorporated into the diet of laying hens (10 g/kg or 20 g/kg for 42 days) to contribute to the pigmentation of the egg yolk; in addition, it has been demonstrated that increased its levels of total saturated fatty acids (C16:0 and C18:0) and decreased monounsaturated fatty acids (C16:1 n-7 and C18:1 n-9) (Altuntaş and Aydin 2014). Therefore, the inclusion of the flour of *Tagetes erecta* L. flowers in the diet of laying hens can be used to reduce cholesterol levels in egg yolk (Altuntaş and Aydin 2014). Regarding their effects on human health, *Tagetes* spp. flowers and their extracts have been shown to have antimicrobial, anti-inflammatory, antiviral, antitumor, and analgesic properties, among others (Chitrakar et al. 2019). In world gastronomy, thanks to its intense colour and its pleasant sweet, minty, and almost citrus flavour, the common marigold flower is used as an ingredient in soups, creams, salads, desserts, and corn-based drinks (atole) (Fig. 13.1a and g) (Chitrakar et al. 2019; Mulík and Ozuna 2020). In these culinary preparations, marigold flowers provide flavour, nutritional properties, and their characteristic yellow-orange colour.

The *Agave* genus groups approximately 160 species distributed in various arid regions, mainly in North America. Agave plants require between 7 and 10 years to flower (Fig. 13.3e). These flowers are an important source of fibre, carbohydrates, and proteins. In addition, these flowers contain compounds with antioxidant activity, such as polyphenols, flavonoids, and carotenoids. However, for their consumption, agave flowers need to be cooked. In traditional Mexican and Central American cuisine, agave flowers are cooked primarily by stir-frying and boiling for 10–20 min (Fig. 13.1e) (Mulík et al. 2022). Among the main dishes that include agave flowers, meat stews, batter-fried flowers stuffed with fresh cheese, and scrambled with eggs are the most common.

Coral flowers (Fig. 13.3j) come from a tree of the genus *Erythrina* spp., which has a height of between 5 and 10 m. The tree blooms in spring, and its flowers are bright red and conically arranged (Mulík and Ozuna 2020). Due to their intense red colour, coral flowers have an ornamental use. However, in various regions in Mexico, these flowers are used to prepare several dishes, such as stews, stuffings, and pancakes. Furthermore, in various traditional cuisine recipes, coral flowers are served with eggs, beans, onion, and tomato (Mulík et al. 2022). These flowers contain alkaloids (hydrosoluble compounds), so they must be boiled, and the cooking water removed, before consumption. Due to their bioactive compound content, coral flowers have been used in traditional medicine as a remedy for chest pain (Mulík and Ozuna 2020).

The roselle flower (*Hibiscus sabdariffa* L.) (Fig. 13.3h), which stands out for its intense red colour, is native to Asia, but its origins are believed to be on the African continent. Worldwide, the main producers of Roselle are China, India, Sudan, Uganda, Indonesia, Malaysia, and Mexico. Roselle flowers are known for having a high content of protein, vitamins, crude fibre, and minerals. They also contain bioactive compounds (phenolic compounds and anthocyanins) with wide functionality in health. Various studies have shown that aqueous and ethanolic extracts of roselle flowers have antioxidant activity. These extracts can act as anticarcinogens and reduce cardiovascular and chronic diseases (diabetes mellitus, dyslipidemia, and hypertension). Internationally, this flower is mainly used in floral infusions and refreshing drinks. However, due to their intense red colour and flavour, they have also been incorporated into vinaigrettes, sauces, desserts, and bread. The flowers that remain after preparing drinks and infusions have also been used as stuffing in certain dishes, including tacos.

13.5 New Technologies and Strategies for the Generation of Added Value in Edible Flowers

Whether eaten raw or cooked, edible flowers are generally purchased fresh. However, due to their high moisture content, edible flowers are highly perishable, which limits their commercialization. Therefore, different conservation technologies have been used to increase the shelf life of flowers and avoid minor effects on their organoleptic and functional characteristics. Drying, for example, has been historically used to increase the shelf life of different food products. On the other hand, drying can deteriorate the functional properties of plant foods. For example, high temperatures can promote the oxidation of the bioactive compounds in the flowers, resulting in a low or null antioxidant capacity. Moreover, outdoor drying processes, such as drying in the shade and drying in the sun, depend largely on weather conditions, limiting their use (Zhao et al. 2019). A study of *Centaurea cyanus* flowers collected in Portugal suggested that drying in the shade is a process that retains most of the bioactive compounds compared to drying with hot or cold air (Fernandes et al. 2018). Although under the right conditions, drying flowers may not affect their functional and nutritional properties, this process changes their appearance and colour compared to their fresh form, which may limit their culinary applications. Moreover, various preparations incorporate fresh flowers. Therefore, conservation processes that preserve the appearance of flowers are necessary. The use of modified atmospheres has been tested in *Cucurbita pepo* L. blossoms. In a study, gaseous nitrogen and carbon dioxide-controlled atmospheres improved sugar retention, total soluble solids, pH, and titratable acidity (Aquino-Bolaños et al. 2013). Furthermore, after 16 days, between 49.5% and 72.8% of bioactive compounds were retained (Aquino-Bolaños et al. 2013).

As previously mentioned, edible flowers are an important component in the diet of people who consume them due to their macronutrient, micronutrient, and mineral content (Sotelo et al. 2007; Pinedo-Espinoza et al. 2020). Moreover, edible flowers also contain components with technological and functional characteristics that could be of great interest to the food industry. Edible flowers contain green (chlorophylls), yellow and orange (carotenoids), and red, violet, and purple (anthocyanins) pigments. This is interesting since the demand for natural foods has given considerable popularity to using natural colourants compared to their synthetic counterparts. This explains the importance of finding new natural sources to extract these compounds. As in other plant matrices, the different compounds in edible flowers can be extracted and then incorporated into food matrices as value-added ingredients.

The extraction processes are a series of operations that allow the recovery of pure compounds from plants or other matrices of commercial interest. The most common extraction processes include the use of solvents, which, in addition to being environmental pollutants, require expensive purification processes that limit their application in the food industry. That is why, in recent decades, interest has increased in implementing or combining with new technologies that are friendly to the environment and less invasive with the components that are intended to be extracted (Zhao et al. 2019). Among these are high-power ultrasound, microwaves, and extraction with supercritical fluids (Zhao et al. 2019; López-Hortas et al. 2022). The compounds extracted from edible flowers can be used for technological purposes in food. For example, some fermented dairy products, such as yogurts, require colouring and flavouring agents. The most required colours are green, red, yellow, and blue. A previous study reported that an aqueous extract obtained from roses (*Rosa damascena* and *Rosa gallica*), cornflowers (*Centaurea cyanus* L.), and dahlias (*Dahlia mignon*) did not change the nutritional characteristics and the sugar profile of yogurts, compared to their counterparts prepared with the E163 colouring (Pires et al. 2018).

Furthermore, edible flower extracts can improve the functional value of foods. A previous study used the extract from *Rosa rugosa* cv. Plena to fortify a functional yogurt. Although the extract modified the physicochemical properties of the yogurt, its phenolic content and antioxidant capacity increased due to the flower's compounds. As another example, lavender (*Lavandula angustifolia*), rose (*Rosa centifolia*), jasmine (*Jasminum officinale*), elder (*Sambucus nigra*), and hop (*Humulus lupulus*) flowers were used as alternative ingredients for the partial substitution of wheat flour in the cookie manufacturing process (Šť'Astná et al. 2021). Flowers increase the nutritional value of the food products that incorporate them. As for the cookies, flowers increased the phenolic compound content and antioxidant capacity.

13.6 Conclusion and Future Perspectives

Based on the information reviewed in this chapter, we can appreciate that there is relevant knowledge about the chemical and functional composition of fresh edible flowers. However, it is important to continue investigating their physicochemical and organoleptic characteristics to increase their use in various dishes and processed products. This could facilitate the selection of edible flowers that are in tune with the other ingredients in a culinary preparation. Furthermore, these flowers could provide contrasting flavours, such as acidity or sweetness, in riskier preparations. As for the edible flowers that, due to their chemical composition, require heat treatment, it is important to continue the research on the effect of the cooking methods on their nutritional and functional properties. This is essential due to the delicate nature of the plant material, which could suffer important structural modifications and compromise its beneficial health effects. Moreover, either by drying or extracting bioactive compounds or pigments, it is important to continue developing functional foods that incorporate edible flowers and their active components. This could open the door to a new type of processed food that is safe, nutritious, and attractive to the consumer.

References

Adler-Nissen J (2007) Continuous wok-frying of vegetables: process parameters influencing scale up and product quality. J Food Eng 83:54–60. https://doi.org/10.1016/j.jfoodeng.2006.11.002

Ahmed FA, Ali RFM (2013) Bioactive compounds and antioxidant activity of fresh and processed white cauliflower. Biomed Res Int 2013:367819. https://doi.org/10.1155/2013/367819

Altuntaş A, Aydin R (2014) Fatty acid composition of egg yolk from chickens fed a diet including Marigold (Tagetes erecta L.). J Lipids 2014:1–4. https://doi.org/10.1155/2014/564851

Aquino-Bolaños EN, Urrutia-Hernández TA, López Del Castillo-Lozano M et al (2013) Physicochemical parameters and antioxidant compounds in edible squash (cucurbita pepo) flower stored under controlled atmospheres. J Food Qual 36:302–308. https://doi.org/10.1111/jfq.12053

Blumfield M, Mayr H, De Vlieger N et al (2022) Should we 'eat a rainbow'? An umbrella review of the health effects of colorful bioactive pigments in fruits and vegetables. Molecules 27:4061. https://doi.org/10.3390/molecules27134061

Bragueto Escher GB, Cardoso Borges LDC, Sousa Santos JS et al (2019) From the field to the pot: phytochemical and functional analyses of Calendula officinalis L. Flower for incorporation in an organic yogurt. Antioxidants 8:559. https://doi.org/10.3390/antiox8110559

Cartea ME, Velasco P (2008) Glucosinolates in brassica foods: bioavailability in food and significance for human health. Phytochem Rev 7:213–229. https://doi.org/10.1007/s11101-007-9072-2

Castañeda-Rodríguez R, Larrea V, Hernando I, Ozuna C (2023) Cooking changes agave flower properties, including the bioaccessibility of bioactive compounds. Int J Gastron Food Sci 32:100749. https://doi.org/10.1016/j.ijgfs.2023.100749

Castañeda-Rodríguez R, Quiles A, Hernando I, Ozuna C (2024a) Cooking methods determine chemical composition and functional properties of squash blossoms: a study of microstructural and bioaccessibility changes. Food Res Int 180:114095. https://doi.org/10.1016/j.foodres.2024.114095

Castañeda-Rodríguez R, Quiles A, Llorca E, Ozuna C (2024b) How to cook Yucca spp. flowers? An analysis of their chemical composition, microstructure, and bioactive compound bioaccessibility. Appl Food Res 4:100414. https://doi.org/10.1016/j.afres.2024.100414

Chitrakar B, Zhang M, Bhandari B (2019) Edible flowers with the common name "marigold": their therapeutic values and processing. Trends Food Sci Technol 89:76–87. https://doi.org/10.1016/j.tifs.2019.05.008

Czarnowska-Kujawska M, Draszanowska A, Starowicz M (2022) Effect of different cooking methods on the folate content, organoleptic and functional properties of broccoli and spinach. LWT 167:113825. https://doi.org/10.1016/J.LWT.2022.113825

Esquivel L (1992) Como Agua para Chocolate. Suma de Letras, México

Fernandes L, Casal S, Pereira JA et al (2018) Effects of different drying methods on the bioactive compounds and antioxidant properties of edible Centaurea (Centaurea cyanus) petals. Braz J Food Technol 21:e2017211. https://doi.org/10.1590/1981-6723.21117

Ferracane R, Pellegrini N, Visconti A et al (2008) Effects of different cooking methods on antioxidant profile, antioxidant capacity, and physical characteristics of artichoke. J Agric Food Chem 56:8601–8608. https://doi.org/10.1021/jf800408w

Ghosh P, Rana SS (2021) Physicochemical, nutritional, bioactive compounds and fatty acid profiling of Pumpkin flower (Cucurbita maxima), as a potential functional food. SN Appl Sci 3:216. https://doi.org/10.1007/s42452-020-04092-0

Gomes S, Belo AT, Alvarenga N et al (2019) Characterization of Cynara cardunculus L. flower from Alentejo as a coagulant agent for cheesemaking. Int Dairy J 91:178–184. https://doi.org/10.1016/J.IDAIRYJ.2018.09.010

Gostin AI, Waisundara VY (2019) Edible flowers as functional food: a review on artichoke (Cynara cardunculus L.). Trends Food Sci Technol 86:381–391. https://doi.org/10.1016/j.tifs.2019.02.015

Hnin KK, Zhang M, Devahastin S, Wang B (2019) Influence of novel infrared freeze drying of rose flavored yogurt melts on their physicochemical properties, bioactive compounds and energy consumption. Food Bioprocess Technol 12:2062–2073. https://doi.org/10.1007/s11947-019-02368-x

Hosseini H, Ghorbani M, Meshginfar N, Mahoonak AS (2016) A review on frying: procedure, fat, deterioration Progress and health hazards. J Am Oil Chem Soc 93:445–466. https://doi.org/10.1007/s11746-016-2791-z

Hwang ES (2019) Effect of cooking method on antioxidant compound contents in cauliflower. Prev Nutr Food Sci 24:210–216. https://doi.org/10.3746/pnf.2019.24.2.210

Inocencio C, Rivera D, Alcaraz F et al (2000) Flavonoid content of commercial capers (Capparis spinosa, C. sicula and C. orientalis) produced in mediterranean countries. Eur Food Res Technol 212:70–74. https://doi.org/10.1007/s002170000220

Kayahan S, Saloglu D (2021) Microwave-assisted extraction of antioxidant phenolic compounds from artichoke (Cynara scolymus L. cv Bayrampasa): optimisation and kinetic modelling. Int Food Res J 28:704–715. https://doi.org/10.47836/ifrj.28.4.07

Kritsi E, Tsiaka T, Ioannou A-G et al (2022) In vitro and in Silico studies to assess edible flowers' antioxidant activities. Appl Sci 12:7331. https://doi.org/10.3390/app12147331

Lara-Cortés E, Osorio-Díaz P, Jiménez-Aparicio A, Bautista-Baños S (2013) Nutritional content, functional properties and conservation of edible flowers. Rev Arch Latinoam Nutr 63:197–208

López-Hernández AA, Ortega-Villarreal AS, Vázquez Rodríguez JA et al (2022) Application of different cooking methods to improve nutritional quality of broccoli (Brassica oleracea var. italica) regarding its compounds content with antioxidant activity. Int J Gastron Food Sci 28:100510. https://doi.org/10.1016/J.IJGFS.2022.100510

López-Hortas L, Rodríguez P, Díaz-Reinoso B et al (2022) Supercritical fluid extraction as a suitable technology to recover bioactive compounds from flowers. J Supercrit Fluids 188:105652. https://doi.org/10.1016/j.supflu.2022.105652

Manzano S, Martínez C, Megías Z et al (2011) The role of ethylene and brassinosteroids in the control of sex expression and flower development in Cucurbita pepo. Plant Growth Regul 65:213–221. https://doi.org/10.1007/s10725-011-9589-7

Mapes C, Basurto F (2016) Biodiversity and edible plants of Mexico. In: Lira R, Casas A, Blancas J (eds) Ethnobotany of Mexico. Ethnobiology. Springer, New York, pp 83–131

Matyjaszczyk E, Śmiechowska M (2019) Edible flowers. Benefits and risks pertaining to their consumption. Trends Food Sci Technol 91:670–674. https://doi.org/10.1016/J.TIFS.2019.07.017

Miękus N, Marszałek K, Podlacha M et al (2020) Health benefits of plant-derived sulfur compounds, Glucosinolates, and organosulfur compounds. Molecules 25:3804. https://doi.org/10.3390/molecules25173804

Mlcek J, Rop O (2011) Fresh edible flowers of ornamental plants—a new source of nutraceutical foods. Trends Food Sci Technol 22:561–569. https://doi.org/10.1016/J.TIFS.2011.04.006

Monalisa K, Bhuiyan JA, Islam MZ, Sayem ASM (2020) Boiling-induced changes on physicochemical, bioactive compounds, color, and texture properties of pumpkin (Cucurbita maxima). Food Sci Technol Int 26:333–343. https://doi.org/10.1177/1082013219894402

Moorthy P, Rajan M, Sathyanarayanan S et al (2020) Effect of different cooking methods of hibiscus surratensis L. leaf vegetable on nutritional, anti-nutritional composition, and antioxidant activities. J Culin Sci Technol 18:13–28. https://doi.org/10.1080/15428052.2018.1502110

Moreno DA, López-Berenguer C, García-Viguera C (2007) Effects of stir-fry cooking with different edible oils on the phytochemical composition of broccoli. J Food Sci 72:S064–S068. https://doi.org/10.1111/j.1750-3841.2006.00213.x

Mulík S, Ozuna C (2020) Mexican edible flowers: cultural background, traditional culinary uses, and potential health benefits. Int J Gastron Food Sci 21:100235. https://doi.org/10.1016/j.ijgfs.2020.100235

Mulík S, Hernández-Carrión M, Pacheco-Pantoja SE et al (2022) Culinary uses of Mexican edible flowers: recipe analysis. Int J Gastron Food Sci 28:100539. https://doi.org/10.1016/j.ijgfs.2022.100539

Murador DC, Mercadante AZ, De Rosso VV (2016) Cooking techniques improve the levels of bioactive compounds and antioxidant activity in kale and red cabbage. Food Chem 196:1101–1107. https://doi.org/10.1016/j.foodchem.2015.10.037

Ng A, Waldron KW (1997) Effect of cooking and pre-cooking on cell-wall chemistry in relation to firmness of carrot tissues. J Sci Food Agric 73:503–512. https://doi.org/10.1002/(SICI)1097-0010(199704)73:4<503::AID-JSFA762>3.0.CO;2-Z

Panyayong C, Srikaeo K (2022) Foods from banana inflorescences and their antioxidant properties: an exploratory case in Thailand. Int J Gastron Food Sci 28:100436. https://doi.org/10.1016/J.IJGFS.2021.100436

Pinedo-Espinoza JM, Gutiérrez-Tlahque J, Santiago-Saenz YO et al (2020) Nutritional composition, bioactive compounds and antioxidant activity of wild edible flowers consumed in semiarid regions of Mexico. Plant Foods Hum Nutr 75:413–419. https://doi.org/10.1007/s11130-020-00822-2

Pires TCSP, Dias MI, Barros L et al (2018) Incorporation of natural colorants obtained from edible flowers in yogurts. LWT 97:668–675. https://doi.org/10.1016/j.lwt.2018.08.013

Pires TCSP, Barros L, Santos-Buelga C, Ferreira ICFR (2019) Edible flowers: emerging components in the diet. Trends Food Sci Technol 93:244–258. https://doi.org/10.1016/J.TIFS.2019.09.020

Putriani N, Perdana J, Meiliana, Nugrahedi PY (2022) Effect of thermal processing on key phytochemical compounds in green leafy vegetables: a review. Food Rev Int 38:783–811. https://doi.org/10.1080/87559129.2020.1745826

Ramu R, Shirahatti PS, Anilakumar KR et al (2017) Assessment of nutritional quality and global antioxidant response of banana (Musa sp. CV. Nanjangud Rasa Bale) pseudostem and flower. Pharm Res 9:S74–S83. https://doi.org/10.4103/pr.pr_67_17

Rodrigues H, Spence C (2023) Looking to the future, by studying the history of edible flowers. Int J Gastron Food Sci 34:100805. https://doi.org/10.1016/j.ijgfs.2023.100805

Santos CSP, Cruz R, Cunha SC, Casal S (2013) Effect of cooking on olive oil quality attributes. Food Res Int 54:2016–2024. https://doi.org/10.1016/j.foodres.2013.04.014

Song J, Park J, Jung J et al (2015) Analysis of trans fat in edible oils with cooking process. Toxicol Res 31:307–312. https://doi.org/10.5487/TR.2015.31.3.307

Sotelo A, López-García S, Basurto-Peña F (2007) Content of nutrient and antinutrient in edible flowers of wild plants in Mexico. Plant Foods Hum Nutr 62:133–138. https://doi.org/10.1007/s11130-007-0053-9

Soundararajan P, Kim JS (2018) Anti-carcinogenic glucosinolates in cruciferous vegetables and their antagonistic effects on prevention of cancers. Molecules 23:2983. https://doi.org/10.3390/molecules23112983

Št'Astná K, Sumczynski D, Yalcin E (2021) Nutritional composition, in vitro antioxidant activity and phenolic profile of shortcrust cookies supplemented by edible flowers. Foods 10:2531. https://doi.org/10.3390/foods10112531

Takahashi JA, Rezende FAGG, Moura MAF et al (2020) Edible flowers: bioactive profile and its potential to be used in food development. Food Res Int 129:108868. https://doi.org/10.1016/j.foodres.2019.108868

Tarhan L, Kayali HA, Urek RO (2007) In vitro antioxidant properties of Cucurbita Pepo L. male and female flowers extracts. Plant Foods Hum Nutr 62:49–51. https://doi.org/10.1007/s11130-006-0038-0

Xiong L, Yang J, Jiang Y et al (2014) Phenolic compounds and antioxidant capacities of 10 common edible flowers from China. J Food Sci 79:517–525. https://doi.org/10.1111/1750-3841.12404

Yang W, Lu X, Zhang Y, Qiao Y (2019) Effect of cooking methods on the health-promoting compounds, antioxidant activity and nitrate of tatsoi (Brassica rapa L. ssp. narinosa). J Food Process Preserv 43:e14008. https://doi.org/10.1111/jfpp.14008

Zhao L, Fan H, Zhang M et al (2019) Edible flowers: review of flower processing and extraction of bioactive compounds by novel technologies. Food Res Int 126:108660. https://doi.org/10.1016/j.foodres.2019.108660

Zheng J, Yu X, Maninder M, Xu B (2018) Total phenolics and antioxidants profiles of commonly consumed edible flowers in China. Int J Food Prop 21:1524–1540. https://doi.org/10.1080/10942912.2018.1494195

Zhou Q, Jia X, Deng Q et al (2019) Quality evaluation of rapeseed oil in Chinese traditional stir-frying. Food Sci Nutr 7:3731–3741. https://doi.org/10.1002/fsn3.1232

Zribi A, Jabeur H, Aladedunye F et al (2014) Monitoring of quality and stability characteristics and fatty acid compositions of refined olive and seed oils during repeated pan- and deep-frying using GC, FT-NIRS, and chemometrics. J Agric Food Chem 62:10357–10367. https://doi.org/10.1021/jf503146f

Chapter 14
Edible Flowers for Food Colouring

Cláudia Novais, Beatriz Helena Paschoalinotto, Lillian Barros, Maria Inês Dias, and Carla Pereira

Abstract The consumption of flowers as a dietary component has been documented across various cultures, serving not only as an ornament but also as a means to enhance traditional cuisine and alternative medicine. As consumers increasingly seek novel flavours and textures, edible flowers have garnered attention within the food industry, offering not only visual appeal and taste, but also unique sensory characteristics. These flowers play a pivotal role in contemporary dietary trends, driven by the pursuit of healthier lifestyles and a growing interest in natural foods. In addition to being rich in essential nutrients such as fibre, protein, and amino acids, they boast a wealth of bioactive compounds, including polyphenols, flavonoids, and anthocyanins, which exhibit significant biological activities. Technological advancements have facilitated the extraction of these compounds paving the way for the development of functional ingredients and potential natural colourants for the food industry. Despite the presence of edible flowers across numerous plant species, only a fraction of them is utilized, underscoring the necessity for comprehensive studies to ensure their safe and effective utilization, including taxonomy classification and toxicological profiling. Scientific exploration of the chemical characteristics of edible flowers aims to bolster understanding and unlock their potential as colourants in the food industry.

Keywords Anthocyanins · Colourants · Ethnobotanics · Extraction · Food industry · Natural colourants · Polyphenols

C. Novais · B. H. Paschoalinotto · L. Barros · M. I. Dias (✉) · C. Pereira (✉)
CIMO, LA SusTEC, Instituto Politécnico de Bragança, Campus de Santa Apolónia, Bragança, Portugal
e-mail: maria.ines@ipb.pt; carlap@ipb.pt

M. M. Jowkar et al. (eds.), *Advances in Research on Edible Flowers*, Advances in Olericulture, https://doi.org/10.1007/978-3-032-04561-4_14

Abbreviations

CO2	Carbon dioxide
EAE	Enzyme-Assisted Aqueous Extraction
EC	Edible coatings
EF	Edible Flowers
EFSA	European Food Safety Authority
FAO	Food and Agriculture Organization of the United Nations
FDA	Food and Drugs Administration
FMASE	Microwave assisted Soxhlet extraction
MAE	Microwave-Assisted Extraction
MAHD	Microwave Hydrodistillation
MPa	Megapascal
MSD	Microwave Steam Distillation
MWHD	Microwave Hydrodistillation
N2	Nitrogen
NPMAE	Nitrogen-protected microwave-assisted extraction
O2	Oxygen
PEFE	Pulse Electric Field Extraction
PUE	Ultrasonic Probe Extraction
RASFF	Rapid Alert System for Food and Feed
S/L	Solid/Liquid
SFE	Supercritical Fluid Extraction
UAE	Ultrasound-Assisted Extraction
UMAE	Ultrasonic microwave assisted extraction
VMAE	Microwave-assisted Vacuum Extraction
WHO	World Health Organization

14.1 Introduction

Edible flowers have long been cherished in various cultures around the world, not just for their aesthetic value, but also for their culinary contributions. From adding a pop of colour to everyday dishes to elevating gourmet creations, these blooms offer a unique sensory experience. Moreover, as consumers increasingly seek natural and sustainable alternatives, the market for edible flowers as food colouring agents is witnessing a remarkable surge, presenting exciting opportunities for chefs, food artisans, and entrepreneurs alike. Nature provides a rich palette of potential colourants, and edible flowers are no exception. From anthocyanins to carotenoids, these compounds not only lend colour but also offer potential health benefits, making them an attractive alternative to synthetic dyes.

The enchanting world of edible flowers extends far beyond mere decoration on a plate. In recent years, there has been a growing interest in utilizing these delicate blossoms not only for their visual appeal but also for their vibrant natural pigments. This comprehensive chapter delves into the realm of edible flowers for food colouring, exploring their cultural significance, traditional uses, extraction methods, and the challenges and opportunities they present in today's culinary landscape.

14.2 Cultural Consumption, Contribution on Daily Dishes, and New Trends and Market Opportunities

The versatility of edible flowers (EF) has been popular since the ancient world since they are traditionally applied in cooking for the attractiveness of dishes and drinks, as well as for their nutritional appeal and in common medicine to combat minor illnesses. The enchantment for EF is due to the multiplicity of aromas, textures, sizes, colours, flavours, and herbal properties; they were used as salad ingredients in medieval France, ancient Greece, and Rome as condiments and flavour enhancers in various dishes and in China as infusions for health benefits, while saffron is used as a food colouring even today (Mlcek and Rop 2011; Lu et al. 2016; Purohit et al. 2021; Shantamma et al. 2021). Figure 14.1 exemplifies some species consumed in different parts of the world.

At the beginning of the twenty-first century, wild plants began to emerge and be praised for gastronomic uses and consequently, among the species, there are edible flowers used both as a decorative element or as an ingredient in recipes, significantly promoting their adoption among the broader population. However, it is worth noting that EF remain relatively unfamiliar in some countries (Santos and Reis 2021). It is estimated that edible flowers are obtained from 97 families, 100 genera, and 180 species worldwide (Lu et al. 2016). This wide diversity is currently present in the so-called Mediterranean Diet (mainly adopted by southern Europe) as salads, starters, desserts, drinks, vegetables, pastry ingredients, and jams (Amrouche et al. 2022). In Costa Rica and Portugal, the prevalent methods of consumption are mainly

Fig. 14.1 Edible flowers with staining properties consumed in different parts of the world. (**a**) *Turmeric,* (**b**) *Calendula officinalis,* (**c**) *Taraxacum officinale* F.G. Wigg, (**d**) Centaurea cyanus, (**e**) Tagetes spp., (**f**) Tropaeolum majus and (**g**) Dahlia spp.

in salads and using them as decorative elements in dishes, such as jellies, and aroma enhancers, with the most frequent way of obtaining the species is by harvesting them in the wild, followed by domestic cultivation and then in supermarkets (Guiné et al. 2020). In India, the main motivation for EF consumption is sociocultural and

Table 14.1 Consumption of edible flowers in different regions of the world: popular species, edible parts, flavour, and traditional uses

Family	Scientific name/ Common mane	Continent	Colour	Common use	References
Asteraceae	*Tagetes patula*/ French marigold	Australia	Orange	Refreshing drinks	Purohit et al. (2021)
	Calendula officinalis/ Calendula	Europe	Yellow/ Orange	Infusions and soups	Purohit et al. (2021)
	Centaurea cyanus/ Cornflower	Europe	Blue	Medicinal + Garnishing	Purohit et al. (2021)
	Matricaria chamomilla/ chamomile	Europe, Egypt, and North America	White	Herbal teas and cakes	Amrouche et al. (2022)
Malvaceae	*Hibisco sabdariffa* L.	Africa	Red	Fermented drinks, ice cream, jam and cakes	Amrouche et al. (2022)
	Hibiscus rosa sinensis L.	Africa	Pink	Infusions	Pires et al. (2019)
Tropaeolaceae	*Tropaeolum majus*/ Nasturtium	Australia	Yellow, orange, red	Medicinal + Garnishing	Purohit et al. (2021)
Boraginaceae	*Borago officinalis* / Borage	Europe and North Africa	Blue	Salads, drinks and desserts	Asadi-Samani et al. (2014)
	Anchusa azurea P. Mill. / Italian bugloss	Europe	Violet to blue	Salads, soups, boil, or fries	Pires et al. (2019)
Rosaceae	*Rosa centifolia* /Rose	Europe	Pink, red, yellow	Tea, cakes, and flavour extracts	Jena et al. (2012)
Violaceae	*Viola x wittrockiana* / Pansy	Europe	Violet, red and yellow	Salads and herbal tea	Mlcek and Rop (2011)
Lamiaceae	*Rosmarinus officinalis* / Rosemary	Europe and North Africa	Purple	Herbal tea and aromatic on meals	Amrouche et al. (2022)
	Lavandula stoechas L / Lavander	Europe	Purple	Herbal tea	
	Thymus vulgaris L / Thyme	Europe	White-violet	Spice	

spiritual, they are consumed raw right after harvesting or cooked; in Brazil, certain flowers have found their way into culinary traditions through the influence of European and Asian immigrants, such as pumpkin, artichoke, broccoli, and cauliflower (Santos and Reis 2021). Table 14.1 shows the popular edible plants consumed in different continents.

Although the current review does not emphasize all phytochemicals abundantly present in edible flowers, it is important to point out the presence of different groups that confer different beneficial health properties. Amrouche et al. (2022) and Lu et al. (2016) explained potential bioactive mechanisms of different edible flowers belonging to several families (mainly Asteraceae and Lamiaceae), highlighting the antioxidant, antidiabetic, anti-inflammatory, antimicrobial, cardiovascular, neuroprotective, and antitumor activities.

Due to cultural, commercial ties, and consumer interest in new foods, there is a renewal in the use of edible flowers. This trend is evidenced by the development of research on nutritional and health effects, accompanied by a rising number of cookbooks and culinary articles. However, the high sensitivity, perishability and seasonality of edible flowers are the biggest obstacles to produce and commercialized these plants, which increases the costs of all logistics. In 2018 in the European market, popular edible flowers such as violas, cornflowers, roses, pansies, tagetes, mixed dried cornflower petals and borages were sold in units from 1.97 to 64.97 euros, with the most expensive product being roses. In North and South America, the market for edible flowers is small, with the most sold species being calendula, hibiscus, viola, and pansy in the range of 5 to 12 euros. Asia has the lowest market value, with EF traded from 1 euro (Fernandes et al. 2019). The aesthetic properties of EF represent a current market niche, relying on the innovations within contemporary gastronomy, particularly when esteemed by chefs (Santos and Reis 2021). In the academic area, studies focus on aromas, essential oils, composition of dishes, preparation of infusions, nutritional profile, and bioactive properties (Fernandes et al. 2017a; Pires et al. 2019; da Silva et al. 2020). However, in addition to the emerging demand for flowers to compose dishes for their flavour and attractiveness, there are new trends for the colouring potential of these natural sources for extraction and use in an extensive list of foods and nutraceuticals (Zhao et al. 2019).

14.3 Traditional Edible Flowers and Common Colouring Uses

Turmeric (Fig. 14.1a) is popularly known as Indian saffron. The rhizomes are valued both as a spice and as a colourant used in food and cosmetic manufacturing. Turmeric comprises approximately 5% volatile oil, resin and vibrant yellow pigments known as curcuminoids. Curcumin, demethoxycurcumin, and bisdemethoxycurcumin are responsible for its intense yellowish colour. It has good microbiological stability; however, it is highly sensitive to light. Curcumin has antioxidant, anti-inflammatory, anticancer, and hepatoprotective potentials. The pharmacological activities of curcuminoids are due to their unique molecular structure (Chengaiah et al. 2010; Brudzyńska et al. 2021). Sowbhagya et al. (2005) studied the replacement of tartrazine by saffron in extruded products and proposed that curcumin could provide a viable alternative to the synthetic colouring agents.

The common marigold (*Calendula officinalis* L.) (Fig. 14.1b) has 29 species and its flowers are widely used as food, dried or fresh, as a garnish for various meals, salads, drinks, desserts, in addition to natural food and textile colouring (Sausserde and Kampuss 2014; Pires et al. 2019). In research conducted by Pintea et al. (2003), using calendula varieties, it was verified that all contained the same pigments, but in different proportions between the individual pigments, prevailing flavoxanthin, luteoxanthin, lutein, rubixanthin, lycopene, β-carotene and γ-carotene. It is known that the concentrations of carotenoids in petals reach 7.7% and in pollens 1.6%, and these pigments can be used in foods, such as dairy products, with a colour intensity very similar to that of conventional colouring (Verma et al. 2018; Rojas-Bedoya et al. 2020).

Dandelions (*Taraxacum officinale* F.H. Wigg, Asteraceae, Fig. 14.1c) is a plant that easily adapts to climatic adversities, which is why it is considered a weed and grows in countries such as Brazil, the United States of America, Asia, and Europa (de Almeida et al. 2022). It is a versatile plant since all its parts (stems, leaves, and flowers) have applicability in traditional medicine or cooking, its flowers serve as additives in the production of wines and desserts, while extracts are employed as flavouring agents in a range of dairy food products. They have proteins, fibres, fats, flavonoids, vitamins, and minerals, including carotenoids that give the yellow colour of the edible flowers of dandelions (Cacak-Pietrzak et al. 2022). The most common carotenoids are luteolin 7-glycosides and luteolin 7-diglycosides (Fernandes et al. 2017b).

Cornflower or bachelors' button (*Centaurea cyanus* L., Fig. 14.1d) is an annual flower typical of Europe and Asia. Its petals are commonly used to prepare infusions with relaxing properties from a slightly sweet to spicy taste, as a natural food colouring, in addition to having antioxidant properties (Fernandes et al. 2017b). The attractiveness and elegance of this edible flower owes much to the uniqueness of the intense bluish colour characteristic of protocyanin. This supramolecular pigment is formed by a complex composed of six anthocyanins (cyanidin-3-O-(6-O-succinylglucoside)-5-O-glucoside), six flavones (7-O-glucuronide-4′-O-(6-O-malonylglucoside)), a magnesium and an iron bound by metal complexation and hydrophobic interactions. However, these bonds are considered unstable, so the blue colour is only stable in concentrated solutions, since, if diluted, the colour fades and at acidic pH it appears reddish-purple (Kondo et al. 1998; Escher et al. 2018). With the emerging problem, optimization studies of extraction of this valuable compound for application as natural colourants are studied and optimized. Escher et al. (2018) optimized the aqueous extract of *C. cyanus* rich in total phenolics, reversible balance of anthocyanins, and antioxidant activity with pH variation, in addition to low cytotoxicity.

Tagetes (Fig. 14.1e), commonly called marigold, belongs to Tagetes family which comprises 56 species, distributed worldwide, the most common being *T. erecta* L. and *T. patula* L. (Riaz et al. 2020). Fresh calendula petals are used in soups, salads, dessert decorations, teas and incorporated into sauces which, in addition to adding flavour, act as a colouring agent (Mulík and Ozuna 2020). Marigold flower extract has been used commercially as poultry feed additive to improve egg fat, skin, and yolk pigmentation (Siriamornpun et al. 2012). The flowers of *T. erecta* are also of great relevance as a natural food colouring, being one of the richest sources of carotenoids, such as α- and β-carotenes, xanthophylls, zeaxanthin and lutein, the latter being an esterified liposoluble carotenoid, the main pigment of the flowers, representing about 90% of the carotenoids contained in the marigold (Skrajda-Brdak et al. 2020). Depending on flower colour, various amounts of lutein esters have been reported, ranging from 4 mg g^{-1} in yellow-green flowers to 800 mg g^{-1} in orange-brown flowers (Sowbhagya et al. 2006; Chitrakar et al. 2019). Other carotenoids such as violaxanthin, zeaxanthin, β-carotene, and lycopene were also isolated (Siriamornpun et al. 2012). *T. erecta* colour can be applied in colouring foods such as edible oils, mayonnaise, mustard, and other salads, in yoghurts, cakes, ice cream, and dairy products (Sowbhagya et al. 2006). *Tagetes patula* L., has a smaller flower than *T. erecta*, with colours ranging from yellow, orange, red and brown. Traditionally considered an aromatic herb, its use varies from one country to another, and its flowers can be used as a spice, salads, flavouring, and food colouring (Purohit et al. 2021). Despite its diversified use, there is still some lack of literature on this plant; however, it is known that T. *patula* flowers are rich in carotenoids, lutein being mostly isolated. This is present in free form in the petals and calyces of *T. patula*, with values of up to 154.96 mg g^{-1} of extract; and as esters of myristate, monomyristate, dimyristate, palmitate, monostearate and palmitate-stearate. Other major carotenoids isolated from the *T. patula* flower are antheraxanthin, violaxanthin and zeaxanthin (Bhattacharyya et al. 2008). Carotenoid content varies between species, cultivars, and plant parts; however, it is known that the petals have higher levels of carotenoids than the calyxes (Riaz et al. 2020).

Nasturtium (*Tropaeolum majus* L.) (Fig. 14.1f) has some varieties that differ in terms of structure, size, and colour of the flowers and is widely cultivated as an ornamental plant and widely used in cooking. Giving dishes a diversity of colours and a characteristic peppery flavour, the flowers are added to soups, meats, pasta, pies, omelets, and pancakes. In Canada, they are consumed with cottage cheese filling and can be used to replace mustard in salads, and some are even incorporated into wines and flavoured liqueurs (Pires et al. 2019; Souza et al. 2020). Regarding pigments, nasturtium is rich in lutein and anthocyanins (cyanidin, delphinidin, and pelargonidin), and its extracts are used as natural colouring in the pharmaceutical and food industries. The content of individual compounds is different depending on the colour of the flowers. Red flowers have a higher concentration of anthocyanins, mainly delfinidin-3-dihexoside (769 ug.g^{-1} fresh weight—fw), yellow flowers have higher levels of lutein (450 μg.g^{-1} fw), which is the only carotenoid found in *T. majus* and orange blossoms have intermediate levels of anthocyanins and carotenoids when compared to yellow and red blossoms (Garzón et al. 2015; Jakubczyk et al. 2018).

One of the most prized ornamental flowers in the world, *Dahlia* spp. (Fig. 14.1g), has 37 species and its flowers have a wide diversity of size, shape, and colour, all imaginable shades from red, purple, pink, lilac, white, yellow, and orange (Martínez-Damián et al. 2021). Its petals are used fresh in salads and drinks; coated with sugar for decorating desserts; or cooked, in soups, stews, pastas, sauces, snacks, fish, purées, jams, and jellies (Mulík and Ozuna 2020). Dahlia flowers are rich in anthocyanins, especially those derived from pelargonidin and cyanidin, for purple, red, and pink flowers; carotenoids for orange flowers; and as far as the yellow flowers are concerned, their colouring results from the formation of chalcones (Halbwirth et al. 2008; Martínez-Damián et al. 2021), therefore relevant as a source of natural food colouring. Pires et al. (2018) used dahlia flower extract in food products, namely yoghurts, comparing it with the additive E163 (commercial anthocyanin colouring) and observed that yogurts prepared with flower extracts had similar nutritional profiles to those prepared with E163.

14.4 Potential Sources of Natural Colourants

The attractiveness of the consumption of edible flowers was undoubtedly boosted by the diversity of colours that a species has, since it plays a fundamental role in the consumer's decision. The primary factors contributing to the colouration of flowers are the levels of carotenoids and anthocyanins. Table 14.2 summarizes the compound/colour ratio of popular edible flowers.

For instance, the chrysanthemum (*Chrysanthemum x morifolium* (Ramat)) has flowers of various colours, the pigments in the yellow flowers consist of flavones and carotenoids, those in the white flowers only in flavones, while those in the red and purple flowers consist of anthocyanins and flavonoids, and the of orange into carotenoids, anthocyanins, and other flavonoids (Park et al. 2015; Chen et al. 2021). Anthocyanins are widely distributed in plants, it is one of the major classes of flavonoids; the water-soluble pigments belong to the class of flavonoids that appear orange, pink, red, purple, and blue (Barani et al. 2022). The search for replacing synthetic colourants with natural pigments is valid, as legislative restrictions on the use of synthetic colourants grow. However, until the legal sanction of a natural pigment comes into force there is an extensive path on chemistry, extraction, processing, stability, and applications, often a complete study is required for each natural pigment molecule. In addition to providing colour, natural pigments have bioactive properties that help maintain long-term health (Sowbhagya et al. 2006).

Table 14.2 Sources of natural colourants

Compound	Colour	Plant/Flower	References
Cyanidin-3-glucoside	Orange-red	*Bauhinia purpurea* L.	Li et al. (2014)
		Cercis canadensis	Perkins-Veazie et al. (2017)
Cyanidin-3-*O*-glucoside	Orange-red	*Bombax malabaricum*	Li et al. (2014)
		Brunfelsia acuminate L.	Li et al. (2014)
		Calliandra haematocephala Hassk	Pires et al. (2019)
Cyanidin-3-(3″-malonoyl) glucoside	Red/intense purple	*Dendranthema grandiflorum* Ramat.	Park et al. (2015)
Cyanidin-3,5-diglucoside	Pink	*Rosa* spp.	Cunningham and Gantt (2005)
Delphinidin-3-glucoside	Purple	*Begônia × tuberhybrida* Voss.	Morais et al. (2020)
		Torenia fournieri F. Lind	
		Clitoria ternatea L.	
Delphinidin-3,7-O-diglucoside		*Crocus sativus L*	Goupy et al. (2013)
Malvidin-3-glucoside	Red wine	*Dianthus chinensis* L.	Morais et al. (2020)
		Cercis canadensis	Perkins-Veazie et al. (2017)
Malvidin-3,5-diglucoside	Blue-red	*Begônia × tuberhybrida* Voss.	Morais et al. (2020)
		Torenia fournieri F. Lind	
		Clitoria ternatea L.	
		Cercis canadensis	Perkins-Veazie et al. (2017)
Pelargonidin 3,5-diglucoside	Pink	*Rosa* spp.	Cunningham and Gantt (2005)
Peonidin—3-glucoside		*Cercis canadensis*	Perkins-Veazie et al. (2017)
Petunidin-3-glucoside	Red	*Cercis canadensis*	Perkins-Veazie et al. (2017)
		Crocus sativus L	Goupy et al. (2013)
		Begônia × tuberhybrida Voss.	Morais et al. (2020)
		Torenia fournieri F. Lind.	
		Clitoria ternatea L.	
Petunidin-3,7-*O*-diglucoside	Intense red	*Crocus sativus L*	Goupy et al. (2013)
Astaxanthin-(3,3′-dihydroxy-4,4′-diketo-*β*,*β*-carotene)	Intense red	*Artemisia annua*	Cunningham and Gantt (2005)
		Adonis aestivalis	
β-carotene	Orange	*Tagetes erecta* L	Siriamornpun et al. (2012)
		Viola wittrockiana	González-Barrio et al. (2018)
		Antirrhinum majus	

(continued)

Table 14.2 (continued)

Compound	Colour	Plant/Flower	References
Lutein	Orange-yellow	*Tagetes erecta* L	Siriamornpun et al. (2012)
		Dendranthema grandiflorum Ramat.	Park et al. (2015)
		Viola wittrockiana	González-Barrio et al. (2018)
		Antirrhinum majus	
Lycopene	Orange-red	*Tagetes erecta* L	Siriamornpun et al. (2012)
Violaxanthin	Light yellow	*Rosa* spp.	Cunningham and Gantt (2005)
		Viola wittrockiana	González-Barrio et al. (2018)
Zeaxanthin	Yellow	*Antirrhinum majus*	González-Barrio et al. (2018)
		Viola wittrockiana	

14.5 Conventional and Non-conventional Methods of Extraction

The growing interest in natural colourants, replacing synthetic colourants, compounds such as anthocyanins and carotenoids are widely used in the food industry. To preserve the colour, stability, and other properties of these molecules, it is crucial to select the most efficient extraction technique, considering the various factors involved. Among these factors, the need for prior elimination of other compounds, such as lipids, proteins, or contaminants, present in the sample matrix, and which may hinder the extraction process, is highlighted. Consequently, decide the most appropriate extraction techniques is the main objective of many industries.

14.5.1 Conventional Methods of Extraction

Conventional extraction techniques primarily involve on the application of various solvents and/or heat and are the most used mainly in the natural colourant industries. The most used techniques are infusion, decoction, maceration (with or without heat), percolation, filtration and Soxhlet. Methanol, ethanol, hexane, water, acetone, or mixtures thereof are the solvents generally used to extract colouring compounds (Picot-Allain et al. 2021). Their compositions, powder particle size, the solvent/solid ratio, the extraction duration, and temperature are parameters that must be optimized according to the matrix and intended compound to guarantee maximum extraction yields. Despite their widespread use, these methods are known for their high consumption of harmful solvents for the environment; high energy cost and long periods of low-yield extraction, low selectivity, and use of considerable temperatures, which can cause the degradation of compounds and change the colour and properties of extracts (Belwal et al. 2018). The fact that they have a low instrumentation cost encourages their continued use on a large scale, however aspects such as solid/liquid (S/L) ratios and amount of solvent to be used are of great relevance for optimization, minimizing the use of contaminants according to the requirements legal, and to reduce evaporation costs.

14.5.2 *Green Methods of Extraction*

Although solvent extraction methods are the most common in the extraction of natural pigments, the purity and quality of these compounds must be considered, which are closely related to the extraction technology and its conditions. In recent years, several novel extraction techniques have developed to address the limitations of traditional extraction methods and align with the increasing public demand for chemical-free compounds, including ultrasound-assisted extraction, microwave-assisted extraction, by supercritical fluid, high-pressure liquid extraction, pulsed electric fields, high-voltage electrical discharge, and enzyme-assisted extraction (Barani et al. 2022). These approaches demonstrate greater ecological sustainability by typically avoiding the use of heat during extraction, resulting in reduced energy consumption and enhanced extract stability. Moreover, they minimize solvent volume, in some cases approaching zero, in contrast to conventional techniques, thereby reducing greenhouse gas emissions. Additionally, these methods hold significant industrial potential by targeting improved extraction efficiency and yield. However, their widespread adoption is hindered by the need for more advanced instrumentation compared to traditional techniques.

14.5.2.1 Microwave-Assisted Extraction (MAE)

Microwave-assisted technology, known as green technology, employs microwave radiation to exert pressure on plant cells. These waves induce elevated pressure within the cell walls and producing heat, which consequently leads to the rupture of the cell wall, thus facilitating easier infiltration of the solvent into the plant cell, with greater transport of its mass to the environment (Belwal et al. 2018). Compared to conventional extractions, microwave-assisted extraction has a lower solvent consumption and time durability, resulting in increased cost-effectiveness and greater extraction yields (Yiğit et al. 2022). However, the heating caused by microwave energy becomes its main limitation, given the thermo-lability of many of the colouring compounds. To address potential constraints associated with the MAE, certain modifications to the technique have already been implemented. MAE techniques that allow for less degradation of thermolabile and oxygen-sensitive phytochemicals found in plants are nitrogen-protected microwave-assisted extraction (NPMAE) and microwave-assisted vacuum extraction (VMAE). These are based on the use of low oxygen content inside the extraction tanks, and moderate or low temperatures. Microwave assisted Soxhlet extraction (FMASE), ultrasonic microwave assisted extraction (UMAE), microwave hydrodistillation (MWHD or MAHD) and microwave steam distillation (MSD) are other examples of alternatives to conventional MAE technique. This technique and its variants are regarded as potential alternatives to traditional approaches and can be applied on an industrial scale (Bagade and Patil 2019).

14.5.2.2 Ultrasound-Assisted Extraction (UAE)

The ultrasound-assisted extraction method harnesses the phenomenon of cavitation, breaking the cell wall of the sample tissue. This disruption facilitates the seamless penetration of the solvent into the solid matrix, thereby enhancing mass transfer between the solid matrix and the solvent, thus decreasing the extraction time (Morata et al. 2021). It can be performed using an ultrasonic bath (BUE), which is more affordable and simpler to manage, but with an uneven energy distribution; or by ultrasonic probe (PUE), which consists of immersing a probe into the extraction vessel, which disperses the ultrasound in the medium, minimizing energy loss by concentrating it in a specific area of the sample. This approach enhances extraction efficiency, being preferably chosen for extract the colouring compounds (Kumar et al. 2021) It is a relatively easy and economical method; however, one must carefully manage and control the extraction conditions to optimize the extraction process (Tan et al. 2022). In addition to being a fast technique, delivering high yields within brief timeframes at low temperatures, this method also minimizes solvent usage, exhibiting a more environmentally conscious approach compared to conventional extraction techniques and it is a straightforward technique, that doesn't involve complex maintenance (Morata et al. 2021; Tena and Asuero 2022).

14.5.2.3 Supercritical Fluid Extraction (SFE)

According to Silva et al. (2016) supercritical fluid extraction has seen extensive use in recent years, earning recognition as one of the most sustainable green technologies, since it frequently employs supercritical CO_2 as a solvent, which is non-toxic, non-flammable, non-corrosive and easy to handle, it is economical and easily available with a high level of purity, minimizing changes to compounds. It effectively dissolves lipophilic substances and can be effortlessly separated from the final products. The fact that this method uses smaller amounts of solvent, compared to those performed under low pressure, exhibits greater efficiency and selectivity, and a shorter extraction time, being also suitable for thermosensitive compounds, makes it very promising for the extraction of colouring compounds. The SFE works with solvents under conditions exceeding their critical points, and although CO_2 is the chosen extraction solvent (for non-polar compounds), a co-solvent (ethanol, methanol, and aqueous solutions of these alcohols) can be employed to facilitate the extraction of polar compounds, such as anthocyanins.

However, more studies should be performed focusing on the optimization of extraction time for various plant materials, co-solvent type, and flow rates to enhance the cost-effectiveness of extraction, one of the drawbacks being the costly implementation that the method needs for its large-scale application (Tan et al. 2022).

14.5.2.4 Pulse Electric Field Extraction (PEFE)

Pulsed electric field extraction has gained increasing attention in recent years, mainly because it is a non-thermal technique, rendering it an excellent method for extracting heat-sensitive compounds, sustainably, thereby augmenting the extraction yields while selectively deactivating microorganisms (Ranjha et al. 2021). PEFE involves applying brief, high-intensity electrical pulses at ambient temperatures. With the increased permeability of cell membranes, through the electroporation process, its structural functionality is lost, facilitating the extraction of the intended compounds. Parameters such as exposure time, electric field intensity, and total pulse specific energy input are critical to the extraction efficiency, as well as the nature of the solvent, sample composition, and chamber size (Gagneten et al. 2019).

14.5.2.5 Enzyme-Assisted Aqueous Extraction (EAE)

Enzyme-assisted aqueous extraction has been a longstanding practice and continues to evolve. It consists in enzymatic action within cell walls, through the hydrolysis of its constituents, thus reaching the colouring and bioactive compounds. Four families of enzymes are commonly used, cellulases, hemicelluloses, pectinases, and proteases, the first three of which act differently on the cell wall and can be used individually or in combination. These enzymes have been used in the extraction of pigments from various plant materials, being able to extract such compounds more efficiently than conventional techniques (Lotfi et al. 2015). The constitution of the enzymatic blend plays a pivotal role in this technique. In addition, it is essential to optimize conditions such as pH or temperature, according to each enzyme. The high selectivity of the enzymes, their ability to degrade the cell wall, and their operation under moderate temperature and pressure conditions, collectively position this technology as a promising tool for industrial applications (Binaschi et al. 2018). EAE can also be combined with other extraction processes, such as UAE, to increase extraction yield. However, the high cost of enzymes and process energy, as well as the high sensitivity to pH, temperature, and nutrient availability, which imply process optimization, limit its application on an industrial scale, requiring further studies for its implementation (Tan et al. 2022).

14.6 Challenges of Ethnobotanical Conservation

Another important perspective on edible flowers is their safe food intake and storage strategies. No food and/or agricultural health regulatory body, including the Food and Agriculture Organization of the United Nations (FAO), Food and Drugs Administration (FDA), World Health Organization (WHO) and the European Food Safety Authority (EFSA) establish an official list of edible and inedible flowers (Fernandes et al. 2017b) as well as there is scarcity on cultivation criteria, limits of pesticides and recommended daily intake of edible flowers, only of phytochemicals present in them (Matyjaszczyk and Śmiechowska 2019). However, food safety issues have already been registered in the European Union Rapid Alarming System for Food and Feed (RASFF).

It is widely documented that the cultivation parameters of any plant material influence the primary and secondary metabolites, which can vary the composition of potential phytochemicals, such as fibre content, phenolic compounds, vitamins, minerals (Petropoulos et al. 2018; Krzymińska et al. 2022; Polyzos et al. 2022). However, the records indicated in the RASFF prove that the cultivation, maintenance, preservation, transport, and storage demand special attention. Safety factors involved in flowers include natural toxic substances, microbial contamination, chemical residues, and heavy metal pollution. The cultivation of the flower can be considered the origin of the various food security issues, and organic cultivation is currently the most appropriate (Fernandes et al. 2017a, b; Fornefeld et al. 2017). There are also flowers collected from wild plants, which have no control over their development. In this case, the greatest concern is centred on possible heavy metals in the soil, especially when the harvest area is not used for agriculture (Matyjaszczyk and Śmiechowska 2019).

There are several challenges for the processing and conservation of edible flowers, since the shelf life is extremely short and there is a need to ensure the preservation of the nutritional, phytochemical, and bioactive value, as well as the sensory characteristics, appearance, and colour, whether for compose dishes or as a natural colourant. Edible flowers are highly perishable due to their elevated water content, volatile compounds, and sensitivity to environmental factors, if improperly stored they may have colour changes due to tissue darkening, flower wilt and/or dehydration (Shantamma et al. 2021). Conventionally applied preservation methods (Fig. 14.2) are hot air drying or low temperature storage (Fernandes et al. 2018; Barani et al. 2022). On the other hand, non-thermal methods emerge (Fig. 14.2) with the appeal of sustainable technologies, capable of changing the minimum of the visual appearance of the food, as well as allowing a better retention of phytochemicals such as anthocyanins, carotenoids and proteins (Shantamma et al. 2021), being that irradiation, ultraviolet light and high hydrostatic pressure techniques extended the shelf life of edible flowers (Matyjaszczyk and Śmiechowska 2019), while edible coating or modified atmosphere packaging also aids in the integrity of EF (Shantamma et al. 2021).

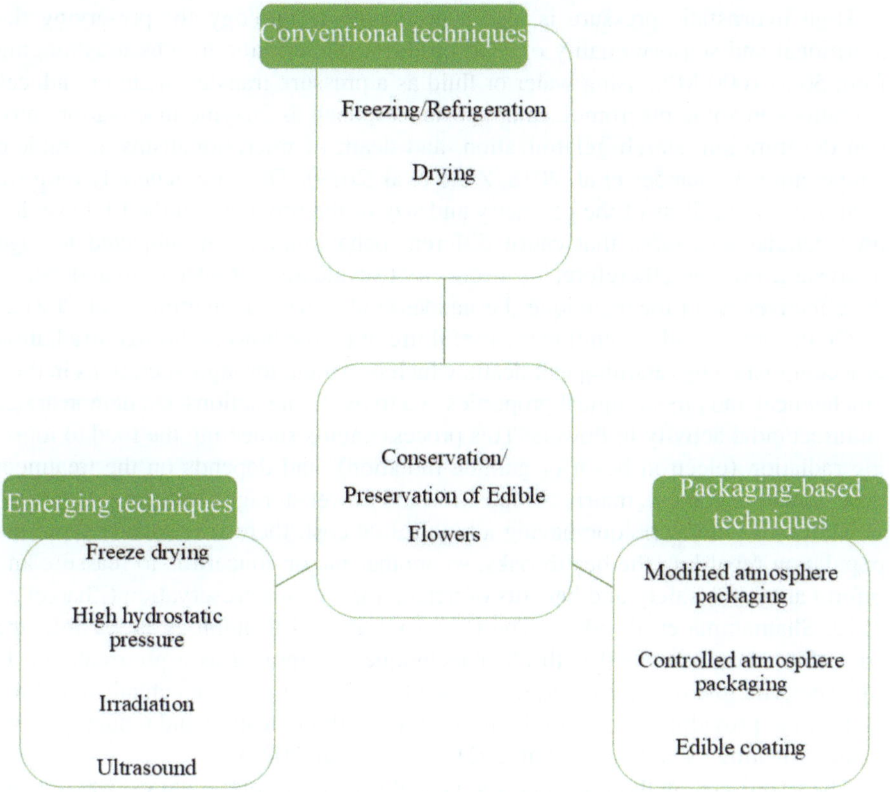

Fig. 14.2 Conventional and emerging techniques for the preservation of edible flowers

Drying is the most applied conventional technique for food preservation, since it extends shelf life, inhibits enzymatic degradation, can prevent microbial growth, and reduce weight for transport and storage with lower costs (Fernandes et al. 2018; Marchioni et al. 2022). There are several drying methods, among them, those adopted by the EF processing industry are those with the application of heat (drying with hot air or in the sun) for easy handling and control. However, this approach can lead to undesirable biochemical and nutritional changes due to heat that affect quality; on the other hand, storage at low temperatures prolongs the shelf life of EF by reducing respiration and internal degradation of tissues by enzymes, water loss and wilting, and slow growth of microorganisms and the disadvantage of this technique is associated with cold injuries in some flowers (Fernandes et al. 2018; Shantamma et al. 2021). One of the most widespread emerging techniques in academia and industry is freeze-drying. The technique involves low processing temperature, lack of oxygen and water sublimation to preserve the nutritional and sensory integrity of the fresh product (Marchioni et al. 2022). The loss of nutrients and aroma compounds from EF is minimal, resulting in a processed product of superior quality compared to heat-based drying methods (Zhao et al. 2019).

High hydrostatic pressure is also a promising technology for preserving the nutritional and sensory quality of food products. Submission to pressures ranging from 50 to 1000 MPa using water or fluid as a pressure transfer medium, induces alterations in some macromolecular substances, such as enzyme inactivation, protein denaturation, starch gelatinization, and death of microorganisms at ambient temperature (Fernandes et al. 2018; Zhao et al. 2019). They are generally uniform treatments, regardless of the geometry and size of the product, but the EF have distinct cellular structures that cause different behaviours when subjected to high hydrostatic pressure, therefore, it is necessary to evaluate each species to understand the effectiveness of the technique (Fernandes et al. 2018; Shantamma et al. 2021).

On the other hand, extending the useful life of edible flowers through irradiation is accomplished by retarding cell death, which is evident through alterations in their biochemical and physiological properties due to oxidizing actions and demonstrates an insecticidal activity in flowers. This process entails subjecting the food to ionizing radiation (electron beam or gamma radiation), and depends on the treatment dose, moisture content, matrix composition and flower storage conditions. However, in addition to the technique having a high initial cost, there is resistance from the population regarding the health risks, so another major concern is to reassure and inform about the safety and benefits of irradiation in food preservation (Zhao et al. 2019; Shantamma et al. 2021). The use of ultrasound technology is feasible and advisable when combined with other techniques or applied as a pre-treatment in EF. The principle of the technique is based on the mechanical mechanism and not on heating, providing a decrease in drying time without heating and reducing enzymatic browning (Shantamma et al. 2021; Barani et al. 2022).

The traditional method of food storage is through low temperatures, which is no different for edible plants. However, when the traditional strategy is not enough, it is necessary to think about modified or controlled atmosphere packaging. Modified and controlled atmosphere packages differ because the parameters of temperature, humidity, and gaseous concentrations of CO_2, O_2 and N_2 are more controlled in the so-called modified atmosphere. The control of these parameters reduces enzymatic, respiratory and senescence activity, therefore, they can provide EF storage without pre-treatments (Shantamma et al. 2021). Alternatively, the use of edible coatings (EC) is already widely used in fruits and vegetables. EC provide a selective gas displacement barrier and provide resistance to mechanical disturbances and are preferably made of polysaccharides, proteins, and lipids, with alginate being the most widespread biopolymer as EC (Fernandes et al. 2018; Shantamma et al. 2021). Table 14.3 shows some examples of emerging edible flower processing and storage technologies.

Therefore, proposing and studying a suitable cultivation system for edible flowers, as well as processing and storage methods, becomes challenging since the uniqueness of each flower, whether in terms of nutritional, chemical or colourants composition, may require different parameters for the best development and preservation of the same, so it is extremely important to understand the purposes of each plant, their concentrations of phytochemicals and harmful compounds to ensure the fresh consumption of edible flowers or quality flowers for the processing of natural colourants.

Table 14.3 Emerging technologies for processing and storing edible flowers

Applied technology	Edible flower	Conditions	References
Drying	*Tagetes erecta*	Freezing for 48 h followed by hot air (60 °C/4 h) and infrared radiation combined with convective hot air	Siriamornpun et al. (2012)
Lyophilization	*Begonia cucullata*	Vacuum freeze-dried for 48 h	Marchioni et al. (2022)
High hydrostatic pressure	*Centaurea cyanus*	75, 100, 200 and 300 MPa, 5 min	Fernandes et al. (2017a, b)
Irradiation	*Viola tricolour*	^{60}Co (0.5, 0.8 and 1 kGy), room temperature	Koike et al. (2015)
Ultrasound	*Brassica oleracea var. Botrytis*	20-40 kHz, 600 W/3 a 10 min	Jambrak et al. (2007)
Low temperatures	*Borago officinalis*	Temperature range: −2.5 to 20 °C in polyethylene bags	Kelley et al. (2003)
Modified atmosphere packaging	*Calendula officianlis* L.	Polypropylene trays covered with three different films, followed by storage at 5 °C for 6 days	Fadda et al. (2020)
Controlled atmosphere packaging	*Brassica oleracea* var. *Botrytis*	Storage at 3% O_2, 5% CO_2 and 0 °C for 56 days	Hodges et al. (2006)
Edible coatings	*Viola* × *wittrockiana*	Immersion in sodium alginate solution (2%) followed by immersion in $CaCl_2$ solution (1%) for crosslinking. Storage for 21 days at 4 °C	Fernandes et al. (2017a)

14.7 Conclusions

As interest in edible flowers continues to soar, it is crucial to address the challenges surrounding their sustainable cultivation and conservation, reflecting on the importance of ethnobotanical conservation efforts and offer concluding insights into the promising future of edible flowers as food colouring agents. This comprehensive overview serves as a gateway to the captivating world of edible flowers, inviting readers to explore their culinary potential while honouring their cultural and botanical significance.

Acknowledgements The authors are grateful to the Foundation for Science and Technology (FCT, Portugal) for financial support through national funds FCT/MCTES (PIDDAC) to CIMO (UIDB/00690/2020 and UIDP/00690/2020) and SusTEC (LA/P/0007/2020). This research also received national funding by FCT, P.I., through the institutional scientific employment program for the contracts of C.P. (10.54499/CEECINST/00016/2018/CP1505/CT0010), M.I.D. (10.54499/CEECINST/00016/2018/CP1505/CT0004), and L.B.; as also for C.N. (2021.05369.BD) and B.H.P. (2023.02731.BD) PhD grants.

References

Amrouche TA, Yang X, Capanoglu E et al (2022) Contribution of edible flowers to the Mediterranean diet: phytonutrients, bioactivity evaluation and applications. Food Front 3:592–630. https://doi.org/10.1002/FFT2.142

Asadi-Samani M, Bahmani M, Rafieian-Kopaei M (2014) The chemical composition, botanical characteristic and biological activities of *Borago officinalis*: a review. Asian Pac J Trop Med 7:22–28. https://doi.org/10.1016/S1995-7645(14)60199-1

Bagade SB, Patil M (2019) Recent advances in microwave assisted extraction of bioactive compounds from complex herbal samples: a review. Crit Rev Anal Chem 51:138–149. https://doi.org/10.1080/10408347.2019.1686966

Barani YH, Zhang M, Mujumdar AS, Chang L (2022) Preservation of color and nutrients in anthocyanin-rich edible flowers: progress of new extraction and processing techniques. J Food Process Preserv 46:e16474. https://doi.org/10.1111/JFPP.16474

Belwal T, Ezzat SM, Rastrelli L et al (2018) A critical analysis of extraction techniques used for botanicals: trends, priorities, industrial uses and optimization strategies. TrAC Trends Anal Chem 100:82–102. https://doi.org/10.1016/J.TRAC.2017.12.018

Bhattacharyya S, Roychowdhury A, Ghosh S (2008) Lutein content, fatty acid composition and enzymatic modification of lutein from marigold (*Tagetes patula* L.) flower petals. J Indian Chem Soc 85:942–944

Binaschi M, Garrido GD, Cirelli C, Spigno G (2018) Biotechnological strategies to valorise grape pomace for food applications. Chem Eng Trans 64:367–372. https://doi.org/10.3303/CET1864062

Brudzyńska P, Sionkowska A, Grisel M (2021) Plant-derived colorants for food, cosmetic and textile industries: a review. Materials 14:3484–3484. https://doi.org/10.3390/MA14133484

Cacak-Pietrzak G, Dziki D, Gawlik-Dziki U et al (2022) Dandelion flowers as an additive to wheat bread: physical properties of dough and bread quality. Appl Sci 13:477–477. https://doi.org/10.3390/APP13010477

Chen S, Liu J, Dong G et al (2021) Flavonoids and caffeoylquinic acids in *Chrysanthemum morifolium* Ramat flowers: a potentially rich source of bioactive compounds. Food Chem 344:128733. https://doi.org/10.1016/J.FOODCHEM.2020.128733

Chengaiah B, Mallikarjuna Rao K, Mahesh Kumar K et al (2010) Medicinal importance of natural dyes - a review. Int J Pharmtech Res 2:144–154

Chitrakar B, Zhang M, Bhandari B (2019) Edible flowers with the common name "marigold": their therapeutic values and processing. Trends Food Sci Technol 89:76–87. https://doi.org/10.1016/J.TIFS.2019.05.008

Cunningham FX, Gantt E (2005) A study in scarlet: enzymes of ketocarotenoid biosynthesis in the flowers of *Adonis aestivalis*. Plant J 41:478–492. https://doi.org/10.1111/J.1365-313X.2004.02309.X

da Silva RPFF, Rocha-Santos TAP, Duarte AC (2016) Supercritical fluid extraction of bioactive compounds. TrAC Trends Anal Chem 76:40–51. https://doi.org/10.1016/J.TRAC.2015.11.013

da Silva LA, Fischer SZ, Zambiazi RC (2020) Proximal composition, bioactive compounds content and color preference of *Viola x Wittrockiana* flowers. Int J Gastron Food Sci 22:100236. https://doi.org/10.1016/J.IJGFS.2020.100236

de Almeida LC, Salvador MR, Pinheiro-Sant'Ana HM et al (2022) Proximate composition and characterization of the vitamins and minerals of dandelion (Taraxacum officinale) from the Middle Doce River region – Minas Gerais, Brazil. Heliyon 8:e11949. https://doi.org/10.1016/J.HELIYON.2022.E11949

de Morais JS, Sant'Ana AS, Dantas AM et al (2020) Antioxidant activity and bioaccessibility of phenolic compounds in white, red, blue, purple, yellow and orange edible flowers through a simulated intestinal barrier. Food Res Int 131:109046. https://doi.org/10.1016/J.FOODRES.2020.109046

dos Santos IC, Reis SN (2021) Edible flowers: traditional and current use. Ornam Hortic 27:438–445. https://doi.org/10.1590/2447-536X.V27I4.2392

Escher GB, Santos JS, Rosso ND et al (2018) Chemical study, antioxidant, anti-hypertensive, and cytotoxic/cytoprotective activities of *Centaurea cyanus* L. petals aqueous extract. Food Chem Toxicol 118:439–453. https://doi.org/10.1016/J.FCT.2018.05.046

Fadda A, Palma A, Azara E, D'Aquino S (2020) Effect of modified atmosphere packaging on overall appearance and nutraceutical quality of pot marigold held at 5 °C. Food Res Int 134:109248–109248. https://doi.org/10.1016/J.FOODRES.2020.109248

Fernandes L, Casal S, Pereira JA et al (2017a) Effect of high hydrostatic pressure on the quality of four edible flowers: *Viola × wittrockiana*, *Centaurea cyanus*, *Borago officinalis* and *Camellia japonica*. Int J Food Sci Technol 52:2455–2462. https://doi.org/10.1111/IJFS.13530

Fernandes L, Casal S, Pereira JA et al (2017b) Edible flowers: a review of the nutritional, antioxidant, antimicrobial properties and effects on human health. J Food Compos Anal 60:38–50. https://doi.org/10.1016/J.JFCA.2017.03.017

Fernandes L, Saraiva JA, Pereira JA et al (2018) Post-harvest technologies applied to edible flowers: a review. Food Rev Int 35:132–154. https://doi.org/10.1080/87559129.2018.1473422

Fernandes L, Casal S, Pereira JA et al (2019) An overview on the market of edible flowers. Food Rev Int 36:258–275. https://doi.org/10.1080/87559129.2019.1639727

Fornefeld E, Schierstaedt J, Jechalke S et al (2017) Interaction between Salmonella and plants: potential hosts and vectors for human infection. In: Current topics in Salmonella and Salmonellosis. InTech, pp 171–191

Gagneten M, Leiva G, Salvatori D et al (2019) Optimization of pulsed electric field treatment for the extraction of bioactive compounds from blackcurrant. Food Bioprocess Technol 12:1102–1109. https://doi.org/10.1007/S11947-019-02283-1/FIGURES/4

Garzón GA, Manns DC, Riedl K et al (2015) Identification of phenolic compounds in petals of nasturtium flowers (*Tropaeolum majus*) by high-performance liquid chromatography coupled to mass spectrometry and determination of oxygen radical absorbance capacity (ORAC). J Agric Food Chem 63:1803–1811. https://doi.org/10.1021/jf503366c

González-Barrio R, Periago MJ, Luna-Recio C et al (2018) Chemical composition of the edible flowers, pansy (*Viola wittrockiana*) and snapdragon (*Antirrhinum majus*) as new sources of bioactive compounds. Food Chem 252:373–380. https://doi.org/10.1016/J.FOODCHEM.2018.01.102

Goupy P, Vian MA, Chemat F, Caris-Veyrat C (2013) Identification and quantification of flavonols, anthocyanins and lutein diesters in tepals of *Crocus sativus* by ultra performance liquid chromatography coupled to diode array and ion trap mass spectrometry detections. Ind Crop Prod 44:496–510. https://doi.org/10.1016/J.INDCROP.2012.10.004

Guiné RPF, Florença SG, Moya KV, Anjos O (2020) Edible flowers, old tradition or new gastronomic trend: a first look at consumption in Portugal versus Costa Rica. Foods 9:977. https://doi.org/10.3390/FOODS9080977

Halbwirth H, Muster G, Stich K (2008) Unraveling the biochemical base of dahlia flower coloration. Nat Prod Commun 8:1259–12666

Hodges DM, Munro KD, Forney CF, McRae KB (2006) Glucosinolate and free sugar content in cauliflower (*Brassica oleracea* var. *botrytis* cv. Freemont) during controlled-atmosphere storage. Postharvest Biol Technol 40:123–132. https://doi.org/10.1016/J.POSTHARVBIO.2005.12.019

Jakubczyk K, Janda K, Watychowicz K et al (2018) Garden nasturtium (*Tropaeolum majus* L.) - a source of mineral elements and bioactive compounds. Rocz Panstw Zakl Hig 69:119–126

Jambrak AR, Mason TJ, Paniwnyk L, Lelas V (2007) Accelerated drying of button mushrooms, Brussels sprouts and cauliflower by applying power ultrasound and its rehydration properties. J Food Eng 81:88–97. https://doi.org/10.1016/J.JFOODENG.2006.10.009

Jena J, Kumar A, Jitendra J et al (2012) *Rosa centifolia*: plant review. Int J Res Pharm Chem 2:794–796

Kelley KM, Cameron AC, Biernbaum JA, Poff KL (2003) Effect of storage temperature on the quality of edible flowers. Postharvest Biol Technol 27:341–344. https://doi.org/10.1016/S0925-5214(02)00096-0

Koike A, Barreira JCM, Barros L et al (2015) Edible flowers of *Viola tricolor* L. as a new functional food: antioxidant activity, individual phenolics and effects of gamma and electron-beam irradiation. Food Chem 179:6–14. https://doi.org/10.1016/J.FOODCHEM.2015.01.123

Kondo T, Ueda M, Isobe M, Goto T (1998) A new molecular mechanism of blue color development with protocyanin, a supramolecular pigment from cornflower, *Centaurea cyanus*. Tetrahedron Lett 39:8307–8310. https://doi.org/10.1016/S0040-4039(98)01858-9

Krzymińska A, Frąszczak B, Gąsecka M et al (2022) The content of phenolic compounds and organic acids in two *Tagetes patula* cultivars flowers and its dependence on light colour and substrate. Molecules 27:527. https://doi.org/10.3390/MOLECULES27020527

Kumar K, Srivastav S, Sharanagat VS (2021) Ultrasound assisted extraction (UAE) of bioactive compounds from fruit and vegetable processing by-products: a review. Ultrason Sonochem 70:105325. https://doi.org/10.1016/J.ULTSONCH.2020.105325

Li AN, Li S, Li HB et al (2014) Total phenolic contents and antioxidant capacities of 51 edible and wild flowers. J Funct Foods 6:319–330. https://doi.org/10.1016/J.JFF.2013.10.022

Lotfi L, Kalbasi-Ashtari A, Hamedi M, Ghorbani F (2015) Effects of enzymatic extraction on anthocyanins yield of saffron tepals (*Crocos sativus*) along with its color properties and structural stability. J Food Drug Anal 23:210–218. https://doi.org/10.1016/J.JFDA.2014.10.011

Lu B, Li M, Yin R (2016) Phytochemical content, health benefits, and toxicology of common edible flowers: a review (2000–2015). Crit Rev Food Sci Nutr 56:S130–S148. https://doi.org/10.1080/10408398.2015.1078276

Marchioni I, Taglieri I, Dimita R et al (2022) Postharvest treatments on sensorial and biochemical characteristics of *Begonia cucullata* Willd edible flowers. Foods 11:1481. https://doi.org/10.3390/FOODS11101481

Martínez-Damián MT, Mejía-Muñoz JM, Colinas-León MT et al (2021) Nutritional value, bioactive compounds and capacity antioxidant in edible flowers of dahlia. Acta Sci Pol Hortorum Cultus 20:63–72. https://doi.org/10.24326/asphc.2021.5.6

Matyjaszczyk E, Śmiechowska M (2019) Edible flowers. Benefits and risks pertaining to their consumption. Trends Food Sci Technol 91:670–674. https://doi.org/10.1016/J.TIFS.2019.07.017

Mlcek J, Rop O (2011) Fresh edible flowers of ornamental plants – A new source of nutraceutical foods. Trends Food Sci Technol 22:561–569. https://doi.org/10.1016/J.TIFS.2011.04.006

Morata A, Escott C, Loira I et al (2021) Emerging non-thermal technologies for the extraction of grape anthocyanins. Antioxidants 10:1863. https://doi.org/10.3390/ANTIOX10121863

Mulík S, Ozuna C (2020) Mexican edible flowers: cultural background, traditional culinary uses, and potential health benefits. Int J Gastron Food Sci 21:100235. https://doi.org/10.1016/J.IJGFS.2020.100235

Park CH, Chae SC, Park SY et al (2015) Anthocyanin and carotenoid contents in different cultivars of *Chrysanthemum* (*Dendranthema grandiflorum* Ramat.) flower. Molecules 20:11090–11102. https://doi.org/10.3390/MOLECULES200611090

Perkins-Veazie P, Ma G, Werner D (2017) Anthocyanin pigments in redbud (*Cercis* spp) flowers. J Hortic Sci Res 2017:13–18. https://doi.org/10.36959/745/393

Petropoulos S, Fernandes Â, Barros L, Ferreira ICFR (2018) Chemical composition, nutritional value and antioxidant properties of Mediterranean okra genotypes in relation to harvest stage. Food Chem 242:466–474. https://doi.org/10.1016/J.FOODCHEM.2017.09.082

Picot-Allain C, Mahomoodally MF, Ak G, Zengin G (2021) Conventional versus green extraction techniques — a comparative perspective. Curr Opin Food Sci 40:144–156. https://doi.org/10.1016/J.COFS.2021.02.009

Pintea A, Bele C, Andrei S, Socaciu C (2003) HPLC analysis of carotenoids in four varieties of *Calendula officinalis* L. flowers. Acta Biol Szeged 47:37–40

Pires TCSP, Dias MI, Barros L et al (2018) Incorporation of natural colorants obtained from edible flowers in yogurts. LWT 97:668–675. https://doi.org/10.1016/J.LWT.2018.08.013

Pires TCSP, Barros L, Santos-Buelga C, Ferreira ICFR (2019) Edible flowers: emerging components in the diet. Trends Food Sci Technol 93:244–258. https://doi.org/10.1016/J.TIFS.2019.09.020

Polyzos N, Paschoalinotto BH, Compocholi M et al (2022) Fertilization of pot-grown *Cichorium spinosum* L.: how it can affect plant growth, chemical profile, and bioactivities of edible parts? Horticulturae 8:890. https://doi.org/10.3390/HORTICULTURAE8100890

Purohit SR, Rana SS, Idrishi R et al (2021) A review on nutritional, bioactive, toxicological properties and preservation of edible flowers. Future Foods 4:100078. https://doi.org/10.1016/J.FUFO.2021.100078

Ranjha MMAN, Kanwal R, Shafique B et al (2021) A critical review on pulsed electric field: a novel technology for the extraction of phytoconstituents. Molecules 26:4893. https://doi.org/10.3390/MOLECULES26164893

Riaz M, Ahmad R, Rahman NU et al (2020) Traditional uses, phytochemistry and pharmacological activities of *Tagetes Patula* L. J Ethnopharmacol 255:112718. https://doi.org/10.1016/J.JEP.2020.112718

Rojas-Bedoya L, Gómez-López C, Marín-Pareja N et al (2020) Extraction of metabolites from *Calendula officinalis* and evaluation of their colorant and antibacterial capacity. Rev Colomb Biotecnol 22:60–69. https://doi.org/10.15446/REV.COLOMB.BIOTE.V22N1.79999

Sausserde R, Kampuss K (2014) Composition of carotenoids in calendula (*Calendula officinalis* L.) flowers. In: Straumite E (ed) 9th Baltic Conference on Food Science and Technology "Food for Consumer Well-Being." Jelgava, pp 13–18

Shantamma S, Vasikaran EM, Waghmare R et al (2021) Emerging techniques for the processing and preservation of edible flowers. Future Foods 4:100094. https://doi.org/10.1016/J.FUFO.2021.100094

Siriamornpun S, Kaisoon O, Meeso N (2012) Changes in colour, antioxidant activities and carotenoids (lycopene, β-carotene, lutein) of marigold flower (*Tagetes erecta* L.) resulting from different drying processes. J Funct Foods 4:757–766. https://doi.org/10.1016/J.JFF.2012.05.002

Skrajda-Brdak M, Dąbrowski G, Konopka I (2020) Edible flowers, a source of valuable phytonutrients and their pro-healthy effects – A review. Trends Food Sci Technol 103:179–199. https://doi.org/10.1016/J.TIFS.2020.06.016

Souza HDA, Nascimento ALAA, Stringheta PC, de Barros FAR (2020) REVISTA PONTO DE VISTA. REVISTA PONTO DE VISTA 9:73–84

Sowbhagya HB, Smitha S, Sampathu SR et al (2005) Stability of water-soluble turmeric colourant in an extruded food product during storage. J Food Eng 67:367–371. https://doi.org/10.1016/J.JFOODENG.2004.05.003

Sowbhagya HB, Sampathu SR, Krishnamurthy N (2006) Natural colorant from marigold-chemistry and technology. Food Rev Int 20:33–50. https://doi.org/10.1081/FRI-120028829

Tan J, Han Y, Han B et al (2022) Extraction and purification of anthocyanins: a review. J Agric Food Res 8:100306. https://doi.org/10.1016/J.JAFR.2022.100306

Tena N, Asuero AG (2022) Up-to-date analysis of the extraction methods for anthocyanins: principles of the techniques, optimization, technical progress, and industrial application. Antioxidants 11:286. https://doi.org/10.3390/ANTIOX11020286

Verma PK, Raina R, Agarwal S, Kaur H (2018) Phytochemical ingredients and pharmacological potential of *Calendula officinalis* Linn. Pharm Biomed Res 4:1–17. https://doi.org/10.18502/PBR.V4I2.214

Yiğit Ü, Turabi Yolaçaner E, Hamzalıoğlu A, Gökmen V (2022) Optimization of microwave-assisted extraction of anthocyanins in red cabbage by response surface methodology. J Food Process Preserv 46:e16120. https://doi.org/10.1111/JFPP.16120

Zhao L, Fan H, Zhang M et al (2019) Edible flowers: review of flower processing and extraction of bioactive compounds by novel technologies. Food Res Int 126:108660. https://doi.org/10.1016/J.FOODRES.2019.108660

Chapter 15
Edible Flowers for Pastry and Candies: Technologies Used for Their Application

Liqing Qiu and Min Zhang

Abstract With people transitioning to healthier lifestyles and modifying their eating habits, edible flowers have become pioneers in the food industry. Due to their flavourful and colourful nature, edible flowers are believed to contain a variety of nutritional and bioactive compounds. Despite the great sensory appeal, high nutritional value, and rich bioactive compounds of edible flowers, they have not been utilized in the food industry as widely as fresh fruits and vegetables. Due to their high moisture content and soft texture, edible flowers have an extremely short shelf life, limiting their commercial application. Appropriate processing methods should be implemented to extend their shelf life and extend their commercial usage. Moreover, the superfine grinding technique can reduce the size of whole flowers and facilitate the release of bioactive substances, playing a vital role in their food applications. Furthermore, flower extracts account for a significant proportion of food additives, enriching the nutritional content of prepared food products. To enhance the quality attributes and expand the commercial usage of edible flowers in pastry and candy preparations, studies on edible flowers have been summarized. This review highlights emerging technologies such as drying, superfine grinding, extraction techniques, and microencapsulation, particularly for edible flowers and flower extracts.

Keywords Confectionery · Drying · Extraction methods · Microencapsulation · Quality attributes · Superfine grinding · Sweets

L. Qiu
School of Perfume and Aroma Technology, Shanghai Institute of Technology, Shanghai, China

M. Zhang (✉)
State Key Laboratory of Food Science and Technology, Jiangnan University, Wuxi, Jiangsu, China
e-mail: min@jiangnan.edu.cn

M. M. Jowkar et al. (eds.), *Advances in Research on Edible Flowers*, Advances in Olericulture, https://doi.org/10.1007/978-3-032-04561-4_15

Abbreviations

CO_2	Carbon dioxide
DMSO	Dimethyl sulphoxide
EAE	Enzyme-assisted extraction
FIR-HA	Combined far-infrared drying and hot air drying
IRFD	Infrared freeze drying
IRPSFD	Infrared pulse-spouted freeze drying
MAE	Microwave-assisted extraction
MFD	Microwave-assisted freeze drying
MFWE	Mohammadi rose flower waste extract
RP	Rose powder
SCFs	Supercritical fluids
SFE	Supercritical fluid extraction
SMP	Superfine marigold flower powder
SW	Subcritical water
SWE	Subcritical water extraction
UAE	Ultrasonic-assisted extraction

15.1 Introduction

Edible flowers, recognized as the nontoxic and safe reproductive organs of plants, offer health-promoting benefits when incorporated into the human diets (Alasalvar et al. 2013). Since ancient times, edible flowers have been a significant part of the human diet, with their use well-documented across the globe, from ancient Rome and Greece, Europe, medieval France, the Middle Eastern region, Victorian England, to Asian countries like China and Japan. (Pires et al. 2019). The consumption of edible flowers has been reported for thousands of years (Pop et al. 2010). Nowadays, consumer awareness and globalization have contributed to a revival of earlier life-styles, where edible flowers play a critical role (Pop et al. 2010). Therefore, more attention has been focused on edible flowers since then. The demand for edible flowers has been increasing recently, which is evidenced by their appearance in cookbooks, television shows, scientific papers, and magazine articles (Mlcek and Rop 2011). The appealing colours of edible flowers not only provide delicate flavour, exotic aroma, and freshness to foods, but also offer health benefits to human body due to their high antioxidant concentration and antioxidative activities (Chitrakar et al. 2019).

Edible flowers can be consumed in various ways, such as in beverages, vegetables, and salads; as stuffing for stir-fried dishes; as a fresh garnish on or in various cooked meals; or as an ingredient in different products, including pastries (cakes, cookies, biscuits, etc.), candies, and jams (cakes, cookies, biscuits etc.), candies, and jams (Fennell 2006). For example, in ancient Rome, rose petals were used as ingredients in omelettes and purees (Melillo 1994). In medieval France, pot marigold (*Calendula officinalis*) flowers were used as an ingredient in preparing salads (Chitrakar et al. 2019). Similarly, based on the assumption that edible flowers would confer medical benefits to human health, various edible flowers were applied as herbal teas in Europe (Alasalvar et al. 2013). In addition, edible flowers were previously reported to be used as herbal medicine for treating chronic diseases. Particularly, edible flowers have a long history of being used as food components and resources for herbal medicine in China (Wongwattanasathien et al. 2010).

Edible flowers are rich in bioactive compounds (e.g., phenolics, flavonoids, phenolic acids, anthocyanins, carotenes, etc.), making them attractive possibilities for the development of functional food products (Gostin and Waisundara 2019). The consumption of functional foods containing bioactive substances in appropriate doses is beneficial for preventing, controlling, or treating acute and chronic diseases. Edible flowers retain proteins, vitamins, essential oils, and antioxidants, especially when consumed in minimally processed or natural forms (Fernandes et al. 2019a). Based on this, various biological activities have recently been associated with edible flowers, including anticholinergic activity, anti-hyperglycemic effects, treatment of ulcerative colitis, inhibition of red blood cell oxidation, and even anticancer activity (Takahashi et al. 2020).

Edible flowers have been widely used in the preparation of food products such as pastry and candies due to their abundant bioactive substances and strong biological activities. However, only a small amount of flowers is used directly in their fresh form, and proper processing steps are necessary depending on the category of the prepared food. Nowadays, there are various food products (including pastry and candies) fortified with edible flowers in many forms (fresh flowers, dried flowers, flower powder, extracts or microcapsules). Compared to fresh flowers, flower products that have undergone proper processing offer versatility in form and greater convenience in handling (Zhao et al. 2019). Additionally, edible flowers that undergo processing are safer to eat because high moisture content in fresh flowers can promote the growth of microorganisms (Braun et al. 2015). Therefore, fresh edible flowers must be sold promptly to avoid the undesirable effects mentioned above. Proper processing steps can extend the shelf life and preserve the quality attributes of edible flowers. In addition, edible flowers with enhanced colour attributes and increased levels of bioactive compounds can be achieved through size reduction technology (grinding technology). Apart from processing edible flowers into stable or powder products, certain bioactive substances in flowers can also be extracted as food additives (Ran et al. 2019). This review focused on various aspects of edible flowers for pastry and candies, including their quality attributes, nutritional aspects, and novel processing technologies such as drying, superfine grinding, extraction, and microencapsulation. In addition, examples of pastry and candy preparations using edible flowers are provided to illustrate the role of these flowers.

15.2 Quality Attributes of Edible Flowers

Appearance, aroma, colour, and nutrition are four key quality attributes of edible flowers that influence their market appeal (Fernandes et al. 2017). These attributes are significantly influenced by maturity stages, flower variety, and harvesting methods, and they can change rapidly during processing and post-harvest storage (Kou et al. 2012).

Figure 15.1 shows some edible flowers used for food production. As shown in Fig. 15.1, appearance is the most crucial attribute for both fresh and processed flowers. Generally, flowers are prone to dehydration, petal abscission, discolouration, wilting, and browning after harvest (Zhao et al. 2019). The main considerations that determine people's purchase and selection of flowers are colour, flower integrity, and petal shape. (Takahashi et al. 2020). Flowers with a defective appearance are less attractive to consumers when used as garnish in dishes. The colour of edible flowers is predetermined by various chemical compounds, such as flavonoids, carotenoids and anthocyanins (Friedman et al. 2010). The aroma of flowers was contributed by the production and emission of fragrant volatile substances, which are sensed by special receptors in the human olfactory system, resulting in a feeling of pleasure. Based on their mode of biosynthesis, aroma compounds can be mainly divided into three categories: phenylpropanoids/benzenoids, aliphatic compounds,

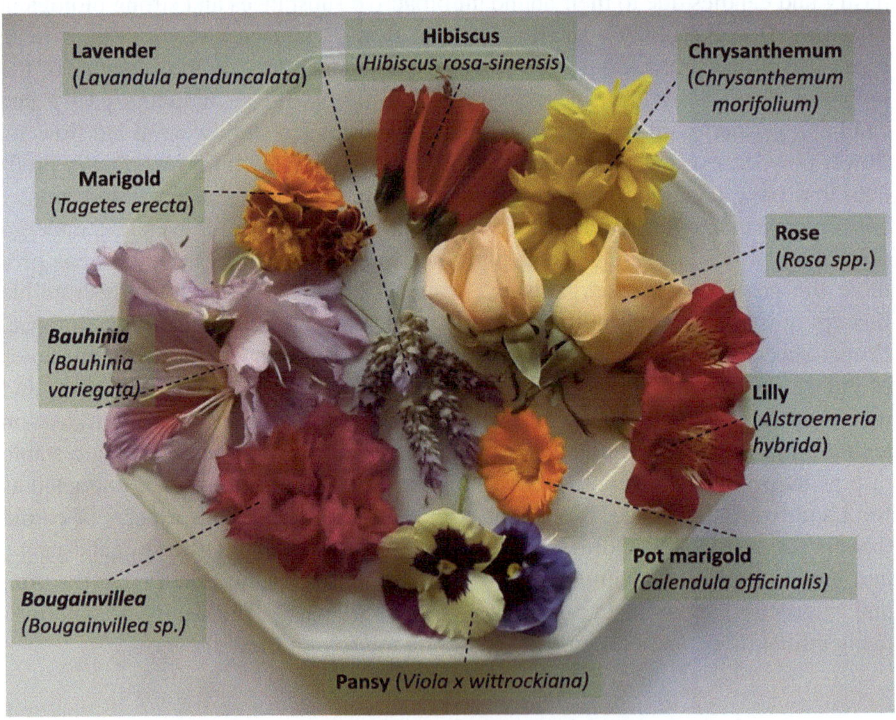

Fig. 15.1 The pictures of some representative edible flowers. (Takahashi et al. 2020)

and terpenes (Friedman et al. 2010). The nutritional value of flowers is closely related to the content of bioactive compounds (e.g., phenolics, flavonoids, anthocyanins, etc.), which play a significant role in promoting body health by inhibiting certain chronic degenerative diseases. (Hegde et al. 2022; Pires et al. 2023; Skrajda-Brdak et al. 2020).

15.3 Nutritional Properties of Edible Flowers

Details on the nutritional properties of edible flowers are described in Chap. 3 (Bioactive Compounds, Nutraceutical aspects and industrial food exploitation). In brief, the flower sample can be divided into three major components, each playing a vital role in human nutrition from a nutritional standpoint. Although pollen constitutes only a small portion of the whole flower, it is rich in carbohydrates, proteins, amino acids, unsaturated and saturated lipids, flavonoids, carotenoids, etc. (Zhao et al. 2019). The taste of pollen is commonly distinctive and often not very delicious. Nectar, a sweet liquid, contains fructose, glucose, sucrose, amino acids, proteins, inorganic ions, lipids, organic acids, phenolic compounds, alkaloids, terpenoids, and more (Nicolson et al. 2007). The petals and other parts of flowers are the third component. They may also provide the aforementioned compounds, as well as vitamins, minerals, antioxidants, and more (Rop et al. 2012).

Edible flowers are rich in phytochemicals, which comprise a vast array of naturally occurring bioactive compounds. These compounds are crucial for aroma, colour, flavour, and, most importantly, provide mechanical and physiological protection to the flowers and their organs against infections, predators, and natural calamities (Wong et al. 2019). Meanwhile, they provide immense health benefits to the human body (Kumar et al. 2021). Edible flowers contain various bioactive phytochemicals, such as carotenoids, phenolic acids, and flavonoids (Kumar et al. 2021). All these phytochemicals can act as agents of antidiabetic, anticancer, anti-inflammatory, cyclooxygenase, antioxidant, and many other therapeutic capacities (Pensamiento-Niño et al. 2023; Skrajda-Brdak et al. 2020; Pinakin et al. 2020). Additionally, the cellulose in flowers can promote gastrointestinal peristalsis, cleanse the intestinal wall, and prevent the occurrence of intestinal malignant tumours (Mlcek and Rop 2011). Therefore, these substances make flowers a superior resource of nutrition.

Considering the quality attributes and nutritional properties listed above, edible flowers used in the production of pastry and candies should possess a colourful appearance, a pleasant aroma, and high nutritional characteristics. As reported, the edible flowers used in pastry and candies commonly in the form of dried products, flower powders, or extract with/without microencapsulation. Therefore, appropriate drying, grinding, or extraction methods should be carefully selected to produce high-quality flower products (dried flowers, powder, extract, and microcapsules) for the preparation of high-quality pastry and candies. Thus, the combination of a satisfactory sensory experience and the pursuit of health benefits makes edible flowers a good option and a potential food that can be incorporated into a balanced diet.

15.4 Processing Technologies of Edible Flowers

As is known to all, fresh edible flowers are perishable due to their high moisture content, making it difficult to achieve long-term storage without proper processing. Additionally, certain technologies should be employed to enhance the quality attributes of edible flowers, thereby broadening their application in pastry and candies. The following sections aim to describe these technologies in detail.

15.4.1 Drying

Drying can extend the shelf life of edible flowers, facilitate their subsequent transportation, and broaden their commercial applications (Zhang et al. 2017). Table 15.1 presents some drying techniques used in edible flowers. In addition, edible flowers used in in pastry and candies are always in the form of dried products, powders, or extracts etc. As reported, flowers are commonly dried using traditional methods such as natural drying (sun drying, solar drying, and shade drying) and hot air drying (Zhao et al. 2019). Natural drying is characterized by low costs and simple operation. In addition, natural drying can be conducted at the place of origin and can also promote the further development of the physiological properties of plant materials that are not yet fully mature. However, the natural drying process is highly affected by climate, and it faces challenges such as contamination, long processing times, and severe nutrient loss. (Zhao et al. 2019). In contrast to natural drying, hot air drying is less dependent on climate compared to natural drying (Afzal and Abe 1999). However, materials being dried experience severe visual changes and nutrient loss due to prolonged exposure to high temperatures (Zhao et al. 2019). For example, Qiu et al. (2020a) reported that hot air drying resulted in the lowest contents of vitamin C, flavonoids, and anthocyanins in *Rosa rugosa* flowers compared to freeze drying and infrared freeze drying. Freeze drying is a common method for the dehydration of high-value materials (e.g., pharmaceuticals, fine chemicals, biotechnological products, and foods with high nutritional and market values) due to its advantages of minimizing the loss of flavour and bioactive compounds and reducing shrinkage (Sadikoglu et al. 2006). Nevertheless, the freeze drying technique faces high operation costs, which impedes its widespread application.

Recently, modern drying methods, including microwave drying and hybrid drying methods, have also been recognized as promising technologies for the dehydration of flowers (Le et al. 2022; Lu et al. 2021). Microwave, an electromagnetic wave with a high frequency range of 300 MHz to 300 GHz, can generate a heating effect through dielectric induction, with its direction constantly changing in space due to the high-frequency electric field. During microwave heating, the external part of the material experiences a slightly lower temperature compared to the internal part due to moisture evaporation, which significantly improves the dehydration rate (Zielinska et al. 2020). Because of the bipolar molecule characteristics of water,

Table 15.1 Some representative studies of novel drying methods for the dehydration process of edible flowers

Sources		Drying methods	Key findings	Reference
Rose	*Rosa rugosa*	Infrared pulse-spouted freeze drying	Infrared pulse-spouted freeze drying shortened the drying time and energy consumption by 8–30% and 15–36%. The bioactive substances and quality of rose were well preserved.	Hnin et al. (2021a)
		Infrared freeze drying	Infrared freeze drying lowered the drying time and energy consumption. The contents of bioactive substances were well retained.	Qiu et al. (2020a)
		Microwave-vacuum drying	Microwave-vacuum drying reduce the drying time of the dehydration process.	Ran et al. (2021)
		Microwave drying	Flower dried by embedding in silica gel during microwave drying yield best quality dried flowers as it scored maximum point for all the quality parameters studied.	Safeena and Patil (2014)
Daylily flower buds	*Hemerocallis Citrine Baroni cv. Mengzihua*	Microwave drying	The microwave energy utilization was more effective when materials' water content was less than 6.8 g/g.	Ding et al. (2012)
Loquat flower tea	*Eriobotrya japonica* (Lindl.) Thunb.	Microwave drying	Microwave drying at 420 W and freeze drying can effectively preserve the effective components compared to other two drying techniques (hot air drying and vacuum drying).	Zheng et al. (2015)
Eucommia male flower	*Eucommia ulmoides*	Microwave drying	Microwave dehydration (560 W and 2–3 min) could maximally preserve the functional constituents of the flowers and increase the drying rate.	Dong et al. (2011)
Marigold flower	*Tagetes erecta* L.	Combined far-infrared drying and hot air drying	The combined drying technology could lead to high levels of lycopene and lutein compared to the far-infrared drying (FIR) and hot air drying (HA) method.	Siriamornpun et al. (2012)

which endow it with a strong polarity. Therefore, a significant amount of heat is generated due to the vibration of water molecules during the drying process. In brief, the higher the moisture content, the greater the microwave absorption ability. Additionally, the affinity between water molecules and the materials being dehydrated can also be reduced, particularly that between bound water and material compounds. Thus, the water molecules can be easily dissociated form material structures and diffused to the external environment. Meanwhile, less microwave energy was absorbed by the material itself, resulting in a relatively low temperature of the processed raw materials, which helps preserve functional components, reduce browning, and shorten the drying period (Bi et al. 2022). Ding et al. (2012) investigated the drying kinetics of fresh daylily flower buds using microwave drying at four different microwave intensities (60, 120, 180 and 240 W). The results indicated that the utilization of microwave energy was more effective when the material's water content was less than 6.8 g/g. Zheng et al. (2015) evaluated the effects of different drying methods (hot air drying, freeze drying, vacuum drying, and microwave drying) on the active components of loquat flower tea. Microwave drying at 420 W and freeze drying can effectively preserve the active components compared to other two drying techniques (hot air drying and vacuum drying). Additionally, microwave drying presented the highest drying rate among four drying technologies. A similar study was conducted by Dong et al. (2011) with *Eucommia ulmoides* flower tea by microwave and conventional drying methods (pan baking drying, and oven ventilation baking drying). Compared to conventional drying methods, microwave dehydration (560 W for 2–3 min) could maximally preserve the functional constituents of the flowers. Although microwave drying can reduce drying duration and preserve functional substances, the tissue damage and charring problems in the final products caused by direct exposure to microwave power may negatively affect the visual appearance of the dried products.

Over the past few decades, hybrid drying techniques (incorporating two or more drying systems or processing units) have garnered increasing interest due to their superior advantages, including high product quality, enhanced drying efficiency, and reduced operational costs (Fathi et al. 2022). To address the disadvantages of microwave drying mentioned above, numerous attempts have been made. It has been reported that microwave-assisted vacuum drying is applied to the dehydration process of edible flowers. For example, edible rose flowers have been successfully dried by microwave vacuum drying (Ran et al. 2021). Recently, microwave-assisted freeze drying (MFD) has emerged as a promising drying technique for the dehydration of plant materials due to its high efficiency, low processing time, and good quality retention characteristics compared to conventional freeze drying (Wang et al. 2023a; Fan et al. 2019). In recent years, MFD has been applied to the dehydration of several categories of vegetable products, fruits, solid soups, and seafood (Fan et al. 2019). Although no literature has been reported on the drying process of flowers using MFD, MFD has been applied to the drying process of several food products with edible flowers, such as restructured rose flower chips (prepared with *Rosa rugosa*) (Qiu et al. 2020b), rose cookies (Hnin et al. 2021b), and rose-flavoured yogurt melts (Hnin et al. 2020). Therefore, MFD could also be tested for the production of dehydrated flowers.

Infrared radiation can generate heat inside the materials being dried through molecular vibration, promoting excited vibrations when the radiation of certain energy and wavelength is absorbed by the molecules. As a result, the dehydrated materials can directly absorb electromagnetic energy, thereby reducing energy loss. In the study by Hnin et al. (2021a), edible rose flowers were dried in an infrared-assisted freeze dryer at different temperature levels (50, 60, and 70 °C), and the bioactive compounds, drying time, and energy consumption were measured. Compared to traditional freeze drying, the drying time and energy consumption of infrared pulse-spouted freeze drying (IRPSFD) can be reduced by 8–30% and 15–36%, respectively. The contents of the bioactive substances (phenolics and anthocyanins) in the IRPSFD sample were close to those of freeze drying. Similarly, in another study by Qiu et al. (2020a), edible rose was processed using infrared freeze drying (IRFD). In comparison to hot air drying, freeze drying and IRFD were more effective in retaining the aroma and bioactive substances. In addition, IRFD could lower the drying time and total energy consumption by 21.43% and 27.78%, respectively, compared to conventional freeze drying. The combination of infrared and hot air drying has also been successfully applied to the dehydration of edible flowers. For instance, Siriamornpun et al. (2012) used combined far-infrared drying and hot air drying (FIR-HA) for the dehydration of marigold flowers (*Tagetes erecta* L.). Results indicated that the FIR-HA technology could lead to higher levels of lycopene and lutein compared to the far-infrared drying (FIR) and hot air drying methods. The marigold processed by FIR-HA also exhibited the highest contents of protocatechuic, syringic, caffeic, and ferulic acids in all dried marigold samples.

15.4.2 Superfine Grinding

Like dried petals, flower powder is also frequently used in the production of pastry and candies such as yogurt melts (Hnin et al. 2019), cookies (Hnin et al. 2021b), biscuits (Kc et al. 2022), and cakes (Gentscheva et al. 2022), etc. According to research, particle size affects the physiochemical and functional properties of food materials, and reducing the particle size to the micron or nano size range can improve the solubility, antioxidant capacity, and bioaccessibility of bioactive substances in the raw materials (Fernandes et al. 2019b; Qiu et al. 2022a). Due to these advantages listed above, the particle size of flowers should be reduced upon applying in foods to enhance the bio-resources utilization.

Superfine grinding is an emerging technology used in functional food processing, which also holds promise as a technique for grinding flowers. Recently, superfine grinding was conducted on Sanchi (*Panax notoginseng*) flowers using a disc-mill at room temperature. The flower powder after superfine grinding exhibited higher total phenolic and flavonoid contents, accompanied by better antioxidant activities, compared to the coarse powder (Wu et al. 2021). Similarly, superfine marigold flower

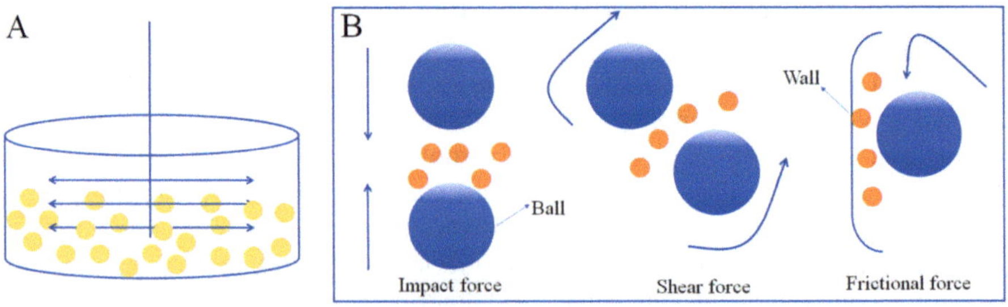

Fig. 15.2 Balling milling (**a**) and its mechanism (**b**)

powder (SMP) was prepared using jet milling technology and applied to sponge cakes as a functional ingredient (Kim et al. 2023). Size reduction increased the lutein content of the marigold flowers, and the sponge cakes with SMP exhibited lower baking loss, hardness, and chewiness compared to the control cake. However, the bioactive substances present in flowers are heat-sensitive and easily degrade during the grounding process due to the heat generated. In order to maintain the quality of flowers to the maximum extent after grinding, ball milling with a cooling system has received increasing attention in the flower grinding process (Chitrakar et al. 2020). The ball milling and its mechanism for size reduction are presented in Fig. 15.2. In the study conducted by Qiu et al. (2022b), rose powder (RP) was superfine grounded using ball milling at 0 °C. They found that superfine grinding at low temperature improved the accessibility of functional substances in RP powders, including anthocyanins, polyphenols, and flavonoids, and also enhanced the antioxidant capacity of RP powders.

15.5 Bioactive Substances in Edible Flowers

Edible flowers are abundant in various bioactive substances, such as carotenoids, anthocyanins, phenolics, flavonoids, dietary fibers, phenolic acids, isothiocyanates, sterols, and prebiotics (e.g., inulin) etc. Bioactive compounds can serve as spice and taste enhancers in pastry and candies, making them more appealing to consumers. To achieve high-quality bioactive extracts for use in food preparation, extraction techniques and encapsulation methods for the extracted bioactive substances are detailed in the following sections.

15.5.1 Traditional Extraction Techniques

Routine extraction technologies include water extraction, steam extraction, and water-steam extraction etc. These methods are the most commonly used and cost-effective ways to obtain the desired bioactive substances. However, the extraction efficiency and product quality produced by these techniques are relatively low, and these methods are often time-consuming (Perović et al. 2024). For example, Vidana Gamage and Choo (2023) extracted anthocyanins from blue pea flowers using hot water extraction, ultrasound, pectinase, and microwave-assisted extraction methods. The results indicated that hot water extraction was the least effective method, while the microwave-assisted method was the most effective, yielding a high anthocyanin content of 37.2%. Similarly, Filly et al. (2016) extracted essential oil from lavender flowers using hydro-distillation, steam distillation, or solvent extraction. They found that microwave-assisted steam distillation could achieve better oil recovery while reducing by-product formation compared to traditional steam distillation and hydro-distillation. Water-steam extraction combines steam and water during the extraction process. During the water-steam extraction process, the plants are initially exposed to boiling water, after which hot steam is immediately introduced into the suspension. As reported, this method is rarely used for the extraction of bioactive substances. In the study by Kiran et al. (2016), water-steam extraction was successfully used for the recovery of *Lavandula angustifolia* essential oil, demonstrating that it provides an alternative extraction method.

Solvent extraction is the most commonly used method for the extraction of bioactive compounds from edible flowers. For example, Kamali et al. (2020) prepared extracts from Mohammadi rose flower waste using the water-ethanol method with 30% (v/v) ethanol solvent and incorporated the extracts into cookies. The addition of Mohammadi rose flower waste extract (MFWE) increased the antioxidant activity and decreased the peroxide value of the samples. In another study, *Impatiens balsamina* L. flowers extract was obtained by immersing in ethanol/water solution (80:20, v/v) and then adding the extract to a Portuguese pastry product as a natural colouring ingredient (Pires et al. 2021). The addition of anthocyanins-rich extract gave the pastry sample a more natural colour and increased its antioxidant capacity, meeting consumers' demand for natural and healthy foods. Soft candies incorporated with *Dendrobium officinale* flower extract obtained through hot water extrusion were prepared by Zhenlin et al. (2022). The soft candy containing 30% flower extract displayed appropriate textural properties, high antioxidant capacity, and the highest sensory scores compared to other formulations. Although solvent extraction has been widely applied to the extrusion of various flower extracts as a functional ingredient or colouring agent in the preparation of pastry and candies, the solvent extraction process faces several disadvantages, including time-consuming operations, high consumption of hazardous chemicals, and poor quality of extracts (Voon et al. 2012). Therefore, other extrusion techniques with advantages such as high efficiency, environmental friendliness, and high retention of quality have attracted increasing attention in recent years.

15.5.2 Supercritical Fluid Extraction

Supercritical fluid extraction (SFE) is an extraction technique that utilizes supercritical fluids (SCFs) as extraction solvents (Fig. 15.3). SCFs are substances at pressures and temperatures above their critical points, at which distinct gas and liquid phases do not exist. Therefore, SCF_S can penetrate solid matrices like gases and subsequently dissolve materials like conventional liquids, thus achieving the purpose of extraction (Ahangari et al. 2021). As reported, SFE possesses a multitude of distinct advantages, such as high selectivity, short processing time, reduced consumption of toxic solvents, and low extraction temperature, thereby enhancing its application in bioactive substance extraction (Zhao et al. 2019; Sagaradze et al. 2017; Chitrakar et al. 2019). However, SCFs encounter difficulty penetrating cell walls with complex structures, which significantly reduces extraction efficiency (Patil et al. 2021). During SFE, carbon dioxide (CO_2) is one of the most widely used supercritical solvents due to its merits of being cheap, non-toxic, and convenient to operate. However, the extraction of bioactive substances from edible flowers using SFE has not been widely developed. Hsu et al. (2016) revealed that the supercritical CO_2 extraction technique effectively separates the bioactive components of *Lonicera*

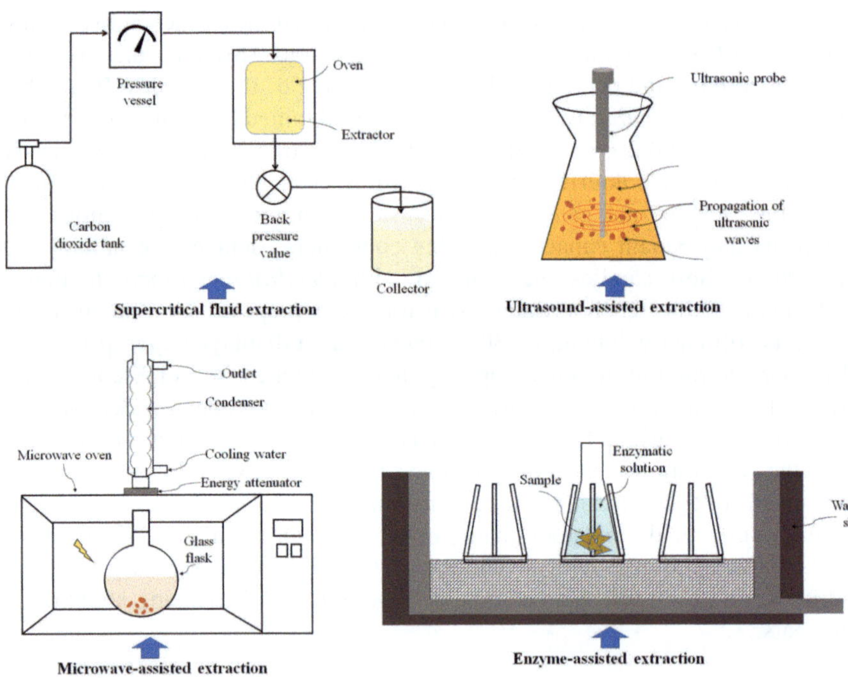

Fig. 15.3 Novel techniques for extraction of bioactive compositions

japonica Thunberg and can enhance the anti-inflammatory activity of the extracts. Previous reports have demonstrated that the experimental conditions (temperature, pressure, time, modifier, and solvent flow rate) significantly affect the extraction process (Dias et al. 2017). Therefore, suitable strategies such as response surface methodology and experimental design should be employed to optimize the process, with the aim of identifying the most influential variables (Pimentel-Moral et al. 2019). In the study conducted by Shaikh et al. (2023), the SFE of lutein from marigold flowers using CO_2 fluid was optimized through the Box-Behnken experimental design. Extraction at 60 °C and 398 bar for 149 min was reported to yield 35.49% (w/w), the highest extraction yield. Furthermore, the extract obtained by SFE demonstrated superior antioxidant capacity, anti-inflammatory properties, and stability. Additionally, the cookies fortified with marigold flower extract showed high organoleptic acceptability (7.19). In another study on the supercritical fluid extraction (SFE) of marigold oleoresin, the operational variables, including solvent flow rate, pressure, and temperature were evaluated (Campos et al. 2005). High pressure resulted in a higher extraction rate, whereas higher solvent flow rates contributed to a lower extraction yield. Additionally, an increase in temperature yielded a higher extraction yield only when the extraction pressure was greater than 15 MPa. Due to the non-polar nature of CO_2, the addition of small amounts of co-solvents and modifiers (e.g., ethanol) is often necessary. For instance, Casas et al. (2007) employed supercritical carbon dioxide to extract bioactive compounds from *Helianthus annuus* L. An increase in extraction efficiency was observed when 5% methanol, water, or dimethyl sulfoxide (DMSO) was added as a modifier to the fluid.

Subcritical water (SW) is also referred to as hot liquid water, superheated water, or high-pressure hot water. The physicochemical properties of subcritical water differ from those at normal pressure and room temperature. As the temperature increases, the hydrogen bonds in subcritical water are either opened or weakened, allowing the water to have sufficient energy to be extracted. By controlling the temperature and pressure, the polarity of water can be adjusted across a wide range, enabling continuous extraction of effective components in natural products from water-soluble to fat-soluble components, and facilitating selective extraction (Zhao et al. 2019). SW extraction (SWE) has been employed in the extraction of bioactive substances from edible flowers. Xu et al. (2015) studied the recovery of phenolics from marigold (*Tagetes erecta* L.) flower residues using SWE. Compared to leaching extraction, SWE enhanced the recovery rate of phenolics and the quercetin, with the quercetin obtained from SWE exhibiting nearly twofold radical scavenging capacities. This technique was also applied to the extraction of bioactive substances from flowers such as *Matricaria chamomilla* L. (Cvetanović et al. 2019) and *Sophora Japonica* (Maksimenko et al. 2018).

15.5.3 Ultrasonic-Assisted Extraction

Recently, ultrasonic-assisted extraction (UAE) techniques have gained widespread attention due to their high efficiency, low energy consumption, short extraction time, mild operation temperature, and reduced damage to active ingredients during the extraction process (Wang et al. 2023b). The principle of ultrasonic-assisted extraction is that the resonance and cavitation generated by ultrasound can disrupt plant cell walls and promote solvent penetration into cells, thus resulting in the dissolution of target extracts and improving extraction efficiency (Zhang et al. 2016). The UAE process is displayed in Fig. 15.3. In addition, some examples using UAE in the extraction of bioactive substances from edible flowers are shown in Table 15.2.

Xu et al. (2017) extracted natural antioxidants from *Limonium sinuatum* flowers using various extraction methods, including maceration extraction, Soxhlet extraction, and ultrasound-assisted extraction. The results showed that ultrasound-assisted extraction not only achieved a higher yield of antioxidants but also significantly reduced the operation time. Sonication studies on the extraction of bioactive substances from *Tagetes erecta* was conducted by Kazibwe et al. (2017). The effects of water bath sonication (300 W/60 Hz) and probe sonication (200 W/20 kHz) on the total phenolic and total flavonoid contents were compared, and it was found that sonication improved the extraction yield of the bioactive substances, with the sample processed by probe sonication showing the highest values. Roriz et al. (2020) extracted betalains from *Gomphrena globosa* L. flowers by ultrasound-assisted extraction and incorporated them as a natural colourant in the preparation of cookies. The betacyanins obtained exhibited a more natural colour compared to the commercially available colouring agents, which may be attributed to the higher yield of colouring molecules. In recent years, the application of multi-frequency ultrasound extraction for the recovery of bioactive substances is a hot topic. For example, Feng et al. (2023) successfully extracted polyphenols from edible roses by multi-frequency ultrasound-assisted extraction. In another study by Xu et al. (2022), the parameters of the multi-frequency power ultrasound green extraction were optimized by Box-Behnken design. The extraction efficiency of polyphenols from Pingyin rose was measured at 187.13 mg/g, which is 19.33% higher than that of the control (without sonication). This method is often combined with a solvent extraction technique to further enhance the extraction rate and yield. The traditional solvents commonly used are organic solvents such as ethanol, methanol, and acetone. Recently, ultrasonic-assisted extraction methods based on ionic liquids and deep eutectic solvents have been developed. feng Wang et al. (2021) combined sonication with an ionic liquid solvent to extract monoterpene glycosides from the flowers of *Paeonia suffruticosa*, achieving satisfactory recoveries ranging from 91.4% to 98.3% for five different monoterpene glycosides. In the study by Li et al. (2023), rose anthocyanins were extracted using an ultrasound-assisted deep eutectic solvent method, and the purity of the final anthocyanin extract was 8.265 mg/g.

Table 15.2 Some representative studies of microwave assisted and/or ultrasound assisted extraction to extract bioactive compounds from edible flowers

Sources	Target compounds	Process condition	Key findings	Reference
Hibiscus sabdariffa	Phenolic compounds	Microwave extraction temperature 50–150 °C Ethanol concentration 15–75% Microwave extraction time 5–20 min	The highest extraction yield of phenolics were 164 °C, 60% ethanol and 22 min.	Pimentel-Moral et al. (2018)
	Anthocyanins.	Ultrasonic power (210–430 w) Ethanol concentration 39.1% Ultrasonic extraction time 26.1 min	Ultrasound-assisted extraction was the most efficient extraction technique yielding 51.76 ± 3.70 mg/g extract.	Pinela et al. (2019)
Osmanthus fragrans	Flavonoids	Ethanol concentration 40–60% Microwave extraction time 5–7 min Ultrasonic extraction time 8–12 min Ultrasonic power 210–430 W	The optimum extraction of the combined microwave- and ultrasound-assisted extraction was ethanol concentration of 48.15%, microwave extraction time of 6.43 min, ultrasonic extraction time 10.09 min and ultrasonic power 370.9 W. The highest extraction yield of flavonoids was 7.86 mg/g.	Yu et al. (2017)
	Antioxidants	Ethanol concentration 39.1% Ultrasonic extraction time 35.2 min Ultrasonic extraction temperature 59.4 °C	The extract extracted by ultrasound-assisted extraction presented the highest Trolox equivalent antioxidant capacity (486.4 ± 12.6 μmol Trolox/g DW) compared to conventional extraction techniques.	Li et al. (2016)

(continued)

Table 15.2 (continued)

Sources	Target compounds	Process condition	Key findings	Reference
Lonicera japonica Thunb	Chlorogenic acid	Microwave extraction temperature 60 °C Ethanol concentration 50% Microwave extraction time 5 min	The yield of chlorogenic acid rapidly reaching 6.14% within 5 min.	Zhang et al. (2008)
Tagetes erecta L.	Xanthophyll	Material to liquid ratio of 1:25 Ultrasonic extraction power 500 W Ultrasonic extraction time 30 min	More than 97% of xanthophyll contents could be achieved under the optimum extraction condition.	Liu et al. (2005)
Tilia amurensis	Polysaccharides	Ultrasonic extraction temperature 25–85 °C Ultrasonic extraction extracting time 30–90 min Ethanol concentration 50%	The optimal ultrasonic temperature was 75 °C and extraction time 52 min. The polysaccharide content was 9.14%.	Mu et al. (2010)
Rosa rugosa	Polyphenols	Ultrasonic extraction temperature 63 °C Ultrasonic extraction time 30 min	The extraction yield of rose polyphenols was 156.81 mg/g.	Xu et al. (2022)
Gomphrena globosa L.	Betacyanins	Ultrasonic power 500 W Ultrasonic extraction time 22 min	The obtained betacyanins displayed a more natural colour compared to the commercially available colouring agents.	Roriz et al. (2020)
Paeonia tenuifolia L.	Polyphenol and flavonoid	Microwave extraction temperature 60 °C Microwave extraction time 10 min	The highest biological activity was observed in the extracts obtained by microwave-assisted drying.	Čutović et al. (2022)

15.5.4 Microwave-Assisted Extraction

Microwave-assisted extraction (MAE) refers to the process of extracting natural bioactive compounds from plants by heating the material with microwave energy or by interacting with the moisture present in the solvents (Fig. 15.3) (Zhao et al. 2019). The application of microwaves in extraction has great potential to shorten the reaction time due to their effective internal heating capability (Kumar Baghel 2023). In recent years, MAE technology has been used to extract bioactive substances from edible flowers (Chitrakar et al. 2019). Cerdá-Bernad et al. (2022) used MAE technique to separate high value-added compounds from saffron (*Crocus sativus* L.), effectively isolating bioactive compounds from saffron flowers with low energy consumption. In the study conducted by Purbowati and Maksum (2019), MAE was applied to extract phenolic compounds from Roselle (*Hibiscus sabdariffa* L.). Experimental results indicated that microwave power and processing time significantly influenced MAE, with the optimal extraction parameters being 325 W and 5 min. Similarly, antioxidant components from *Osmanthus fragrans* (Lour) flower were effectively extracted by MAE under optimized conditions (497.12 W, 59.03 s). Čutović et al. (2022) compared the MAE and UAE of polyphenol and flavonoid from *Paeonia tenuifolia* L., employing various extraction solvents. The MAE was proved to be the most desirable extraction procedure to obtain extracts in aqueous solvents with the highest biological activity. Recently, MAE using deep eutectic solvents has been widely applied in the extraction of edible flowers. For instance, Liu et al. (2022) compared the effects of deep eutectic solvents and conventional solvents (60% ethanol and water) on the extraction yield of natural antioxidants (organic acids and flavonoids) from *Hibiscus manihot* L. They found that the antioxidants extracted using deep eutectic solvents improved by 1.42–1.94 times compared to those in conventional solutions. Similar results were also observed by Pan et al. (2021), who extracted antioxidant compounds from *Osmanthus fragrans* (Lour) flowers by MAE method with deep eutectic solvents. More examples concerning the extraction of bioactive substances via MAE from edible flowers are presented in Table 15.2.

15.5.5 Other Extraction Techniques

15.5.5.1 Enzyme-Assisted Extraction

The solvent extraction method has been associated with problems such as solvent loss, leading to low extraction yields (Ling and Hadinoto 2022). Therefore, enzyme-assisted extraction (EAE) technique was introduced (Fig. 15.3). Enzyme pre-treatment of saffron (*Crocus sativus* L.) has been reported to enhance the extraction of polyphenols (Vardakas et al. 2021). Rose petals, roselle calyces, and camellia flowers pretreated with cellulase and pectinase facilitated the effective recovery of sugars and phenolic compounds, affirming that enzymatic hydrolysis is an environmentally-friendly method for bioactive substance recovery (Trinh et al. 2018). Similarly, Barzana et al. (2002) extracted carotenoids by EAE method, with the highest extraction yield achieved by adding 1% (w/w) Viscozyme/HT-Proteolytic/Pectinex enzyme at a solid-to-liquid ratio of 1:4. Additionally, a study on enzyme-assisted (pectinase) two-phase (ammonium sulphate/ethanol) extraction of marigold (*Tagetes erecta* L.) was carried out by Fu et al. (2019). In comparison to the one-phase extraction, this two-phase method exhibited the highest extraction of lutein and polyphenols, demonstrating a relatively high recovery rate ranging from 95% to 99%.

15.5.5.2 Combined Extrusion Methods

Furthermore, several examples of MAE combined with other technologies for the recovery of bioactive substances from flowers are provided (Zhao et al. 2019; Barani et al. 2022). For example, Li et al. (2020) successfully employed the ultrasound-microwave synergistic extraction method to extract phenolics from peony flowers. Likewise, Boukerche et al. (2023) used the combined ultrasonic- and microwave-assisted extraction to recover anthocyanins and phenolics from roselle (*Hibiscus Sabdariffa*. L.). The combined extraction method significantly ($p < 0.05$) increased the extraction yield and reduced the processing time from 24 h to 15 min. Jiang et al. (2020) used radio frequency-assisted enzymatic extraction method to prepare anthocyanins from *Akebia trifoliata* (Thunb.) Koidz. flowers. Radio frequency treatment for 10 min at 40 °C with a 5 cm electrode gap, an ethanol concentration of 50%, a liquid-solid ratio of 50 mL/g, pH 4, and a cellulase to pectinase ratio of 1:1 resulted in the highest crude extraction yield of 26.55%, compared with hot water, acidified ethanol, and enzymatic extraction techniques. Additionally, no significant degradation of bioactive compounds was observed in the extracts prepared by the novel combined method, which was also more efficient and rapid compared to the other two methods. In addition, Fu et al. (2018) successfully extracted polyphenol and lutein from marigold (*Tagetes erecta* L.) flowers via a novel microwave and enzyme co-assisted aqueous two-phase extraction method. The combination of microwave heating and enzyme-assisted aqueous two-phase extraction improved the yields by 71.14% compared to Soxhlet extraction.

Based on the examples listed above, the extraction methods and parameters should be carefully selected according to the species of flowers and the relevant tissues, which can ensure the extraction yield and quality of the extracts. In summary, the most suitable extraction methods and parameters should be selected based on the production target.

15.5.6 Encapsulation

Bioactive substances extracted from flowers are degradative; therefore, they are usually encapsulated in shell materials to enhance their stability. Additionally, micro-encapsulation can protect the extract from degradation during processing; therefore, encapsulated extracts are preferred and applied in various food preparations. Micro-encapsulation is a common means of protecting sensitive compounds; it refers to the embedding of solid, liquid, or gas into polymer walls to form solid particles. Microcapsules process the advantages of release control and bioactive substance protection, enabling resistance to subsequent processing steps and external environmental changes (e.g., light, heat, and oxygen) as well as intelligent release in food products (Desai and Jin Park 2005). In a recent study, phenolic compounds derived from *Hibiscus sabdariffa* L. extracts were encapsulated via spray-drying, utilizing maltodextrin and gum arabic as the shell materials (Navidad-Murrieta et al. 2023). Improved stability of microcapsules was observed under simulated gastrointestinal conditions. Similarly, the extracts from *Hibiscus sabdariffa* L. can also be encapsulated in calcium alginate spheres as a novel means of preserving them against oxidative damage (Bevan et al. 2023). In another study, microcapsules containing anthocyanin extract from *Hibiscus sabdariffa* L. were prepared by the ionic gelation method (de Moura et al. 2019). The microencapsulation of *Hibiscus sabdariffa* extract enhanced the enteric protection of bioactive compounds, and the microcapsules containing anthocyanins demonstrated their feasibility in jelly candy applications (Fig. 15.4).

In recent years, nano-encapsulation has emerged as one of the widely adopted and rapidly expanding technologies in the food industry, leveraging nanotechnology. It may significantly advance the nanoscale qualities of food components, including sensory characteristics, colour, stability of formulated particles, and many other factors (Tahir et al. 2021). Additionally, nanotechnology can improve the fluid solubility, physical stability, and, most importantly, the bioavailability of bioactive compounds (Ezhilarasi et al. 2013). Calliari et al. (2020) encapsulated *Hibiscus sabdariffa* extract within zein nanoparticles using liquid antisolvent precipitation.

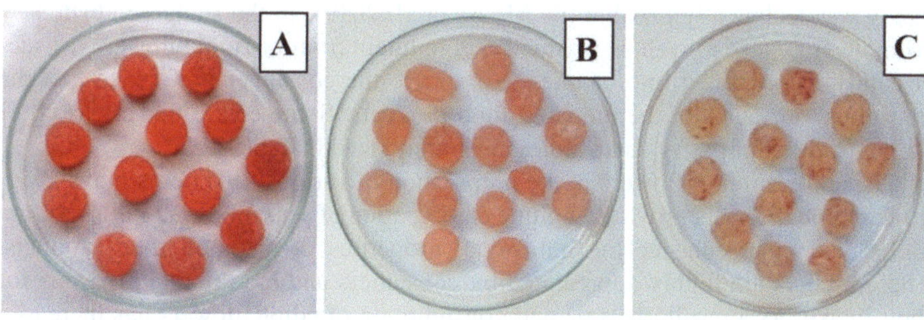

Fig. 15.4 Jelly candies incorporated with free *Hibiscus sabdariffa* extracts (**a**), and microcapsules containing *Hibiscus sabdariffa* extracts (**b**, **c**) (de Moura et al. 2019)

Results revealed that the liquid antisolvent precipitation method provides high extract encapsulation efficiency, good particle size control, and excellent particle stability. Gali et al. (2022) investigated the encapsulation of a rutin-rich extract extracted from *Ruta chalepensis* L. into zein nanoparticles prepared by antisolvent precipitation and stabilized with gum Arabic. The addition of gum Arabic improved the stability of the zein nanoparticles and increased the encapsulation efficiency of rutin.

15.6 Examples of Pastry and Candies Prepared Using Edible Flowers

Pastry and candies occupied a large portion of food products due to their appealing appearance and delicious taste. Recently, various flowers have been successfully incorporated into the formulations of pastry and candies to enhance their taste, aroma, and nutritional profiles (Roriz et al. 2020; Carocho et al. 2015). For instance, freeze-dried chestnut (*Castanea sativa* Mill.) flowers (Carocho et al. 2015) and *Castanea sativa* male flowers powder/extract (Fig. 15.5) (Caleja et al. 2020) were employed in cake preparation. The *Rosa rugosa* flowers, in the form of dried petals or powders, were used as a nutritional enhancer for preparing pastry products such as chips (Qiu et al. 2020b) and yogurt melts in (Fig. 15.6) (Hnin et al. 2019). *Gomphrena globosa* L. flower extracts, prepared by spray drying and freeze drying, were incorporated into cookies as natural colouring agents (Roriz et al. 2020). Results showed that the cookies enriched with freeze-dried pigments displayed a more colourful appearance compared to those incorporating spray-dried extracts (Fig. 15.7). In addition, extracts from *Impatiens balsamina* L. flowers (Pires et al. 2021) and *Dendrobium officinale* flowers (Zhenlin et al. 2022) were used as dye agents in candies. Furthermore, free anthocyanin pigments and microcapsules containing anthocyanin extract from *Hibiscus sabdariffa* L. were used in jelly candies (Fig. 15.4) (de Moura et al. 2019). Results showed that pigment encapsulation improved the enteric protection of bioactive compounds compared to free extracts. Additionally, this study demonstrated that the application of microcapsules in jelly candy was technically feasible (de Moura et al. 2019). In the study conducted by Alim un et al. (2018), yellow pigment rich in lutein from marigold flower was successfully incorporated into candies to improve the consumer preference and the nutritional attributes of the prepared food products. Some edible flowers used for the preparation of pastry and candies are summarized in Table 15.3.

The application of flowers in the preparation of various food items plays a significant role. Flowers provide a unique combination of sensations, including aroma and flavour, and enhance the nutritional properties of prepared food products. These aspects are highly appreciated by consumers and are the main reasons for their intake. As a result, the sale and consumption of edible flowers has increased worldwide (Baldissera et al. 2022). In the past few years, there has been growing

Fig. 15.5 External appearance of "pastel de nata" with different formulations Caleja et al. (2020)

evidence that edible flowers are considered natural products of great interest in scientific research, as they contain important nutrients for human health. Additionally, edible flowers have attracted significant interest from both the food industry and consumers as a nutritional component in various food products, such as pastries and candies.

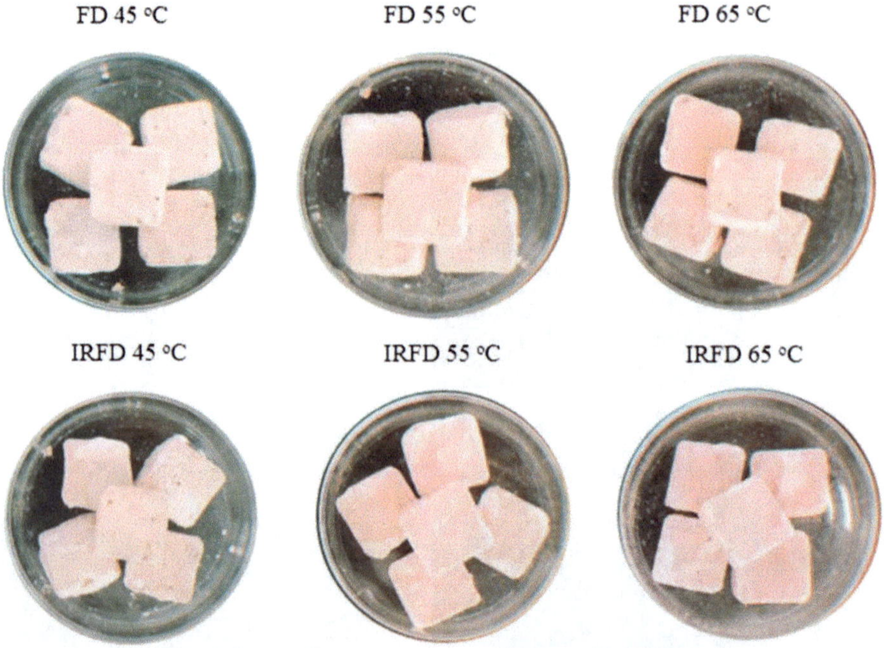

Fig. 15.6 Yogurt melts incorporated with rose powder dehydrated by freeze and infrared freeze drying at different conditions (Hnin et al. 2019)

15.7 Conclusion and Future Perspective

Edible flowers are increasingly used to enhance the visual appearance, nutritional properties, and functional characteristics of dishes, such as pastry and candies. They are gaining increasing attention from consumers seeking healthier and more attractive alternatives. However, due to their high moisture content, edible flowers are prior to microbial reproduction. In addition, whole flowers face issues related to uneven distribution, large size, and low bioactive substance release. Therefore, to extend the shelf life and improve the quality attributes of edible flowers, it is essential to apply proper processing technologies that are highly efficient and have a reduced environmental impact. These technologies include novel drying, superfine grinding, extraction, and encapsulation techniques, which will ultimately benefit their subsequent use in the food preparation process.

Nevertheless, processing technology should be carefully selected, taking into account the characteristics of edible flowers, as the choice of method depends on resource categories, the quality of the end product, consumer demand, and the selling price of the final product. Therefore, future studies could focus on determining the chemical compositions of different flower resources and selecting appropriate

Fig. 15.7 Colours of the cookies enriched with *Gomphrena globosa* L. flower extracts dried by different techniques Roriz et al. (2020)

Table 15.3 Examples of edible flowers used for the preparation of pastry and candies

Kind of flower		Food category	References
Rose Balsam	*Impatiens balsamina* L.	Portuguese pastry product "*bombocas*"	Pires et al. (2021)
Chestnut male flower	*Castanea sativa* Mill.	Portuguese pastry delicacy "*pastel de nata*"	Caleja et al. (2020)
		Portuguese pastry "*económicos*"	Carocho et al. (2015)
Buddy Rose	*Gomphrena globosa* L.	Cookies	Roriz et al. (2020)
Elder Flower	*Sambucus nigra L.*	Sponge cakes	Gentscheva et al. (2022)
Pinying Rose	*Rosa rugosa*	Restructured chips	Qiu et al. (2020b)
Rose	*Rosa damascena*	Yogurt melts	Hnin et al. (2019)
Orchidaceae	*Dendrobium officinale*	Soft candy	Zhenlin et al. (2022)
Hibiscus	*Hibiscus sabdariffa* L.	Jelly candy	de Moura et al. (2019)
Marigold	*Tagetes erecta*	Sponge cakes	Kim et al. (2023)
		Candy	Alim un et al. (2018)

processing technologies to achieve maximum retention of nutrients. In addition, the quality of the end product and consumer demand should also be taken into consideration. As a result, the role of the edible flower (garnish, spice and taste modifiers, or spice and taste modifiers) should be clarified firstly. Finally, a careful compromise between the quality and price of the product should be made to broaden the application of edible flowers in the food industry.

Acknowledgment The author acknowledges the National Key Research and Development Program of China (Contract No. 2022YFF1101600), National Natural Science Foundation of China (No. 31872902), Jinan City Science & Technology Innovation Project of Ten Agricultural Characteristic Industries, Jiangsu Province Key Laboratory Project of Advanced Food Manufacturing Equipment and Technology (No. FMZ202003), enabling us to carry out this study.

References

Afzal TM, Abe T (1999) Some functional attributes of far infrared radiation drying of potato. Dry Technol 17(1–2):138–155. https://doi.org/10.1080/07373939908917522

Ahangari H, King JW, Ehsani A et al (2021) Supercritical fluid extraction of seed oils-a short review of current trends. Trends Food Sci Tech 111:249–260. https://doi.org/10.1016/j.tifs.2021.02.066

Alasalvar C, Pelvan E, Özdemir KS et al (2013) Compositional, nutritional, and functional characteristics of instant teas produced from low- and high-quality black teas. J Agric Food Chem 61(31):7529–7536. https://doi.org/10.1021/jf4015137

Alim un N, Hina S, Mazhar S et al (2018) Stability of lutein content in colour extracted from Marigold flower and its application in candies. Pak J Agric Res 31(1):15–23. https://doi.org/10.17582/journal.pjar/2018/31.1.15.23

Baldissera C, Hoppe A, Carlini NRBS et al (2022) Factors influencing consumers' attitudes towards the consumption of grape pomace powder. Appl Food Res 2(1):100103. https://doi.org/10.1016/j.afres.2022.100103

Barani YH, Zhang M, Mujumdar AS et al (2022) Preservation of colour and nutrients in anthocyanin-rich edible flowers: Progress of new extraction and processing techniques. J Food Process Preserv 46(9):e16474. https://doi.org/10.1111/jfpp.16474

Barzana E, Rubio D, Santamaria RI et al (2002) Enzyme-mediated solvent extraction of carotenoids from Marigold flower (*Tagetes erecta*). J Agric Food Chem 50(16):4491–4496. https://doi.org/10.1021/jf025550q

Bevan P, Pastor MV, Almajano MP et al (2023) Antioxidant and antiradical activities of *Hibiscus sabdariffa* L. Extracts Encapsulated in Calcium Alginate Spheres. Polymers 15(7):1740. https://doi.org/10.3390/polym15071740

Bi L, Liu B, Yang Z et al (2022) Analysis of heat and moisture transfer in the microwave drying of potatoes. Dry Technol 41:1397–1410. https://doi.org/10.1080/07373937.2022.2155972

Boukerche H, Malki F, Saidji N et al (2023) Combination of ultrasound, microwave and conventional extraction techniques for roselle (*Hibiscus Sabdariffa.* L.) total anthocyanins and phenolics recovery: effect on antioxidant and structural properties. Biomass Convers Biorefinery 14:18051–18063. https://doi.org/10.1007/s13399-023-04029-8

Braun S, Prakash V, Martin-Belloso O et al (eds) (2015) Regulating safety of traditional and ethnic foods. Academic, New York

Caleja C, Barros L, Barreira J et al (2020) *Castanea sativa* male flower extracts as an alternative additive in the Portuguese pastry delicacy "pastel de nata". Food Funct 11(3):2208–2217. https://doi.org/10.1039/c9fo03050f

Calliari CM, Campardelli R, Pettinato M et al (2020) Encapsulation of Hibiscus sabdariffa extract into Zein nanoparticles. Chem Eng Technol 43(10):2062–2072. https://doi.org/10.1002/ceat.202000194

Campos LMAS, Michielin EMZ, Danielski L et al (2005) Experimental data and modelling the supercritical fluid extraction of marigold (*Calendula officinalis*) oleoresin. J Supercrit Fluid 34(2):163–170. https://doi.org/10.1016/j.supflu.2004.11.010

Carocho M, Barreira JCM, Barros L, Bento A et al (2015) Traditional pastry with chestnut flowers as natural ingredients: An approach of the effects on nutritional value and chemical composition. J Food Compos Anal 44:93–101. https://doi.org/10.1016/j.jfca.2015.08.003

Casas L, Mantell C, Rodríguez M et al (2007) Effect of the addition of cosolvent on the supercritical fluid extraction of bioactive compounds from *Helianthus annuus* L. J Supercrit Fluid 41(1):43–49. https://doi.org/10.1016/j.supflu.2006.09.001

Cerdá-Bernad D, Baixinho JP, Fernández N et al (2022) Evaluation of microwave-assisted extraction as a potential green Technology for the Isolation of bioactive compounds from saffron (*Crocus sativus* L.) floral by-products. Foods 11(15):2335. https://doi.org/10.3390/foods11152335

Chitrakar B, Zhang M, Bhandari B (2019) Edible flowers with the common name "marigold": their therapeutic values and processing. Trends Food Sci Tech 89:76–87. https://doi.org/10.1016/j.tifs.2019.05.008

Chitrakar B, Zhang M, Zhang X et al (2020) Bioactive dietary fiber powder from asparagus leaf by-product: effect of low-temperature ball milling on physico-chemical, functional and microstructural characteristics. Powder Technol 366:275–282. https://doi.org/10.1016/j.powtec.2020.02.068

Čutović N, Marković T, Kostić M et al (2022) Chemical profile and skin-beneficial activities of the petal extracts of *Paeonia tenuifolia* L. from Serbia. Pharmaceuticals 15(12):1537. https://doi.org/10.3390/ph15121537

Cvetanović A, Švarc-Gajić J, Zeković Z et al (2019) The influence of the extraction temperature on polyphenolic profiles and bioactivity of chamomile (*Matricaria chamomilla* L.) subcritical water extracts. Food Chem 271:328–337. https://doi.org/10.1016/j.foodchem.2018.07.154

de Moura SCSR, Berling CL, Garcia AO et al (2019) Release of anthocyanins from the hibiscus extract encapsulated by ionic gelation and application of microparticles in jelly candy. Food Res Int 121:542–552. https://doi.org/10.1016/j.foodres.2018.12.010

Desai KGH, Jin Park H (2005) Recent developments in microencapsulation of food ingredients. Dry Technol 23(7):1361–1394. https://doi.org/10.1081/DRT-200063478

Dias AMA, da Silva ACS, Botelho JRS et al (2017) Temperature and density effects of the $scCO_2$ extraction of spilanthol from *Spilanthes acmella* flowers. J Supercrit Fluid 121:32–40. https://doi.org/10.1016/j.supflu.2016.11.004

Ding S, You J, An K et al (2012) Effective diffusivities and energy consumption of daylily in microwave drying. Int J Food Sci Tech 47(12):2648–2654. https://doi.org/10.1111/j.1365-2621.2012.03147.x

Dong J, Ma X, Fu Z et al (2011) Effects of microwave drying on the contents of functional constituents of *Eucommia ulmoides* flower tea. Ind Crop Prod 34(1):1102–1110. https://doi.org/10.1016/j.indcrop.2011.03.026

Ezhilarasi PN, Karthik P, Chhanwal N et al (2013) Nanoencapsulation techniques for food bioactive components: a review. Food Bioprocess Technol 6(3):628–647. https://doi.org/10.1007/s11947-012-0944-0

Fan K, Zhang M, Mujumdar AS (2019) Recent developments in high efficient freeze-drying of fruits and vegetables assisted by microwave: a review. Crit Rev Food Sci Nutr 59(8):1357–1366

Fathi F, Ebrahimi SN, Matos LC et al (2022) Emerging drying techniques for food safety and quality: a review. Compr Rev Food Sci Food Saf 21(2):1125–1160. https://doi.org/10.1111/1541-4337.12898

feng Wang Q, Qiu Zhao Y, Bing Sun J et al (2021) Simultaneous separation and determination of five monoterpene glycosides in Paeonia suffruticosa flower samples by ultra-high-performance liquid chromatography with a novel reinforced cloud point extraction based on ionic liquid. Microchem J 168:106457. https://doi.org/10.1016/j.microc.2021.106457

Feng M, Xu B, Nahidul Islam M et al (2023) Individual and synergistic effect of multi-frequency ultrasound and electro-infrared pretreatments on polyphenol accumulation and drying characteristics of edible roses. Food Res Int 163:112120. https://doi.org/10.1016/j.foodres.2022.112120

Fennell CW (2006) Food plants of the world. Identification, culinary uses and nutritional value. S Afr J Bot 72(3):484–485

Fernandes L, Casal S, Pereira JA et al (2017) Edible flowers: a review of the nutritional, antioxidant, antimicrobial properties and effects on human health. J Food Compos Anal 60:38–50. https://doi.org/10.1016/j.jfca.2017.03.017

Fernandes L, Pereira JA, Saraiva JA et al (2019a) Phytochemical characterization of *Borago officinalis* L. and *Centaurea cyanus* L. during flower development. Food Res Int 123:771–778. https://doi.org/10.1016/j.foodres.2019.05.014

Fernandes L, Saraiva JA, Pereira JA et al (2019b) Post-harvest technologies applied to edible flowers: a review. Food Rev Int 35(2):132–154. https://doi.org/10.1080/87559129.2018.1473422

Filly A, Fabiano-Tixier AS, Louis C, Fernandez X et al (2016) Water as a green solvent combined with different techniques for extraction of essential oil from lavender flowers. C R Chim 19(6):707–717. https://doi.org/10.1016/j.crci.2016.01.018

Friedman H, Agami O, Vinokur Y et al (2010) Characterization of yield, sensitivity to *Botrytis cinerea* and antioxidant content of several rose species suitable for edible flowers. Sci Hortic 123(3):395–401. https://doi.org/10.1016/j.scienta.2009.09.019

Fu X-Q, Ma N, Sun W-P et al (2018) Microwave and enzyme co-assisted aqueous two-phase extraction of polyphenol and lutein from marigold (*Tagetes erecta* L.) flower. Ind Crop Prod 123:296–302. https://doi.org/10.1016/j.indcrop.2018.06.087

Fu XQ, Zhang GL, Deng L et al (2019) Simultaneous extraction and enrichment of polyphenol and lutein from marigold (*Tagetes erecta* L.) flower by an enzyme-assisted ethanol/ammonium sulfate system. Food Funct 10(1):266–276. https://doi.org/10.1039/c8fo01865k

Gali L, Bedjou F, Ferrari G et al (2022) Formulation and characterization of zein/gum arabic nanoparticles for the encapsulation of a rutin-rich extract from *Ruta chalepensis* L. Food Chem 367:129982. https://doi.org/10.1016/j.foodchem.2021.129982

Gentscheva G, Milkova-Tomova I, Buhalova D et al (2022) Incorporation of the dry blossom flour of *Sambucus nigra* L. in the production of sponge cakes. Molecules 27(3):1124. https://doi.org/10.3390/molecules27031124

Gostin A-I, Waisundara VY (2019) Edible flowers as functional food: a review on artichoke (*Cynara cardunculus* L.). Trends Food Sci Technol 86:381–391. https://doi.org/10.1016/j.tifs.2019.02.015

Hegde AS, Gupta S, Sharma S et al (2022) Edible rose flowers: a doorway to gastronomic and nutraceutical research. Food Res Int 162:111977. https://doi.org/10.1016/j.foodres.2022.111977

Hnin KK, Zhang M, Devahastin S et al (2019) Influence of novel infrared freeze drying of rose flavoured yogurt melts on their physicochemical properties, bioactive compounds and energy consumption. Food Bioprocess Technol 12(12):2062–2073. https://doi.org/10.1007/s11947-019-02368-x

Hnin KK, Zhang M, Devahastin S et al (2020) Combined infrared freeze drying and infrared drying of rose-flavoured yogurt melts—effect on product quality. Food Bioprocess Technol 13(8):1356–1367. https://doi.org/10.1007/s11947-020-02486-x

Hnin KK, Zhang M, Ju R et al (2021a) A novel infrared pulse-spouted freeze drying on the drying kinetics, energy consumption and quality of edible rose flowers. LWT 136:110318. https://doi.org/10.1016/j.lwt.2020.110318

Hnin KK, Zhang M, Wang B (2021b) Development of nutritional properties in cookies with the incorporation of different levels of rose flower powder by microwave-vacuum drying. Dry Technol 39(9):1136–1148. https://doi.org/10.1080/07373937.2020.1725035

Hsu H-F, Hsiao P-C, Kuo T-C et al (2016) Antioxidant and anti-inflammatory activities of *Lonicera japonica* Thunb. var. *sempervillosa Hayata* flower bud extracts prepared by water, ethanol and supercritical fluid extraction techniques. Ind Crop Prod 89:543–549. https://doi.org/10.1016/j.indcrop.2016.05.010

Jiang Y, Ding Y, Wang D, Deng Y et al (2020) Radio frequency-assisted enzymatic extraction of anthocyanins from *Akebia trifoliata* (Thunb.) Koidz. flowers: process optimization, structure, and bioactivity determination. Ind Crop Prod 149:112327. https://doi.org/10.1016/j.indcrop.2020.112327

Kamali Z, Moazzezi S, Labbeiki G (2020) An investigation on the possibility of production of cookie containing sunflower seed meal flour and *Rosa damascena* waste extract. J Food Process Eng 3(2):147–159

Kazibwe Z, Kim D-H, Chun S et al (2017) Ultrasonication assisted ultrafast extraction of *Tagetes erecta* in water: cannonading antimicrobial, antioxidant components. J Mol Liq 229:453–458. https://doi.org/10.1016/j.molliq.2016.12.044

Kc Y, Bhattarai S, Shiwakoti LD et al (2022) Sensorial and chemical analysis of biscuits prepared by incorporating Moringa flower powder and leaf powder. Int J Food Prop 25(1):894–906. https://doi.org/10.1080/10942912.2022.2069807

Kim S-Y, Hong S-Y, Choi H-S et al (2023) Superfine Marigold powder improves the quality of sponge cake: lutein fortification, texture, and sensory properties. Foods 12:508. https://doi.org/10.3390/foods12030508

Kiran GD, Sharma A, Singh B, Babu GDK (2016) Composition of *lavandula angustifolia* produced by different extraction techniques. J Essent Oil Res 28(6):489–500. https://doi.org/10.1080/10412905.2016.1162210

Kou L, Turner ER, Luo Y (2012) Extending the shelf life of edible flowers with controlled release of 1-Methylcyclopropene and modified atmosphere packaging. J Food Sci 77(5):S188–S193. https://doi.org/10.1111/j.1750-3841.2012.02683.x

Kumar Baghel P (2023) Application of microwave in manufacturing technology: a review. Mater Today Proc 46:2686–2690. https://doi.org/10.1016/j.matpr.2023.02.008

Kumar A, Kaur A, Tomer V et al (2021) Effect of rose syrup and marigold powder on the physicochemical, phytochemical, sensorial and storage properties of nutricereals and milk-based functional beverage. J Am Coll Nutr 40(2):133–140. https://doi.org/10.1080/07315724.2020.1744487

Le X-N, Hu S-C, Zheng J-L et al (2022) The influence of different drying methods on bioactive components of *Eucommia ulmoides* Oliver male flower and the comprehensive assessment for industrial application. Ind Crop Prod 177:114469. https://doi.org/10.1016/j.indcrop.2021.114469

Li A-N, Li S, Li Y et al (2016) Optimization of ultrasound-assisted extraction of natural antioxidants from the *Osmanthus fragrans* flower. Molecules 21(2):218. https://doi.org/10.3390/molecules21020218

Li C, Jiang C, Jing H et al (2020) Effects of different extraction methods on yield, purity, composition, antioxidant and antimicrobial activities of phenolics from peony flowers. J Food Meas Charact 14(2):716–724. https://doi.org/10.1007/s11694-019-00318-x

Li J, Guo X, Wang R, Geng Z et al (2023) Ultrasonic assisted extraction of anthocyanins from rose flower petal in DES system and enzymatic acylation. LWT 180:114693. https://doi.org/10.1016/j.lwt.2023.114693

Ling JKU, Hadinoto K (2022) Deep eutectic solvent as green solvent in extraction of biological macromolecules: a review. Int J Mol Sci 23(6):3381. https://doi.org/10.3390/ijms23063381

Liu D, Liu C, Wang Z (2005) Parameter optimization for ultrasonic extraction of xanthophyll from marigold flower (*Tagetes erecta*). Jiangsu J Agric Sci 21(4):374–377. https://doi.org/10.56042/ijftr.v47i4.47319

Liu J-Z, Lyu H-C, Fu Y-J et al (2022) Simultaneous extraction of natural organic acid and flavonoid antioxidants from *Hibiscus manihot* L. flower by tailor-made deep eutectic solvent. LWT 163:113533. https://doi.org/10.1016/j.lwt.2022.113533

Lu J, Wang Z, Qin L, Shen J et al (2021) Drying methods affect bioactive compound contents and antioxidant capacity of *Bletilla striata* (Thunb.) Reichb.f. flower. Ind Crop Prod 164:113388. https://doi.org/10.1016/j.indcrop.2021.113388

Maksimenko EV, Lekar AV, Borisenko SN et al (2018) The development of a one-step method for production of the antioxidant quercetin from flower buds of the Sophora Japonica (*Sophora japonica* L.) in a subcritical water medium. Russ J Phys Chem B 12(8):1269–1275. https://doi.org/10.1134/S1990793118080092

Melillo L (1994) Diuretic plants in the paintings of Pompeii. Am J Nephrol 14(4–6):423–425. https://doi.org/10.1159/000168758

Mlcek J, Rop O (2011) Fresh edible flowers of ornamental plants—a new source of nutraceutical foods. Trends Food Sci Technol 22(10):561–569. https://doi.org/10.1016/j.tifs.2011.04.006

Mu L-q, Liu L-p, Ma D-l (2010) Optimal ultrasonic extraction condition and determination of polysaccharides in *Tilia amurensis* flowers. J For Res 21(1):77–80. https://doi.org/10.1007/s11676-010-0013-3

Navidad-Murrieta MS, López-Muñoz HA, Torres LM et al (2023) In vitro study on the effect of the appropriate encapsulating agent for the release of phenolic compounds from *Hibiscus sabdariffa* Linn extracts. ACS Food Sci Technol 3(4):608–615. https://doi.org/10.1021/acsfoodscitech.2c00349

Nicolson SW, Nepi M, Pacini E (2007) Nectaries and nectar. Springer, Berlin

Pan C, Zhao L, Zhao D (2021) Microwave-assisted green extraction of antioxidant components from *Osmanthus fragrans* (Lour) flower using natural deep eutectic solvents. J Appl Res Med Aromat Plants 20:100285. https://doi.org/10.1016/j.jarmap.2020.100285

Patil PD, Patil SP, Kelkar RK et al (2021) Enzyme-assisted supercritical fluid extraction: an integral approach to extract bioactive compounds. Trends Food Sci Technol 116:357–369. https://doi.org/10.1016/j.tifs.2021.07.032

Pensamiento-Niño CA, Castañeda-Ovando A, Añorve-Morga J et al (2023) Edible flowers and their relationship with human health: biological activities. Food Rev Int 40(1):620–639. https://doi.org/10.1080/87559129.2023.2182885

Perović AB, Karabegović IT, Krstić MS et al (2024) Novel hydrodistillation and steam distillation methods of essential oil recovery from lavender: a comprehensive review. Ind Crops Prod 211:118244. https://doi.org/10.1016/j.indcrop.2024.118244

Pimentel-Moral S, Borrás-Linares I, Lozano-Sánchez J et al (2018) Microwave-assisted extraction for *Hibiscus sabdariffa* bioactive compounds. J Pharm Biomed Anal 156:313–322. https://doi.org/10.1016/j.jpba.2018.04.050

Pimentel-Moral S, Borrás-Linares I, Lozano-Sánchez J et al (2019) Supercritical CO_2 extraction of bioactive compounds from *Hibiscus sabdariffa*. J Supercrit Fluid 147:213–221. https://doi.org/10.1016/j.supflu.2018.11.005

Pinakin DJ, Kumar V, Suri S et al (2020) Nutraceutical potential of tree flowers: a comprehensive review on biochemical profile, health benefits, and utilization. Food Res Int 127:108724. https://doi.org/10.1016/j.foodres.2019.108724

Pinela J, Prieto MA, Pereira E, Jabeur I, Barreiro MF, Barros L, Ferreira ICFR (2019) Optimization of heat- and ultrasound-assisted extraction of anthocyanins from *Hibiscus sabdariffa* calyces for natural food colourants. Food Chem 275:309–321. https://doi.org/10.1016/j.foodchem.2018.09.118

Pires EO Jr, Pereira E, Carocho M et al (2021) Study on the potential application of *Impatiens balsamina* L. flowers extract as a natural colouring ingredient in a pastry product. Int J Environ Res Public Health 18(17):9026. https://doi.org/10.3390/ijerph18179062

Pires TCSP, Barros L, Santos-Buelga C et al (2019) Edible flowers: emerging components in the diet. Trends Food Sci Technol 93:244–258. https://doi.org/10.1016/j.tifs.2019.09.020

Pires EO, Di Gioia F, Rouphael Y et al (2023) Edible flowers as an emerging horticultural product: a review on sensorial properties, mineral and aroma profile. Trends Food Sci Technol 137:31–54. https://doi.org/10.1016/j.tifs.2023.05.007

Pop M, Lupea AX, Popa S et al (2010) Colour of Bilberry (*Vaccinium Myrtillus* fruits) extracts. Int J Food Prop 13(4):771–777. https://doi.org/10.1080/10942910902894898

Purbowati ISM, Maksum A (2019) The antioxidant activity of Roselle (*Hibiscus sabdariffa* Linii) phenolic compounds in different variations microwave-assisted extraction time and power. IOP Conf Ser Earth Environ Sci 406(1):012005. https://doi.org/10.1088/1755-1315/406/1/012005

Qiu L, Zhang M, Bhandari B et al (2020a) Effects of infrared freeze drying on volatile profile, FTIR molecular structure profile and nutritional properties of edible rose flower (*Rosa rugosa* flower). J Sci Food Agr 100(13):4791–4800. https://doi.org/10.1002/jsfa.10538

Qiu L, Zhang M, Ju R et al (2020b) Effect of different drying methods on the quality of restructured rose flower (*Rosa rugosa*) chips. Dry Technol 38(12):1632–1643. https://doi.org/10.1080/07373937.2019.1653318

Qiu L, Zhang M, Mujumdar AS et al (2022a) Recent developments in key processing techniques for oriental spices/herbs and condiments: a review. Food Rev Int 38(8):1791–1811. https://doi.org/10.1080/87559129.2020.1839492

Qiu L, Zhang M, Xu B et al (2022b) Effects of superfine grinding on the physicochemical properties, antioxidant capacity, and hygroscopicity of *Rosa rugosa* cv. Plena powders. J Sci Food Agric 102(10):4192–4199. https://doi.org/10.1002/jsfa.11768

Ran X-l, Zhang M, Wang Y et al (2019) Novel technologies applied for recovery and value addition of high value compounds from plant byproducts: a review. Crit Rev Food Sci Nutr 59(3):450–461. https://doi.org/10.1080/10408398.2017.1377149

Ran X-L, Zhang M, Liu W et al (2021) Effects of hibiscetin pretreatment on the colour and anthocyanin level of microwave vacuum dried edible roses. Dry Technol 39(9):1231–1239. https://doi.org/10.1080/07373937.2021.1880433

Rop O, Mlcek J, Jurikova T et al (2012) Edible flowers--a new promising source of mineral elements in human nutrition. Molecules 17(6):6672–6683. https://doi.org/10.3390/molecules17066672

Roriz CL, Heleno SA, Carocho M et al (2020) Betacyanins from *Gomphrena globosa* L. flowers: incorporation in cookies as natural colouring agents. Food Chem 329:127178. https://doi.org/10.1016/j.foodchem.2020.127178

Sadikoglu H, Ozdemir M, Seker M (2006) Freeze-drying of pharmaceutical products: research and development needs. Dry Technol 24(7):849–861. https://doi.org/10.1080/07373930600734018

Safeena SA, Patil VS (2014) Studies on the effect of microwave oven drying on flower quality of dried Dutch rose flowers. Natl Acad Sci Lett 37(1):19–24. https://doi.org/10.1007/s40009-013-0197-0

Sagaradze VA, Babaeva EY, Kalenikova EI (2017) HPLC-UV method for determing flavonoids in hawthorn flowers and leaves. Pharm Chem J 51(4):277–280. https://doi.org/10.1007/s11094-017-1597-0

Shaikh AA, Ray A, Singhal RS (2023) Co-extraction of marigold flowers (*Tagetes erecta* L.) and dried coconut (*Cocos nucifera* L.) shreds using supercritical carbon dioxide: characterization and functional food formulations. Food Chem Adv 2:100189. https://doi.org/10.1016/j.focha.2023.100189

Siriamornpun S, Kaisoon O, Meeso N (2012) Changes in colour, antioxidant activities and carotenoids (lycopene, β-carotene, lutein) of marigold flower (*Tagetes erecta* L.) resulting from different drying processes. J Funct Foods 4(4):757–766. https://doi.org/10.1016/j.jff.2012.05.002

Skrajda-Brdak M, Dąbrowski G, Konopka I (2020) Edible flowers, a source of valuable phytonutrients and their pro-healthy effects-a review. Trends Food Sci Tech 103:179–199. https://doi.org/10.1016/j.tifs.2020.06.016

Tahir A, Shabir Ahmad R, Imran M et al (2021) Recent approaches for utilization of food components as nano-encapsulation: a review. Int J Food Prop 24(1):1074–1096. https://doi.org/10.1080/10942912.2021.1953067

Takahashi JA, Rezende FAGG, Moura MAF et al (2020) Edible flowers: bioactive profile and its potential to be used in food development. Food Res Int 129:108868. https://doi.org/10.1016/j.foodres.2019.108868

Trinh LTP, Choi Y-S, Bae H-J (2018) Production of phenolic compounds and bio-sugars from flower resources via several extraction processes. Ind Crop Prod 125:261–268. https://doi.org/10.1016/j.indcrop.2018.09.008

Vardakas AT, Shikov VT, Dinkova RH, Mihalev KM (2021) Optimisation of the enzyme-assisted extraction of polyphenols from saffron (*Crocus Sativus* L.) temals. Acta Sci PolonTechnol Aliment 20(3):359–367. https://doi.org/10.17306/J.AFS.2021.0954

Vidana Gamage GC, Choo WS (2023) Hot water extraction, ultrasound, microwave and pectinase-assisted extraction of anthocyanins from blue pea flower. Food Chem Adv 2:100209. https://doi.org/10.1016/j.focha.2023.100209

Voon HC, Bhat R, Rusul G (2012) Flower extracts and their essential oils as potential antimicrobial agents for food uses and pharmaceutical applications. Compr Rev Food Sci Food Saf 11(1):34–55. https://doi.org/10.1111/j.1541-4337.2011.00169.x

Wang D, Zhang M, Ju R et al (2023a) Novel drying techniques for controlling microbial contamination in fresh food: a review. Dry Technol 41(2):172–189. https://doi.org/10.1080/07373937.2022.2080704

Wang Y, Xiong X, Huang G (2023b) Ultrasound-assisted extraction and analysis of maidenhair tree polysaccharides. Ultrason Sonochem 95:106395. https://doi.org/10.1016/j.ultsonch.2023.106395

Wong F-C, Chai T-T, Xiao J (2019) The influences of thermal processing on phytochemicals and possible routes to the discovery of new phytochemical conjugates. Crit Rev Food Sci Nutr 59(6):947–952. https://doi.org/10.1080/10408398.2018.1479681

Wongwattanasathien O, Kangsadalampai K, Tongyonk L (2010) Antimutagenicity of some flowers grown in Thailand. Food Chem Toxicol 48(4):1045–1051. https://doi.org/10.1016/j.fct.2010.01.018

Wu Z, Ameer K, Jiang G (2021) Effects of superfine grinding on the physicochemical properties and antioxidant activities of Sanchi (*Panax notoginseng*) flower powders. J Food Sci Technol 58(1):62–73. https://doi.org/10.1007/s13197-020-04514-2

Xu H, Wang W, Jiang J et al (2015) Subcritical water extraction and antioxidant activity evaluation with on-line HPLC-ABTS$\cdot^+$ assay of phenolic compounds from marigold (*Tagetes erecta* L.) flower residues. J Food Sci Tech 52(6):3803–3811. https://doi.org/10.1007/s13197-014-1449-9

Xu D-P, Zheng J, Zhou et al (2017) Ultrasound-assisted extraction of natural antioxidants from the flower of *Limonium sinuatum*: optimization and comparison with conventional methods. Food Chem 217:552–559. https://doi.org/10.1016/j.foodchem.2016.09.013

Xu B, Feng M, Tiliwa ES et al (2022) Multi-frequency power ultrasound green extraction of polyphenols from Pingyin rose: optimization using the response surface methodology and exploration of the underlying mechanism. Lwt 156:113037. https://doi.org/10.1016/j.lwt.2021.113037

Yu J, Lou Q, Zheng X et al (2017) Sequential combination of microwave- and ultrasound-assisted extraction of total flavonoids from *Osmanthus fragrans* Lour. Flowers. Molecules 22(2):2216. https://doi.org/10.3390/molecules22122216

Zhang B, Yang R, Liu C-Z (2008) Microwave-assisted extraction of chlorogenic acid from flower buds of *Lonicera japonica* Thunb. Sep Purif Technol 62(2):480–483. https://doi.org/10.1016/j.seppur.2008.02.013

Zhang D-Y, Wan Y, Xu J-Y et al (2016) Ultrasound extraction of polysaccharides from mulberry leaves and their effect on enhancing antioxidant activity. Carbohyd Polym 137:473–479. https://doi.org/10.1016/j.carbpol.2015.11.016

Zhang M, Chen H, Mujumdar AS et al (2017) Recent developments in high-quality drying of vegetables, fruits, and aquatic products. Crit Rev Food Sci Nutr 57(6):1239–1255. https://doi.org/10.1080/10408398.2014.979280

Zhao L, Fan H, Zhang M, Chitrakar B, Bhandari B, Wang B (2019) Edible flowers: review of flower processing and extraction of bioactive compounds by novel technologies. Food Res Int 126:108660. https://doi.org/10.1016/j.foodres.2019.108660

Zheng M, Xia Q, Lu S (2015) Study on drying methods and their influences on effective components of loquat flower tea. LWT Food Sci Technol 63(1):14–20. https://doi.org/10.1016/j.lwt.2015.03.090

Zhenlin Z, Yunjiang M, Qingqing X et al (2022) Optimization of Dendrobium officinale flower soft candy preparation and evaluation of its antioxidant activity. JSFA reports 2(11):548–555. https://doi.org/10.1002/jsf2.88

Zielinska M, Ropelewska E, Xiao H-W et al (2020) Review of recent applications and research progress in hybrid and combined microwave-assisted drying of food products: quality properties. Crit Rev Food Sci Nutr 60(13):2212–2264. https://doi.org/10.1080/10408398.2019.1632788

Index

M. M. Jowkar et al. (eds.), *Advances in Research on Edible Flowers*, Advances in Olericulture, https://doi.org/10.1007/978-3-032-04561-4

D

E

F

T

U

V

MIX
Papier aus verantwortungsvollen Quellen
Paper from responsible sources
FSC® C105338
FSC
www.fsc.org

If you have any concerns about our products,
you can contact us on
ProductSafety@springernature.com

In case Publisher is established outside the EU,
the EU authorized representative is:
Springer Nature Customer Service Center GmbH
Europaplatz 3, 69115 Heidelberg, Germany

Printed by Libri Plureos GmbH
in Hamburg, Germany